T0297796

Software Defined Networks

Software Defined Networks

A Comprehensive Approach

Second Edition

Paul Göransson

Chuck Black

Timothy Culver

AMSTERDAM • BOSTON • HEIDELBERG • LONDON
NEW YORK • OXFORD • PARIS • SAN DIEGO
SAN FRANCISCO • SINGAPORE • SYDNEY • TOKYO
Morgan Kaufmann is an imprint of Elsevier

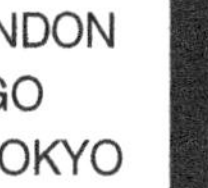

Morgan Kaufmann is an imprint of Elsevier
50 Hampshire Street, 5th Floor, Cambridge, MA 02139, United States

Library of Congress Cataloging-in-Publication Data
A catalog record for this book is available from the Library of Congress

British Library Cataloguing-in-Publication Data
A catalogue record for this book is available from the British Library

ISBN: 978-0-12-804555-8

For information on all Morgan Kaufmann publications
visit our website at https://www.elsevier.com/

ELSEVIER
Book Aid International
Working together
to grow libraries in
developing countries
www.elsevier.com • www.bookaid.org

Publisher: Todd Green
Acquisition Editor: Brian Romer
Editorial Project Manager: Amy Invernizzi
Production Project Manager: Priya Kumaraguruparan
Cover Designer: Matthew Limbert

Typeset by SPi Global, India

This book is dedicated to our families.

Contents

About the Authors

Paul Göransson is a serial entrepreneur who has led two boot-strap start-up companies through successful acquisitions by industry giants—Qosnetics by Hewlett Packard (1999) and Meetinghouse by Cisco (2006). Göransson held management positions with Agilent Technology's Advanced Networks Division and Cisco's Wireless Networking Business Unit. He serves on the Board of Directors of Tallac Networks. Göransson received a BA in Psychology from Brandeis University in 1975, an MS in Computer Engineering from Boston University in 1981, and a PhD in Computer Science from the University of New Hampshire in 1995. He is an avid marathoner, mountaineer, triathlete, scuba diver, and outdoor enthusiast. He has completed over 100 marathons, several Ironman triathlons, numerous ultramarathons including the Leadville Trail 100, and is a PADI Scuba Instructor. He completed a Thru-Hike of the 2189 mile long Appalachian Trail in 2015 on his recently installed artificial right ankle. He has lived, studied, and worked for extensive periods in France, Algeria, Venezuela, Mexico, and Sweden. Göransson co-authored the book *Roaming Securely in 802.11 Networks* as well as articles in technical journals related to computer networking. He has often been an invited speaker at technical conferences. Göransson owns and manages a 220-acre Polled Hereford beef and hay farm in southern Maine in his spare time.

Chuck Black has over 35 years of experience in the field of computer networking, working in research and development labs for Hewlett-Packard for most of that time before becoming co-founder of Tallac Networks, a Software Defined Networking startup. Most recently he has been training engineering staff and customers of major networking vendors in the areas of developing SDN applications at SDN Essentials. He has been the innovator and creator of multiple networking products for HP in the area of Network Access Control and Security, and holds eleven patents in these areas. Prior to this work, he developed products in the field of Network Management for HP's software organization. In the early days of local area networking, he was author of some of the first network topology discovery applications in the industry. Black holds a BS and MS in Computer Science from California Polytechnic State University, San Luis Obispo.

Timothy Culver is a seasoned technology executive and university faculty member who has a broad base of experience in engineering, information technology, business development, sales, and marketing. He is successful and passionate about emerging technologies and has a proven track record in creating, building, and delivering global technology teams. Culver has been involved in 3 start-ups and built businesses spanning 14 countries. As professor at the University of Texas at Dallas, he has created and taught graduate courses in computer science, software engineering, and spearheaded their Software Defined Networking curriculum and program. Prior to joining the University of Texas at Dallas in the 1990s, Culver taught computer language and introduction to computer science classes for the Dallas Community College District. Culver has won the US Presidential Volunteer Service Award from 2011 through 2014 for his volunteer activities in the community. This includes working on STEM (Science Technology Engineering and Math) education efforts in middle and high schools. He actively and enthusiastically promotes these careers to students and serves on the Mentor Advisory Committee for WeTeachScience.org. In his personal life, Culver is a father of four children and has been married to his wife for 30 years. After playing soccer for 25 years, Culver also earned his USSF National Coaching and US Youth Soccer National Coaching Licenses as he coached his children, Culver has refereed soccer matches in the United States and Europe. He lives on a farm in Sunnyvale, Texas, where he raises Black Angus cattle. Culver has four patents in the areas of VoIP, Internet Teleconferencing, and LDAP. He is an honor graduate of Baylor University where he earned an MIS degree. He then went on to Southern Methodist University where he earned a Master of Science in Engineering and an MBA. He has conducted post graduate research on Cloud Computing at Baylor and Walden Universities. Culver is also a Research Associate for the Open Networking Foundation.

Foreword

What is Software Defined Networking and why has it created so much hype in the networking industry? For a term that was coined more than 5 years ago, it seems like it should be a simple question to answer by now. However, as with other technology terms like *cloud*, after marketing teams have years to reshape the term for their own purposes, it becomes nearly impossible to discern its meaning. To add to the confusion, a variety of related terms have popped up such as open networking, programmable networks, and software-driven networks. This leaves many people wondering what SDN really is and if there is actually anything of substance behind the hype.

For people like myself who have been knee-deep in SDN since the beginning, it is easy to separate hype from reality and see what is really happening and why. However, I regularly teach classes on SDN and find that many people who are trying to learn about SDN today are finding it extremely difficult to wade through the various angles and spins on the topic to figure out the core of what is really happening.

In July 2010 when I gave my first public talk on OpenFlow, the protocol at the heart of SDN, virtually no one had heard of OpenFlow or Software Defined Networking. Since then I have explained SDN to literally thousands of people from students to network engineers and C-level executives. Some people immediately "got it" after just a few minutes of explanation, often before I could get to slide 3 in the presentation. With some people it took a full-day class before the "light bulb" went on. Almost universally, the people who were quickest to understand SDN's potential to impact the industry were those who had been in networking long enough to remember a time before the Internet, who had actually built networking products and who understood the business of networking.

This is why Paul and Chuck are perfectly suited for explaining the value of SDN. They both have tremendous experience in the networking industry, dating back longer than they probably care to admit. They have built networking solutions through multiple generations of technology. Having led two successful startups, Paul is also intimately familiar with the business of the networking industry. This depth and breadth of experience is absolutely invaluable when explaining SDN, positioning SDN in the context of the last 30+ years of computer networking and forecasting the potential impacts on networking in the coming years.

There is no shortage of information on the Internet today about SDN in the form of blog posts, whitepapers, podcasts, and videos. There are also several books available now. However, for someone who has not been directly involved with SDN up to this point and wants to get up to speed, I have found no comprehensive source of information on SDN, written from an unbiased viewpoint, like the authors have created with this book.

The book is extremely approachable for anyone interested in networking and Software Defined Networking including undergraduate and graduate students, networking professionals, and IT managers. It is very much self-contained and does not require extensive knowledge of networking in order to provide value to the reader. It provides the reader with valuable context by presenting a brief history of networking and describing the technologies and industry landscape leading up to the creation of OpenFlow and SDN. If you are a network architect or an IT manager trying to compare multiple solutions that claim to be based on SDN, but appear to be based on very different technology, the authors provide a solid basis for understanding and evaluating the various competing approaches to SDN in the market.

The truth is that a massive shift is happening in the networking industry and the drivers are pretty simple. You can draw a straight line from the rise of cloud computing and mobility directly to SDN, or open networking or programmable networks or whatever you choose to call it. If you want to cut through the marketing and the hype and get a comprehensive understanding of SDN and how it is helping reshape the networking industry, I definitely recommend reading this book.

Matt Davy
Tallac Networks, Inc.
April 26, 2016

Matt Davy *is a world-renowned expert in Software Defined Networking technology. At Indiana University, he served as Executive Director of InCNTRE, the SDN Interoperability Lab, network research, and internships and training. He was the lead architect for an enterprise network with 120,000+ users, 100,000+ Ethernet ports, and 5000+ wireless access points. He has 19 years of experience designing and operating large service provider and enterprise networks.*

Preface

When we initially conceived of the idea of writing this book, we were motivated in part by the lack of a single good reference for a comprehensive overview of SDN. Although we were involved professionally with SDN technologies, even we found information related to SDN to be largely unavailable in any single comprehensive source. We realized that for the very large numbers of professionals that were *not* directly working with SDN but who needed to learn about it, this was a big problem. Thus, our broad-brush goal in writing this book is to describe the setting that gave rise to SDN, the defining characteristics that distinguish it from competing technologies, and to explain the numerous significant commercial impacts that this nascent technology is already having. One of the challenges in writing an early book about such a rapidly evolving technology is that it is a moving target.

The preceding paragraph, written for our first edition published 3 years ago, still rings true today. So many technologies that were off our first edition radar are now commonly considered part of SDN. We have selected to use the words *A Comprehensive Approach* as part of our title. There are many competing ideas in use today, all of which wish to jump on the SDN bandwagon. Indeed, the size of that bandwagon seems ever-expanding. Whatever aspect or type of SDN technology with which our reader may be required to work, we at least hope that he or she will be able to place it into the broader SDN context through the reading of this book. For this purpose, we try to discuss a variety of definitions of SDN. We hope that no reader takes offense that we were not dogmatic in our application of the definition of SDN in this work.

Individuals interested in learning about Software Defined Networks or having a general interest in any of the following topics:

- networking
- switching
- Software Defined Networks
- OpenFlow
- OpenStack
- OpenDaylight
- Network Virtualization
- Network Functions Virtualization

will find this book useful.

Software Defined Networking is a broad field that is rapidly expanding. While we have attempted to be as comprehensive as possible, the interested reader may need to pursue certain technical topics via the references provided. We do not assume the reader has any special knowledge other than a basic understanding of computer concepts. Some experience in computer programming and computer networking will be helpful for understanding the material presented. The book contains a large number of figures and diagrams to explain and to illustrate networking concepts that are being defined or discussed. These graphics help the reader to continue straight through the text without feeling the need to reach for other references.

The first edition of this work was very well received. Over the 3 years since its initial publication, we have received numerous inquiries from university faculty using the text as the basis for a course.

Our motivation in producing this second edition is twofold: first, to bring the text up to date with changes in SDN and second, to target it more squarely for use in a graduate course on SDN. To this end, each chapter includes text boxes that include discussion questions relevant to the adjacent material. These questions can either be used to facilitate in-class discussion or as the basis for quiz questions. In addition, our publisher will maintain a faculty-accessible website where ancillary, course-related materials such as lecture notes and lab exercises may be found.

SUGGESTIONS AND CORRECTIONS

Although we have tried to be as careful as possible, the book may still contain some errors and certain topics may have been omitted that readers feel are especially relevant for inclusion. In anticipation of possible future printings, we would like to correct any mistakes and incorporate as many suggestions as possible. Please send comments via email to: chuck.a.black@gmail.com.

Acknowledgments

Many thanks to our families for their tremendous support throughout the years and especially while we were writing this book.

This book would not have been possible without the unwavering support of Bill Johnson, Matt Davy, and Dr. Paul Congdon, co-founders of Tallac Networks. Their deep technical understanding of SDN, their painstaking reviews of the drafts, and their many direct contributions to the text were invaluable.

We thank Dr. Niels Sluijs, Internet Communication and Service Engineer at Sicse in the Netherlands for his refreshingly unbiased view and comments on the manuscript.

Our great appreciation is extended to Nopadon Juneam of Chiang Mai University whose help rendering the final versions of many of the figures in this book was invaluable.

We are also grateful to Ali Ezzet from Tallac Networks for his careful reading of draft chapters of the manuscript. His technical expertise caught a number of bugs and greatly helped to improve the manuscript. We extend our appreciation to Anthony Delli Colli who provided important input and advice on the business and marketing aspects presented in this work. These contributions to this work were indispensable.

A special thanks to Helen Göransson for her meticulous reviews of the many drafts of the manuscript. This work has certainly been made more readable by her efforts.

In addition to all the valuable contributions acknowledged above for our first edition of this work, we want to add our thanks to two individuals whose careful review of this second edition have improved it immeasurably. Hence, a heartfelt thanks to Giles Heron of Cisco and Dr. Alberto Schaeffer-Filho of the Federal University of Rio Grande do Sul (UFRGS).

Finally, thanks to Brian Romer and Amy Invernizzi at Elsevier for their encouragement and support during this project.

The authors gratefully acknowledge Tallac Networks for the use of selected illustrations that appear throughout the book.

Paul Göransson
Chuck Black
Timothy Culver

CHAPTER

INTRODUCTION

1

It is not often that an author of a technology text gets to read about his subject matter in a major story in a current issue of a leading news magazine. The tempest surrounding *Software Defined Networking* (SDN) is indeed intense enough to make mainstream news [1]. The modern computer network has evolved into a complex beast that is challenging to manage and which struggles to scale to the requirements of some of today's environments. SDN represents a new approach to computer networking that attempts to address these weaknesses of the current paradigm. SDN is a fundamentally novel way to program the switches utilized in modern data networks. SDN's move to a highly scalable and centralized network control architecture is better suited to the extremely large networks prevalent in today's mega-scale data centers. Rather than trying to crowbar application-specific forwarding into legacy architectures ill-suited to the task, SDN is designed from the outset to perform fine-grained traffic forwarding decisions. Interest in SDN goes far beyond the research and engineering communities intrigued by this new Internet switching technology. Another of the early drivers of SDN was to facilitate experimentation with and the development of novel protocols and applications designed to address problems related to an Internet bogged down by the weight of three decades of incremental technological fixes—a problem sometimes called *Internet Ossification*. If SDN's technological promise is realized, this will represent nothing short of a tectonic shift in the networking industry, as long-term industry incumbents may be unseated and costs to consumers may plummet. Along with this anticipation, though, surely comes a degree of overhype, and it is important that we understand not only the potentials of this new networking model, but also its limitations. In this work we will endeavor to provide a technical explanation of how SDN works, an overview of those networking applications for which it is well-suited and those for which it is not, a tutorial on building custom applications on top of this technology, and a discussion of the many ramifications it has on the networking business itself.

This introductory chapter provides background on the fundamental concepts underlying current state-of-the-art Internet switches, where data plane, control plane and management plane will be defined and discussed. These concepts are key to understanding how SDN implements these core functions in a substantially different manner than the traditional switch architecture. We will also present how forwarding decisions are made in current implementations and the limited flexibility this offers network administrators to tune the network to varying conditions. At a high level, we provide examples of how more flexible forwarding decisions could greatly enhance the business versatility of existing switches. We illustrate how breaking the control plane out of the switch itself into a separate, open-platform controller can provide this greater flexibility. We conclude by drawing parallels between how the Linux operating system has enjoyed rapid growth by leveraging the open source development community and how the same efficiencies can be applied to the control plane on Internet switches.

Software Defined Networks. http://dx.doi.org/10.1016/B978-0-12-804555-8.00001-6

We next look at some basic packet switching terminology that will be used throughout the text, and following that we provide a brief history of the field of packet switching and its evolution.

1.1 BASIC PACKET SWITCHING TERMINOLOGY

This section defines much of the basic packet switching terminology used throughout the book. Our convention is to italicize a new term on its first use. For more specialized concepts that are not defined in this section, they will be defined on their first use. Many packet switching terms and phrases have several and varied meanings to different groups. Throughout the book we try to use the most-accepted definition for terms and phrases. Acronyms are also defined and emphasized on their first use, and the appendix on acronyms provides an alphabetized list of all of the acronyms used in this work. An advanced reader may decide to skip over this section. Others may want to skim this material, and later look back to refer to specific concepts.

This terminology is an important frame of reference as we explain how SDN differs from traditional packet switching. To some degree, though, SDN does away with some of these historic concepts or changes their meaning in a fundamental way. Throughout this book, we encourage the reader to look back at these definitions and consider when the term's meaning is unchanged in SDN, when SDN requires a nuanced definition, and when a discussion of SDN requires entirely new vocabulary.

A *Wide Area Network* (WAN) is a network that covers a broad geographical area, usually larger than a single metropolitan area.

A *Local Area Network* (LAN) is a network that covers a limited geographical area, usually not more than a few thousand square meters in area.

A *Metropolitan Area Network* (MAN) is a network that fills the gap between LANs and WANs. This term came into use because LANs and WANs were originally distinguished not only by their geographical areas of coverage, but also by the transmission technologies and speeds that they used. With the advent of technologies resembling LANs in terms of speed and access control, but with the capability of serving a large portion of a city, the term MAN came into use to distinguish these networks as a new entity distinct from large LANs and small WANs.

A *Wireless Local Area Network* (*WLAN*) is a LAN in which the transmission medium is air. The typical maximum distance between any two devices in a wireless network is on the order of 50 m. While it is possible to use transmission media other than air for wireless communication, we will not consider these in our use of this term in this work.

The *Physical Layer* is the lowest layer of the seven layer *Open Systems Interconnection* (OSI) model of computer networking [2]. It consists of the basic hardware transmission technology to move bits of data on a network.

The *Data Link Layer* is the second lowest layer of the OSI model. This is the layer that provides the capability to transfer data from one device to another on a single network segment. For clarity, here we equate a LAN network segment with a collision domain. A strict definition of LAN network segment is an electrical or optical connection between network devices. For our definition of data link layer we will consider multiple segments linked by repeaters as a single LAN segment. Examples of network segments are a single LAN, such as an Ethernet, or a point-to-point communications link between adjacent nodes in a WAN. The link layer includes: (1) mechanisms to detect sequencing errors or bit-errors that may occur during transmission, (2) some mechanism of flow control between the sender and receiver across that network segment, and (3) a multiplexing ability that allows multiple network protocols to use the same communications medium. These three functions are considered to be part of the *logical link control* (LLC) component of the data link layer. The remaining functions of the data link layer are part of the *Media Access Control* (MAC) component, described separately below.

The *MAC* layer is the part of the data link layer that controls when a shared medium may be accessed and provides addressing in the case that multiple receivers will receive the data yet only one should process it. For our purposes in this book, we will not distinguish between data link layer and MAC layer.

The *Network Layer* provides the functions and processes that allow data to be transmitted from sender to receiver across multiple intermediate networks. To transit each intermediate network involves the data link layer processes described above. The network layer is responsible for stitching together those discrete processes such that the data correctly makes its way from the sender to the intended receiver.

Layer one is the same as the physical layer defined above.

Layer two is the same as the data link layer defined above. We will also use the term *L2* synonymously with layer two.

Layer three is the same as the network layer defined above. *L3* will be used interchangeably with layer three in this work.

A *port* is a connection to a single communications medium, including the set of data link layer and physical layer mechanisms necessary to correctly transmit and receive data over that link. This link may be of any feasible media type. We will use the term *interface* interchangeably with port throughout this text. Since this book will also deal with virtual switches, the definition of port will be extended to include virtual interfaces, which are the endpoints of tunnels.

A *frame* is the unit of data transferred over a layer two network.

A *packet* is the unit of data transferred over a layer three network. Sometimes this term is used more generally to refer to the units of data transferred over either a layer two network (frames) as well, without distinguishing between layers two and three. When the distinction is important, a packet is always the payload of a frame.

A *MAC address* is a unique value that globally identifies a piece of networking equipment. While these addresses are globally unique, they serve as layer two addresses, identifying a device on a layer two network topology.

An *IP Address* is a nominally unique value assigned to each host in a computer network that uses the Internet Protocol for layer three addressing.

An *IPv4 Address* is an IP address that is a 32-bit integer value conforming to the rules of Internet Protocol Version 4. This 32-bit integer is frequently represented in *dotted* notation, with each of the 4 bytes comprising the address represented by a decimal number from 0 to 255, separated by periods (e.g., 192.168.1.2).

An *IPv6 Address* is an IP address that is a 128-bit integer conforming to the rules of Internet Protocol Version 6, introducing a much larger address space than IPv4.

A *switch* is a device that receives information on one of its ports and transmits that information out one or more of its other ports, directing this information to a specified destination.

A *circuit switch* is a switch where contextual information specifying where to forward the data belonging to a circuit (i.e., connection) is maintained in the switch for a prescribed duration, which may span lapses of time when no data belonging to that connection is being processed. This context is established either by configuration or by some *call set-up* or *connection set-up* procedure specific to the type of circuit switch.

A *packet switch* is a switch where the data comprising the communication between two or more entities is treated as individual packets that each make their way independently through the network toward the destination. Packet switches may be of the connection-oriented or connectionless type.

The *connection-oriented* model is when data transits a network where there is some context information residing in each intermediate switch that allows the switch to forward the data toward its destination. The circuit switch described above is a good example of the connection-oriented paradigm.

The *connectionless* model is when data transits a network and there is sufficient data in each packet such that each intermediate switch can forward the data toward its destination without any a priori context having been established about that data.

A *router* is a packet switch used to separate subnets. A subnet is a network consisting of a set of hosts that share the same network prefix. A network prefix consists of the most significant bits of the IP address. The prefix may be of varying lengths. Usually all of the hosts on a subnet reside on the same LAN. The term router is now often used interchangeably with *layer three switch*. A home wireless access point typically combines the functionality of WiFi, layer two switch, and router into a single box.

To *flood* a packet is to transmit it on all ports of a switch except for the port on which it was received.

To *broadcast* a packet is the same as flooding it.

Line rate refers to the bandwidth of the communications medium connected to a port on a switch. On modern switches this bandwidth is normally measured in *megabits per second* (Mbps) or *gigabits per second* (Gbps). When we say that a switch handles packets at line rate, this means it is capable of handling a continuous stream of packets arriving on that port at that bandwidth.

WiFi is common name for wireless communications systems that are based on the IEEE 802.11 standard.

1.2 HISTORICAL BACKGROUND

The major communications networks around the world in the first half of the 20th century were the telephone networks. These networks were universally circuit switched networks. Communication between endpoints involved the establishment of a communications path for that dialogue and the tearing down of that path at the dialogue's conclusion. The path on which the conversation traveled was static during the call. This type of communications is also referred to as connection-oriented. In addition to being based on circuit switching, the world's telephone networks were quite centralized, with large volumes of end-users connected to large switching centers. Paul Baran, a Polish immigrant who became a researcher working at Rand Corporation in the United States in the 1960s, argued that in the event of enemy attack networks like the telephone network were very easy to disrupt [3,4]. The networks had poor survivability characteristics in that the loss of a single switching center could remove phone capability from a large swath of the country. Mr. Baran's proposed solution was to transmit the voice signals of the phone conversations in packets of data that could travel autonomously through the network, finding their own way toward their destination. This concept included the notion that if part of the path being used for a given conversation was destroyed by enemy attack, the communication would *survive* by automatically rerouting over alternative paths to the same destination. He demonstrated that the national voice communications system could still function even if fifty per cent of the forwarding switches were destroyed, greatly reducing the vulnerability characteristics of the centralized, circuit switching architecture prevalent at the time.

When Mr. Baran did his work at Rand, he never could have envisioned the dominance his idea would ultimately achieve in the area of data networking. While certainly not the universally accepted networking paradigm at the time, the history that followed is now part of Internet folklore. Mr. Baran's ideas became embodied in the *Department of Defense's* (DOD's) experimental ARPANET network that began to operate in 1969. The ARPANET connected academic research institutions, military departments and defense contractors. This decentralized, connectionless network grew over the years until bursting upon the commercial landscape around 1990 in the form of the Internet known and loved around the world today.

For decades after the emergence of the ARPANET, networking professionals waged battles over the advantages of connection-based vs. connectionless architectures and centralized vs. distributed architectures. The tides in this struggle seemed to turn one way, only to reverse some years later. The explosive growth of the Internet in the 1990s seemed to provide a conclusion to these arguments, at least as far as computer networking was concerned. The Internet was in ascendance, and was unequivocally a distributed, connectionless architecture. Older connection-oriented protocols like X.25 [5] seemed destined to become a thing of the past. Any overly centralized design was considered too vulnerable, whether to malicious attacks or to simple acts of nature. Even the new kid on the block, *Asynchronous Transfer Mode* (ATM) [5] hyped in the mid-1990s to be capable of handling line rates greater than the Internet could ever handle, would eventually be largely displaced by the ever-flexible Internet that somehow managed to handle line rates in the tens of gigabits per second range, once thought only to be possible using cell-switching technology like ATM.

DISCUSSION QUESTION:

Describe why Paul Baran's notion of a connectionless network had potential appeal to the United States Department of Defense.

1.3 THE MODERN DATA CENTER

The Internet came to dominate computer networking to a degree rarely seen in the history of technology, and the Internet was at its core connectionless and distributed. During this period of ascendance, the *World Wide Web* (WWW), an offspring of the Internet, spawned by Sir Tim Berners-Lee of the United Kingdom, gave rise to ever-growing data centers, hosting ever more complex and more heavily subscribed web services. These data centers served as environmentally protected warehouses housing large numbers of computers that served as compute and storage servers. The warehouses themselves were protected against environmental entropy as much as possible by being situated in disaster-unlikely geographies, with redundant power systems, and ultimately with duplicate capacity at disparate geographical locations.

Because of the large numbers of these servers, they were physically arranged into highly organized rows of racks of servers. Over time, the demand for efficiency drove the migration of individual server computers into server *blades* in densely packed racks. Racks of compute-servers were hierarchically

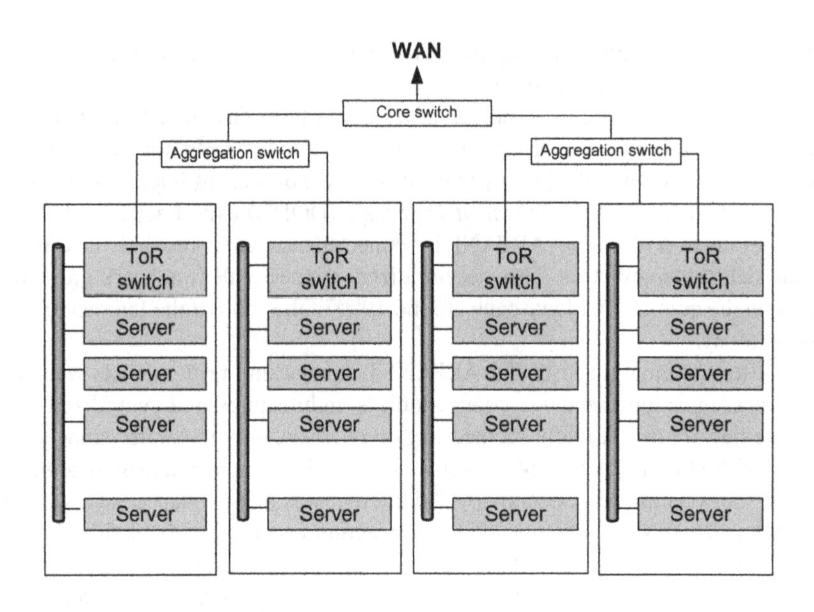

FIG. 1.1

Typical data center network topology.

organized such that *Top-of-Rack* (ToR) switches provided the networking within the rack and the inter-rack interface capability. We depict the rigid hierarchy typical in today's data center in Fig. 1.1.

During this evolution, the sheer numbers of computers in the data center grew rapidly. Data centers are being built now that will accommodate over 120,000 physical servers [6]. State-of-the-art physical servers can conceivably host twenty *virtual machines* (VMs) per physical server. This means that the internal network in such a data center would interconnect 2,400,000 hosts! This growing number of host computers within a data center all communicate with each other via a set of communications protocols and networking devices that were optimized to work over a large, disparate geographical area with unreliable communications links. Obviously, tens of thousands of computers sitting side by side in an environmental cocoon with highly reliable communications links is a different nature of network altogether. It gradually emerged that the goals of survivability in the face of lost communications links or a bombed-out network control center were not highly relevant in this emerging reality of the data center. Indeed, an enormous amount of complexity invented to survive precisely those situations was making the operation of the ever-larger and more complex data center untenable.

Beyond the obvious difference in the stability of the network topology, the sheer scale of these mega-data centers in terms of nodes and links creates a network management challenge different than those encountered previously. Network management systems designed for carrier public networks or large corporate intranets simply cannot scale to these numbers. A new network management paradigm was needed.

Furthermore, while these data centers exist in order to support interaction with the turbulent world outside of their walls, studies [7,8] indicate that the majority of the traffic in current data centers is *East-West* traffic. East-West traffic is composed of packets sent by one host in a data center to another host in that same data center. Analogously, *North-South* traffic is traffic entering (leaving) the data center from (to) the outside world. For example, a user's web browser's query about the weather might be processed by a web server (North-South) which retrieves data from a storage server in the same data center (East-West) before responding to the user (North-South). The protocols designed to achieve robustness in the geographically dispersed wide-area Internet today require that routers spend more than thirty percent of their CPU cycles [6] rediscovering and recalculating routes for a network topology in the data center that is highly static and only changed under strict centralized control. This increasing preponderance of East-West traffic does not benefit from the overhead and complexities that have evolved in traditional network switches to provide just the de-centralized survivability that Mr. Baran so wisely envisioned for the WANs of the past.

The mega-data centers discussed in this section differ from prior networks in a number of ways: stability of topology, traffic patterns, and sheer scale. In addition, since the services provided by these data centers require frequent reconfiguration, these networks demand a level of agility not required in the past. Traditional networking methods are simply insufficiently dynamic to scale to the levels being required. SDN is a technology designed explicitly to work well with this new breed of network, and represents a fundamental transformation from traditional Internet switching. In order to understand how SDN does differ, in Sections 1.4 and 1.5 we review how legacy Internet switches work to establish a baseline for comparison with SDN.

Before continuing, we should emphasize that while the modern data center is the premier driver behind the current SDN fervor, by no means is SDN only applicable to the data center. When we review SDN applications in Chapter 12 we will see that SDN technology can bring important innovations in domains such as mobility that have little to do with the data center.

DISCUSSION QUESTION:

Provide three examples of east-west traffic and three examples of north-south traffic in a modern data center.

1.4 TRADITIONAL SWITCH ARCHITECTURE

We will now look at what the traditional Internet switch looks like from an architectural perspective, and how and why it has evolved to be the complex beast it is today. The various switching functions are traditionally segregated into three separate categories. Since each category may be capable of *horizontal* communication with peer elements in adjacent entities in a topology, and also capable of *vertical* communication with the other categories, it has become common to represent each of these categories as a layer or *plane*. Peer communications occur in the same plane, and cross-category messaging occurs in the third dimension, between planes.

1.4.1 DATA, CONTROL, AND MANAGEMENT PLANES

We show the control, management, and data planes of the traditional switch in Fig. 1.2. The vast majority of packets handled by the switch are only touched by the *data plane*. The data plane consists of the various ports that are used for the reception and transmission of packets and a forwarding table with

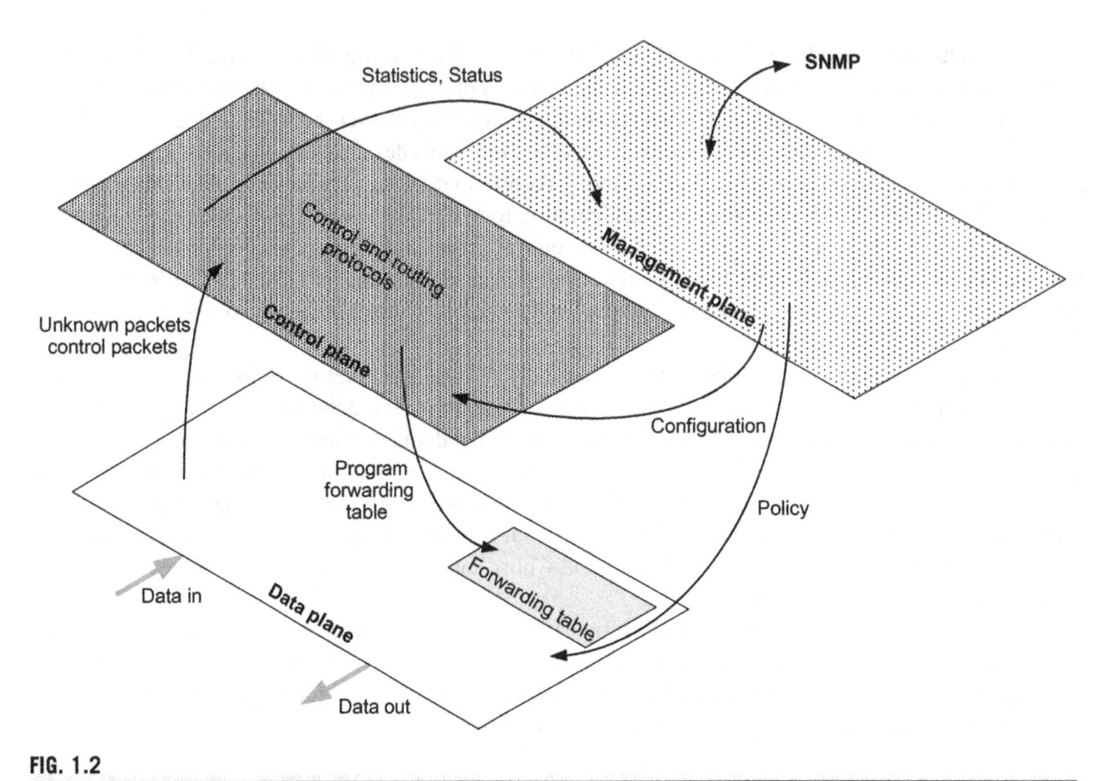

FIG. 1.2

Roles of the control, management, and data planes.

its associated logic. The data plane assumes responsibility for packet buffering, packet scheduling, header modification, and forwarding. If an arriving data packet's header information is found in the forwarding table it may be subject to some header field modifications and then will be forwarded without any intervention of the other two planes. Not all packets can be handled in that way, sometimes simply because their information is not yet entered into the table, or because they belong to a control protocol that must be processed by the *control plane*. The control plane, as shown in Fig. 1.2, is involved in many activities. Its principal role is to keep current the information in the forwarding table so that the data plane can independently handle as high a percentage of the traffic as possible. The control plane is responsible for processing a number of different control protocols that may affect the forwarding table, depending on the configuration and type of switch. These control protocols are jointly responsible for managing the active topology of the network. We review some of these control protocols below in Section 1.5. These control protocols are sufficiently complex as to require the use of general purpose microprocessors and accompanying software in the control plane, whereas we will see in Section 1.4.3 that the data plane logic can today be incarnated entirely in silicon.

The third plane depicted in Fig. 1.2 is the management plane. Network administrators configure and monitor the switch through this plane, which in turn extracts information from or modifies data in the control and data planes as appropriate. The network administrators use some form of network management system to communicate with the management plane in a switch.

1.4.2 SOFTWARE-BASED ROUTING AND BRIDGING

In the late 1980s the commercial networking world was undergoing the cathartic process of the emergence of the Internet as a viable commercial network. New start-ups such as Cisco Systems and Wellfleet Communications had large teams of engineers building special-purpose, commercial versions of the *routers* that had previously been relegated to use in research and military networks. While the battle between connection-oriented packet switching, such as X.25 and ATM, and the connectionless Internet architecture would continue to be waged for a number of years, ultimately the connectionless router would dominate the packet switching world.

Before this time, most routers were just general purpose Unix computers running software that inspected a packet that arrived on one interface and looked up the destination IP address in some efficiently searched data type such as a hierarchical tree structure. This data structure was called the *routing table*. Based on the entry found during this search, the packet would be transmitted on the indicated outbound interface. Control packets would be shunted to the appropriate control processes on the Unix system rather than being processed through that routing table.

In addition to layer three routers, many companies marketed layer two *bridges*. Bridges create a bridged LAN which is a topology of interconnected LAN segments. As there were multiple competing layer two technologies prevalent, including Token Ring, *Fiber Distributed Data Interface* (FDDI), and different speeds and physical media versions of the original Ethernet, these bridges also served as a mechanism to interface between disparate layer two technologies. The then-current terminology was to use the terms *bridging* of *frames* for layer two and *routing* of *packets* for layer three. We will see in this book that as the technologies have evolved, these terms have become blended such that a modern networking device capable of handling both layer two and three is commonly referred to simply as a *packet switch*.

Demand for increased speed brought about concomitant advances in the performance of these network devices. In particular, rapid evolution occurred both in optical and twisted copper pair physical media. Ten Mbps Ethernet interfaces were commonplace by 1990 and one hundred Mbps fiber and ultimately twisted pair were on the horizon. Such increases in media speed made it necessary to continually increase the speed of the router and bridge platforms. At first, such performance increases were achieved by distributing the processing over ever-more parallel implementations involving multiple *blades*, each running state-of-the-art microprocessors with independent access to a distributed forwarding table. Ultimately, though, the speed of the interfaces reached a point where performing the header inspection and routing table look-up in software simply could not keep up.

1.4.3 HARDWARE LOOK-UP OF FORWARDING TABLES

The first major use of hardware acceleration in packet switching was via the use of *Application-Specific Integrated Circuits* (ASICs) to perform high-speed hashing functions for table look-ups. In the mid-1990s advances in *Content-Addressable Memory* (CAM) technology made it possible to perform very high speed look-up using destination address fields to find the output interface for high-speed packet forwarding. The networking application of this technology made its commercial debut in *Ethernet switches*.

At that time, the term *switch* became used to distinguish a hardware-look-up-capable layer two bridge from the legacy software-look-up devices discussed above in Section 1.4.2. The term router still referred to a layer three device, and initially these were still based upon software-driven address look-up. Routers were still software-based for a number of reasons. First, the packet header modifications required for layer three switching were beyond the capability of the ASICs used at the time. Also, the address look-up in layer two switching was based on the somewhat more straightforward task of looking up a 48-bit *MAC* address. Layer three address look-up is more complicated since the devices look up the *closest* match on a network address, where the match may only be on the most significant bits of a network address. Any of the destination addresses matching that network address will be forwarded out the same output interface to the same *next-hop* address. The capabilities of CAMs steadily improved, however, and within a few years there were layer three routers that were capable of hardware look-up of the destination layer three address. At this point, the distinction between router and switch began to blur, and the terms *layer two switch* and *layer three switch* came into common use. Today, where the same device has the capability to simultaneously act as both a layer two and layer three switch, the use of the simple term switch has become commonplace.

1.4.4 GENERICALLY PROGRAMMABLE FORWARDING RULES

In reality there are many packets that need to be processed by a switch that need more complicated treatment than simply being forwarded on another interface different than the one on which it arrived. For example, the packet may belong to one of the control protocols covered in Sections 1.5.1 and 1.5.2, in which case it needs to be processed by the control plane of the switch. The brute-force means of handling these exceptions by passing all exception cases to the control plane for processing rapidly became unrealistic as forwarding rates increased. The solution was to push greater intelligence down into the forwarding table such that as much of the packet processing as possible can take place at line rates in the hardware *forwarding engine* itself.

Early routers only had to perform limited packet header field modifications. This included decrementing the *Time-To-Live* (TTL) field and swapping the MAC header to the source-destination MAC headers for the next hop. In the case of multicast support, they had to replicate the packet for transmission out multiple output ports. As the switch's features grew to support advanced capabilities like *Virtual Local Area Networks* (VLANs) and *Multiprotocol Label Switching* (MPLS), more packet header fields needed ever more complex manipulation. To this end, the idea of *programmable rules* came into being whereby some of this more complex processing could be encoded in rules and carried out directly in the forwarding engines at line rate. Such rules could be communicated to the switch by network managers using a *Command Line Interface* (CLI) or some other primitive configuration mechanism. While most control protocol packets still needed to be shunted to the control plane processor for handling, this capability allowed the evolving switches to implement ever more complex protocols and features at ever increasing line rates. This very programmability of the hardware was one of the early seeds that gave life to the concept of SDN.

While there are many variations in hardware architecture among modern switches, we attempt to depict an idealized state-of-the-art switch in Fig. 1.3. The figure shows the major functional blocks that a packet will transit as it is being forwarded by the switch. In the figure we see that the packet may transit the *packet receive*, *ingress filter*, *packet translation*, *egress filter*, and *packet transmit* functions or be consumed by the *Switch OS*. We can see that the forwarding database is influenced by the arrival

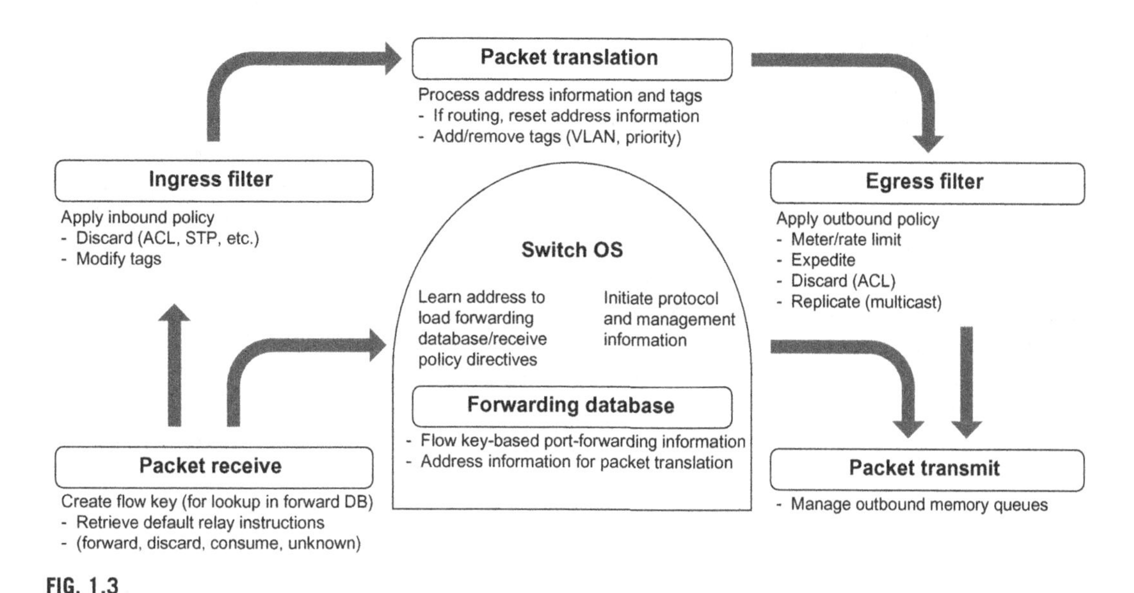

FIG. 1.3

A packet's journey through switching hardware.

of certain packets, such as when a new address is learned. We also see that the hardware generates a flow key to serve as a look-up key into the forwarding database. The forwarding database will provide basic forwarding instructions as well as policy directives, if relevant in this switch. The ingress filter applies policy, perhaps resulting in the packet being dropped. Packet header modifications take place both at the ingress filter as well as in the packet translation function. Outbound policy is implemented at the egress filter, where again the packet may be dropped, if, for example, established *Quality of Service* (QoS) rates are being exceeded. Finally, the packet passes through the packet transmit function where additional priority queue processing occurs prior to its actual transmission on the communications medium.

DISCUSSION QUESTION:

Discuss two major reasons why hardware-based packet forwarding was achievable in layer two bridges before layer three routers.

1.5 AUTONOMOUS AND DYNAMIC FORWARDING TABLES

In order to respond to unstable communications links and the possible disappearance of an entire switching center, protocols were developed so that bridges and routers could dynamically and autonomously build and update their forwarding tables. In this section we provide background on some of the major protocols that were developed to provide this capability.

1.5.1 LAYER TWO CONTROL

A layer two forwarding table is a look-up table indexed by destination MAC address, where the table entry indicates which port on the switch should be used to forward the packet. In the case of early, stand-alone bridges, when a frame was first received from a device on a given interface, the switch could *learn* the sender's MAC address and location and populate its forwarding table with this new information. When a frame arrived destined for that address, it could then use the forwarding table to send that frame out the correct port. In the case that an *unknown* destination address was received, the bridge could flood this out all interfaces, knowing that it would be dropped on all attached networks except the one where the destination actually resided. This approach no longer worked once bridges were interconnected into networks of bridges as there were multiple paths to reach that destination, and unless a single one was chosen predictably, an infinite loop could be created. While a static mapping of MAC addresses to ports was an alternative, protocols were developed that allowed the bridge to learn MAC addresses dynamically, and to learn how to assign these to ports in such a way as to automatically avoid creating switching loops yet predictably be able to reach the destination MAC address, even when that required transiting multiple intermediate bridges. We discuss some of these MAC-learning protocols below to illustrate ways that the legacy layer two switch's forwarding table can be built up automatically.

As we explained earlier, the Ethernet switch's role is to bridge multiple layer two networks. The most simplistic concept of such a device includes only the ability to maintain a forwarding table of MAC addresses such that when a frame reaches the switch it understands on which attached layer two network that destination MAC address exists, and to forward it out on that layer two network. Clearly, if the frame arrived on a given port, and if the destination is reachable via that same port, the Ethernet switch need not be involved in forwarding that frame to its destination. In this case, the frame was dropped. The task of the Ethernet switch is considerably more complex when multiple Ethernet switches are connected into complicated topologies containing loops. Without taking care to prevent such situations, it was possible to accidentally construct forwarding loops where a frame is sent back to a switch that had forwarded it earlier ad infinitum. The performance impact of such switching loops is more exaggerated in layer two switching than layer three, since the MAC header has no analog to the layer three *TTL* field, which limits the number of times the packet can transit routers before it is dropped. Thus, an unintentional loop in a layer two network can literally bring the network to a halt, where the switch appears frozen, but is actually one hundred percent occupied forwarding the same packet over and over through the loop.

The prevalence of broadcast frames in the layer two network exacerbates this situation. Since many layer two control frames must be broadcast to all MAC addresses on that layer two topology, it was very easy for a broadcast frame to return to a switch that had already broadcast that frame. This latter problem is known as *broadcast radiation*. The *Spanning Tree Protocol* (STP), specified in IEEE 802.1D [9], was designed specifically to prevent loops and thus addresses the problem of broadcast radiation. The protocol achieves this by having all switches in the topology collectively compute a *spanning tree*, a computer science concept whereby from a single root (source MAC address) there is exactly one path to every leaf (destination MAC address) on the tree. STP has been superceded, from a standards perspective at least, by protocols offering improvements on the original protocol's functionality, most recently the *Shortest Path Bridging* (SPB) [10] and the *Transparent Interconnection of Lots of Links* (TRILL) protocols [11]. In practice, though, STP is still prevalent in networks today.

1.5.2 LAYER THREE CONTROL

The router's most fundamental task when it receives a packet is to determine if the packet should be forwarded out one of its interfaces or if the packet needs some exception processing. In the former, normal case, the layer three destination address in the packet header is used to determine over which output port the packet should be forwarded. Unless the destination is directly attached to this router, the router does not need to know exactly where in the network the host is located, it only needs to know the next-hop router to which it should forward the packet. There may be large numbers of hosts that reside on the same destination network, and all of these hosts will share the same entry in the router's *routing table*. This is a table of layer three *networks* and sometimes layer three host addresses, not a table of just layer two destination host addresses as is the case with the layer two forwarding table. Thus, the look-up that occurs in a router is to match the destination address in the packet with one of the networks in the forwarding table. Finding a match provides the next-hop router, which maps to a specific forwarding port, and the forwarding process can continue. In a traditional router, this routing table is built primarily through the use of *routing protocols*. While there is a wide variety of these protocols and they are specialized for different situations, they all serve the common purpose of allowing the router to autonomously construct a layer three forwarding table that can automatically and dynamically change in the face of changes elsewhere in the network. Since these protocols are the basic mechanism by which the traditional layer three switch builds its forwarding table, we provide a brief overview of them here.

The *Routing Information Protocol* (RIP) [12] is a comparatively simple routing protocol that was widely deployed in small networks in the 1990s. Each router in the RIP-controlled routing domain periodically broadcasts its entire routing table on all of its interfaces. These broadcasts include the hop count from the broadcasting router to each reachable network. The weight of each hop can be tailored to accommodate relevant differences between the hops, such as link speed. Neighboring routers can integrate this reachability and hop count information into their own routing tables, which will in turn be propagated to their neighbors. This propagation pervades the entire routing domain or *autonomous system* (AS). The routers in an AS share detailed routing information allowing each member router in the AS to compute the shortest path to any destination within that AS. In a stable network, eventually all of the routing tables in that routing domain will converge, and the hop count information can be used to determine a least-cost route to reach any network in that domain. RIP is a distance-vector protocol, where each router uses the weighted hop count over one interface as its distance to that destination network, and is able to select the best next-hop router as the one presenting the least total distance to the destination. RIP does not need to have a complete picture of the network topology in order to make this distance-vector routing computation. This differs from the next two protocols, *Open Shortest Path First* (OSPF) [13] and *Intermediate System to Intermediate System*, (IS-IS) [14] both of which maintain a complete view of the network topology in order to compute best routing paths. OSPF is an Internet Engineering Task Force project, whereas IS-IS has its origins in OSI.

Both OSPF and IS-IS are link-state dynamic routing protocols. This name derives from the fact that they both maintain a complete and current view of the state of each link in the AS and can thus determine its topology. The fact that each router knows the topology permits them to compute the best routing paths. Since both protocols know the topology, they can represent all reachable networks as leaf nodes on a graph. Internal nodes represent the routers comprising that autonomous system. The edges on the graph represent the communication links joining the nodes. The *cost* of edges may be assigned by any metric, but an obvious one is bandwidth. The standard shortest-path algorithm, known as *Dijkstra's algorithm* is used to compute the shortest path to each reachable network from each router's perspective. For this class of algorithm, the shortest path is defined as the one traversed with the smallest total cost. If a link goes down, this information is propagated via the respective protocol, and each router in the AS will shortly converge to have identical pictures of the network topology. OSPF is currently widely used in large enterprise networks. Large service provider networks tend more to use IS-IS for this purpose. This is primarily due to IS-IS having been a more mature and thus more viable alternative than OSPF in the early service provider implementations. Although OSPF and IS-IS are essentially equally capable at this point in time, there is little incentive for the service providers to change.

While the three routing protocols discussed above are *Interior Gateway Protocols* (IGPs) and therefore concerned with optimizing routing within an AS, the Internet is comprised of a large set of interconnected autonomous systems, and it is obviously important to be able to determine the best path transiting these ASs to reach the destination address. This role is filled by an *Exterior Gateway Protocol* (EGP). The only EGP used in the modern Internet is the *Border Gateway Protocol* (BGP) [15]. BGP routers primarily reside at the periphery of an AS and learn about networks that are reachable via peer edge routers in adjacent ASs. This network routing information can be shared with other routers in the same AS either by using the BGP protocol between routers in the same AS, which is referred to as Internal BGP, or by redistributing the external network routing information into the IGP routing protocol. The most common application of BGP is to provide the routing *glue* between ASs run by different service providers, allowing communication from a host in one to transit through other providers' networks to reach the ultimate destination. Certain private networks may also require the use of BGP to interconnect multiple internal ASs that cannot grow larger due to reaching the maximum scale of the IGP they are using. BGP is called a *path vector protocol*, which is similar to a distance vector protocol like RIP discussed above, in that it does not maintain a complete picture of the topology of the network for which it provides routing information. Unlike RIP, the metric is not a simple distance (i.e., hop count), but involves parameters such as network policies and rules as well as the distance to the destination network. The policies are particularly important when used by service providers as there may be business agreements in place that need to dictate routing choices that may trump simple routing cost computations.

1.5.3 PROTOCOL SOUP OR (S)WITCH'S BREW?

The control protocols listed by name above are only a small part of a long list of protocols that must be implemented in a modern layer two or layer three switch in order to make it a fully functional participant in a state-of-the-art Internet. In addition to those we have mentioned thus far, a modern router will likely also support LDP [16] for exchanging information about MPLS labels, as well as IGMP [17], MSDP [18], and PIM [19] for multicast routing. We see in Fig. 1.4 how the control plane of the switch is bombarded with a constant barrage of control protocol traffic. Indeed, Ref. [6] indicates that in large scale data center networks *thirty percent* of the router's control plane capacity today is spent tracking network topology. We have explained above that the Internet was conceived as and continues to evolve as an organism that is extremely robust in the face of flapping communications links and switches that may experience disruption in their operation. The Internet-family protocols were also designed to gracefully handle routers joining and leaving the topology whether intentionally or otherwise.

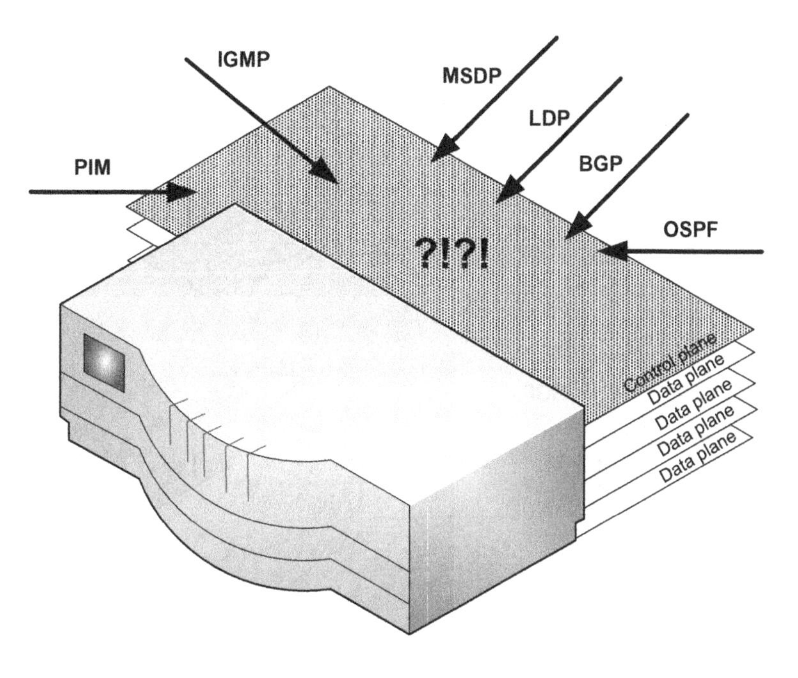

FIG. 1.4

Control plane consternation in the switch.

Such chaotic conditions still exist to varying degrees in the geographically diverse Internet around the globe, though switches and links are generally much more predictable than at the Internet's inception. In particular, the data center is distinct in that it has virtually no unscheduled downtime. While there are links and nodes in the data center that fail occasionally, the base topology is centrally provisioned. Physical servers and VMs do not magically appear or move around in the data center; rather, they only change when the central orchestration software dictates so. Despite the fact that the data center's topology is static and the links are stable, the individual router's travails shown in Fig. 1.4 multiply at data center scale, as depicted in Fig. 1.5. Indeed, since these control protocols are actually distributed protocols, when something is intentionally changed, such as a server being added or a communications link taken out of service, these distributed protocols take time to converge to a set of consistent forwarding tables, wreaking temporary havoc inside the data center. Thus, the overhead experienced by switches shown in Fig. 1.5 is not merely that of four independent units experiencing the problems of the switch shown in Fig. 1.4. Rather, the overhead is exacerbated by the added processing of this convergence. Thus, we reach a situation where most topology change is done programmatically and intentionally, reducing the benefit of the autonomous and distributed protocols, and the scale of the networks is larger than these protocols were designed for, making convergence times unacceptably long.

So, while this array of control protocols was necessary to develop autonomous switches that can dynamically and autonomously respond to rapidly changing network conditions, these conditions do not exist in the modern data center and, by using them in this overkill fashion, we have created a network management nightmare that is entirely avoidable. Instead of drinking this *(s)witch's brew of*

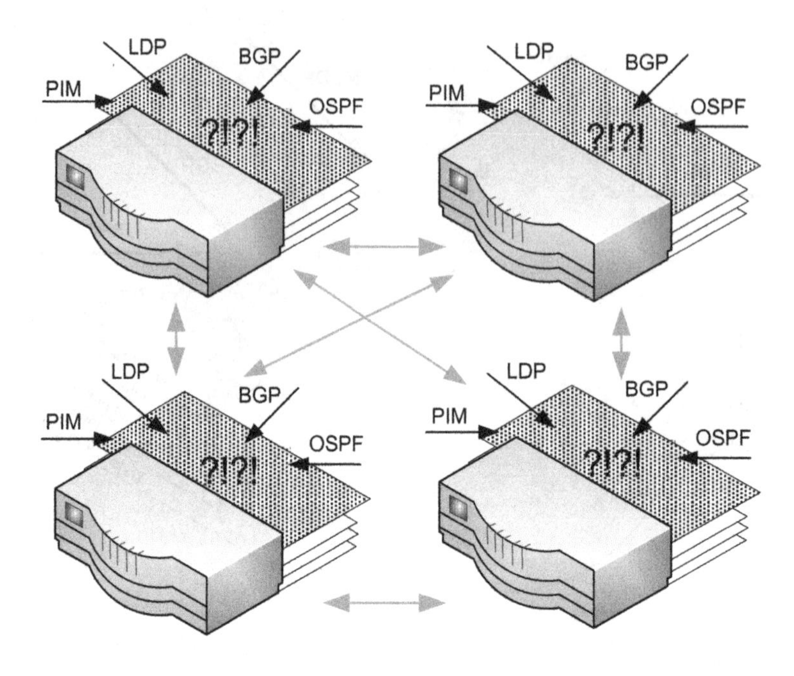

FIG. 1.5

Overhead of dynamic distributed route computation.

protocols, is there a more simple approach? Is there hope for network sobriety by conceiving of a less distributed, less autonomous, but simpler approach of removing the control plane from the switches to a centralized control plane, serving the entire network? This centralized control plane would program the forwarding tables in all the switches in the data center. The simplicity of this concept arises from the facts that (1) the topology inside the data center is quite stable and under strict local administrative control, (2) knowledge of the topology is already centralized and controlled by the same administrators, and (3) when there are node or link failures, this global topology knowledge can be used by the centralized control to quickly reprovision a consistent set of forwarding tables. Thus, it should follow that a framework for programming the switches' forwarding tables like that depicted in Fig. 1.6 might be feasible. This more simple, centralized approach is one of the primary drivers behind SDN. This approach is expected to help address the issue of instability in the control plane which results in long convergence times when changes occur in the network.

There are other important justifications for the migration to SDN. For example, there is a strong desire to drive down the cost of networking equipment. SDN also opens up the possibility of experimentation and innovation by allowing researchers and engineers to build custom protocols in software, which can be easily deployed in the network. We will review these and other arguments underpinning SDN in the next chapter. First, though, we will consider a more fine-grained sort of packet switching that has been pursued for a number of years, and for which SDN has been specifically tailored.

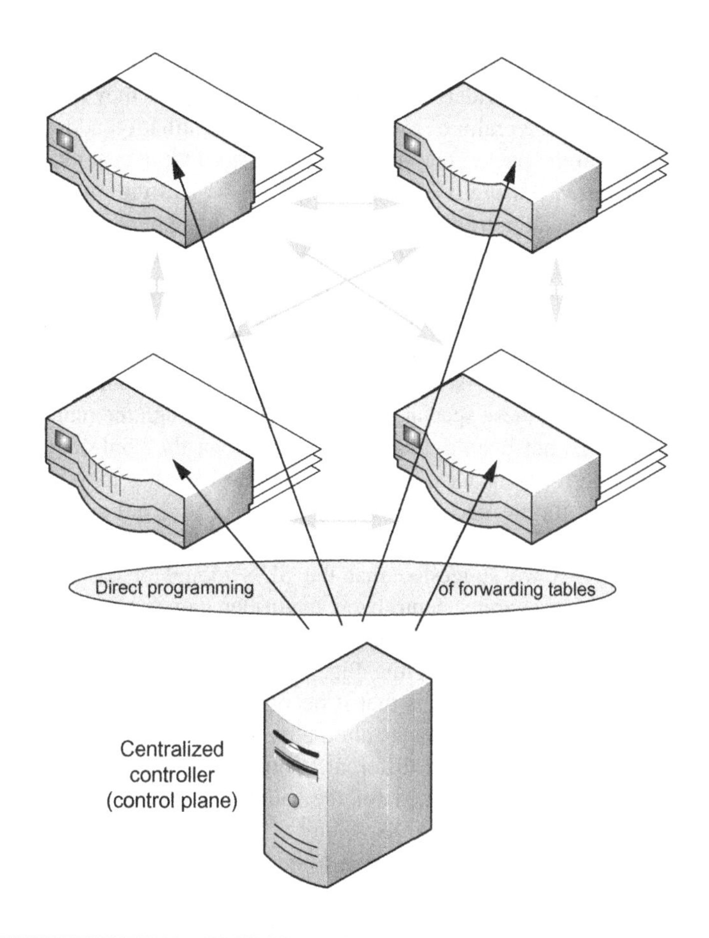

FIG. 1.6

Centralized programming of forwarding tables.

DISCUSSION QUESTION:

A premise of SDN is that the plethora of control plane protocols depicted in Fig. 1.4 creates complexity that is not necessary in all networks. Give an example of a network where the complexity is unnecessary and explain why this is so.

1.6 CAN WE INCREASE THE PACKET FORWARDING IQ?

Whether centrally programmed or driven by distributed protocols, the forwarding tables we have discussed thus far have contained either layer two or layer three addresses. For many years, now, network designers have implemented special processes and equipment that permit more fine-grained forwarding decisions. For example, it may be desirable to segregate traffic destined to different TCP ports on the same IP address such that they travel over separate paths. We may wish to route different QoS classes, such as data, voice, and video, over different paths even if they are directed to the same destination host. Once we leave the realm of routing on just the destination address, we enter a world where many different fields in the packet may be used for forwarding decisions. In general, this kind of routing is called *policy-based routing* (PBR). At this point, we will introduce a new term, *flow*, to describe a particular set of application traffic between two endpoints that receives the same forwarding treatment. On legacy commercial routers, PBR is normally implemented by using *Access Control Lists* (ACLs) to define a set of criteria that determines if an incoming packet corresponds to a particular flow. Thus, if, for example, we decide to forward all video traffic to a given IP address via one port and all email traffic via another port, the forwarding table will treat the traffic as two different flows with distinct forwarding rules for each. While PBR has been in use for a number of years, this has only been possible by configuring these special forwarding rules through the management plane. Such more sophisticated routing has not been dynamically updated via the control plane in the way that simple layer two and three destinations have been. The fact that PBR has long been implemented via central configuration provides further underpinning for SDN's centralized approach of programming forwarding tables.

At the end of Section 1.5.3 we suggested that the SDN model of breaking the control plane out of the switch itself into a separate, centralized controller can provide greater simplicity and flexibility in programming the forwarding tables of the switches in a large and tightly controlled network, such as the modern data center. While Fig. 1.6 hints at an architecture that might allow us such programming capability, the reality is that it becomes possible to do more than just replace the distributed routing algorithms with a cleaner alternative. In particular, we can identify particular user application traffic flows and give them different treatment, routing their traffic along different paths in some cases, even if they are destined for the same network endpoint. In this manner, we can incorporate PBR at the ground level of SDN. Many of the specialized network appliances, such as firewalls, load balancers, and *Intrusion Detection Systems* (IDS), have long relied on performing deeper packet inspection than just the destination address. By looking at source and destination address, and source and destination TCP port number, these devices have been able to segregate and quarantine traffic in a highly useful manner. If we consider that the network's forwarding tables are to be programmed from a centralized controller with global network knowledge, it follows naturally that the kind of information utilized by these special appliances could be programmed directly into the switches themselves, possibly obviating the need for the separate network appliances and simplifying the network topology. If the switches' forwarding tables have the capability to filter, segregate, and quarantine flows, the power of the network itself could grow exponentially. From its birth, SDN has been based on the notion of constructing forwarding tables defining actions to take on flows, rather than having forwarding tables merely map destination address to output port.

SDN purports to incorporate a richer set of information in its processing of individual flows, including fields like the TCP port number which provide clues as to which application that flow corresponds. Nonetheless, we emphasize that the SDN intelligence added at the switching layers will not require any modifications to user applications. SDN is intended to affect how layer two and layer three forwarding decisions are made. As currently defined, SDN can inspect and potentially modify layer two, three, and four protocol header information, but is not intended to make decisions on, nor modify, higher layer protocol fields.

DISCUSSION QUESTION:

In what way is policy-based routing a precursor of SDN?

1.7 OPEN SOURCE AND TECHNOLOGICAL SHIFTS

The SDN paradigm that this book examines represents a major technological deviation from the legacy network protocols and architectures that we have reviewed in this chapter. Those legacy technologies required countless man-years of engineering effort to develop and refine. A paradigm-shifting technology such as SDN will also require a huge engineering effort. While some of this effort will come from commercial enterprises that expect to follow the SDN star to financial success, it is widely expected that SDN will be able to rapidly provide matching functionality to legacy systems by counting on the open source community to develop much of the SDN control plane software.

The open source model has revolutionized the way that software is developed and delivered. Functionality that used to be reinvented in every organization is now often readily available and being incorporated into enterprise-class applications. Often software development organizations no longer have to devote time and energy to implement features that have already been developed.

Probably the most famous and influential body of open source code today is the Linux operating system. This complex and high-quality body of code is still developed by an amorphous and changing team of volunteer software developers around the world. It has changed the markets of commercial operating system vendors and improved quality of both open source and competitive retail products. Numerous businesses have been spawned by innovating on top of this body of work. Small technology start-ups with minimal funding have had free source access to a world-class operating system, allowing them to innovate without incurring onerous licensing fees that may have prevented them from ever escaping their garage.

Another important example of open source is the OpenSSL implementation [20]. Encryption software is notoriously complex, yet is required in more and more products in today's security-conscious world. This very complete body of asymmetrical and symmetrical encryption protocols and key algorithms has been a boon to many researchers and technology start-ups that could never afford to develop or license this technology, yet need it to begin pursuing their security-related ideas.

In the networking world, open-source implementations of routing protocols such as BGP, OSPF, RIP, and switching functionality such as Spanning Tree, have been available for many years now. Unlike the PC world, where a common *Wintel* hardware platform has made it possible to develop software trivially ported to numerous PC platforms, the hardware platforms from different *Network Equipment Manufacturers* (NEMs) remain quite divergent. Thus, porting the open source implementations of these complex protocols to the proprietary hardware platforms of network switches has always been labor-intensive. Moving the control plane and its associated complex protocols off of the proprietary switch hardware and onto a centralized server built on a PC platform makes the use of open source control plane software much more feasible.

It is not possible to fully grasp the origins of SDN without understanding that the early experimentation with SDN was done in a highly collaborative and open environment. Specifications and software were exchanged freely between participating university groups. An impressive amount of pioneering engineering work was performed without any explicit commercial funding, all based on the principle that something was being created for the greater good. It truly was and is a kind of network-world reflection of the Linux phenomenon. With few exceptions, the most vocal supporters of SDN have been individuals and institutions that did not intend to extract financial profit from the sales of the technology. This made a natural marriage with open source possible. We will see in later chapters of this book that SDN has begun to have such a large commercial impact that an increasing number of proprietary implementations of SDN-related technology are reaching the market. This is an unsurprising companion on almost any journey where large sums of money can be made. Nevertheless, there remains a core SDN community that believes that openness should remain an abiding characteristic of SDN.

DISCUSSION QUESTION:

While open source implementations of many control plane protocols exist today, it remains challenging to port these to traditional switches. What characteristic(s) of SDN could potentially make open source control plane software more ubiquitous?

1.8 ORGANIZATION OF THE BOOK

The first two chapters of the book define terminology about packet switching and provide a historical context for evolution of the data, control, and management planes extant in today's switches. We describe some of the limitations of the current technology and explain how these motivated the development of SDN technology. In Chapter 3 we explain the historical origins of SDN. Chapter 4 presents the key interfaces and protocols of SDN and explains how the traditional control plane function is implemented in a novel way. Up to this point in the book our references to SDN are fairly generic. Also in Chapter 4 we begin to distinguish the original or *Open* SDN from proprietary alternatives that also fall under the SDN umbrella. When we use the term *Open* SDN in this work, we refer to SDN as it was developed and defined in research institutions starting around 2008. Open SDN is tightly coupled with the *OpenFlow* protocol presented in Chapter 5 and also strongly committed to the openness of the protocols and specifications of SDN. Alternative definitions of SDN exist that share some but not all of the basic precepts of open SDN. In Chapter 6 we provide more details about these alternative SDN implementations, as well as a discussion about limitations and potential drawbacks of using SDN. Since this book was originally published, commonly accepted definitions of what constitutes SDN have been stretched considerably. SDN control protocols and controller models that formerly lay on the sidelines have become mainstream. We discuss the major new protocols and controllers in Chapter 7. In Chapter 8 we present use cases related to the earliest driver of interest in SDN, the exploding networking demands of the data center. In Chapter 9 we discuss additional use cases that contribute to the current frenzy of interest in this new technology, most notably the application of SDN in wide-area networks. One of the most significant aspects of these use cases is that of *Network Functions Virtualization* (NFV) wherein physical appliances such as *IDS* can be virtualized via SDN. This particular class of SDN use cases is significant enough to merit treatment separately in Chapter 10. The book provides in Chapter 11 an inventory of major enterprises and universities developing SDN products today. Chapter 12 presents several real-world SDN applications, including a tutorial on writing one's own SDN application. Chapter 13 provides a survey of open source initiatives that will contribute to the standardization and dissemination of SDN. In Chapter 14 we examine the impact that this new technology is already making in the business world, and consider what likely future technological and business consequences may be imminent. The final chapter delves into current SDN research directions and considers several novel applications of SDN. There is an appendix of SDN-related acronyms, source code from a sample SDN application and a comprehensive user-friendly index.

REFERENCES

[1] Network effect. The Economist 2012;405(8815):67–8.
[2] Bertsekas D, Gallager R. Data networks. Englewood Cliffs, NJ: Prentice Hall; 1992.
[3] Internet Hall of Fame. Paul Baran. Retrieved from: http://internethalloffame.org/inductees/paul-baran.
[4] Christy P. OpenFlow and Open Networking: an introduction and overview. Ethernet Technology Summit, February 2012, San Jose, CA; 2012.
[5] de Prycker M. Asynchronous transfer mode—solution for broadband ISDN. London: Ellis Horwood; 1992.
[6] Gahsinsky I. Warehouse scale datacenters: the case for a new approach to networking. Open Networking Summit, October 2011. Stanford University; 2011.
[7] Guis I. Enterprise Data Center Networks. Open Networking Summit, April 2012, Santa Clara, CA; 2012.
[8] Kerravala Z. Arista launches new security feature to cover growing East-to-West data center traffic. Network World, October 6; 2015. Retrieved from: http://www.networkworld.com/article/2989761/cisco-subnet/arista-launches-new-security-feature-to-cover-growing-east-to-west-data-center-traffic.html.
[9] Information technology—telecommunications and information exchange between systems—local and metropolitan area networks—common specifications—Part 3: Media Access Control (MAC) Bridges. IEEE P802.1D/D17, May 25, 1998.
[10] Draft Standard for Local and Metropolitan Area Networks—Media Access Control (MAC) Bridges and Virtual Bridged Local Area Networks—Amendment XX: Shortest Path Bridging. IEEE P802.1aq/D4.6, February 10, 2012.
[11] Perlman R, Eastlake D, Dutt D, Gai S, Ghanwani A. Routing Bridges (RBridges): Base Protocol Specification, RFC 6325. Internet Engineering Task Force; 2011.
[12] Hendrick C. Routing Information Protocol, RFC 1058. Internet Engineering Task Force; 1988.
[13] Moy J. OSPF Version 2, RFC 2328. Internet Engineering Task Force; 1998.
[14] Oran D. OSI IS-IS Intra-domain Routing Protocol, RFC 1142. Internet Engineering Task Force; 1990.
[15] Meyer D, Patel K. BGP-4 protocol analysis, RFC 4274. Internet Engineering Task Force; 2006.
[16] Andersson L, Minei I, Thomas B. LDP specification, RFC 3036. Internet Engineering Task Force; 2007.
[17] Cain B, Deering S, Kouvelas I, Fenner B, Thyagarajan A. Internet Group Management Protocol, Version 3, RFC 3376. Internet Engineering Task Force; 2002.
[18] Fenner B, Meyer D. Multicast Source Discovery Protocol (MSDP), RFC 3618. Internet Engineering Task Force; 2003.
[19] Fenner B, Handley M, Holbrook H, Kouvelas I. Protocol Independent Multicast—Sparse Mode (PIM-SM): Protocol Specification (Revised), RFC 2362. Internet Engineering Task Force; 2006.
[20] OpenSSL. Retrieved from: https://www.openssl.org.

CHAPTER

WHY SDN?

2

Networking devices have been successfully developed and deployed for several decades. Repeaters and bridges, followed by routers and switches, have been used in a plethora of environments, performing their functions of filtering and forwarding packets throughout the network toward their ultimate destination. Despite the impressive track record of these traditional technologies, the size and complexity of many modern deployments leaves them lacking. The reasons for this include the ever-increasing costs of owning and operating networking equipment, the need to accelerate innovation in networking and, in particular, the increasing demands of the modern data center. This chapter investigates these trends and describes how they are nudging networking technology away from traditional methods and protocols toward the more open and innovation-friendly paradigm of SDN.

2.1 EVOLUTION OF SWITCHES AND CONTROL PLANES

We begin with a brief review of the evolution of switches and control planes that has culminated in a fertile playing field for SDN. This complements the material presented in Sections 1.4 and 1.5. The reader may find it useful to employ Fig. 2.1 as a visual guide through the following sections, as it provides a graphical summary of this evolution and allows the reader to understand the approximate timeframes when different components of switching moved from software to hardware.

2.1.1 SIMPLE FORWARDING AND ROUTING USING SOFTWARE

In Chapter 1 we discussed the early days of computer networking, where almost everything other than the physical layer (layer one) was implemented in software. This was true for end-user systems as well as for networking devices. Whether the devices were bridges, switches, or routers, software was used extensively inside the devices in order to perform even the simplest of tasks, such as MAC-level forwarding decisions. This remained true even through the early days of the commercialized Internet in the early 1990s.

2.1.2 INDEPENDENCE AND AUTONOMY IN EARLY DEVICES

Early network device developers and standards-creators wanted each device to perform in an autonomous and independent manner, to the greatest extent possible. This was because networks were generally small and fixed, with large shared domains. A goal also was to simplify rudimentary management tasks and to make the networks as *plug and play* as possible. Their relatively static

Software Defined Networks. http://dx.doi.org/10.1016/B978-0-12-804555-8.00002-8

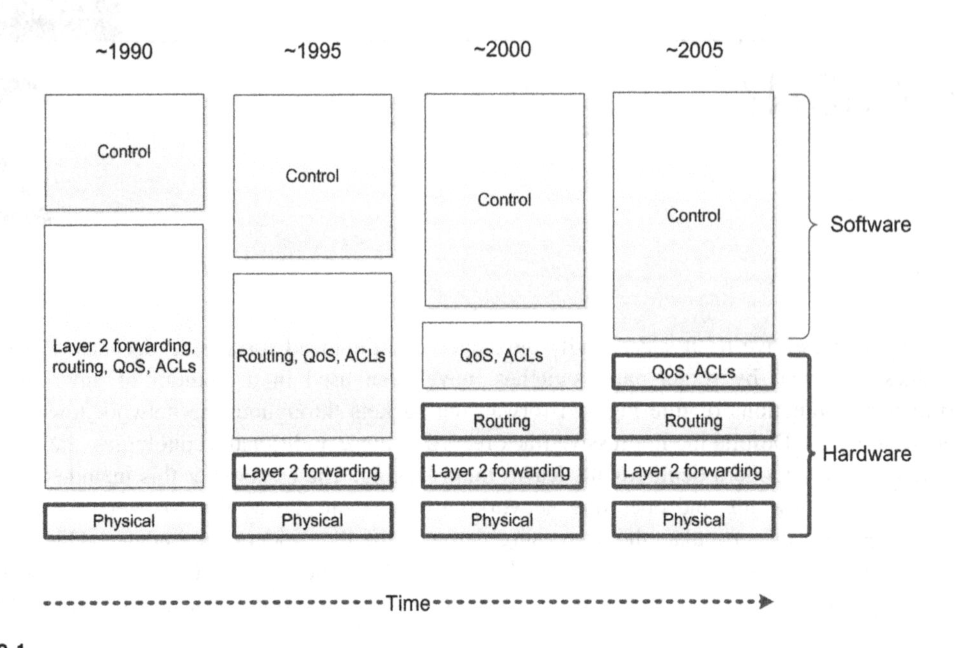

FIG. 2.1

Networking functionality migrating to hardware.

configuration needs were performed manually. Developers went to great lengths to implement this distributed environment with intelligence resident in every device. Whenever coordination between devices was required, collective decisions could be made through the collaborative exchange of information between devices.

Interestingly, many of the goals of this distributed model, such as simplicity, ease-of-use, and automatic recovery, are similar to the goals of SDN, but as the scale and complexity of networks grew, the current distributed model has become increasingly dysfunctional.

Examples of this distributed intelligence are the layer two (bridging) and layer three (routing) protocols, which involved negotiation between the devices in order to reach a consensus of how forwarding and routing would be performed. We introduced these protocols in Chapter 1 and provide more details on them below.

- **Spanning Tree Protocol**
  Basic layer two forwarding, also known as *transparent bridging*, can be performed independently by each switch in the network. However, certain topologies require an imposition of a hierarchy on the network in order to prevent loops which would cause broadcast radiation. The *Spanning Tree Protocol* (STP) is an example of the operation of autonomous devices participating in a distributed decision-making process in order to create and enforce a hierarchy on the network. The result is the correct operation of transparent bridging throughout the domain, at the expense of convergence latency and possibly arbitrary configuration. This solution was a trade-off between cost and complexity. Multiple paths could have been supported but at greater cost. While STP was adequate

when networks were of smaller scale, as networks grew the spanning tree solution has become problematic. These problems manifest themselves in a striking fashion when networks reach the scale of the modern data center. For example, IEEE 802.1D specifies the following default timers for STP: fifteen seconds for listening, fifteen seconds for learning, and twenty seconds for max-age timeout. In older networks, convergence times of thirty to fifty seconds were common. Such delays are not acceptable in today's data centers. As the scale of the layer two network grows, the likelihood of greater delays increases. The *Rapid Spanning Tree Protocol* (RSTP) protocol, specified in IEEE 802.1D-2004 [1], improves this latency significantly but unfortunately is not deployed in many environments.

- **Shortest Path Bridging**
  STP allowed only one active path to a destination, suffered from relatively slow convergence times, and was restricted to small network topologies. While the newer implementations of STP have improved the convergence times, the single active path shortcoming has been addressed in a new layer two protocol, SPB, introduced in Section 1.5.1. SPB is a mechanism for allowing multiple concurrent paths through a layer two fabric through collaborative and distributed calculation of shortest and most efficient paths, and sharing that information amongst the participating nodes in the meshed network. This characteristic is called *multipath*. SPB accomplishes this by utilizing IS-IS to construct a graph representing the layer two link-state topology. Once this graph exists, shortest path calculations are straightforward, though more complex than with spanning tree.
  To elaborate on what we mean by shortest path calculations, in Fig. 2.2 we depict a simple graph that can be used for calculating shortest paths in a network with six switches. The costs assigned to the various links may be assigned their values according to different criteria. A simple criterion is

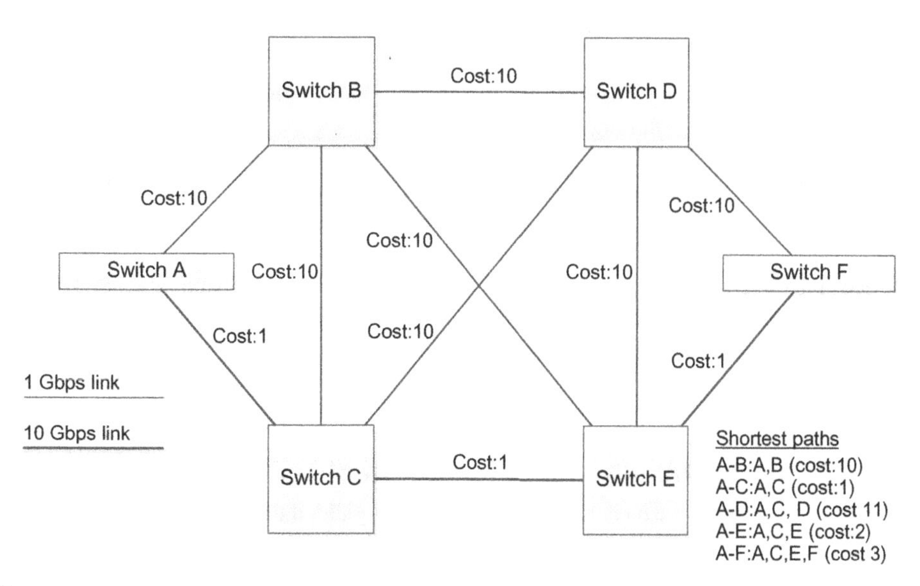

FIG. 2.2

Example of graph of network for shortest path calculation.

to make the cost of a network link inversely proportional to its bandwidth. Thus, the cost of transiting a ten Gbps link is one-tenth that of transiting a one Gbps link. When the shortest path calculation is complete, the node performing the calculation knows the least-cost path to any of the other nodes in the network. The least-cost path is considered the shortest path. For the sake of clarity, we should point out that IS-IS is used in the SPB context strictly for layer two path calculation. This differs from its classical application in calculating layer three routes, as described below. In the trivial example of Fig. 2.2, there is a single shortest path from node *A* to every other node. In real life networks it is common for there to be more than one least cost path between two nodes. The multipath characteristic of SPB would allow the traffic to be distributed across those multiple paths.

- **RIP, BGP, OSPF, and IS-IS**
  Routing at layer three requires cooperation between devices in order to know which routers are attaching which subnets to the network. In Chapter 1 we provided background on four routing protocols: RIP, BGP, OSPF, and IS-IS. These routing protocols involve the sharing of local routing information by each device, either at the edge of the network or as an intermediate node. Their collective sharing of information allows the routing state to converge as devices share their information with each other. Each router remains autonomous in terms of its ability to make routing decisions as packets arrive. This process is one of peers sharing and negotiating amongst themselves, without a centralized entity aiding in the decision.

2.1.3 SOFTWARE MOVES INTO SILICON

Vendors originally had to write their own software to implement even the basic functions like layer two forwarding and routing. Fig. 2.1 shows that over time these more basic functions moved from software into hardware. We now see most forwarding and filtering decisions implemented entirely in hardware. These decisions are driven by configured tables, set by the control plane software above. This shift of the lower-level decision-making from software to hardware has yielded tremendous improvements in the performance/cost ratio of switching equipment.

Today, switching devices are typically composed of hardware components such as *Application-Specific Integrated Circuits* (ASICs), *Field-Programmable Gate Arrays* (FPGAs), and *Ternary Content-Addressable Memories* (TCAMs). The combined power of these integrated circuits allows for the forwarding decisions to be made entirely in the hardware at line rate. This has become more critical as network speeds have increased from one Gbps to ten Gbps, to forty Gbps, and beyond. The hardware is now capable of handling all forwarding, routing, *Access Control List* (ACL), and QoS decisions. Higher-level control functions, responsible for network-wide collaboration with other devices, are implemented in software. This control software runs independently in each network device.

2.1.4 HARDWARE FORWARDING AND CONTROL IN SOFTWARE

The network device evolution we have recounted thus far has yielded the following current situation:

- **Bridging** (Layer Two Forwarding)
  Basic layer two MAC forwarding of packets is handled in the hardware tables.
- **Routing** (Layer Three Forwarding)

In order to keep up with today's high-speed links and to route packets at link speeds, layer three forwarding functionality is also implemented in hardware tables.

- **Advanced Filtering and Prioritization**
  General traffic management rules, such as ACLs, which filter, forward, and prioritize packets, are handled via hardware tables located in the hardware (e.g., in TCAMs), and accessed through low-level software.
- **Control**
  The control software used to make broader routing decisions and to interact with other devices in order to converge on topologies and routing paths is implemented in software that runs autonomously inside the devices. Since the current control plane software in networking devices lacks the ability to distribute policy information about such things as security, QoS and ACLs, these features must still be provisioned through relatively primitive configuration and management interfaces.

Given this landscape of (1) layer two and layer three hardware handling most forwarding tasks, (2) software in the device providing control plane functionality, and (3) policy implemented via configuration and management interfaces, an opportunity presents itself to simplify networking devices and move forward to the next generation of networking.

2.1.5 THE GROWING NEED FOR SIMPLIFICATION

In [2] the authors state that one of the major drivers for SDN is simplification. As time has passed, networking devices have become increasingly more complex. This is due in part to the existing independent and autonomous design of devices that make it necessary that so much intelligence be placed inside each device. Placing more functionality in hardware in some ways simplifies the device, but in other ways makes it more complicated because of the difficult handshakes and tradeoffs between handling packets in hardware versus software.

Attempting to provide simplicity by adding features to legacy devices tends to complicate implementations rather than simplifying them. An analogy to the evolution of the *Central Processing Unit* (CPU) can be made here. Over time CPUs became highly complex as they attempted to support more and more functions. Ultimately, another simpler, easier to use CPU model emerged which was called the *Reduced Instruction Set Computing* (RISC) model. In the same way the RISC architecture served as a reset to CPU architecture, so, too, SDN may serve as a simplifying reset for network equipment design.

In addition to simplifying the devices themselves, there is an opportunity to simplify the management of the networks of these devices. Rather than using primitive network management tools such as SNMP and CLI, network operators would prefer to use policy-based management systems. SDN may enable such solutions [3].

DISCUSSION QUESTION:

Explain the analogy between RISC CPU architecture in computation and SDN in networking.

2.1.6 MOVING CONTROL OFF OF THE DEVICE

We remind the reader that control software in our context is the intelligence that determines optimal paths and responds to outages and new networking demands. At its core, SDN is about moving that control software off of the device and into a centrally located compute resource which is capable of seeing the entire network and making decisions which are optimal, given a complete understanding of the situation. While we will discuss this in much greater detail in the chapters that follow, basically, SDN attempts to segregate network activities in the following manner:

- **Forwarding, Filtering, and Prioritization**
  Forwarding responsibilities, implemented in hardware tables, remain on the device. In addition, features such as filtering based on ACLs and traffic prioritization, are enforced locally on the device as well.
- **Control**
  Complicated control software is removed from the device and placed into a centralized controller, which has a complete view of the network and the ability to make optimal forwarding and routing decisions. There is a migration to a programming paradigm for the control plane. The basic forwarding hardware on the networking device is available to be programmed by external software on the controller. The control plane is no longer embedded, closed, closely coupled with the hardware, or optimized for particular embedded environments.
- **Application**
  Above the controller is where the network applications run, implementing higher-level functions and, additionally, participating in decisions about how best to manage and control packet forwarding and distribution within the network.

Subsequent chapters will examine in greater detail how this can be achieved, with a minimum of investment and change by networking vendors, while providing the maximum control and capability by the controller and its applications. The next section of this chapter will discuss another major reason why SDN is needed today—the cost of networking devices.

2.2 COST

Arguments related to the need for SDN often include cost as a driving factor for this shift [4,5]. In this section we consider the impact of the status quo in networking on the cost of designing, building, purchasing, and operating network equipment.

2.2.1 INCREASED COST OF DEVELOPMENT

Today's autonomous networking devices must store, manage, and run the complicated control plane software that we discussed in the previous section. Over time, the result of this increased control plane sophistication is an increase in the amount of control plane software in the device, as can be seen in Fig. 2.1. Despite the overall downward trend in the cost of networking hardware, this growing complexity acts as an upward pressure on the hardware component costs due to the processing power required to run that advanced software as well as the storage capacity to hold it.

In Chapter 1 we described how software development outside the networking realm benefits greatly from the readily available open source software. For example, application server frameworks provide platforms which allow software developers to reuse code provided by those common frameworks, and, therefore to concentrate on solving domain-specific problems. Without the ability to leverage software functionality in this manner, each vendor has to develop, test, and maintain large amounts of redundant code, which is not the case in an open software environment. With the closed networking environment that is prevalent today, little such leverage is available, and, consequently, each vendor must implement all of the common functionality required by their devices. Common network functionality and protocols must be developed by every device vendor. This clearly increases the costs attributable to software development.

In recent years, silicon vendors have been producing *common off-the-shelf* (COTS) ASICs which are capable of speeds and functionality that rivals or surpasses the proprietary versions developed by networking hardware vendors. However, given the limited software leverage mentioned above, vendors are often unable to efficiently make use of these *merchant silicon* chips, since software must be re-engineered for each product line. So while their products may be quite profitable, NEMs must write and support larger amounts of software than would otherwise be necessary if networking devices were developed in truly open environments. The fact that such a large body of software must run on each and every network device serves to further increase this cost. There is additional overhead resulting from the requirement to support multiple versions of legacy protocols as well as keeping up with the latest protocols being defined by standards bodies.

DISCUSSION QUESTION:

Even with a wholesale migration to SDN there would still be complex control plane software needed to program the data plane. Surely, there would be improvements in this software over time that would result in multiple versions of this software. Why, then, would this represent less complexity than the current paradigm?

2.2.2 CLOSED ENVIRONMENTS ENCOURAGE VENDOR LOCK-IN

It is true that over the years standards have been developed in the networking space for most relevant protocols and data that are used by switches and routers. For the most part, vendors do their best to implement these standards in a manner that allows heterogeneous networks of devices from multiple vendors to co-exist with one another.

However, in spite of good intentions by vendors, enhancements are often added to these standard implementations, which attempt to allow a vendor's product to outperform its competition. With many vendors adding such enhancements, the end result is that each vendor product will have difficulty interoperating smoothly with products from another vendor. Adherence to standards helps alleviate the issues associated with attempting to support multiple vendor types in a network, but problems with interoperability and management often far outweigh the advantages that might be gained by choosing another vendor. As a result, customers frequently become effectively married to a vendor they chose years or even decades before. This sort of vendor lock-in alleviates downward pressure on cost as the vendor is largely safe from competition and can thus preserve high profit margins.

2.2.3 COMPLEXITY AND RESISTANCE TO CHANGE

Quite often in networking we arrive at a point of having made the network operational and the normal impulse from that point on is to just leave things as they are, to not disturb it lest it break and we must start all over again. Others may have been burned by believing the latest vendor who proposed a new solution and, when the dust settled, their closed, proprietary, vendor-specific solution was just as complex as that of the previous vendor.

Unfortunately, in spite of efforts at standardization, there is still a strong argument to stay with that single-vendor solution. Often, that closed, complex solution may be easier to deploy precisely because there is only one vendor involved, and that vendor's accountability is not diluted in any way. By adopting and embracing a solution that works, we believe we lower our short-term risk. That resistance to change results in long-term technological stagnation and sluggishness. The ideal would be a simpler, more progressive world of networking, with open, efficient, and less expensive networking devices. This is a goal of SDN.

2.2.4 INCREASED COST OF OPERATING THE NETWORK

As networks become ever-larger and more complex, the *Operational Expense* (OPEX) of the network grows. This component of the overall costs is increasingly seen to be more significant than the corresponding *Capital Expense* (CAPEX) component. SDN has the capacity to accelerate the automation of network management tasks in a multivendor environment [6,7]. This, combined with the fact that SDN will permit faster provisioning of new services and provides the agility to switch equipment between different services [8] should lead to lower OPEX with SDN. In Section 15.3.6, we will examine proposals where SDN may be used in the future to reduce the power consumption of the networking equipment in a data center, which is another major contributor to network OPEX.

2.3 SDN IMPLICATIONS FOR RESEARCH AND INNOVATION

Networking vendors have enjoyed an enviable position for over two decades. They control networking devices from the bottom up: the hardware, the low level firmware, and the software required to produce an intelligent networking device. This platform on which the software runs is closed, and, consequently, only the networking vendors themselves can write the software for their own networking devices.

The reader should contrast this to the world of software and computing, where you can have many different hardware platforms created by multiple vendors, consisting of different and proprietary capabilities. Above that hardware reside multiple layers of software, which ultimately provide a common and open interface to the application layer. The *Java Virtual Machine* and the *Netbeans Integrated Development Environment* provide a good example of cross-platform development methods. Using such tools, one can develop software that may be ported between Windows PC, Linux, or an Apple MAC. Analogously, tablets and smartphones, such as iPhones or Android-based phones, can share application software and can run on different platforms with relative ease. Imagine if only Apple were able to write software for Apple products and only Microsoft was able to write software to run on

PCs or Windows-based servers? Would the technological advances we have enjoyed in the last decade have taken place? Probably not.

In [9] the authors explain that this status quo has negatively impacted innovation in networking. In the next section, we examine this relationship and, how, on the contrary, the emergence of SDN is likely to accelerate such innovation.

2.3.1 STATUS QUO BENEFITS INCUMBENT VENDORS

The collective monopolies extant in the networking world today are advantageous to networking vendors. Their only legitimate competition comes from their fellow established NEMs. Although, periodically, new networking companies emerge to challenge the status quo, that is the exception rather than the rule. The result is that the competitive landscape evolves at a much slower pace than it would in a truly open environment. This is due to limited incentives to invest large amounts of money when the current status quo is generating reasonable profits, due primarily to the lack of real competition and the resulting high margins they are able to charge for their products.

Without a more competitive environment, the market will naturally stagnate to some degree. The incumbent NEMs will continue to behave as they have in the past. The small players will struggle to survive, attempting to chip away at the industry giants, but with limited success, especially since the profit margins of those giants are so large. This competition-poor environment is unhealthy for the market and for the consumers of its products and services. Consider the difference in profit margins for a server vendor versus that for a networking vendor. Servers are sold at margins close to five percent or below. Networking device margins can be as high as thirty percent or more for the established vendors.

2.3.2 SDN PROMOTES RESEARCH AND INNOVATION

Universities and research labs are focal points of innovation. In technology, innovations by academia and other research organizations have accelerated the rate of change in numerous industries. Open software environments such as Linux have helped to promote this rapid pace of advancement. For instance, if researchers are working in the area of operating systems, they can look at Linux and modify its behavior. If they are working in the area of server virtualization or databases, they can look at KVM or Xen and MySQL or Postgres. All of these open-source packages are used in large-scale commercial deployments. There is no equivalent in networking today. Unfortunately, the current closed nature of networking software, network protocols, network security, and network virtualization is such that it has been challenging to experiment, test, research, and innovate in these areas. This, in fact, is one of the primary drivers of SDN [4]. In that case, a number of universities collaborated to propose a new standard for networking called OpenFlow, which would allow for this free and open research to take place. This makes one wonder if SDN will ultimately be to the world of networking what Linux has become to the world of computing.

General innovation, whether from academia or by entrepreneurs, is stifled by the closed nature of networking devices today. How can a creative researcher or entrepreneur develop a novel mechanism for forwarding traffic through the Internet? That would be nearly impossible today. Is it reasonable for a start-up to produce a new way of providing hospitality-based networking access in airports, coffee shops, and malls? Perhaps, but it would be required to run across network vendor equipment such as *wireless access points* (APs) and access switches and routers, that are themselves closed systems. The more that the software and hardware components of networking are commoditized, the lower their cost to customers. SDN promises both the hardware commoditization as well as openness, and both of these factors contribute to innovation.

To be fair, though, one must keep in mind that innovation is also driven by the prospect of generating wealth. It would be naive to imagine that a world of low-cost networking products will have a purely positive impact on the pace of innovation. For some companies, the lower product margins presaged by SDN will reduce their willingness to invest in innovation. We will examine this correlation and other business ramifications of SDN in Chapter 14.

2.4 DATA CENTER INNOVATION

In Section 1.3 we explained that in the last few years, server virtualization has caused both the capacity and the efficiency of data centers to increase exponentially. This unbounded growth has made possible new computing trends such as the cloud, which is capable of holding massive amounts of computing power and storage capacity. The whole landscape of computing has changed as a result of these technological advances in the areas of compute and storage virtualization.

2.4.1 COMPUTE AND STORAGE VIRTUALIZATION

Virtualization technology has been around for decades. The first commercially available VM technology for IBM mainframes was released in 1972 [10], complete with the *hypervisor* and the ability to abstract the hardware below, allowing multiple heterogeneous instances of other operating systems to run above it in their own space. In 1998 VMware was established and began to deliver software for virtualizing desktops as well as servers.

Use of this *compute virtualization* technology did not explode until data centers became prevalent, and the need to dynamically create and tear down servers, as well as moving them from one physical server to another, became important. Once this occurred, however, the state of data center operations immediately changed. Servers could be instantiated with a mouse click, and could be moved without significantly disrupting the operation of the server being moved.

Creating a new VM, or moving a VM from one physical server to another, is straightforward from a server administrator's perspective, and may be accomplished very rapidly. Virtualization software, such as VMware, Hyper-V, KVM, and XenServer, are examples of products that allow server administrators to readily create and move virtual machines. This has reduced the time needed to start up a new instance of a server to a matter of minutes or even seconds. Fig. 2.3 shows the simple creation of a new instance of a virtual machine on a different physical server.

Likewise, *storage virtualization* has existed for quite some time, as has the concept of abstracting storage blocks and allowing them to be separated from the actual physical storage hardware. As with servers, this achieves efficiency in terms of speed (e.g., moving frequently used data to a faster device), as well as in terms of utilization (e.g., allowing multiple servers to share the same physical storage device).

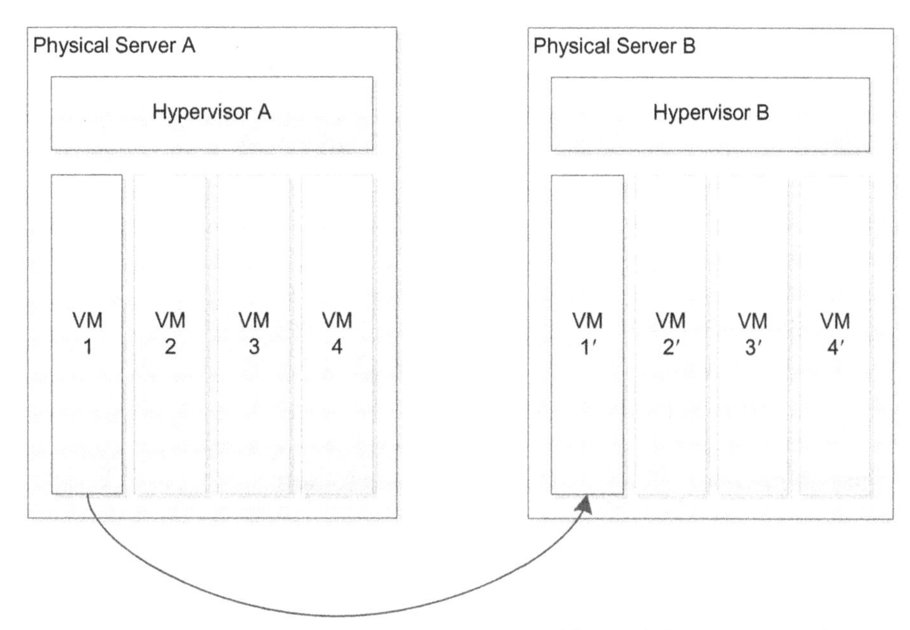

FIG. 2.3

Server virtualization: creating new VM instance.

These technological advancements allow servers and storage to be manipulated quickly and efficiently. While these advances in computer and storage virtualization have been taking place, the same has not been true in the networking domain [11].

2.4.2 INADEQUACIES IN NETWORKS TODAY

In Chapter 1 we discussed the evolution of networks which allowed them to survive catastrophic events such as outages and hardware or software failures. In large part, networks and networking devices have been designed to overcome these rare but severe challenges. However, with the advent of data centers, there is a growing need for networks to not only recover from these types of events, but also to be able to respond quickly to frequent and immediate changes.

While the tasks of creating a new network, moving a new network, and removing a network are similar to those performed for servers and storage, doing so requires work orders, coordination between server and networking administrators, physical or logical coordination of links, *Network Interface Cards* (NICs), and ToR switches, to name a few. These time-consuming tasks are reflected in Fig. 2.4, which illustrates the difference in elapsed time between creating a new instance of a VM, which is in the order of minutes, compared to the multiple days that it may take to create a new instance of a network. This disparity is due to the fact that the servers are virtualized yet the network is still purely a physical network. Even when we are configuring something virtual for the network, such as a VLAN, changes

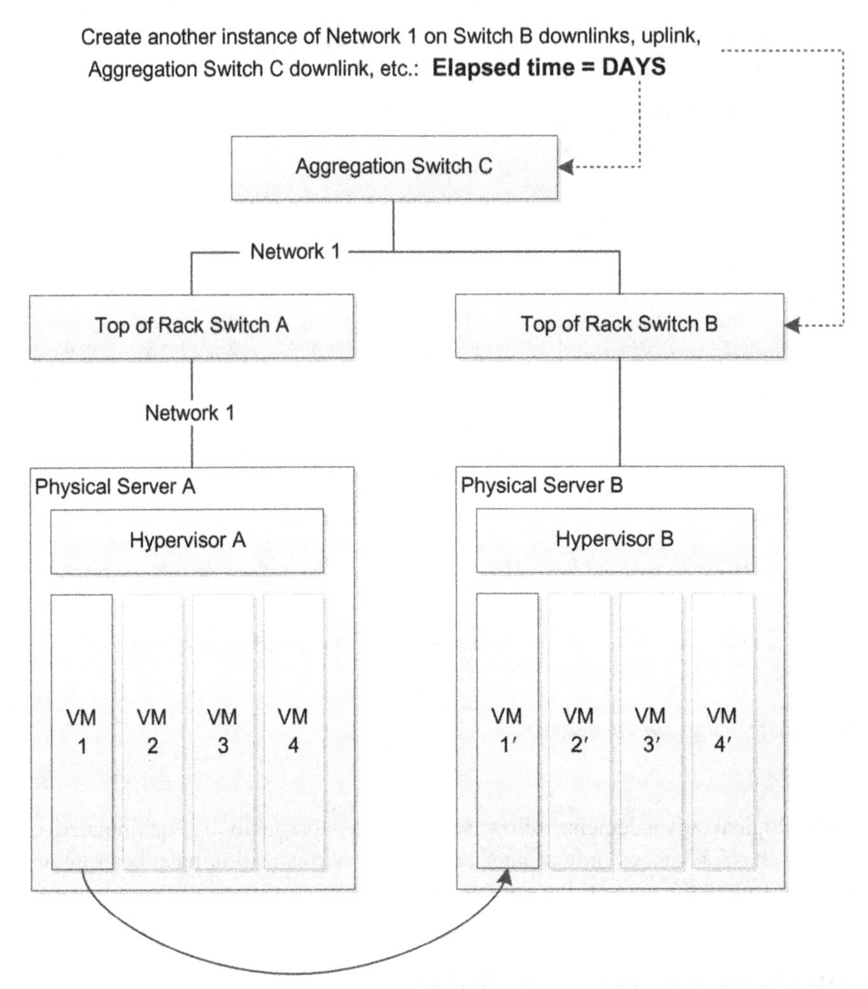

FIG. 2.4

Creating a new network instance in the old paradigm.

are more cumbersome than in their server counterparts. In Chapter 1 we explained that while the control plane of legacy networks had sophisticated ways of autonomously and dynamically distributing layer two and layer three state, no corresponding protocols exist for distributing the policies that are used in policy-based routing. Thus, configuring security policy, such as ACLs or virtualization policy, such as to which VLAN a host belongs, remains static and manual in traditional networks. Thus, the task of reconfiguring a network in a modern data center does not take minutes, but, rather, days. Such inflexible networks are hindering IT administrators in their attempts to automate and streamline their virtualized data center environments. SDN holds the promise that the time required for such network reconfiguration be reduced to the order of minutes, such as is already the case for reconfiguration of VMs.

DISCUSSION QUESTION:

Discuss why in Fig. 2.4, creating an instance of network 1 on switch B may take days yet creating another instance of VM1 need only take minutes.

2.5 DATA CENTER NEEDS

The explosion of the size and speed of data centers has strained the capabilities of traditional networking technologies. We discuss these needs briefly below and cover them in greater detail in Chapter 8. The sections below serve as an indication of new requirements emerging from the technological advances taking place now in data center environments.

2.5.1 AUTOMATION

Automation allows networks to come and go at will, following the movements of servers and storage as needs change. This characteristic is sometimes referred to as *agility*, the ability to dynamically instantiate networks and to disable them when they are no longer needed. This must happen fast, efficiently, and with a minimum of human intervention. Not only do networks come and go, but they also tend to expand and contract. Supporting such agility is only possible through automation.

2.5.2 SCALABILITY

With data centers and cloud environments, the sheer number of end stations that connect to a single network has grown exponentially. The limitations of MAC address table sizes and number of VLANs have become impediments to network installations and deployments. The large number of physical devices present in the data centers also poses a *broadcast control* problem. The use of tunnels and virtual networks can contain the number of devices in a broadcast domain to a reasonable number.

2.5.3 MULTIPATHING

Accompanying the large demands placed on the network by the scalability requirement stated above, is the need for the network to be efficient and reliable. That is, the network must make optimal use of its resources, and it must be resistant to failures of any kind; and, if failures do occur, the network must be able to recover immediately. Fig. 2.5 shows a simple example of multipath through a network, both for choosing the shortest path as well as for alternate or redundant paths. Legacy layer two control plane software would block some of the alternate and redundant paths shown in the figure in order to eliminate forwarding loops. Because we are living with network technology invented years ago, the network is forced into a hierarchy, which results in links which could have provided shortest-path routes between nodes lying entirely unused and dormant. In cases of failure, the current hierarchy can reconfigure

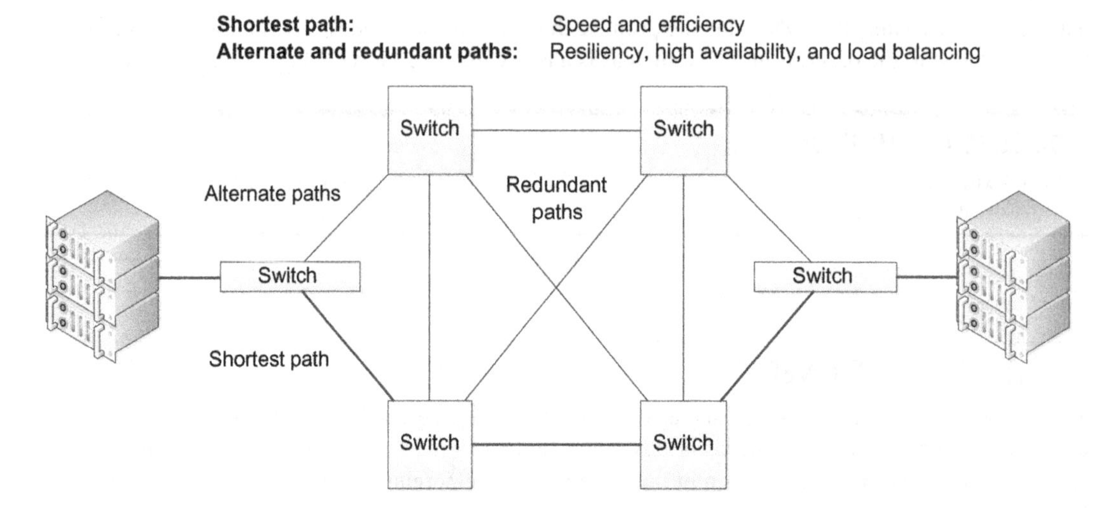

FIG. 2.5

Multipath.

itself in a nondeterministic manner and with unacceptable latency. The speed and high-availability requirements of the modern data center mandate that multiple paths not be wasted by being blocked and, instead, be put into use to improve efficiency as well as to achieve resiliency and load-balancing.

2.5.4 MULTITENANCY

With the advances in data center technology described above and the subsequent advent of *cloud computing*, the idea of hosting dozens, or even hundreds or thousands of customers or *tenants* in the same physical data center has become a requirement. One set of physical hardware hosting multiple tenants has been feasible for some time in the server and storage area. Multitenancy implies that the data center has to provide each of its multiple tenants with their own (virtual) network that they can manage in a manner similar to the way that they would manage a physical network.

2.5.5 NETWORK VIRTUALIZATION

The urgency for automation, multitenancy, and multipathing has increased as a result of the scale and fluidity introduced by server and storage virtualization. The general idea of virtualization is that you create a higher-level abstraction that runs on top of the actual physical entity you are abstracting. The growth of compute and storage server virtualization has created demand for network virtualization. This means having a virtual abstraction of a network running on top of the actual physical network. With virtualization the network administrator should be able to create a network anytime and anywhere (s)he chooses, as well as expand and contract networks that are already in existence. Intelligent virtualization software should be capable of this, without requiring the upper virtualized layer to be aware of what is occurring at the physical layer.

Server virtualization has caused the scale of networks to increase as well, and this increased scale has put pressure on layer two and layer three networks as they exist today. Some of these pressures can be alleviated to some degree by tunnels and other type of technologies, but fundamental network issues remain, even in those situations. Consequently, the degree of network virtualization required to keep pace with data center expansion and innovation is not possible with the network technology that is available today.

To summarize, advances in data center technology have caused weaknesses in the current networking technology to become more apparent. This situation has spurred demand for better ways to construct and manage networks [12], and that demand has driven innovation around SDN [13].

DISCUSSION QUESTION:

Explain how network virtualization can support multitenancy.

2.6 CONCLUSION

The issues of reducing cost and the speed of innovation as motivators for SDN will be recurring themes in the balance of this work. The needs of the modern data center are so tightly linked with the demand for SDN that we dedicate all of Chapter 8 to the examination of use cases in the data center that benefit from SDN technology. These needs did not appear overnight nor did SDN simply explode onto the networking scene in 2009, however. There were a number of tentative steps over many years that formed the basis for what ultimately appeared as the SDN revolution. In the next chapter we will review this evolution and examine how it culminated in the birth of what we now call SDN. We will also discuss the context in which SDN has matured and the organizations and forces that continue to mold its future.

REFERENCES

[1] IEEE Standard for Local and metropolitan area networks—Media Access Control (MAC) Bridges; IEEE 802.1D-2004, New York, June 2004.
[2] Shenker S. The future of networking, and the past of protocols. Open Networking Summit, October 2011. Stanford University; 2011.
[3] Kim H, Feamster N. Improving network management with software defined networking. IEEE Commun Mag 2013;51(2).
[4] McKeown N, Parulkar G, et al. OpenFlow: enabling innovation in campus networks. ACM SIGCOMM Comput Commun Rev Arch 2005;35(3).
[5] Kirkpatrick K. Software-defined networking. Commun ACM 2013;56(9).
[6] Malim G. SDN's value is in operational efficiency, not capex control. Global Telecoms Business; 2013. Retrieved from: http://www.globaltelecomsbusiness.com/article/3225124/SDNs-value-is-in-operational-efficiency-not-capex-control.html.
[7] Wilson C. NTT reaping opex rewards of SDN; 2013. Retrieved from: http://www.lightreading.com/carrier-sdn/ntt-reaping-opex-rewards-of-sdn/d/d-id/705306.
[8] Yegulalp S. Five SDN benefits enterprises should consider. Network Computing; 2013. Retrieved from: http://www.networkcomputing.com/next-generation-data-center/commentary/networking/five-sdn-benefits-enterprises-should-con/240158206.

[9] Greenberg A, Hjalmtysson G, Maltz D, Myers A, Rexford J, Xie G, et al. A clean slate 4D approach to network control and management. ACM SIGCOMM Comput Commun Rev 2005;35(3).
[10] Witner B, Wade B. Basics of z/VM virtualization. IBM. Retrieved from: http://www.vm.ibm.com/devpages/bkw/vmbasics.pdf.
[11] Bari MF, Boutaba R, et al. Data center network virtualization: a survey. IEEE Commun Surv Tutorials 2013;15(2).
[12] Narten T, Sridharan M, et al. Problem statement: overlays for network virtualization. Internet Draft. Internet Engineering Task Force; 2013.
[13] Brodkin J. Data center startups emerging to solve virtualization and cloud problems. Network World; 2011. Retrieved from: http://www.networkworld.com/news/2011/061411-data-center-startups.html.

CHAPTER

GENESIS OF SDN

3

We have now seen a number of the reasons that have led to the emergence of SDN. This included the fact that there were too many disparate control plane protocols attempting to solve many different networking problems in a distributed fashion, which has resulted in relative confusion and inertia in network switching evolution as compared to compute and storage technologies. Overall, the system that has evolved is far too closed, and there is a strong desire to migrate to a model closer to that of open source development and open source initiatives (such as Linux with respect to computer operating systems). We have also noted that for a variety of reasons the *network equipment manufacturers* (NEMs) are bogged down in their current technologies and are unable to innovate at the rate required by the modern data center. This is partly attributable to the fact that the current business model for networking equipment is a very high margin business and it is not obviously in the incumbents' interest for the status quo to change drastically [1]. Even the standards bodies generally move too slowly [2]. Thus, the need for an SDN-like transformative technology has become apparent to many users of network equipment, especially in large data centers. In this chapter we will explore how the SDN movement began. In order to understand the base from which these SDN pioneers started, we begin with a brief review of the evolution of networking technology up to the advent of SDN.

3.1 THE EVOLUTION OF NETWORKING TECHNOLOGY

In the earliest days of computing there was no networking—each individual computer occupied an entire room or even an entire floor. As mainframes emerged and began to find their way into our technological consciousness, these computing powerhouses still operated as islands, and any sharing of data took place using physical media such as magnetic tapes.

3.1.1 MAINFRAME NETWORKING: REMOTE TERMINALS

Even in the age of mainframes, remote connectivity to the mainframe was needed. This was provided in the form of remote terminal controllers and card readers, which operated as subservient devices, known as peripherals, with control residing entirely in the central mainframe. Network connections in this case were simple point-to-point or point-to-multipoint links. Communication was solely between a few connected entities, in this case the mainframe and a small set of remote terminal controllers or card readers. This communication was controlled by the central mainframe.

Software Defined Networks. http://dx.doi.org/10.1016/B978-0-12-804555-8.00003-X

3.1.2 PEER-TO-PEER POINT-TO-POINT CONNECTIONS

As computer technology began to move from solely mainframes to the addition of mini-computers, these machines had a greater need to share information in a quick and efficient manner. Computer manufacturers began to create protocols for sharing data between two peer machines. The network in this case was also point-to-point, although the nature of the connection was peer-to-peer, in that the two machines (e.g., mini-computers) would communicate with each other and share data as relative equals, at least compared to the earlier mainframe-to-terminal-controller type of connections.

Of course, in these point-to-point connections, the network was trivial, with only the two parties communicating with each other. Control for this communication resided not in any networking device, but in the individual computers participating in this one-to-one communication.

3.1.3 LOCAL AREA NETWORKS

Eventually, with further evolution of computing toward smaller, independent systems, the need arose for a way to connect these devices in order to allow them to share information and collaborate in a manner that wasn't required when everything ran on one large mainframe or even on a powerful minicomputer. Hence, *Local Area Networking* (LAN) technology arrived, with various battles being fought between technologies (e.g., Ethernet/IEEE 802.3 versus Token Ring/IEEE 802.5). Notably, these early LANs were running on shared media, so all traffic was seen by every device attached to the network.

IEEE 802.3 emerged as the more popular of these technologies. It uses *Carrier-Sense Multiple-Access/Collision Detect* (CSMA/CD) technology, which exhibits poor aggregate throughput when the number of active devices reached a certain level. The exact number of devices where this would occur was dependent on the amount of data attempted to be transmitted by each. This decrease in performance resulted from the backing off and waiting to transmit done by each device when a collision occurs, as stipulated by CSMA/CD. The number of collisions reaches a critical point as a function of the number of nodes and their respective transmission activity. Once this critical point is reached the network becomes very inefficient, with too much time spent either in the midst of or recovering from collisions.

These flat, shared-media networks were quite simple, and the *repeater* devices provided physical extensions to the shared media by merely forwarding all frames to the extended medium. The greatest impact of these early devices was a new topology created by wire concentration. This made layer two networks more deployable and reliable than the former method of snaking a coaxial cable from one node to the next. This early networking required minimum control plane intelligence, if any. Simple repeaters basically just did one thing: forward to everybody. More advanced, *managed* repeaters did have limited control planes where segmentation and error monitoring occurred. Managed repeaters did perform functions such as removing erroneous traffic from the network and also isolated ports that were causing problems. Segmentable repeaters were able to split themselves into multiple repeaters via configuration. Different groups of ports could be configured to reside in different collision domains. These separate collision domains would be connected by bridges. This feature provided more control over the size of the collision domains to optimize performance.

DISCUSSION QUESTION:

Explain why shared medium networks often cease to work efficiently above a certain level of utilization.

3.1.4 BRIDGED NETWORKS

Eventually these shared-media networks needed to scale in physical extent as well as number of nodes. As explained in the previous section, shared-media networks do not scale well as the number of hosts grows. It became desirable to split the shared-media network into separate segments. In so doing, and since not all the nodes are transmitting all the time, *spatial reuse* occurs and the aggregate bandwidth available actually increases due to the locality of transmissions in each segment. The first devices to perform this functionality were called *bridges*—forerunners of today's switches, but much simpler. They typically had only two ports, connecting two shared domains. These bridges possessed an incipient control plane in that they were able to actually learn the location and MAC address of all devices, and created forwarding tables which allowed them to make decisions about what to do with incoming packets. As we mentioned in the previous chapter, these forwarding tables were implemented entirely in software.

Furthermore, these bridges were implemented in such a way that each device was able to operate independently and autonomously without requiring any centralized intelligence. The goal was to facilitate expansion of the network without a lot of coordination or interruption across all the devices in the network. One of the first manifestations of this distributed intelligence paradigm was the *Spanning Tree Protocol*, (STP) which allowed a group of bridges to interact with each other and converge on a topology decision (the spanning tree), which eliminated loops and superimposed a hierarchical loop-free structure on the network.

The important point here is that this greater scale in terms of the number of nodes as well as physical extent drove the need for this new model of individually autonomous devices, with distributed protocols implementing the required control functionality. If we translate this period's realities into today's terminology, there was no centralized controller, merely a collector of statistics. *Policies*, if indeed one can use that term, were administered by setting specific parameters on each device in the network. We need to keep in mind that at the time networks in question were small, and this solution was entirely acceptable.

3.1.5 ROUTED NETWORKS

In the same manner that bridged and switched networks dealt with layer two domains with distributed protocols and intelligence, similar strategies were employed for layer three routing. Routers were directly connected locally to layer two domains and interconnected over large distances with point-to-point WAN links. Distributed routing protocols were developed to allow groups of interconnected routers to share information about those networks to which they were directly connected. By sharing this information amongst all the routers, each was able to construct a routing table allowing it to route packets to remote layer three IP subnets using the optimal forwarding ports. This was another application of autonomous devices utilizing distributed protocols in order to allow each to make appropriate forwarding decisions.

This has led to the current state of affairs, with networking intelligence distributed in the networking devices themselves. During this evolution, however, the growth of network size and complexity was unrelenting. The size of MAC forwarding tables grew, control plane protocols became more complex, and network overlays and tunneling technology became more prevalent. Making major changes to these implementations was a continual challenge. Since the devices were designed to operate independently, centrally administered large-scale upgrades were challenging. In addition, the fact that the actual control plane implementations were from many different sources, and not perfectly matched, created a sort of *lowest common denominator* effect, where only those features that were perfectly aligned between the varied implementations could truly be relied upon. In short, existing solutions were not scaling well with this growth and complexity. This led network engineers and researchers to question whether this evolution was headed in the right direction. We describe some of the innovations and research that this led to in the following section.

3.2 FORERUNNERS OF SDN

Prior to OpenFlow, and certainly prior to the birth of the term Software Defined Networking, forward-thinking researchers and technologists were considering fundamental changes to today's world of autonomous, independent devices and distributed networking intelligence and control. This section considers some of those early explorations of SDN-like technology. There is a steady progression of ideas around advancing networking technology toward what we now know as SDN. Table 3.1 shows this progression, which will be discussed in more detail in the following subsections. The timeframes shown in the table represent approximate time periods when these respective technologies were developed or applied in the manner described.

Table 3.1 Precursors of SDN

Project	Description
Open Signaling	Separating the forwarding and control planes in ATM switching (1999, see Section 3.2.1)
Active Networking	Separate control and programmable switches (late 1990s, see Section 3.2.1)
DCAN	Separating the forwarding and control planes in ATM switching (1997, see Section 3.2.1)
IP Switching	Control layer two switches as a layer three routing fabric (late 1990s, see Section 3.2.1)
MPLS	Separate control software establishes semi-static forwarding paths for flows in traditional routers (late 1990s, see Section 3.2.1)
RADIUS, COPS	Use of admission control to dynamically provision policy (2010, see Section 3.2.2)
Orchestration	Use of SNMP and CLI to help automate configuration of networking equipment (2008, see Section 3.2.3)
Virtualization Manager	Use of plug-ins to perform network reconfiguration to support server virtualization (2011, see Section 3.2.4)
ForCES	Separating the forwarding and control planes (2003, see Section 3.2.5)
4D	Control plane intelligence located in a centralized system (2005, see Section 3.2.6)
Ethane	Complete enterprise and network access and control using separate forwarding and control planes, and utilizing a centralized controller (2007, see Section 3.2.7)

3.2.1 EARLY EFFORTS

Good surveys of early programmable networks that were stepping-stones on the path to SDN can be found in [3,4]. Some of the earliest work in programmable networks began not with Internet routers and switches, but with ATM switches. Two notable examples were *Devolved Control of ATM Networks* (DCAN) and *Open Signaling*.

As its name indicates, DCAN [5] prescribed the separation of the control and management of the ATM switches from the switches themselves. This control would be assumed by an external device that is similar to the role of the controller in SDN networks.

Open Signaling [6] proposed a set of open, programmable interfaces to the ATM switching hardware. The key concept was to separate the control software from the switching hardware. This work led to the IETF effort which resulted in the creation of the *General Switch Management Protocol* (GSMP) [7]. In GSMP, a centralized controller is able to establish and release connections on an ATM switch, as well as a multitude of other functions that may otherwise be achieved via distributed protocols on a traditional router. Tag switching was Cisco's version of label switching, and both of these terms refer to the technology that ultimately became known as *Multiprotocol Label Switching* (MPLS). Indeed, MPLS and related technologies are a deviation from the autonomous, distributed forwarding decisions characteristic of the the traditional Internet router and in that sense were a small step toward a more SDN-like Internet switching paradigm. In the late 1990s, Ipsilon Networks utilized the GSMP protocol to set up and tear down ATM connections internally in their IP Switch product. The Ipsilon IP Switch [8] presented normal Internet router interfaces externally, but its internal switching fabric could utilize ATM switches for persistent flows. Flows were defined as a relatively persistent set of packets between the same two endpoints, where an endpoint was determined by IP address and TCP/UDP port number. Since there was some overhead in establishing the ATM connection that would carry that flow, this additional effort would only be expended if the IP Switch believed that a relatively large number of packets would be exchanged between the endpoints in a short period of time. This definition of flow is somewhat consistent with the notion of flow within SDN networks.

The *Active Networking* project [9,10] also included the concept of switches that could be programmed by out-of-band management protocols. In the case of Active Networking, the switching hardware was not ATM switches but Internet routers. Active Networking also included a very novel proposal for small downloadable programs called *capsules* that would travel in packets to the routers and could reprogram the router's behavior on the fly. These programs could be so fine-grained as to prescribe the forwarding rules for a single packet, even possibly the payload of the packet that contained the program. The concept of Active Networking was embraced by a number of projects [11–13].

3.2.2 NETWORK ACCESS CONTROL

Network Access Control (NAC) products control access to a network based on policies established by the network administration. The most basic control is to determine whether or not to admit the user onto the network. This is usually accomplished by some exchange of credentials between the user and the network. If the user is admitted, his rights of access will also be determined and restricted in accordance with those policies. There were early efforts where network policy beyond admission control was dynamically provisioned via NAC methods, such as *Remote Authentication Dial In User*

Service (RADIUS) [14] and *Common Open Policy Service* (COPS) [15]. As RADIUS was far more widely implemented, we discuss it in more detail here.

RADIUS has been used to provide the automatic reconfiguration of the network. RADIUS can be viewed as a simple, proactive, policy-based precursor to SDN. The idea is that via RADIUS, networking attributes would change based on the identity of the compute resource that had just appeared and required connectivity to the network. While RADIUS was originally designed for the *Authentication, Authorization and Accounting* (AAA) related to granting a user access to a network, that original paradigm maps well to our network reconfiguration problem. The identity of the resource connecting to the network serves to identify the resource to the RADIUS database, and the authorization attributes returned from that database could be used to change the networking attributes described above. Solutions of this nature achieved automatic reconfiguration of the network, but never gained the full trust of IT administrators and, thus, never became mainstream. While this RADIUS solution did an adequate job of automatically reconfiguring the edge of the network, the static, manually configured core of the network remained the same. This problem still awaited a solution.

Fig. 3.1 shows an overview of the layout and operation of a RADIUS-based solution for automating the process of creating network connectivity for a virtual server. In the figure, the process would be that a VM is moved from physical server A to physical server B, and as a result, the RADIUS server becomes aware of the presence of this VM at a new location. The RADIUS server is then able to automatically configure the network based on this information, using standard RADIUS mechanisms.

3.2.3 ORCHESTRATION

Early attempts at automation involved applications which were commonly labeled *Orchestrators* [16]. Just as a conductor can make a harmonious whole out of the divergent instruments of an orchestra, such applications could take a generalized command or goal, and apply it across a wide range of often heterogeneous devices. These orchestration applications would typically utilize common device *Application Programming Interfaces* (APIs), such as the *Command Line Interface* (CLI) or *Simple Network Management Protocol* (SNMP).[1]

Fig. 3.2 shows an overview of the layout and operation of an orchestration management solution for automating the process of creating network connectivity for a virtual server. The figure shows SNMP/CLI plug-ins for each vendor's specific type of equipment; an orchestration solution can then have certain higher-level policies which are in turn executed at lower levels by the appropriate plug-ins. The vendor-specific plug-ins are used to convert the higher-level policy requests into the corresponding native SNMP or CLI request specific to each vendor.

Such orchestration solutions alleviated the task of updating device configurations. But since they were really limited to providing a more convenient interface to existing capabilities, and since no capability existed in the legacy equipment for network-wide coordination of more complex policies such as security and virtual network management, tasks such as configuring VLANs remained hard.

[1]The reader should note that while in this work we classify CLI and SNMP as APIs, we acknowledge that they are much more primitive than most of the APIs we discuss later in the book and there is not universal consensus that they should even be considered APIs.

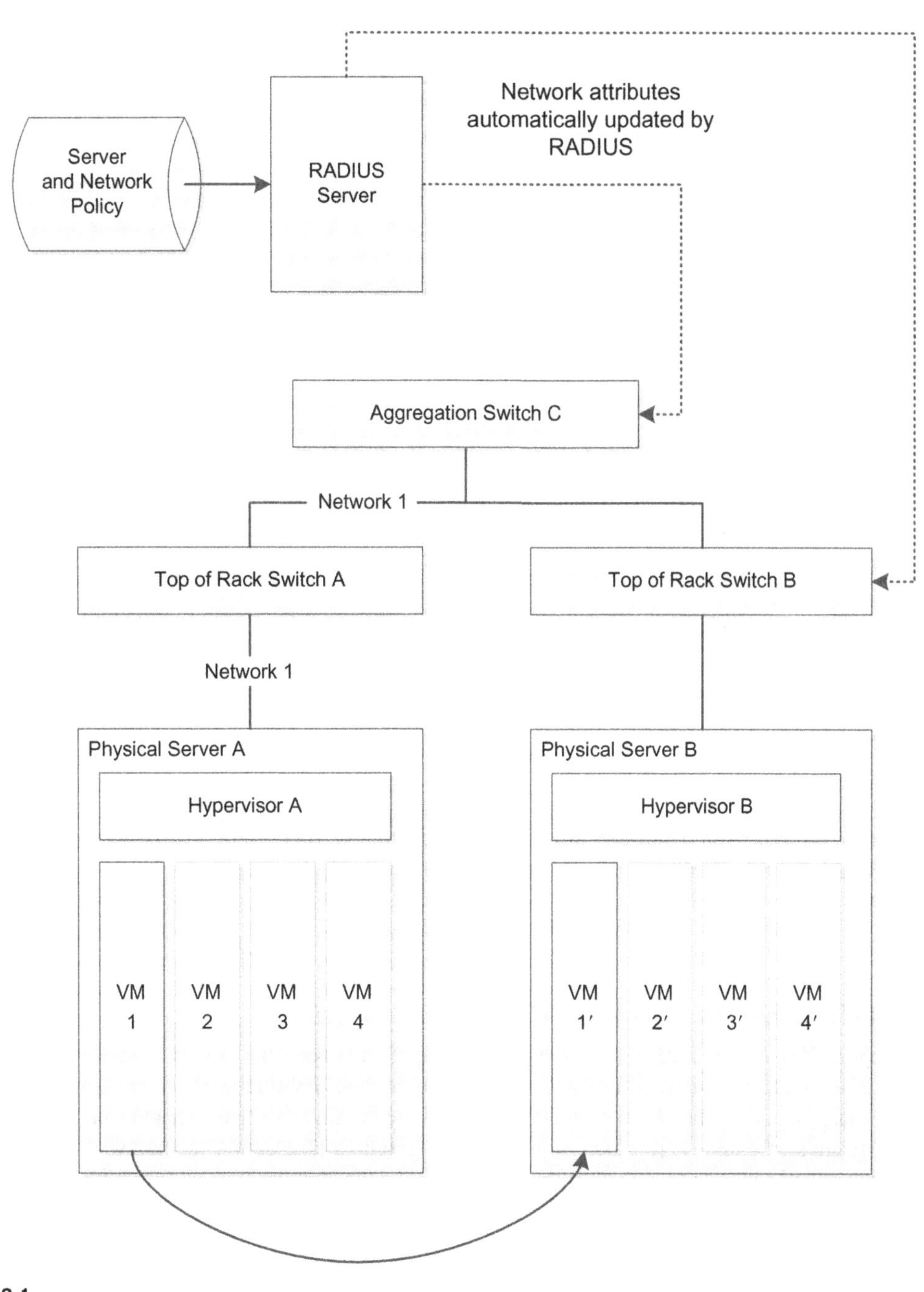

FIG. 3.1

Early attempts at SDN: RADIUS.

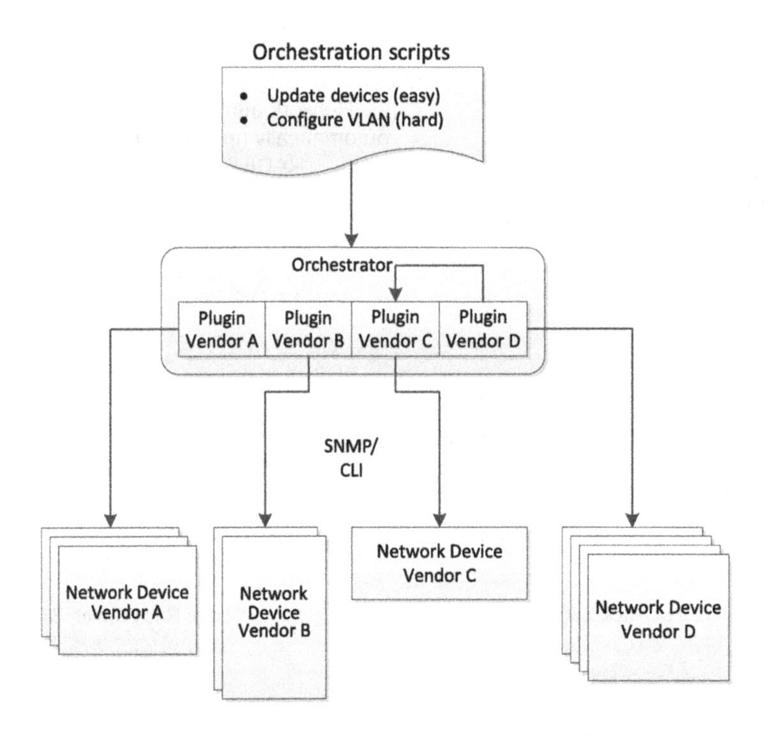

FIG. 3.2

Early attempts at SDN: Orchestration.

Consequently, orchestration management applications continued to be useful primarily for tasks such as software and firmware updates, but not for tasks which would be necessary in order to automate today's data centers.

3.2.4 VIRTUALIZATION MANAGER NETWORK PLUG-INS

The concept of *Virtualization Manager Network Plug-ins* builds upon the notion of orchestration and attempts to automate the network updates that are required in a virtualization environment. Tools specifically targeted at the data center would often involve virtualization manager plug-ins (e.g., plug-ins for VMware's vCenter [17]) which would be configured to take action in the event of a server change, such as a vMotion [18]. The plug-in would then take the appropriate actions on the networking devices they controlled in order to make the network follow the server and storage changes with changes of its own. Generally the mechanism for making changes to the network devices would be SNMP or CLI commands. These plug-ins can be made to work, but, since they must use the static configuration capabilities of SNMP and CLI, they suffer from being difficult to manage and prone to error.

Fig. 3.3 shows an overview of the layout and operation of a virtualization manager plug-in solution for automating the process of creating network connectivity for a virtual server. This figure reflects a

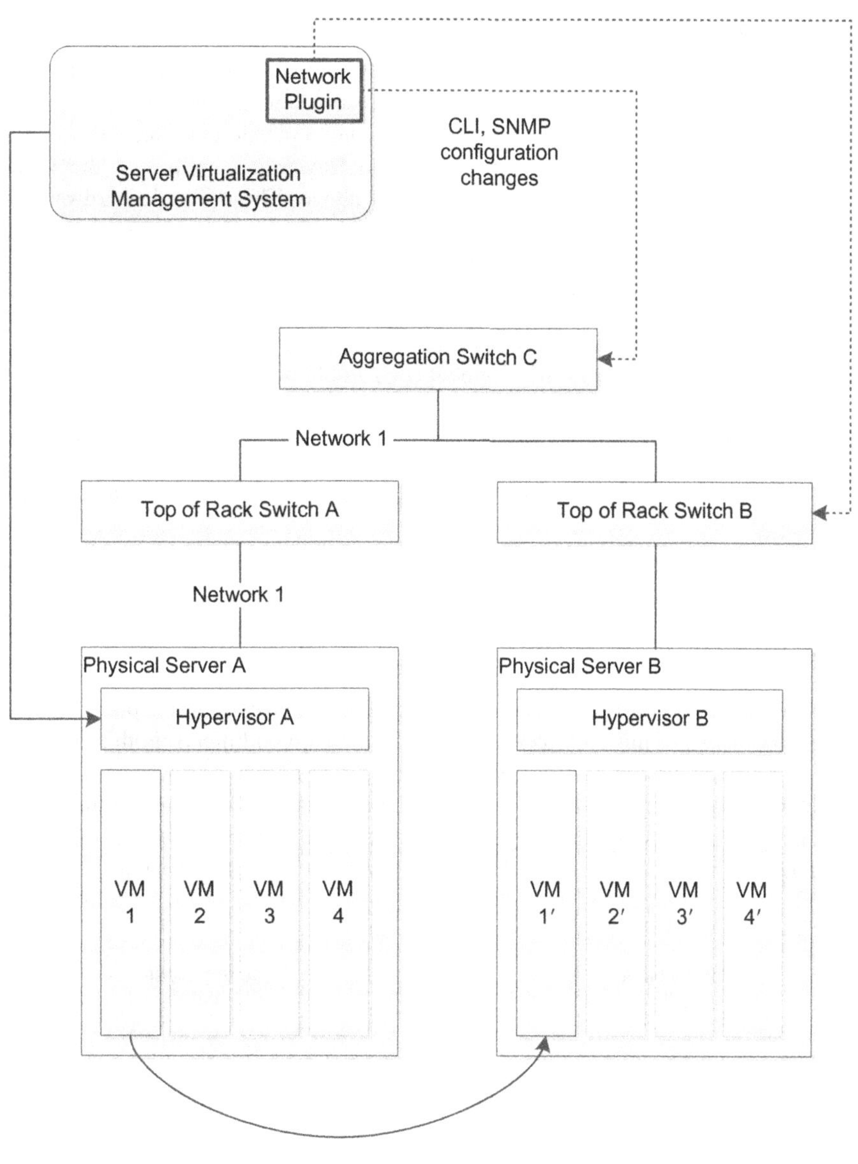

FIG. 3.3

Early attempts at SDN: Plug-ins.

similar use of SNMP/CLI plug-ins to that which we saw in Fig. 3.2. The most notable difference is that this application starts from a base that supports virtualization. The nature of the reconfiguration managed by this application is oriented to support the networking of virtual machines via VLANs and tunnels.

3.2.5 FORCES: SEPARATION OF FORWARDING AND CONTROL PLANES

The *Forwarding and Control Element Separation* (ForCES) [19] work produced in the IETF began around 2003. ForCES was one of the original proposals recommending the decoupling of forwarding and control planes as well as a standard interface for communication between them. The general idea of ForCES was to provide simple hardware-based forwarding entities at the foundation of a network device, and software-based control elements above. These simple hardware forwarders were constructed using cell switching or tag switching technology. The software-based control had responsibility for the broader tasks often involving coordination between multiple network devices (e.g., Border Gateway Protocol routing updates).

The functional components of ForCES are as follows:

- **Forwarding Element**: The *Forwarding Element* (FE) would be typically implemented in hardware and located in the network. The FE is responsible for enforcement of the forwarding and filtering rules that it receives from the *controller*.
- **Control Element**: The *Control Element* (CE) is concerned with the coordination between the individual devices in the network, and for communication of forwarding and routing information to the FEs below.
- **Network Element**: The *Network Element* (NE) is the actual network device which consists of one or more FEs and one or more CEs.
- **ForCES Protocol**: The ForCES protocol is used to communicate information back and forth between FEs and CEs.

ForCES proposes the separation of the forwarding plane from the control plane, and it suggests two different embodiments of this architecture. In one of these embodiments, both the forwarding and control elements are located within the networking device. The other embodiment speculates that it would be possible to actually move the control element(s) off of the device and to locate them on an entirely different system. Although the suggestion of a separate controller thus exists in ForCES, the emphasis is on the communication between CE and FE over a switch backplane, as shown in Fig. 3.4.

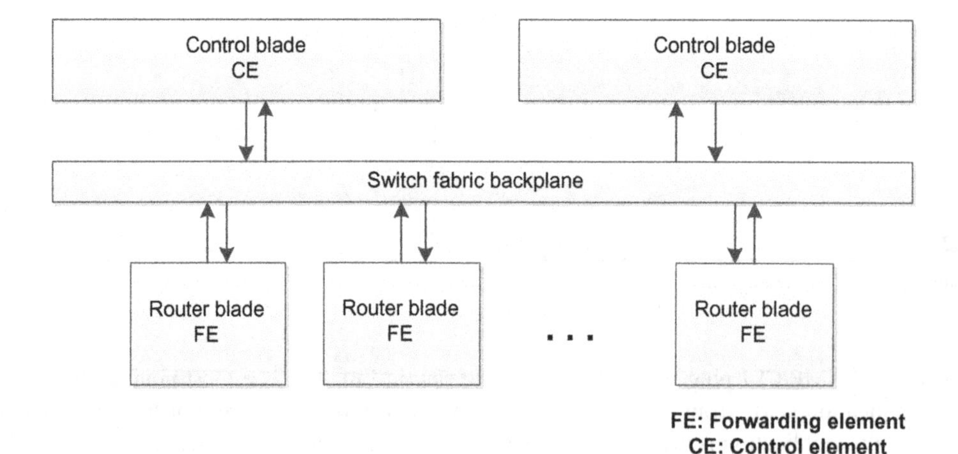

FIG. 3.4

ForCES design.

3.2.6 4D: CENTRALIZED NETWORK CONTROL

Seminal work on the topic of moving networking technology from distributed networking elements into a centralized controller appeared in the 4D proposal [20], *A Clean Slate 4D Approach to Network Control and Management*. 4D, named after the architecture's four planes—*decision*, *dissemination*, *discovery*, and *data*—proposes a complete refactoring of networking away from autonomous devices and toward the idea of concentrating control plane operation in a separate and independent system dedicated to that purpose.

4D argues that the state of networking today is fragile and, therefore, often teeters on the edge of failure because of its current design based on distributed, autonomous systems. Such systems exhibit a defining characteristic of unstable, complex systems: a small local event such as a misconfiguration of a routing protocol can have a severe, global impact. The proposal argues that the root cause is the fact that the control plane is running on the network elements themselves.

4D centers around three design principles:

- **Network-level Objectives**: In short, the goals and objectives for the network system should be stated in network-level terms based on the entire network domain, separate from the network elements, rather than in terms related to individual network device performance.
- **Network-wide View**: There should be a comprehensive understanding of the whole network. Topology, traffic, and events from the entire system should form the basis on which decisions are made and actions are taken.
- **Direct Control**: The control and management systems should be able to exert direct control over the networking elements, with the ability to program the forwarding tables for each device, rather than only being able to manipulate some remote and individual configuration parameters, as is the case today.

Fig. 3.5 shows the general architecture of a 4D solution, with centralized network control via the control and management system.

One aspect of 4D that is actually stated in the title of the paper is the concept of a *clean slate*, meaning the abandoning of the current manner of networking in favor of this new method as described by the three aforementioned principles. A quote from the 4D proposal states that "*We hope that exploring an extreme design point (the clean slate approach) will help focus the attention of the research and industrial communities on this crucially important and intellectually challenging area.*"

The 4D proposal delineates some of the challenges faced by the centralized architecture proposed. These challenges continue to be relevant today in SDN. We list a few of them here:

- **Latency**: Having a centralized controller means that a certain (hopefully small) number of decisions will suffer nontrivial round-trip latency as the networking element requests policy directions from the controller. How this delay impacts the operation of the network, and to what extent, remains to be determined. Also, with the central controller providing policy advice for a number of network devices, it is unknown whether the conventional servers on which the controller runs will be able to service these requests at sufficient speed to have minimal or no impact on network operation.
- **Scale**: Having a centralized controller means that responsibility for the topological organization of the network, determination of optimal paths, and responses to changes, must be handled by the controller. As has been argued, this is the appropriate location for this functionality; however, as more and more network devices are added to the network, questions arise of scale and the ability of a single controller to handle all those devices. It is difficult to know how well a centralized system

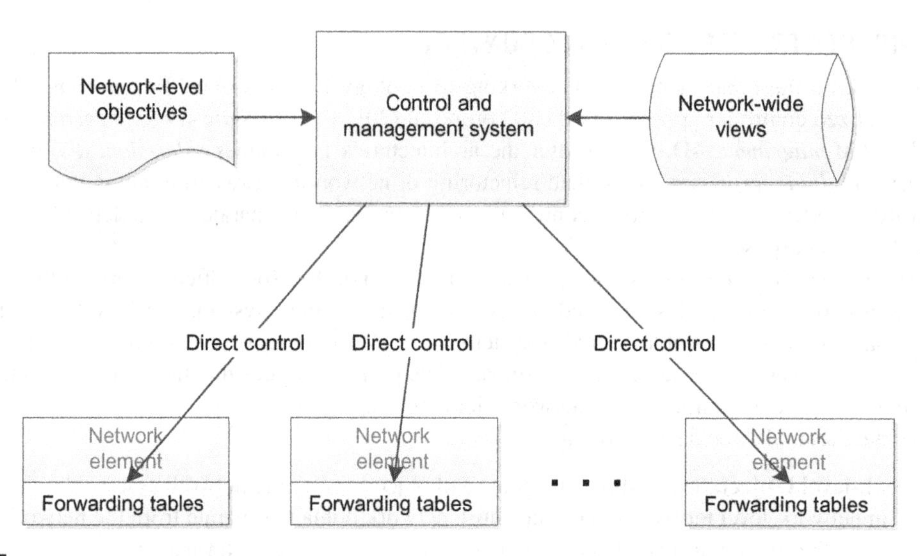

FIG. 3.5

4D principles.

can handle hundreds, thousands, or tens of thousands of network devices, nor what is the solution when the number of network devices outgrows the capacity of the controller to handle them. If we attempt to scale by adding additional controllers, how do they communicate, and who orchestrates coordination amongst the controllers? Section 6.1.3 will address these questions.

- **High Availability (HA)**: The centralized controller must not constitute a *single point of failure* for the network. This implies the need for redundancy schemes in a number of areas. First, there must be redundant controllers such that processing power is available in the event of failure of a single controller. Secondly, the actual data used by the set of controllers needs to be mirrored such that they can program the network devices in a consistent fashion. Thirdly, the communication paths to the different controllers need to be redundant to ensure that there is always a functioning communications path between a switch and at least one controller. We further discuss high availability in the context of a modern SDN network in Section 6.1.2.
- **Security**: Having a centralized controller means that security attacks are able to focus on that one point of failure, and, thus, the possibility exists that this type of solution is more vulnerable to attack than a more distributed system. It is important to consider what extra steps must be taken to protect both the centralized controller and the communication channels between it and the networking devices.

ForCES and 4D contribute greatly toward the evolution of the concepts that underlie SDN—separation of forwarding and control planes (ForCES), and having a centralized controller responsible for overall routing and forwarding decisions (4D). However, both of these proposals suffer from a lack of actual implementations. Ethane, examined below, benefited from the learnings that can only come from a real-life implementation.

3.2.7 ETHANE: CONTROLLER-BASED NETWORK POLICY

Ethane was introduced in a paper entitled *Rethinking Enterprise Network Control* [21] in 2007. Ethane is a policy-based solution which allows network administrators to define policies pertaining to network-level access for users, which includes authentication and quarantine for misbehaving users. Ethane was taken beyond the proposal phrase. Multiple instances have been implemented and shown to behave as suggested in [21].

Ethane is built around three fundamental principles:

- **The network should be governed by high-level policies**: Similar to 4D, Ethane espouses the idea that the network be governed by policies defined at high levels, rather than on a per-device basis. These policies should be at the level of services and users and the machines through which users can connect to the network.
- **Routing for the network should be aware of these policies**: Paths that packets take through the network are to be dictated by the higher-level policies described in the previous bullet point, rather than as in the case of today's networks, in which paths are chosen based on lower-level directives. For example, *guest* packets may be required to pass through a filter of some sort, or certain types of traffic may require routing across lightly loaded paths. Some traffic may be highly sensitive to packet loss, while other traffic (e.g., *Voice over IP* (VoIP)) may tolerate dropped packets but not latency and delay. These higher-level policies are more powerful guidelines for organizing and directing traffic flows than low-level and device-specific rules.
- **The network should enforce binding between packets and their origin**: If policy decisions rely on higher-level concepts, such as the concept of a *user*, then the packets circulating in the network must be traceable back to their point of origin (i.e., the user or machine which is the actual source of the packet).

Fig. 3.6 illustrates the basics of the Ethane solution. As will become apparent in the following chapters, there are many similarities between this solution and OpenFlow.

In order to test Ethane, the researchers themselves had to develop switches in order to have them implement the protocol and the behavior of such a device. That is, a device that allows control plane functionality to be determined by an external entity (the controller), and which communicates with that external entity via a protocol that allows flow entries to be configured into its local flow tables, which then perform the forwarding functions as packets arrive.

The behavior of the Ethane switches is generally the same as OpenFlow switches today which forward and filter packets based on the flow tables that have been configured on the device. If the switch does not know what to do with the packet, it forwards it to the controller and awaits further instructions.

In short, Ethane is basically a software defined networking technology, and its components are the antecedents of OpenFlow, which we describe in detail in Chapter 5.

DISCUSSION QUESTION:

A key aspect of SDN is segregation of the control plane to a centralized device. Which of the technologies listed in Table 3.1 used this approach?

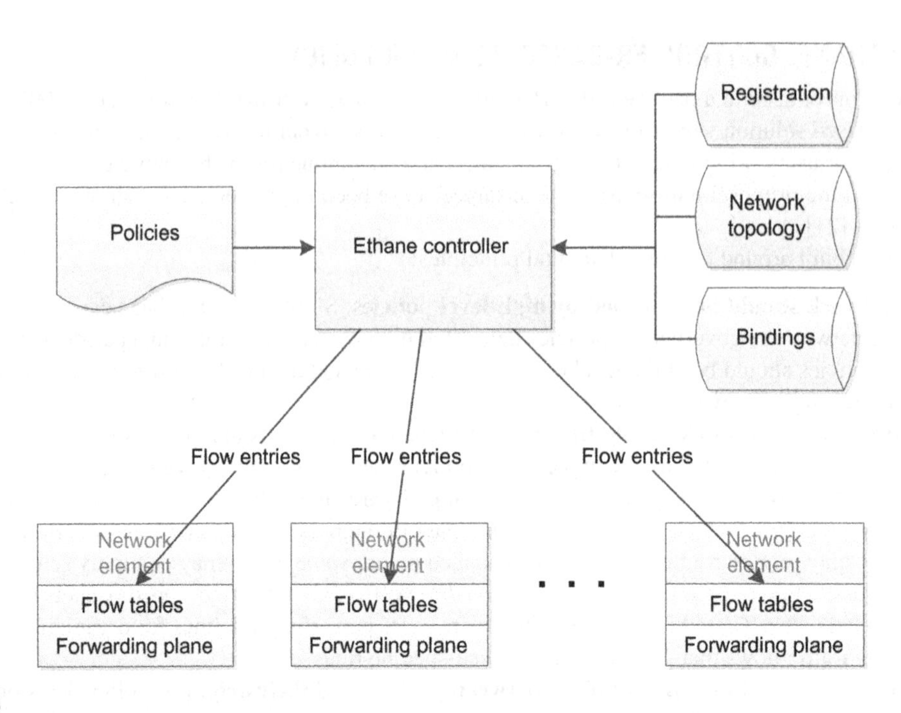

FIG. 3.6

Ethane architecture.

3.3 LEGACY MECHANISMS EVOLVE TOWARD SDN

The urgent need for SDN we described in Chapter 2 could not wait for a complete new network paradigm to be fleshed out through years of research and experimentation. It is not surprising, then, that there were early attempts to achieve some SDN-like functionality within the traditional networking model. One such example was Cisco's *Policy Based Routing* that we described in Chapter 1. The capabilities of legacy switches were sometimes extended to support detailed policy configuration related to security, QoS and other areas. Old APIs were extended to allow centralized programming of these features. Some SDN providers have based their entire SDN solution on a rich family of extended APIs on legacy switches, orchestrated by a centralized controller. In Chapter 6 we will examine how these alternative solutions work and, whether or not they genuinely constitute an SDN solution.

3.4 SOFTWARE DEFINED NETWORKING IS BORN

3.4.1 THE BIRTH OF OPENFLOW

OpenFlow is a protocol specification that describes the communication between OpenFlow switches and an OpenFlow controller. Just as the previous sections presented standards and proposals which were precursors to SDN, seeing SDN through a gestation period, then the arrival of OpenFlow is the point at which SDN was actually born. In reality, the term SDN did not come into use until a year after OpenFlow made its appearance on the scene in 2008, but the existence and adoption of OpenFlow by research communities and networking vendors marked a sea change in networking, one that we are still witnessing even now. Indeed, while the term SDN was in use in the research community as early as 2009, SDN did not begin to make a big impact in the broader networking industry until 2011.

For reasons identified in the previous chapter, OpenFlow was developed and designed to allow researchers to experiment and innovate with new protocols in everyday networks. The OpenFlow specification encouraged vendors to implement and enable OpenFlow in their switching products for deployment in college campus networks. Many network vendors have implemented OpenFlow in their products.

The OpenFlow specification delineates both the protocol to be used between the controller and the switch as well as the behavior expected of the switch. Fig. 3.7 illustrates the simple architecture of an OpenFlow solution.

The following list describes the basic operation of an OpenFlow solution:

- The controller populates the switch with flow table entries.
- The switch evaluates the header(s) of incoming packets and finds a matching flow, then performs the associated action. Depending on the match criteria, this evaluation would begin with the layer two header, and then potentially continue to the layer three and even layer four headers in some cases.
- If no match is found, the switch forwards the packet to the controller for instructions on how to deal with the packet.
- Typically the controller will update the switch with new flow entries as new packet patterns are received, so that the switch can deal with them locally. It is also possible that the controller will program *wildcard rules* that will govern many flows at once.

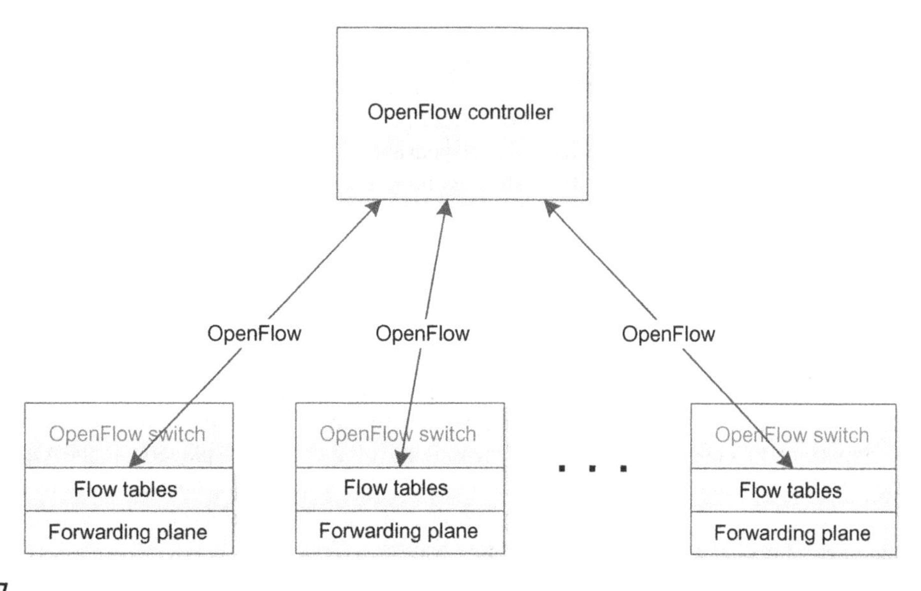

FIG. 3.7

General OpenFlow design.

OpenFlow will be examined in detail in Chapter 5. For now, though, the reader should understand that OpenFlow has been adopted by both the research community and by a number of networking vendors. This has resulted in a significant number of network devices supporting OpenFlow on which researchers can experiment and test new ideas.

DISCUSSION QUESTION:

What similarities exist between Ethane and OpenFlow?

3.4.2 OPEN NETWORKING FOUNDATION

OpenFlow began with the publication of the original proposal in 2008. By 2011 OpenFlow had gathered enough momentum that the responsibility for the standard itself moved to the *Open Networking Foundation* (ONF). The ONF was established in 2011 by Deutsche Telekom, Facebook, Google, Microsoft, Verizon, and Yahoo!. It is now the guardian of the OpenFlow standard, and consists of a number of councils, areas and working groups. We depict the interrelationships between these bodies in Fig. 3.8 [27].

One novel aspect of the ONF is that corporate members of the Board of Directors consist of major network *operators*, and not the networking *vendors* themselves. As of this writing, the ONF board is composed of CTOs, Technical Directors, professors from Stanford and Princeton, and Fellows from companies such as Google, Yahoo!, Facebook, Deutsche Telekom, Verizon, Microsoft, NTT, and Goldman Sachs among others. This helps to prevent the ONF from supporting the interests of one major networking vendor over another. It also helps to provide a *real-world* perspective on what should be investigated and standardized. Conversely, it runs the risk of defining a specification that is difficult for NEMs to implement. Our previous comments about NEMs being locked into their status quo notwithstanding, it is also true that there is a wealth of engineering experience resident in the NEMs regarding how to actually design and build high performance, high reliability switches. While the real-world experience of the users is indeed indispensable, it is imperative that the ONF seek input from the vendors to ensure that the specifications they produce are in fact implementable.

3.5 SUSTAINING SDN INTEROPERABILITY

At the current point in the evolution of SDN, we have a standard which has been accepted and adopted by academia and industry alike, and we have an independent standards body to shepherd OpenFlow forward and to ensure it is independent and untethered to any specific institution or organization. It is now important to ensure that the implementations of the various players in the OpenFlow space adhere to the standards as they are defined, clarify and expand the standards where they are found to be incomplete or imprecise, and in general guarantee interoperability amongst OpenFlow implementations. This goal can be achieved in a few ways:

- **Plugfests**: Plugfests, staged normally at conferences, summits, and congresses, are environments where vendors can bring their devices and software in order to test them with devices and software

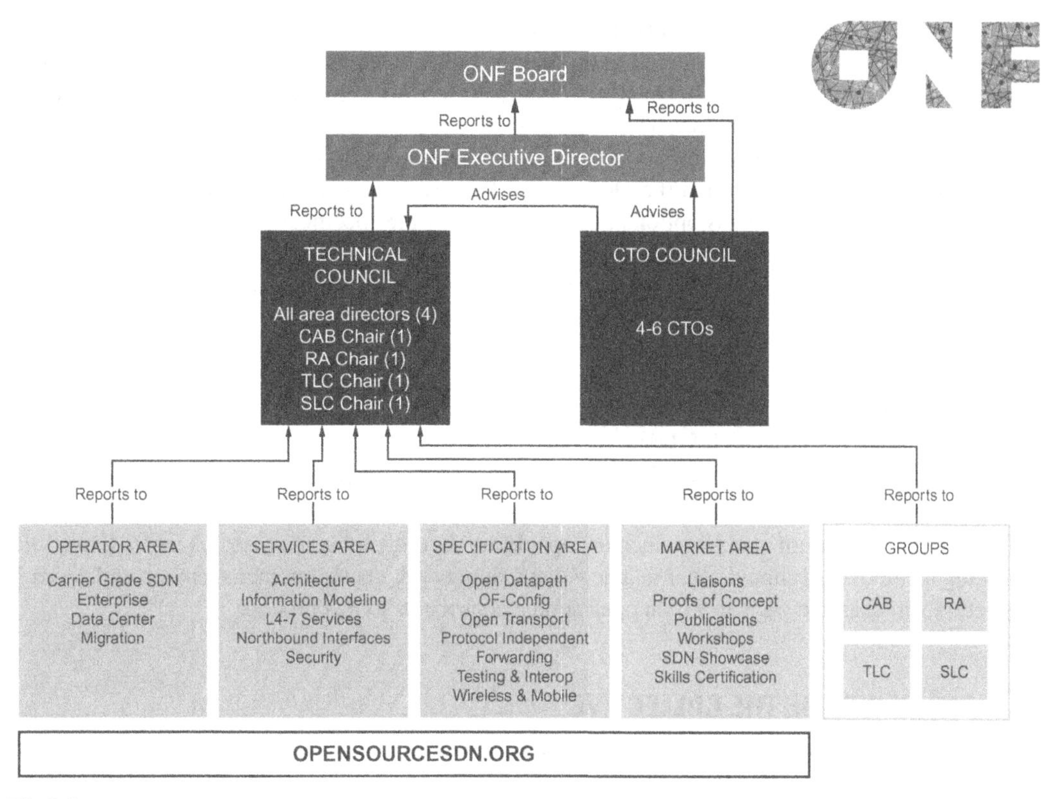

FIG. 3.8

ONF orgchart.

Reproduced with permission from https://www.opennetworking.org/certification/158-module-content/technical-communities-modules/1908-org-chart.

from other vendors. These are rich opportunities to determine where implementations may be lacking, or where the standard itself is unclear and needs to be made more precise and specific.

- **Interoperability Labs**: Certain institutions have built dedicated test labs for the purpose of testing the interoperability of equipment from various vendors and organizations. One such lab is the *Indiana Center for Network Translational Research and Education* (InCNTRE) at Indiana University, which hosts a large collection of vendor devices and controllers, as well as experimental devices and controllers from open source contributors. We discuss open source contributions to SDN in the next section.
- **Certification Programs**: There is a need for certification of switches so buyers can know they are getting a switch that is certified to support a particular version(s) of OpenFlow. The ONF has now implemented such a program [22].
- **Education and Consulting**: A complex, game-changing technological shift such as that represented by SDN will not easily permeate a large industry without the existence of an infrastructure to train and advise networking staff about the migration. It is important that a cadre of highly qualified yet vendor-neutral organizations address this need.

Initially, many SDN interoperability tests revealed dissimilarities and issues. While existing implementations of OpenFlow are increasingly interoperable, challenges remain. For example, as of this writing, it remains difficult for an OpenFlow controller to work consistently across multiple, varied switch implementations of OpenFlow. This is largely due to limitations in existing ASICs that lead to switches only supporting different subsets of OpenFlow consistent with their particular hardware. Other problems emanate from the fact that the OpenFlow 1.0 specification is vague in terms of specifying which features are required versus optional. Later versions of the specification attempt to clarify this, but even the latest specifications suffer from a large number of optional features for which support is far from universal. As we will see in Chapter 5, the OpenFlow standard is not static, and the goalposts of interoperability are moved with each new release.

3.6 OPEN SOURCE CONTRIBUTIONS

One of the basic rationales for SDN is that innovation and advancement have been stifled as a result of the closed and inflexible environment that exists in networking today. Thus, the openness that results from the creation of standards such as OpenFlow should encourage researchers to dissect old networking methods, and should usher in a new dawn of network operation, management, and control. This section will examine ways in which open source contributes to this process.

3.6.1 THE POWER OF THE COLLECTIVE

Technological advancement sometimes arises from the efforts of corporations and major organizations, quite often due to the fact that they are the only ones in the position to make contributions in their domains. In the world of software, however, it is occasionally possible for small players to develop technology and make it freely available to the general public. Some examples are:

- **Operating Systems**: The Linux operating system was developed as open source and is used today to control countless devices that we use every day, from *Digital Video Recorders* (DVRs) to smartphones.
- **Databases**: Many of the websites we visit for news reports or to purchase products store their information and product data in databases that have been developed, at least initially, by the open source community. MySQL is an example of such an open source data base.
- **Servers**: When we access locations on the Internet, many of those servers that we are accessing are running application server software and using tools which have been developed over time by the open source community. The open source Apache Web Server is used in countless applications worldwide.
- **Security**: Applying the open source model is also often considered to deliver more secure environments [23]. Open source can be more secure because of the peer review and white box evaluation that naturally occur in the open source development paradigm. Proprietary protocols may be less secure because they are not open and evaluated. Many security products providing antivirus protection and maintaining lists of malicious sites and programs are running open source software to accomplish their tasks. OpenSSL is probably the foremost example of a widely used open source encryption toolkit.

- **File Sharing**: The BitTorrent protocol is an example of a hugely successful [24] open protocol used for file sharing. BitTorrent works as a *P2P/Overlay* network that achieves high-bandwidth file downloading by performing the download of a file piecemeal and in parallel from multiple servers.

These are just a few examples. Imagine our world today without those items listed above. Would private institutions have eventually implemented the software solutions required in order to provide that functionality? Most likely. Would these advancements in technology have occurred at the velocity that we have witnessed in the past ten years, without current and ongoing contributions from open source? Almost certainly not.

3.6.2 THE DANGER OF THE COLLECTIVE

Of course, with an endeavor being driven by individuals who are governed not only by their own desire to contribute but also by their free will, whims and other interests, there is bound to be some risk. The areas of risk include quality, security, timeliness, and support. We explain below how these risks can be mitigated.

Open source software must undergo tests and scrutiny by even larger numbers of individuals than its commercial counterpart. This is due to the fact that an entire world of individuals has access to and can test those contributions. For any given feature being added or problem being solved, the open source community may offer a number of competing approaches to solve the problem. Even Open Source initiatives are subject to release cycles and there are key individuals involved in deciding what code will make its way into the next release. Just because an open source developer creates a body of code for an open source product does not mean that it makes it into a release. The fact that competing approaches may be assessed by the community and that admission into a release is actually controlled by key individuals associated with the open source effort manage to keep quality high. These same factors serve to minimize the risk of security threats. Since the source code is open for all to view, it is more difficult to hide malicious threats such as back doors.

The issue of timeliness is more nebulous. Unlike a project sponsored and fully funded by one's own organization, only indirect influence can be applied to exactly when a particular feature appears in an open source product. The prudent approach is not to depend on future deliverables, but to use existing functionality.

There seem to be two fundamental solutions to the issue of support. If you are a business intending to utilize open source in a product you are developing, you either need to feel that you have the resources to support it on your own, or that there is such a large community that has been using that code for a long enough time that you simply trust that the bad bugs have already surfaced and that you are using a stable code base.

It is important to remember that open source licensing models differ greatly from one model to the next. Some models severely restrict how contributions can be made and are used. We discuss in more detail in Section 13.4 some of the different major open source licenses and the issues with them.

3.6.3 OPEN SOURCE CONTRIBUTIONS TO SDN

Based on the previous discussions, it is easy to see the potential value of open source contributions in the drive toward SDN. Huge advances in SDN technology are attributable to open source projects.

Multiple open source implementations of SDN switches, controllers and applications are available. In Chapter 13 we provide details of the open source projects that have been specifically targeted to accelerate innovation in SDN. In that chapter we will also discuss other open source efforts which, while not as directly related to SDN, are nonetheless influential on the ever-growing acceptance of the SDN paradigm.

3.7 NETWORK VIRTUALIZATION

In Chapter 2 we discussed how *network virtualization* lagged behind its compute and storage counterparts and how this has resulted in a strong demand for network virtualization in the data center. Network virtualization, in essence, provides a network service that is decoupled from the physical hardware below that offers a feature set identical to the behavior of its physical counterpart. An important and early approach to such network virtualization was the *Virtual Local Area Network* (VLAN). VLANs permitted multiple virtual local area networks to co-reside on the same layer two physical network in total isolation from one another. While this technical concept is very sound, the provisioning of VLANs is not particularly dynamic, and they only scale to the extent of a layer two topology. Layer three counterparts based on tunneling scale better than VLANs to larger topologies. Complex systems have evolved to use both VLAN as well as tunneling technologies to provide network virtualization solutions.

One of the most successful commercial endeavors in this space was Nicira, now part of VMware. Early on, Nicira claimed that there were seven properties of network virtualization [25]:

1. Independence from network hardware
2. Faithful reproduction of the physical network service model
3. Follow operational model of compute virtualization
4. Compatibility with any hypervisor platform
5. Secure isolation between virtual networks, the physical networks, and the control plane
6. Cloud performance and scale
7. Programmatic network provisioning and control

Several of these characteristics closely resemble what we have said is required from an SDN solution. SDN promises to provide a mechanism for automating the network and abstracting the physical hardware below from the software defined network above. Network virtualization for data centers has undoubtedly been the largest commercial driver behind SDN. This momentum has become so strong that to some network virtualization has become synonymous with SDN. Indeed, VMware's (Nicira) standpoint [26] on this issue is that SDN is simply about abstracting control plane from data plane, and, therefore, network virtualization is SDN. Well, is it SDN or not?

3.8 MAY I PLEASE CALL MY NETWORK SDN?

If one were to ask four different attendees at a networking conference in 2016 what they thought qualified a network to be called SDN, they would likely provide divergent answers. Based on the genesis of SDN as presented in Section 3.4, in this book we will define an SDN network as characterized by five fundamental traits: *plane separation, a simplified device, centralized control, network automation and virtualization*, and *openness*. We will call an SDN solution possessing these five traits an *Open SDN* technology. We acknowledge that there are many competing technologies offered today that claim that their solution is an SDN solution. Some of these technologies have had larger economic impact in terms of the real-life deployments and dollars spent by customers than those that meet all five of our criteria. In some respects they may address customers' needs better than Open SDN. For example, a network virtualization vendor such as Nicira has had huge economic success and widespread installations in data centers, but does this without simplified devices. We define these five essential criteria not to pass judgment on these other SDN solutions, but in acknowledgement of what the SDN pioneers had in mind when they coined the term SDN in 2009 to refer to their work on OpenFlow. We will provide details about each of these five fundamental traits in Section 4.1, and compare and contrast competing SDN technologies against these five as well as other criteria in Chapter 6.

DISCUSSION QUESTION:

Looking back at Table 3.1, after Ethane, which two of the technologies listed there most closely approach what we have defined in this book as SDN? Why?

3.9 CONCLUSION

With the research and open source communities clamoring for an open environment for expanded research and experimentation, and the urgent needs of data centers for increased agility and virtualization, networking vendors have been forced into the SDN world. Some have moved readily into that world, while others have dragged their feet or have attempted to redefine SDN. In the next chapter we will examine what SDN is and how it actually works.

REFERENCES

[1] Ferro G. Networking vendors should step up with an SDN strategy. Network Computing, June 7, 2012. Retrieved from: http://www.networkcomputing.com/next-gen-network-tech-center/networking-vendors-should-step-up-with-a/240001600.

[2] Godbole A. Data communications and networks. New Delhi: Tata McGraw-Hill; 2002.

[3] Mendonca M, Astuto B, Nunes A, Nguyen X, Obraczka K, Turletti T. A survey of software-defined networking: past, present and future of programmable networks. IEEE Commun Surv Tutorials 2014;16(3):1617–34.

[4] Feamster N, Rexford J, Zegura E. The road to SDN: an intellectual history of programmable networks. ACM Mag 2013;11(12).

[5] Devolved control of ATM networks. Retrieved from: http://www.cl.cam.ac.uk/research/srg/netos/old-projects/dcan/.

[6] Campbell A, Katzela I, Miki K, Vicente J. Open signalling for ATM, Internet and Mobile Networks (OPENSIG'98). ACM SIGCOMM Comput Commun Rev 1999;29(1):97–108.

[7] Doria A, Hellstrand F, Sundell K, Worster T. General Switch Management Protocol (GSMP) V3, RFC 3292. Internet Engineering Task Force; 2002.

[8] A comparison of IP switching technologies from 3Com, Cascade, and IBM. CSE 588 Network Systems, Spring, 1997. University of Washington; 1997. Retrieved from: http://www.cs.washington.edu/education/courses/csep561/97sp/paper1/paper11.txt.

[9] Tennehouse D, Smith J, Sincoskie W, Wetherall D, Minden, G. A survey of active network research. IEEE Commun Mag 1997;35(1):80–6.

[10] Tennehouse D, Wetherall D. Towards an active network architecture. In: Proceedings of the DARPA Active networks conference and exposition, 2002, IEEE; 2002. p. 2–15.

[11] Bhattacharjee S, Calvert K, Zegura E. An architecture for active networking. In: High performance networking VII: IFIP TC6 Seventh international conference on High Performance Networks (HPN '97), April–May 1997, White Plains, NY, USA; 1997.

[12] Schwartz B, Jackson A, Strayer W, Zhou W, Rockwell R, Partridge C. Smart packets for active networks. In: Open architectures and network programming proceedings, IEEE OPENARCH '99, March 1999, New York, NY, USA; 1999.

[13] da Silva S, Yemini Y, Florissi D. The NetScript active network system. IEEE J Sel Areas Commun 2001;19(3):538–51.

[14] Rigney C, Willens S, Rubens A, Simpson W. Remote Authentication Dial In User Service (RADIUS), RFC 2865. Internet Engineering Task Force; 2000.

[15] Durham D, Boyle J, Cohen R, Herzog S, Rajan R, Sastry A. The COPS (Common Open Policy Service) Protocol, RFC 2748. Internet Engineering Task Force; 2000.

[16] Ferro G. Automation and orchestration. Network Computing, September 8, 2011. Retrieved from: http://www.networkcomputing.com/private-cloud-tech-center/automation-and-orchestration/231600896.

[17] VMWare vCenter. Retrieved from: http://www.vmware.com/products/vcenter-server/overview.html.

[18] VMWare vMotion. Retrieved from: http://www.vmware.com/files/pdf/VMware-VMotion-DS-EN.pdf.

[19] Doria A, Hadi Salim J, Haas R, Khosravi H, Wang W, Dong L, et al. Forwarding and Control Element Separation (ForCES) protocol specification, RFC 5810. Internet Engineering Task Force; 2010.

[20] Greenberg A, Hjalmtysson G, Maltz D, Myers A, Rexford J, Xie G, et al. A clean slate 4D approach to network control and management. ACM SIGCOMM Comput Commun Rev 2005;35(3).

[21] Casado M, Freedman M, Pettit J, McKeown N, Shenker S. Ethane: taking control of the enterprise. ACM SIGCOMM Comput Commun Rev 2007;37(4):1–12.

[22] Lightwave Staff. Open Networking Foundation launches OpenFlow certification program. Lightwave, July 2013. Retrieved from: http://www.lightwaveonline.com/articles/2013/07/open-networking-foundation-launches-openflow-certification-program.html.

[23] Benefits of Open Source Software. Open Source for America. Retrieved from: http://opensourceforamerica.org/learn-more/benefits-of-open-source-software/.

[24] BitTorrent and μTorrent software surpass 150 million user milestone; announce new consumer electronics partnerships. BitTorrent, January 2012. Retrieved from: http://www.bittorrent.com/intl/es/company/about/ces_2012_150m_users.

[25] Gourley B. The seven properties of network virtualization. CTOvision, August 2012. Retrieved from: http://ctovision.com/2012/08/the-seven-properties-of-network-virtualization/.

[26] Metz C. What is a virtual network? It's not what you think it is. Wired, May 9, 2012. Retrieved from: http://www.wired.com/wiredenterprise/2012/05/what-is-a-virtual-network.

[27] Org Chart. Open Networking Foundation; 2016. Retrieved from: https://www.opennetworking.org/certification/158-module-content/technical-communities-modules/1908-org-chart.

CHAPTER

HOW SDN WORKS

4

In previous chapters we have seen why SDN is necessary and what preceded the actual advent of SDN in the research and industrial communities. In this chapter we provide an overview of how SDN actually works, including the basic components of a software defined networking system, their roles, and how they interact with one another. In the first part of this chapter we focus on the methods used by Open SDN. We also examine how some *alternate* forms of SDN work. As SDN has gained momentum, some networking vendors have responded with alternate definitions of SDN, which better align with their own product offerings. Some of these methods of implementing SDN-like solutions are new (some are not), and are innovative in their approach. We group the most important of these alternate SDN implementations in two categories, *SDN via APIs* and *SDN via Hypervisor-Based Overlay Networks*, and discuss them separately in the latter half of this chapter.

4.1 FUNDAMENTAL CHARACTERISTICS OF SDN

As introduced in Chapter 3, software defined networking, as it evolved from prior proposals, standards, and implementations such as ForCES, 4D, and Ethane, is characterized by five fundamental traits: *plane separation*, *a simplified device*, *centralized control*, *network automation and virtualization*, and *openness*.

4.1.1 PLANE SEPARATION

The first fundamental characteristic of SDN is the separation of the forwarding and control planes. Forwarding functionality, including the logic and tables for choosing how to deal with incoming packets, based on characteristics such as MAC address, IP address, and VLAN ID, reside in the forwarding plane. The fundamental actions performed by the forwarding plane can be described by how it dispenses with arriving packets. It may *forward*, *drop*, *consume*, or *replicate* an incoming packet. It may also transform the packet in some manner before taking further action. For basic forwarding, the device determines the correct output port by performing a lookup in the address table in the hardware ASIC. A packet may be dropped due to buffer overflow conditions or due to specific *filtering* resulting from a QoS rate-limiting function, for example. Special-case packets that require processing by the control or management planes are consumed and passed to the appropriate plane. Finally, a special case of forwarding pertains to multicast, where the incoming packet must be replicated before forwarding the various copies out different output ports.

Software Defined Networks. http://dx.doi.org/10.1016/B978-0-12-804555-8.00004-1

The logic and algorithms that are used to program the forwarding plane reside in the control plane. Many of these protocols and algorithms require global knowledge of the network. The control plane determines how the forwarding tables and logic in the data plane should be programmed or configured. Since in a traditional network each device has its own control plane, the primary task of that control plane is to run routing or switching protocols so that all the distributed forwarding tables on the devices throughout the network stay synchronized. The most basic outcome of this synchronization is the prevention of loops.

While these planes have traditionally been considered logically separate, they co-reside in legacy Internet switches. In SDN, the control plane is moved off of the switching device and onto a centralized controller. This is the inspiration behind Fig. 1.6 in Chapter 1.

4.1.2 SIMPLE DEVICE AND CENTRALIZED CONTROL

Building on the idea of separation of forwarding and control planes, the next characteristic is the simplification of devices, which are then controlled by a centralized system running management and control software. Instead of hundreds of thousands of lines of complicated control plane software running on the device and allowing the device to behave autonomously, that software is removed from the device and placed in a centralized controller. This software-based controller may then manage the network based on higher-level policies. The controller provides primitive instructions to the simplified devices when appropriate in order to allow them to make fast decisions about how to deal with incoming packets.

4.1.3 NETWORK AUTOMATION AND VIRTUALIZATION

Three basic abstractions forming the basis for SDN are defined in [1]. This asserts that SDN can be derived precisely from the abstractions of *distributed state*, *forwarding*, and *configuration*. They are derived from decomposing the actual complex problem of network control faced by networks today into simplifying abstractions. For a historical analogy, note that today's high-level programming languages represent an evolution from their machine language roots through the intermediate stage of languages like C, where today's languages allow great productivity gains by allowing the programmer to simply specify complex actions through programming abstractions. In a similar manner, Shenker [1] purports that SDN is a similar natural evolution for the problem of network control. The distributed state abstraction provides the network programmer with a global network view that shields the programmer from the realities of a network that is actually comprised of many machines, each with its own state, collaborating to solve network-wide problems. The forwarding abstraction allows the programmer to specify the necessary forwarding behaviors without any knowledge of vendor-specific hardware. This implies that whatever language(s) emerges from the abstraction needs to represent a sort of lowest common denominator of forwarding capabilities of network hardware. Finally, the configuration abstraction, which is sometimes called the *specification* abstraction, must be able to express the desired goals of the overall network without getting lost in the details of how the physical network will implement those goals. To return to the programming analogy, consider how unproductive software developers would be if they always needed to be aware of what is actually involved in writing a block of data to a hard disk when they are instead happily productive with the abstraction of file input and output.[1] Working with the network through this configuration abstraction is really network virtualization at its most basic level. This kind of virtualization lies at the heart of how we will define Open SDN in this work.

The centralized software-based controller in SDN provides an open interface on the controller to allow for automated control of the network. In the context of Open SDN, the terms *northbound* and *southbound* are often used to distinguish whether the interface is to the applications or to the devices. These terms derive from the fact that in most diagrams the applications are depicted above (i.e., to the north) the controller while devices are depicted below (i.e., to the south) the controller. An example of a southbound API is the OpenFlow interface that the controller uses to program the network devices. The controller offers a northbound API, allowing software applications to be plugged into the controller, and thereby allowing that software to provide the algorithms and protocols that can run the network efficiently. These applications can quickly and dynamically make network changes as the need arises. The northbound API of the controller is intended to provide an abstraction of the network devices and topology. There are three key benefits that the application developer should derive from the northbound API: (1) it converts to a syntax that is more familiar to developers (e.g., REST or JSON are more convenient syntaxes than are TLVs); (2) it provides abstraction of the network topology and network layer allowing the application programmer to deal with the network as a whole rather than individual nodes; and (3) it provides abstraction of the network protocols themselves, hiding the application developer from the details of OpenFlow or BGP. In this way, applications can be developed that work over a wide array of manufacturers' equipment that may differ substantially in their implementation details.

One of the results of this level of abstraction is that it provides the ability to virtualize the network, decoupling the network service from the underlying physical network. Those services are still presented to host devices in such a way that those hosts are unaware that the network resources they are using are virtual and not the physical ones for which they were originally designed.

4.1.4 OPENNESS

A characteristic of Open SDN is that its interfaces should remain standard, well documented, and not proprietary. The APIs that are defined should give software sufficient control to experiment with and control various control plane options. The premise is that by keeping open both the northbound and southbound interfaces to the SDN controller, this will allow for research into new and innovative methods of network operation. Research institutions as well as entrepreneurs can take advantage of this capability in order to easily experiment with and test new ideas. Hence, the speed at which network technology is developed and deployed is greatly increased, as a much larger pool of individuals and organizations are able to apply themselves to today's network problems, resulting in better and faster technological advancement in the structure and functioning of networks. The presence of these open interfaces also encourages SDN-related open source projects. As discussed in Section 1.7 and Section 3.6, harnessing the power of the open source development community should greatly accelerate

[1]Granted, programmers who *are* aware of technical details of device driver interactions with hardware usually make better programmers and are better able to optimize performance. While the abstractions discussed here will increase the efficiency of future network programmers, we are not suggesting that deep technical knowledge of how packets move through networking switches and communication links will cease to be needed by elite network programmers.

innovation in SDN [2]. In 2014, Lloyd Carney, CEO of Brocade made the following statement on openness [3]: "*Vertical architectures are giving way to open solutions that allow customers to select best-of-breed technologies that enable new services and networks optimized for the customer's own specific requirements. I can say without hesitation that openness is an integral part of nearly every discussion I have with customers, and I firmly believe it is creating meaningful business opportunities for Brocade.*"

In addition to facilitating research and experimentation, open interfaces permit equipment from different vendors to interoperate. This normally produces a competitive environment which lowers costs to consumers of network equipment. This reduction in network equipment costs has been part of the SDN agenda since its inception.

DISCUSSION QUESTION:

Of the five fundamental traits of SDN we define here, not all are viewed as essential to SDN within the networking industry. Which do you think are more controversial for certain industry players and why?

4.2 SDN OPERATION

At a conceptual level, the behavior and operation of an SDN is straightforward. In Fig. 4.1 we provide a graphical depiction of the operation of the basic components of SDN: the SDN devices, the controller and the applications. The easiest way to understand the operation is to look at it from the bottom up, starting with the SDN device. As can be seen in Fig. 4.1, the SDN devices contain forwarding

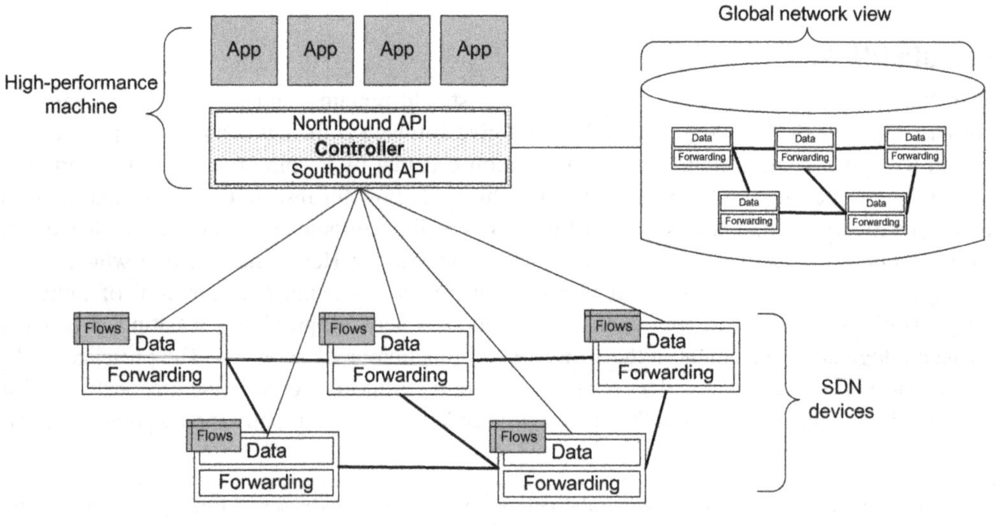

FIG. 4.1

SDN operation overview.

functionality for deciding what to do with each incoming packet. The devices also contain the data which drives those forwarding decisions. The data itself is actually represented by the flows defined by the controller, as depicted in the upper left portion of each device.

A flow describes a set of packets transferred from one network endpoint (or set of endpoints) to another endpoint (or set of endpoints). The endpoints may be defined as IP address-TCP/UDP port pairs, VLAN endpoints, layer three tunnel endpoints, and input port, among other things. One set of rules describes the actions that the device should take for all packets belonging to that flow. A flow is unidirectional in that packets flowing between the same two endpoints in the opposite direction could each constitute a separate flow. Flows are represented on a device as a flow entry.

A flow table resides on the network device and consists of a series of flow entries and the actions to perform when a packet matching that flow arrives at the device. When the SDN device receives a packet it consults its flow tables in search of a match. These flow tables had been constructed previously when the controller downloaded appropriate flow rules to the device. If the SDN device finds a match, it takes the appropriate configured action, which usually entails forwarding the packet. If it does not find a match, the switch can either drop the packet or pass it to the controller, depending on the version of OpenFlow and the configuration of the switch. We will describe flow tables and this packet matching process in greater detail in Section 4.3 and Section 5.3.

The definition of a flow is a relatively simple programming expression of what may be a very complex control plane calculation previously performed by the controller. For the reader less familiar with traditional switching hardware architecture, it is important to understand that this complexity is such that it simply cannot be performed at line rates and instead must be digested by the control plane and reduced to simple rules that can be processed at that speed. In Open SDN, this digested form is the flow entry.

The SDN controller is responsible for abstracting the network of SDN devices it controls and presenting an abstraction of these network resources to the SDN applications running above. The controller allows the SDN application to define flows on devices and to help the application to respond to packets which are forwarded to the controller by the SDN devices. In Fig. 4.1 we see on the right-hand side of the controller that it maintains a view of the entire network that it controls. This permits it to calculate optimal forwarding solutions for the network in a deterministic, predictable manner. Since one controller can control a large number of network devices, these calculations are normally performed on a high performance machine, with an order-of-magnitude performance advantage over the CPU and memory capacity than is typically afforded to the network devices themselves. For example, a controller might be implemented on an eight-core two-GHz CPU versus the single-core one-GHz CPU more typical on a switch.

SDN applications are built on top of the controller. These applications should not be confused with the application layer defined in the seven-layer OSI model of computer networking. Since SDN applications are really part of network layers two and three, this concept is orthogonal to that of applications in the tight hierarchy of OSI protocol layers. The SDN application interfaces with the controller, using it to set *proactive* flows on the devices and to receive packets which have been forwarded to the controller. Proactive flows are established by the application; typically the application will set these flows when the application starts up, and the flows will persist until some configuration change is made. This kind of proactive flow is known as a *static* flow. Another kind of proactive flow is where the controller decides to modify a flow based on the traffic load currently being driven through a network device.

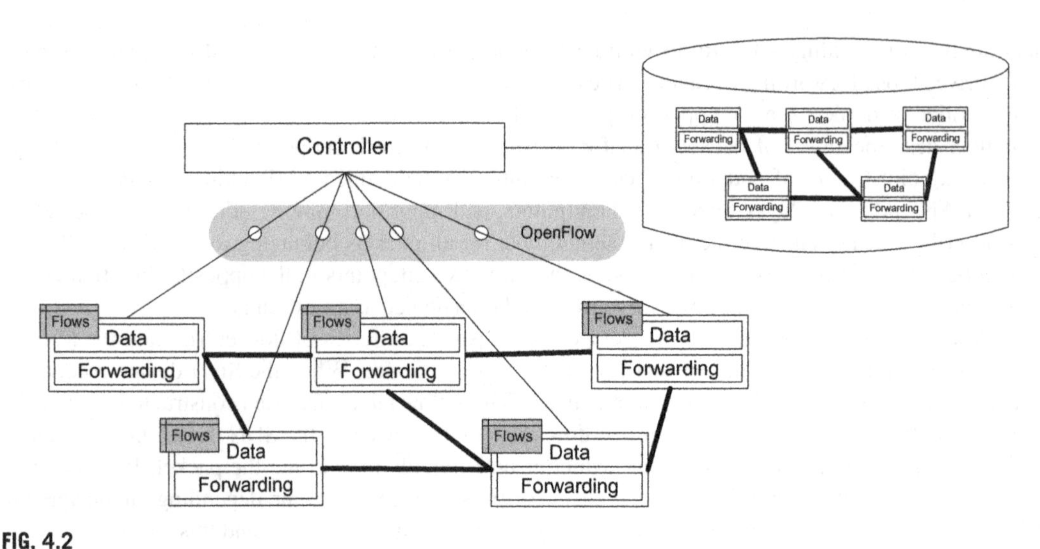

FIG. 4.2

Controller to device communication.

In addition to flows defined proactively by the application, some flows are defined in response to a packet forwarded to the controller. Upon receipt of incoming packets that have been forwarded to the controller, the SDN application will instruct the controller as to how to respond to the packet, and, if appropriate, will establish new flows on the device in order to allow that device to respond locally the next time it sees a packet belonging to that flow. Such flows are called *reactive* flows. In this way, it is now possible to write software applications which implement forwarding, routing, overlay, multipath, and access control functions, among others.

There are also reactive flows that are defined or modified as a result of stimuli from sources others than packets from the controller. For example, the controller can insert flows reactively in response to other data sources such as *Intrusion Detection Systems* (IDS) or the NetFlow traffic analyzer [4].

Fig. 4.2 depicts the OpenFlow protocol as the means of communication between the controller and the device. While OpenFlow is the defined standard for such communication in Open SDN, there are alternative SDN solutions discussed later in this chapter which may use vendor-specific proprietary protocols. The next sections discuss in greater detail SDN devices, controllers and applications.

4.3 SDN DEVICES

An SDN device is composed of an API for communication with the controller, an abstraction layer, and a packet processing function. In the case of a virtual switch, this packet processing function is packet processing software as shown in Fig. 4.3. In the case of a physical switch, the packet processing function is embodied in the hardware for packet processing logic, as shown in Fig. 4.4.

The abstraction layer embodies one or more flow tables, which we discuss in Section 4.3.1. The packet processing logic consists of the mechanisms to take actions based on the results of evaluating

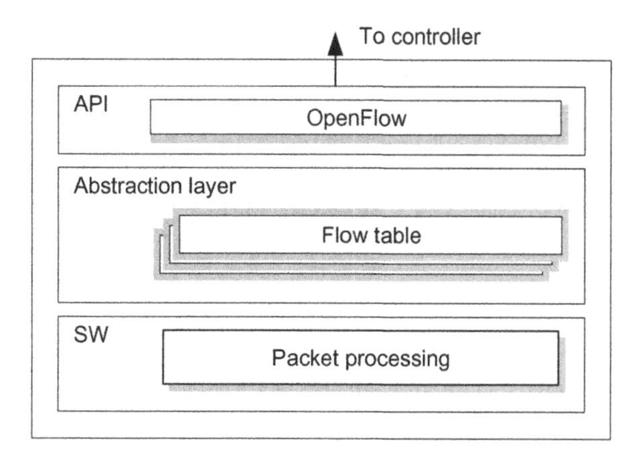

FIG. 4.3

SDN software switch anatomy.

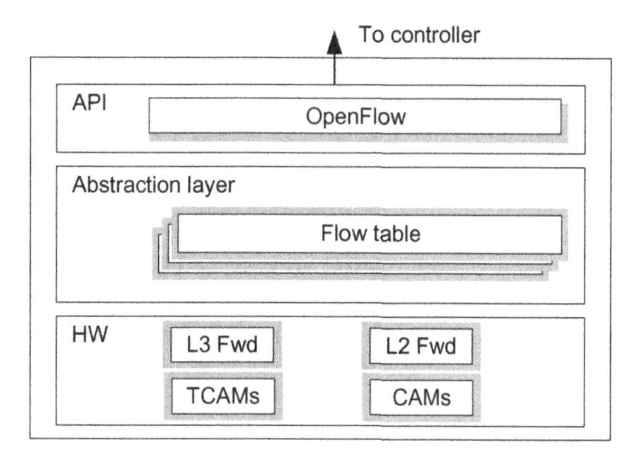

FIG. 4.4

SDN hardware switch anatomy.

incoming packets and finding the highest priority match. When a match is found, the incoming packet is processed locally, unless it is explicitly forwarded to the controller. When no match is found, the packet may be copied to the controller for further processing. This process is also referred to as the controller *consuming* the packet. In the case of a hardware switch, these mechanisms are implemented by the specialized hardware we discuss in Section 4.3.3. In the case of a software switch, these same functions are mirrored by software. Since the case of the software switch is somewhat simpler than the hardware switch, we will present that first in Section 4.3.2. Some readers may be confused by the distinction between a hardware switch and a software switch. The earliest routers that we described in Chapter 1 were indeed just software switches. Later, as we discussed in Section 2.1, we explained that over time the actual packet forwarding logic migrated into hardware for switches that needed to process packets arriving at ever-increasing line rates. More recently, a role has reemerged in the data center for the pure software switch. Such a switch is implemented as a software application usually running in conjunction with a hypervisor in a data center rack. Like a VM, it may be instantiated or moved under software control. It normally serves as a virtual switch and works collectively with a set of other such virtual switches to constitute a virtual network. We will discuss this concept in greater depth in Chapter 8.

4.3.1 FLOW TABLES

Flow tables are the fundamental data structures in an SDN device. These flow tables allow the device to evaluate incoming packets and take the appropriate action based on the contents of the packet that has just been received. Packets have traditionally been received by networking devices and evaluated based on certain fields. Depending on that evaluation, actions are taken. These actions may include forwarding the packet to a specific port, dropping the packet, and flooding the packet on all ports, among others. An SDN device is not fundamentally different, except that this basic operation has been rendered more generic and more programmable via the flow tables and their associated logic.

Flow tables consist of a number of prioritized flow entries, each of which typically consists of two components, *match fields* and *actions*. Match fields are used to compare against incoming packets. An incoming packet is compared against the match fields in priority order, and the first complete match is selected. Actions are the instructions that the network device should perform if an incoming packet matches the match fields specified for this flow entry.

Match fields can have wildcards for fields that are not relevant to that particular match. For example, when matching packets based just on IP address or subnet, all other fields would be wildcarded. Similarly, if matching on only MAC address or UDP/TCP port, the other fields are irrelevant, and, consequently, those fields are wildcarded. Depending on the application needs, all fields may be important, in which case there would be no wildcards. The flow table and flow entry constructs allow the SDN application developer to have a wide range of possibilities for matching packets and taking appropriate actions.

Given this general description of an SDN device, we now look at two embodiments of an SDN device: first, the more simple software SDN device, and then a hardware SDN device.

4.3.2 SDN SOFTWARE SWITCHES

In Fig. 4.3 we provide a graphical depiction of a purely software-based SDN device. Implementation of SDN devices in software is the simplest means of creating an SDN device because the flow tables, flow entries, and match fields involved are easily mapped to general software data structures, such as sorted arrays and hash tables. Consequently, it is more probable that two software SDN devices produced by different development teams will behave consistently than is the case for two different hardware implementations. Conversely, implementations in software are likely to be slower and less efficient than those implemented in hardware, as they do not benefit from any hardware acceleration. Consequently, for network devices that must run at high speeds, such as 100Gbs, only hardware implementations are feasible with current technology.

Due to the use of wildcards in matching, which poses a problem for typical hash tables, the packet processing function depicted in Fig. 4.3 uses sophisticated software logic in order to implement efficient match field lookups. Hence, in the early days of SDN, there was a wide variance in the performance of different software implementations, based on the efficiency with which these lookups are accomplished. Fortunately, software SDN device implementations have matured. The fact that there are two widely recognized software reference implementations (see Section 4.3.4), both of which use sophisticated and efficient methods of performing these lookups, has resulted in greater uniformity in software SDN device performance.

Software device implementations also suffer less from resource constraints, since considerations such as processing power and memory size are not an issue in typical implementations. Thus, whereas a hardware SDN device implementation will support only a comparatively limited number of flow entries, the ceiling on the number of flow entries on a software device may be orders of magnitude larger. As software device implementations have more flexibility to implement more complex actions, we expect to see a richer set of actions available on software SDN device implementations than on the hardware SDN devices that we examine in the next section.

Software SDN device implementations are most often found in software-based network devices, such as the hypervisors of a virtualization system. These hypervisors often incorporate a software switch implementation which connects the various virtual machines to the virtual network. The virtual switch working with a hypervisor is a natural fit for SDN. In fact, the whole virtualization system is often controlled by a centralized management system, which also meshes well with the centralized controller aspect of the SDN paradigm.

4.3.3 HARDWARE SDN DEVICES

Hardware implementations of SDN devices hold the promise of operating much faster than their software counterparts and, thus, are more applicable to performance-sensitive environments, such as in data centers and network cores. In order to understand how SDN objects such as flow tables and flow entries can be translated into hardware, we will briefly review some of the hardware components of today's networking devices.

Currently, network devices utilize specialized hardware designed to facilitate the inspection of incoming packets and the subsequent decisions that follow, based on the packet matching operation. We see in Fig. 4.4 that the packet processing logic shown in Fig. 4.3 has been replaced by this specialized hardware. This hardware includes the layer two and layer three forwarding tables, often implemented using *Content-Addressable Memories* (CAMs) and *Ternary Content-Addressable Memories* (TCAMs). The price-performance gains enjoyed by RAM technology makes its use more popular for layer two and layer three Forwarding Tables in contemporary designs. The layer three Forwarding Table is used for making IP-level routing decisions. This is the fundamental operation of a router. It matches the destination IP address against entries in the table, and, based on the match, takes the appropriate routing action (e.g., forward the packet out interface B3). The layer two Forwarding Table is used for making MAC-level forwarding decisions. This is the fundamental operation of a switch. It matches the destination MAC address against entries in the table, and, based on the match, takes the appropriate forwarding action (e.g., forward out interface 15).

The layer two forwarding table is typically implemented using regular CAMs or hardware-based hashing. These kinds of associative memories are used when there are precise indices, such as a forty-eight bit MAC address. TCAMs, however, are associated with more complex matching functions. TCAMs are used in hardware to check not only for an exact match, but also for a third state, which uses a mask to treat certain parts of the match field as wild cards. A straightforward example of this is matching an IP destination address against networks where a *longest prefix match* is performed. Depending on subnet masks, multiple table entries may match the search key and the goal is to determine the closest match. A more important and innovative use of TCAMs is for potentially matching some but not all header fields of an incoming packet. These TCAMs are thus essential for functions such as *policy-based routing* (PBR).

This hardware functionality allows the device to both match packets and then take actions at a very high rate. However, it also presents a series of challenges to the SDN device developer. Specifically:

- How best to translate from flow entries to hardware entries; for example, where and when should the hardware developer utilize CAMs, TCAMs or more straightforward RAM-based hash tables?
- Which of the flow entries to handle in hardware, versus how many to fall back to using software. Most implementations are able to use hardware to handle some of the lookups, but others are handed off to software to be handled there. Obviously, hardware will handle the flow lookups much faster than software, but hardware tables have limitations on the number of flow entries they can hold at any time, and software tables could be used to handle the overflow.
- How to deal with hardware action limitations which may impact whether to implement the flow in hardware versus software; for example, certain actions such as packet modification may be limited or not available if handled in hardware.
- How to track statistics on individual flows. When using devices such as TCAMs which may match multiple flows, it is not possible to use that device to count individual flows separately. Also, gathering statistics across the various tables can be problematic because the tables may count something twice or not at all.

These and other factors will impact the quality, functionality, and efficiency of the SDN device being developed, and must be considered during the design process. For example, hardware table sizes may limit the number of flows, and hardware table capabilities may limit the breadth and depth of special features supported. The limitations presented by a hardware SDN device might require adaptations to SDN applications in order to interoperate with multiple heterogeneous types of SDN devices.

While the challenges to the SDN device designer are no doubt formidable, the range of variables confronting the SDN application developer is vast. The first generation SDN device developer is corralled into basically retrofitting existing hardware to SDN, and thus does not have many choices and indeed may be unable to implement all specified features. The SDN application developer, on the other hand, must deal with inconsistencies across vendor implementations, with scaling performance on a network-wide basis, and a host of other more nebulous issues. We discuss some of these application-specific issues in Section 4.5.

This section provided an overview of the composition of SDN devices and what considerations must be taken into account during their development and their use as part of an SDN application. We provide further specifics on flow tables, flow entries and actions in Chapter 5.

DISCUSSION QUESTION:

Why is the use of devices like TCAMs more challenging in an SDN environment than in a traditional switch?

4.3.4 EXISTING SDN DEVICE IMPLEMENTATIONS

There are a number of SDN device implementations available today, both commercial and open source. Software SDN devices are predominantly open source. Currently, two main alternatives are available, *Open vSwitch* (OVS) [5] from Nicira and *Indigo* [6] from Big Switch. Incumbent *Network Equipment Manufacturers* (NEMs), such as Cisco, HP, NEC, IBM, Juniper, and Extreme, have added OpenFlow support to some of their legacy switches. Generally, these switches may operate in both legacy mode as well as OpenFlow mode. There is also a new class of device called *white box switches*, which are minimalist in that they are built primarily from merchant silicon switching chips and a commodity CPU and memory by a low-cost *Original Device Manufacturer* (ODM) lacking a well-known brand name. One of the premises of SDN is that the physical switching infrastructure may be built from OpenFlow-enabled white box switches at far less direct cost than switches from established NEMs. Most legacy control plane software is absent from these devices, since this is largely expected to be provided by a centralized controller. Such white box devices often use the open source OVS or Indigo switch code for the OpenFlow logic, and then map the packet processing part of those switch implementations to their particular hardware. Chuck Robbins, Cisco CEO, recognizes the impact of SDN when he states "*Everything we build will be programmable*" [7].

4.3.5 SCALING THE NUMBER OF FLOWS

The granularity of flow definitions will generally be more fine as the device holding them approaches the edge of the network and more general the more the device approaches the core. At the edge, flows will permit different policies to be applied to individual users and even different traffic types of the same user. This will imply in some cases multiple flow entries for a single user. This level of flow granularity would simply not scale if it were applied closer to the network core, where large switches deal with the traffic for tens of thousands of users simultaneously. In those core devices, the flow definitions will be generally more coarse, with a singe aggregated flow entry matching the traffic from a large number of users whose traffic is aggregated in some way such as a tunnel, a VLAN or a MPLS LSP. Policies applied deep into the network will likely not be user-centric policies but policies that apply to these aggregated flows. One positive result of this is that there will not be an explosion of the number of flow entries in the core switches due to handling the traffic emanating from thousands of flows in edge switches.

4.4 SDN CONTROLLER

We have noted that the controller maintains a view of the entire network, implements policy decisions, controls all the SDN devices that comprise the network infrastructure, and provides a northbound API for applications. When we have said that the controller implements policy decisions regarding routing, forwarding, redirecting, load balancing, and the like, these statements referred both to the controller and to the applications that make use of that controller. Controllers often come with their own set of common application modules, such as a learning switch, a router, a basic firewall, and a simple load balancer. These are really SDN applications, but they are often bundled with the controller. Here we focus strictly on the controller.

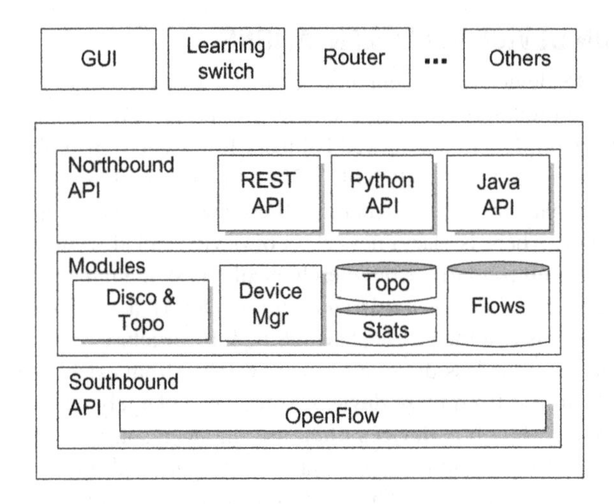

FIG. 4.5

SDN controller anatomy.

Fig. 4.5 exposes the anatomy of an SDN controller. The figure depicts the modules that provide the core functionality of the controller, both a northbound and southbound API, and a few sample applications that might use the controller. As we described earlier, the southbound API is used to interface with the SDN devices. This API is OpenFlow in the case of Open SDN or some alternative such as BGP in other SDN solutions. It is worth noting that in some product offerings both OpenFlow and alternatives coexist on the same controller. Early work on the southbound API has resulted in more maturity of that interface with respect to its definition and standardization. OpenFlow itself is the best example of this, but de facto standards like the Cisco CLI and SNMP also represent standardization in the southbound-facing interface. OpenFlow's companion protocol OF-Config [8] and Nicira's *Open vSwitch Database Management Protocol* (OVSDB) [9] are both open protocols for the southbound interface, though these are limited to configuration roles.

Currently, the Open Source SDN (http://www.opensourcesdn.org) group is proposing a northbound counterpart to the southbound OpenFlow standard. While the lack of a standard for the controller-to-application interface is considered a current deficiency in SDN, organizations like the OpenSource SDN group are developing proposals to standardize this. The absence of a standard notwithstanding, northbound interfaces have been implemented in a number of disparate forms. For example, the Floodlight controller [10] includes a Java API, and a *Representational State Transfer* (RESTful) [11] API. The OpenDaylight controller [12] provides a RESTful API [13] for applications running on separate machines. The northbound API represents an outstanding opportunity for innovation and collaboration amongst vendors and the open source community.

4.4.1 SDN CONTROLLER CORE MODULES

The controller abstracts the details of the SDN controller-to-device protocol so that the applications above are able to communicate with those SDN devices without knowing the nuances of those devices.

Fig. 4.5 shows the API below the controller, which is OpenFlow in Open SDN, and the interface provided for applications. Every controller provides core functionality between these raw interfaces. Core features in the controller will include:

- **End-user Device Discovery**: Discovery of end-user devices, such as laptops, desktops, printers, mobile devices, etc.
- **Network Device Discovery**: Discovery of network devices which comprise the infrastructure of the network, such as switches, routers, and wireless access points.
- **Network Device Topology Management**: Maintain information about the interconnection details of the network devices to each other, and to the end-user devices to which they are directly attached.
- **Flow Management**: Maintain a database of the flows being managed by the controller and perform all necessary coordination with the devices to ensure synchronization of the device flow entries with that database.

The core functions of the controller are device and topology discovery and tracking, flow management, device management and statistics tracking. These are all implemented by a set of modules internal to the controller. As shown in Fig. 4.5, these modules need to maintain local databases containing the current topology and statistics. The controller tracks the topology by learning of the existence of switches (SDN devices) and end-user devices and tracking the connectivity between them. It maintains a *flow cache* which mirrors the flow tables on the different switches it controls. The controller locally maintains per-flow statistics that it has gathered from its switches. The controller may be designed such that functions are implemented via pluggable modules such that the feature set of the controller may be tailored to an individual network's requirements.

Many companies implementing SDN look to these core modules to help them model their network and create abstraction layers. Level 3's CTO Jack Waters described his company's efforts with SDN: "*The combined company (Level 3 and Time Warner Telecom) has done a number of network trials. With respect to SDN, we're in the development phase around a few NFV betas so it's early but we have use cases up and running already. The combined company is laser focused on how to leverage the technology really to help build a network abstraction layer for things like provisioning, configuration management, and provisioning automation is something we think will be great for us and our customers.*" [14]

4.4.2 SDN CONTROLLER INTERFACES

For SDN applications, a key function provided by the SDN controller is the API for accessing the network. In some cases, this northbound API is a low level interface, providing access to the network devices in a common and consistent manner. In this case, that application is aware of individual devices, but is shielded from their differences. In other instances the controller may provide high level APIs that provide an abstraction of the network itself, so that the application developer need not be concerned with individual devices, but with the network as a whole.

Fig. 4.6 takes a closer look at how the controller interfaces with applications. The controller informs the application of *events* that occur in the network. Events are communicated from the controller to the application. Events may pertain to an individual packet that has been received by the controller or some state change in the network topology such as a link going down. Applications use different *methods* to

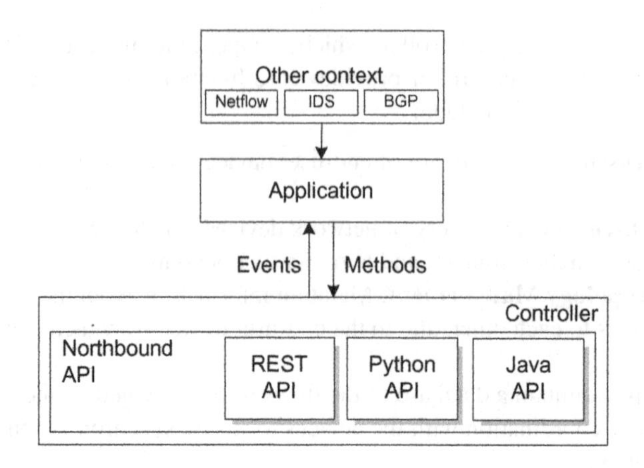

FIG. 4.6

SDN controller northbound API.

affect the operation of the network. Such methods may be invoked in response to a received event and may result in a received packet being dropped, modified and/or forwarded, or the addition, deletion or modification of a flow. The applications may also invoke methods independently without the stimulus of an event from the controller, as we explain in Section 4.5.1. Such inputs are represented by the *other context* box in Fig. 4.6.

As of this writing, a search of the web reveals more than twenty SDN controllers available on the market. Interestingly, there are no standards for the northbound API, also called the *northbound interface* (NBI). Why is the standardization of the NBI important? From a customer perspective, a standard NBI would allow for the SDN application market to materialize. SDN application developers no longer would have to modify their application to multiple proprietary interfaces across SDN controllers. If one acquires an SDN application, the ability to plug-n-play different controllers into the environment without big changes to code would become a reality. In addition, investments made by the enterprise or carrier in creating custom SDN applications to differentiate their own end services would be more flexible because of the ability to plug-n-play SDN controllers.

4.4.3 EXISTING SDN CONTROLLER IMPLEMENTATIONS

There are a number of implementations of SDN controllers available on the market today. They include both open source SDN controllers and commercial SDN controllers. Open source SDN controllers come in many forms, from basic C-language controllers, such as NOX [15], to Java-based versions, such as Beacon [16] and Floodlight [10]. There is even a Ruby-based [17] controller called Trema [18]. Interfaces to these controllers may be offered in the language in which the controller is written or other alternatives, such as REST or Python. The open source controller OpenDaylight (ODL) [12] has been built by a consortium of vendors. An important open source controller, *Open Network Operating System* (ONOS) [19] is rapidly gaining traction in the carrier market. Other vendors offer their own commercial version of an SDN controller. Vendors such as NEC, IBM, and HP offer controllers that are primarily OpenFlow implementations. Most other NEMs offer vendor-specific and proprietary SDN controllers that include some level of OpenFlow support, though these controllers vary considerably in terms of which version of the OpenFlow specification they support. In general, these commercial controllers do not support the latest version of the specification, which evolves more quickly than the NEMs' ability and motivation to keep up. As of this writing, ODL and ONOS have assumed dominant roles for commercial applications. For this reason, we provide deeper treatment of these two controllers in Section 7.3.

There are pros and cons to the proprietary alternative controllers. Although proprietary controllers are more closed than the nominally open systems, they do offer some of the automation and programmability advantages of SDN while at the same time providing a *buck stops here* level of support for the network equipment. They permit SDN-like operation of legacy switches, obviating the need to replace older switching equipment in order to begin the migration to SDN. They do constitute closed systems, however, which ostensibly violates one of the original tenets of SDN. They also may do little to offload control functionality from devices, resulting in the continued high cost of network devices. These proprietary alternative controllers are generally a component of the alternative SDN methodologies we introduce in Section 4.6.

4.4.4 POTENTIAL ISSUES WITH THE SDN CONTROLLER

In general the Open SDN controller suffers from the birthing pains common to any new technology. While many important problems are addressed by the concept and architecture of the controller, there have been comparatively few large-scale commercial deployments thus far. As more commercial deployments scale, more real-life experience in large, demanding networks will be needed. In particular, experience with a wider array of applications with a more heterogeneous mix of equipment types is needed before widespread confidence in this architecture is established. Achieving success in these varied deployments will require that a number of potential controller issues be adequately addressed. In some cases, these solutions will come in multiple forms from different vendors. In other cases, a standards body such as the ONF will have to mandate a standard. In Section 3.2.6 we stated that a centralized control architecture is needed to grapple with the issues of latency, scale, high availability, and security. In addition to these more general issues, the centralized SDN controller will need to confront the challenges of *coordination between applications*, *the lack of a standard northbound API*, and *flow prioritization*.

Scale will be a very important aspect of SDN efforts at large carriers like AT&T and Verizon. John Donovan, Chief Infrastructure Executive at AT&T recently said [20]: "*No part of our network will be unaffected. We've catalogued the hundreds of network functions that we manage, and decided which will be relevant in the future and which are becoming obsolete. Of those 200 critical future functions, we plan to move 75% to this new software defined architecture in the next five years. The 5% of network functions that we plan to virtualize in 2015 includes elements of our mobile packet core, the heart of our mobile network.*" AT&T is not alone in its challenge to scale. In November of 2014, Vodafone Australia's CTO indicated that the company's goal is to virtualize most of its core network within five years [21].

DISCUSSION QUESTION:

What might be some solutions for dealing with issues of scale in an Open SDN environment?

There may be more than one SDN application which is running on a single controller. When this is the case, issues related to application prioritization and flow handling become important. Which application should receive an event first? Should the application be required to pass along this event to the next application in line, or can it deem the processing complete, in which case no other applications get a chance to examine and act on the received event?

An emerging standard for the northbound API holds promise for efforts to develop applications that will be reusable across a wide range of controllers. Early standardization efforts for OpenFlow generally assumed such a northbound counterpart would emerge, and much of the efficiency gain assumed to come from a migration to SDN will be lost without it. Late in 2013 the ONF formed a workgroup that focuses on the standardization of the northbound API. Today, in collaboration with OpenSource SDN, that effort has produced an *intents-based* NBI, called *Boulder*. Boulder has created the semantics and information models that support applications in communicating the network changes while abstracting away the underlying implementation [22]. The hope is that significant momentum will be gained in SDN via the adoption of this unifying NBI by multiple SDN controllers.

The reader should note that there is not universal agreement that a single, standard NBI is always needed across all the mainstream controllers. An alternative view reveals an ecosystem growing around each controller. This is reminiscent of how Microsoft Windows and Apple OSX each have their own ecosystems with some applications remaining unique to each platform while other applications are ported across multiple platforms. Moreover, an application written inside a particular controller will likely retain a performance advantage over applications written northbound of that controller.

Flows in an SDN device are processed in priority order. The first flow which matches the incoming packet is acted upon. Within a single SDN application, it is critical for the flows on the SDN device to be prioritized correctly. If they are not, the resulting behavior will be incorrect. For example, the designer of an application will put more specific flows at a higher priority (e.g., match all packets from IP address 10.10.10.2 and TCP port 80), and the most general flows at the lowest priority (e.g., match everything else). This is relatively easy to do for a single application. However, when there are multiple SDN applications, flow entry prioritization becomes more difficult to manage. How does the controller appropriately interleave the flows from all applications? This is a challenge and requires special coordination between the applications.

4.5 SDN APPLICATIONS

SDN applications run above the SDN controller, interfacing to the network via the controller's northbound API. SDN applications are ultimately responsible for managing the flow entries that are programmed on the network devices, using the controller's API to manage flows. Through this API the applications are able to: (1) configure the flows to route packets through the best path between two endpoints; (2) balance traffic loads across multiple paths or destined to a set of endpoints; (3) react to changes in the network topology such as link failures and the addition of new devices and paths, and (4) redirect traffic for purposes of inspection, authentication, segregation, and similar security-related tasks.

Fig. 4.5 includes some standard applications, such as a GUI for managing the controller, a learning switch and a routing application. The reader should note that even the basic functionality of a simple layer two learning switch is not obtained by simply pairing an SDN device with an SDN controller. Additional logic is necessary to react to the newly seen MAC address and update the forwarding tables in the SDN devices being controlled in such a way as to provide connectivity to that new MAC address throughout the network while avoiding switching loops. This additional logic is embodied in the learning switch application in Fig. 4.5. One of the perceived strengths of the SDN architecture is the fact that switching decisions can be controlled by an ever richer family of applications that control the controller. In this way, the power of the SDN architecture is highly expandable. Other applications that are well-suited to this architecture are load balancers and firewalls, among many others.

These examples represent some typical SDN applications which have been developed by researchers and vendors today. Applications such as these demonstrate the promise of SDN—being able to take complex functionality that formerly resided in each individual network device or appliance, and allowing it to operate in an Open SDN environment.

4.5.1 SDN APPLICATION RESPONSIBILITIES

The general responsibility of an SDN application is to perform whatever function for which it was designed, be it load-balancing, firewalling, or some other operation. Once the controller has finished initializing devices and has reported the network topology to the application, the application spends most of its processing time responding to events. While the core functionality of the application will vary from one application to another, application behavior is driven by events coming from the controller as well as external inputs. External inputs could include network monitoring systems, such as Netflow, IDS, or BGP peers. The application affects the network by responding to the events as modeled in Fig. 4.6. The SDN application registers as a *listener* for certain events, and the controller will invoke the application's callback *method* whenever such an event occurs. This invocation will be accompanied by the appropriate details related to the event. Some examples of events handled by an SDN application are: *End-user Device Discovery*, *Network Device Discovery*, and *Incoming Packet*. In the first two cases, events are sent to the SDN application upon the discovery of a new end-user device (i.e., MAC address) or a new network device (e.g., switch, router, or wireless access point), respectively. Incoming Packet events are sent to the SDN application when a packet is received from an SDN device, due to either a flow entry instructing the SDN device to forward the packet to the controller, or because there is no matching flow entry at the SDN device. When there is no matching flow entry the default action is usually to forward the packet to the controller though it could be to drop the packet, depending on the nature of the applications.

There are many ways in which an SDN application can respond to events that have been received from the SDN controller. There are simple responses, such as downloading a set of default flow entries to a newly discovered device. These default or *static* flows typically are the same for every class of discovered network device, and, hence, little processing is required by the application. There are also more complex responses which may require state information gathered from some other source apart from the controller.

This can result in variable responses depending on that state information. For example, based on a user's state, an SDN application may decide to process the current packet in a certain manner, or else it may take some other action, such as downloading a set of user-specific flows.

4.6 ALTERNATE SDN METHODS

Thus far in this chapter we have been examining what we consider the original definition of SDN, which we distinguish from other alternatives by the term *Open* SDN. Open SDN most certainly has the broadest support in the research community and amongst the operators of large data centers such as Google, Yahoo! and their ilk. Nevertheless, there are other proposed methods of accomplishing at least some of the goals of SDN. These methods are generally associated with a single networking vendor or a consortium of vendors. We define here two alternate categories of SDN implementations, *SDN via APIs* and *SDN via Hypervisor-Based Overlay Networks*. The first of these consists of employing functions that exist on networking devices that can be invoked remotely, typically via traditional methods, such as SNMP, CLI, or NETCONF. Alternatively, newer mechanisms such as RESTful APIs may be used. In SDN via Hypervisor-Based Overlay Networks the details of the underlying network infrastructure are not relevant. Virtualized *overlay* networks are instantiated across the top of the physical network. We make the distinction that the overlays be hypervisor-based since network overlays exist in other, nonhypervisor-based forms as well. One early example of this sort of network virtualization is VLAN technology. Another type of overlay network that is not related to our use of the term are *P2P/Overlay* networks such as Napster and BitTorrent. We will further clarify the distinction between SDN and P2P/Overlay networks later in Section 9.7.

In the following sections we provide an introduction of how these two alternate SDN methods work, and defer a more detailed treatment of these as well as the introduction of some others to Chapter 6.

4.6.1 SDN VIA APIs

If a basic concept of SDN is to move control functionality from devices to a centralized controller, then this can be accomplished in other ways than the OpenFlow-centric approach coupled to Open SDN. Looking at the value provided by SDN as described in earlier chapters, a couple of key functional improvements are *programmability* and *centralized control*. Programmability provides the opportunity for applications to be written in order to automate processes and to provide a level of dynamic control which goes beyond what is available today with static, CLI-based networking updates. Centralized control provides the ability for applications running on a controller or independently to have views of the entire network, and hence increases the ability to make optimal policy-based forwarding and routing decisions.

Fig. 4.7 depicts a number of levels at which APIs can be provided for SDN applications. These levels are *Device*, *Controller*, and *Policy*, for which we provide detail in the following sections.

SDN via Device APIs

In *SDN via Device APIs*, the strategy is to provide a richer set of control points on the devices, so that centrally located software can manipulate those devices and provide the intelligent and predictable behavior that is expected in an SDN-controlled network. Consequently, many vendors offer SDN solutions by improving the means of affecting configuration changes on their network devices.

We depict *SDN via Device APIs* graphically in Fig. 4.8. The diagram shows a set of centralized applications communicating with devices via device-level APIs. Often with SDN via Device APIs solutions, vendors provide an enhanced level of APIs on their devices rather than just the traditional CLI and SNMP. The architecture shown in the figure is reminiscent of the earlier diagrams in this chapter. The most obvious difference with the earlier diagrams is that we here depict the controller as optional. This is because even when the controller is present in this paradigm, it adds little or no value to the communication between the application and device. That is, it serves as a pass-through

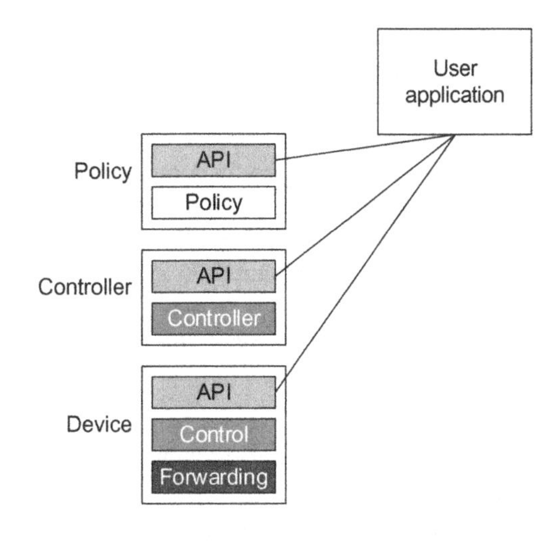

FIG. 4.7

SDN via different levels of APIs.

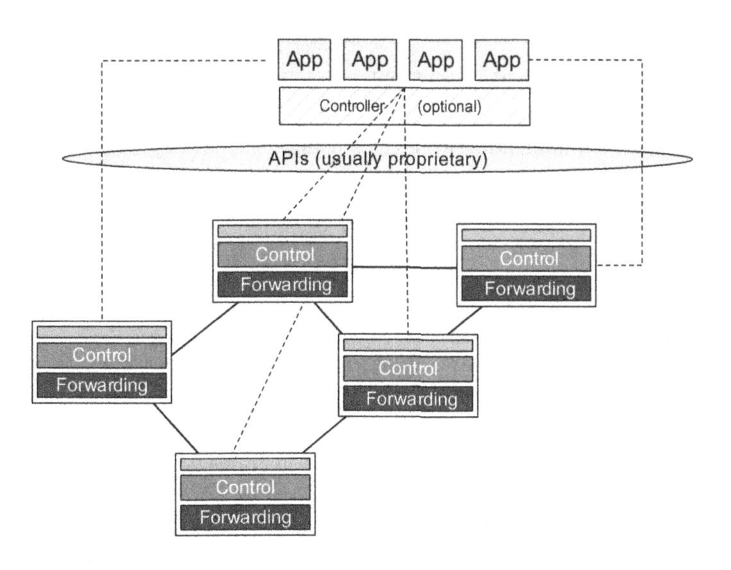

FIG. 4.8

SDN via Device APIs.

only. Indeed, the applications may sometimes communicate with the devices directly without using a controller at all.

Since the early days of the commercial Internet, it has been possible to set configuration parameters on devices using methods such as CLI and SNMP. While these mechanisms have long been available, they can be cumbersome and difficult to maintain. Furthermore, they are geared toward relatively rare static management tasks, not the dynamic, frequent, and automated tasks required in environments such as today's data centers. Newer methods of providing the means to make remote configuration changes have been developed in the last few years.

One such approach used by some vendors is the use of RESTful APIs. REST has become the dominant method of making API calls across networks for computational tasks. REST uses *HyperText Transfer Protocol* (HTTP), the protocol commonly used to pass web traffic. RESTful APIs are simple and extensible and have the advantage of using a standard TCP port and, thus, require no special firewall configuration to permit the API calls to pass through firewalls. We provide a more detailed discussion of RESTful APIs in Section 6.2.4.

Another related API is the IETF's *Interface to the Routing System* (I2RS) [23] I2RS provides an interface between routing protocols and the *Routing Information Base* (RIB). The four key drivers inspiring this new protocol are:

- The need for an interface that is programmatic, asynchronous, and offers fast, interactive access for atomic operations.
- To provide access to structured routing information and state that is frequently not directly configurable or modeled in existing implementations or configuration protocols.
- To provide the ability for network management and other applications to subscribe to structured, filterable event notifications from the routing system.
- To facilitate extensibility and provide standard data-models to be used by network applications.

While I2RS seems to be very synergistic with SDN, it remains to be seen whether this new architecture will gain traction in commercial use.

Note in the figure that communication to devices using these APIs (rather than OpenFlow) can be done either directly, or through a controller. In either case, the solution is using improved APIs which provide the ability to control the forwarding plane through mechanisms made available on the device by the manufacturer.

SDN via Controller APIs

In order to promote programmability from a centralized location, SDN via Controller-level APIs can provide a platform from which to build SDN applications. Solutions using this type of SDN utilize the set of APIs provided by the controller. These APIs are open and available for application developers. Since there is no standard NBI, applications written for one controller's set of APIs may not run on a different controller.

Fig. 4.9 depicts the SDN via Controller APIs model. The reader should note that this model does not appear to contrast fundamentally from those we presented earlier in Section 4.4 in our discussion about Open SDN. The fundamental distinction comes in the nature of the southbound APIs the controller uses. In the case of Open SDN, the southbound protocol is OpenFlow. In the case of SDN via Controller APIs, the southbound protocol consists of one or more legacy protocols that are being re-purposed to provide SDN-like functionality via the centralized controller.

One goal of implementing APIs at the controller level is to provide a level of abstraction between the devices and the application, abstracting details of each specific device and each protocol, and making it possible for the application to interact with the controller at a higher level. One example of this is the case of gathering topology information about devices in the network and presenting

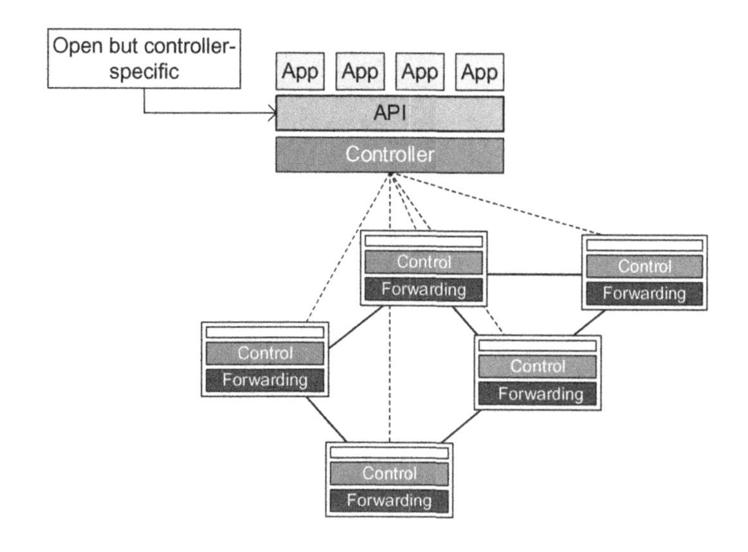

FIG. 4.9

SDN via Controller APIs.

it via the controller API. In this case, the application benefits from the work done by the controller and is able to read information about nodes and links from the controller without having to perform the actual device and host discovery and topology. This is precisely how existing routing protocols such as *Border Gateway Protocol* (BGP), *Border Gateway Protocol Link State* (BGP-LS), and *Path Computation Element Protocol* (PCE-P) are being used today. Implementations of SDN using these device protocols make use of technology existing in devices in order to provide the ability to create SDN applications which control routes and paths throughout the network. The use of these protocols with SDN will be described in detail in Section 6.2.3.

In other cases, the controller performs only light processing of the messages between the application and device. An example of this is found in one of the typical uses of NETCONF as an SDN API. NETCONF is one of the device management and configuration protocols most favored by vendors and customers today. It is used especially in higher-level devices such as core routers. These devices expose their configurable information via data models defined using the YANG data definition language. (NETCONF and YANG will be discussed in greater detail in Section 6.2.2.) In our context of SDN via Controller APIs, the application typically makes RESTCONF requests to the controller. RESTCONF is a protocol wherein NETCONF requests are bundled in easier-to-use REST messages. The controller in turn translates the RESTCONF request into the NETCONF protocol, with the contents of the request remaining fundamentally the same.

One SDN controller that fits squarely in the *SDN via Controller APIs* category is ODL. A prime feature of ODL is its ability to support multiple southbound protocols. Included in the long list of supported protocols are OpenFlow (giving support for Open SDN), NETCONF (supporting SDN via APIs), and BGP-LS/PCE-P (supporting SDN via APIs). Fig. 4.10 shows ODL support for multiple southbound protocol plugins. ODL evangelists argue that northbound APIs can be developed that would allow applications to manage hybrid sets of devices, some managed on the southbound side

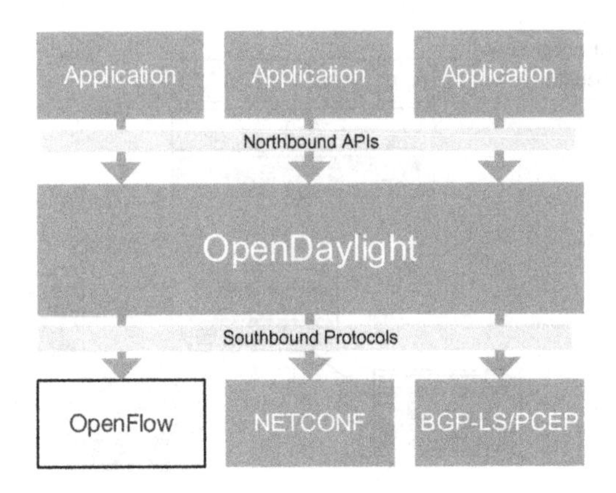

FIG. 4.10

OpenDaylight controller APIs.

via OpenFlow and some managed by legacy APIs. While this would not afford the fine-grained control over all flows afforded by a pure OpenFlow solution, it offers some intriguing possibilities for intelligent control of hybrid networks. Fig. 4.10 shows that the distinction between an Open SDN controller as described earlier in this chapter and SDN via Controller APIs can become blurred.

SDN via Controller level APIs remains a popular choice for SDN application developers, since it provides SDN functionality in the form of programmability and centralized control, as well as supporting legacy devices.

SDN via Policy-level APIs

Another approach is to provide a level of APIs which reside at a layer above the controller level. These APIs are created at a level of abstraction such that they address *policies*, rather than merely individual device or network capabilities. These APIs come in different flavors and may be directed at different target domains, but they all attempt to address network configuration from a *declarative* perspective, rather than an *imperative* one.

These concepts are defined as follows:

- *Imperative:* Imperative systems and APIs require the user to input exactly *how* to do a particular task.
- *Declarative:* Declarative systems and APIs request the user to input exactly *what* is to be accomplished. It is the responsibility of the system to determine *how* to do it.

Fig. 4.11 shows SDN via Policy APIs; notice the policy layer running on top of the controller, providing functionality and a level of abstraction which shields the SDN application from the nuances and details of interacting with a specific device.

These policy-based API solutions are gaining traction not only in the SDN via APIs category, but also in the Open SDN category as well. An example of this type of API would be the NBI concept of *intents*, which raises the API level of abstraction such that requests are truly declarative. The notion of

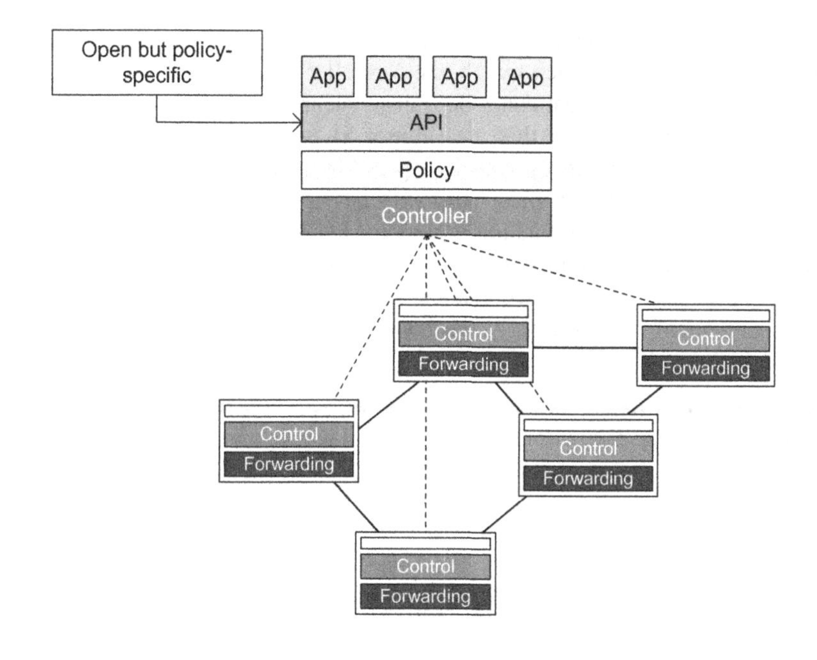

FIG. 4.11

SDN via Policy APIs.

controlling network behavior via policy is an active area of research in SDN. We cover relevant policy languages in Section 7.6.

4.6.2 BENEFITS AND LIMITATIONS OF SDN VIA APIS

There are a number of benefits of SDN via APIs. One distinct advantage of this approach is that, because it uses legacy management interfaces, it therefore works with legacy switches. Thus, this solution does not require upgrading to OpenFlow-enabled switches. Another benefit of this approach is that it allows for some improvement in agility and automation. These APIs also make it easier to write software such as orchestration tools which can respond quickly and automatically to changes in the network (e.g., the movement of a virtual machine in a data center). A third advantage is that it allows for some amount of centralized control of the devices in the network. Therefore, it is possible to build an SDN solution using the provided APIs on the distributed network devices. Finally, there is potential for increased openness in the SDN via APIs approach. Although the individual interfaces may be proprietary to individual vendors, when they are exposed to the applications, they are made open for exploitation by applications. The degree of openness will vary from one NEM to another.

Of course, the API-based SDN methods have their limitations. First, in some cases *there is no controller at all*. The network programmer needs to interact directly with each switch. Second, even when there is a controller, it may not provide an abstract, *network-wide* view to the programmer. This is certainly true when dealing with NETCONF-based APIs on the controller. In these situations, the programmer needs to think in terms of individual switches. (Note that it is possible to develop a layer above NETCONF that abstracts away device details, so it would not be fair to conclude that the use of NETCONF or other similar APIs precludes an abstraction layer above.) Third, since there is still a control plane operating on each switch, the controller and, more importantly, the programmer developing applications on top of that controller must synchronize with what the distributed control plane is doing. Another drawback is that the solution is often proprietary. When these APIs are nonstandard (as opposed to a protocol such as OpenFlow), SDN-like software applications using this type of API-based approach will only work with devices from that specific vendor, or a small group of compatible vendors. This limitation is sometimes circumvented by extending this approach to provide support for multiple vendors' APIs. This masks the differences between the device APIs to the application developer, who will see a single northbound API despite the incompatible device interfaces on the southbound side. Obviously, this homogeneity on the northbound interface is achieved by increased complexity within the controller.

DISCUSSION QUESTION:

What do you see as the major advantages of the SDN via APIs solution? Do you see these as advantages going forward into the future of networking, or are they short-term benefits?

4.6.3 SDN VIA HYPERVISOR-BASED OVERLAY NETWORKS

Another more innovative alternate SDN method is what we refer to as *Hypervisor-Based Overlay* networks. Under this concept the current physical network is left as it is, with networking devices and their configurations remaining unchanged. Above that network, however, *hypervisor-based virtualized* networks are erected. The systems at the edges of the network interact with these virtual networks, which obscure the details of the physical network from the devices that connect to the overlays.

We depict such an arrangement in Fig. 4.12, where we see the virtualized networks *overlaying* the physical network infrastructure. The SDN applications making use of these overlay network resources

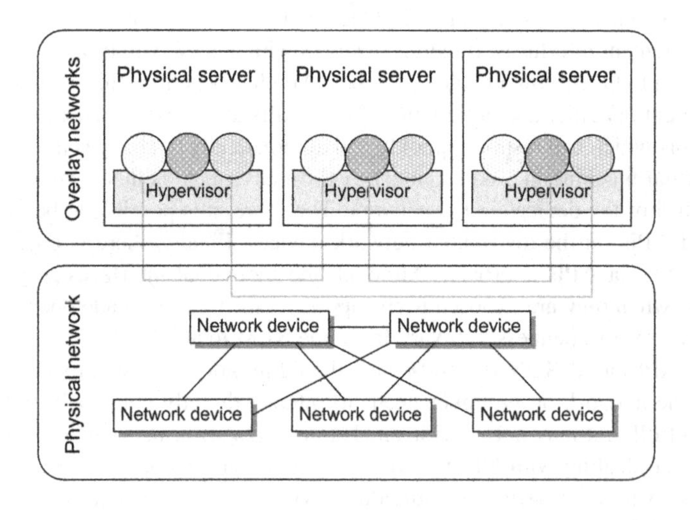

FIG. 4.12

Virtualized networks.

are given access to virtualized networks and ports, which are abstract in nature and do not necessarily relate directly to their physical counterparts below.

As shown in Fig. 4.12, conceptually the virtual network traffic runs *above* the physical network infrastructure. The hypervisors inject traffic into the virtual network and receive traffic from it. The traffic of the virtual networks is passed through those physical devices, but the endpoints are unaware of the details of the physical topology, how routing occurs or other basic network functions. Since these virtual networks exist above the physical infrastructure, they can be controlled entirely by the devices at the very edge of the network. In data centers, these would typically be the hypervisors of the VMs that are running on each server.

The mechanism that makes this possible is tunneling, which uses encapsulation. When a packet enters the edge of the virtual network at the source, the networking device (usually the hypervisor) will take the packet in its entirety and encapsulate it within another frame. This is shown in Fig. 4.13. Note that the edge of the virtual network is called a *Tunnel Endpoint* or *Virtual Tunnel Endpoint* (VTEP).

The hypervisor then takes this encapsulated packet and, based on information programmed by the controller, sends it to the destination's VTEP. This VTEP decapsulates the packet and forwards it to the destination host. As the encapsulated packet is sent across the physical infrastructure, it is being sent from the source's VTEP to the destination's VTEP. Consequently, the IP addresses are those of the source and destination VTEP. Normally, in network virtualization, the VTEPs are associated with hypervisors.

This tunneling mechanism is referred to as *MAC-in-IP* tunneling, because the entire frame, from MAC address inwards, is encapsulated within this unicast IP frame, as shown in Fig. 4.13. Different vendors have established their own methods for MAC-in-IP tunneling. Specifically, Cisco offers VXLAN [24], Microsoft uses NVGRE [25], and Nicira's is called STT [26].

This approach mandates that a centralized controller be in charge of making sure there is always a mapping from the actual destination host to the destination hypervisor which serves that host.

Fig. 4.14 shows the roles of these VTEPs as they serve the source and destination host devices. The virtual network capability is typically added to a hypervisor by extending it with a virtual switch. We introduced the notion of virtual switch in Section 4.3.2, and it is well suited to the overlay network concept. The virtual network has a virtual topology, consisting of the virtual switches interconnected by virtual point-to-point links. The virtual switches are depicted as the VTEPs in Fig. 4.14 and the

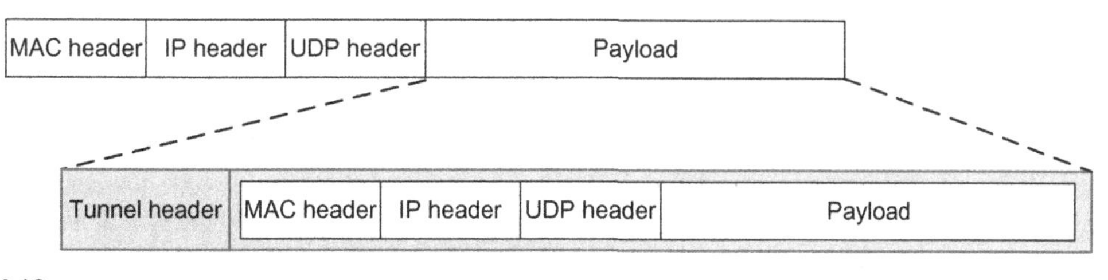

FIG. 4.13

Encapsulated frames.

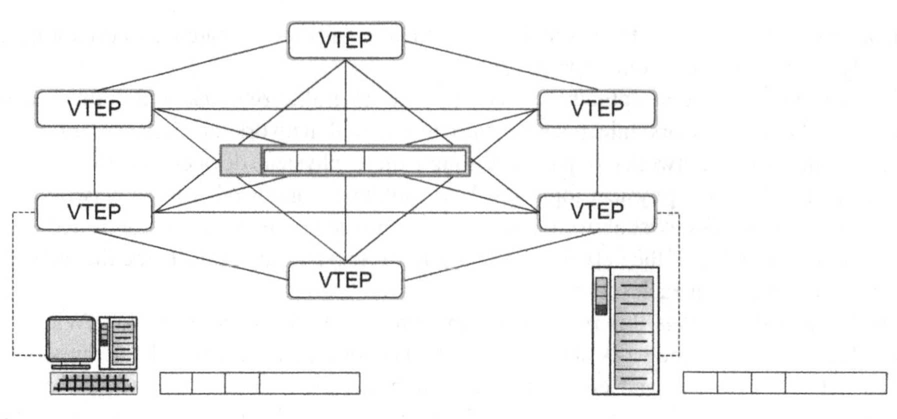

FIG. 4.14

Virtual tunnel endpoints.

virtual links are the tunnels interconnecting them. All the traffic on each virtual network is encapsulated as shown in Fig. 4.13 and sent VTEP-to-VTEP. The reader should note that the tunnels depicted in Fig. 4.14 are the same as the links interconnecting hypervisors in Fig. 4.12. As Fig. 4.12 indicates, there can be multiple overlay networks existing independently and simultaneously over the same physical network.

In summary, SDN via Hypervisor-Based Overlay Networks is well-suited to environments such as data centers already running compute and storage virtualization software for their servers. It does address a number of the needs of an SDN solution. First, it addresses MAC address explosion in data centers and cloud environments, because all those host MAC addresses are hidden within the encapsulated frame. Second, it addresses VLAN limitations, because all traffic is tunneled and VLANs are not required for supporting the isolation of multiple tenants. Third, it addresses agility and automation needs, because it is implemented in software and these virtual networks can be constructed and taken down in a fraction of the time that would be required to change the physical network infrastructure.

Nevertheless, these overlay networks do not solve all the problems that can be addressed by an Open SDN solution. In particular it does not address existing issues within the physical infrastructure, which still requires manual configuration and maintenance. Moreover, it fails to address traffic prioritization and efficiency in the physical infrastructure, so confronting STP's blocked links and QoS settings continue to challenge the network engineer. In the introduction to this section we said that with overlays the current physical network is left as it is, with networking devices and their configurations remaining unchanged. Under this assumption, hypervisor-based overlays do not directly address the desire to open up network devices for innovation and simplification, since those physical network devices have not changed at all. Indeed, while one of the benefits of overlays is the ability to build the virtual network on top on an existing physical one, the overlay strategy can also be used over a *new* physical network that incorporates, for instance, more simple switches. Hence, the overlay strategy does not itself preclude device simplification.

DISCUSSION QUESTION:

What do you find most compelling about the SDN via Overlays solution? Why do you think this solution is popular in certain domains?

4.7 CONCLUSION

This chapter has described the basic functionality related to the manner in which an SDN solution actually works. It is important to realize that there is no fundamental incompatibility between the hypervisor-based overlay network approach to SDN and Open SDN. In fact, some implementations use OpenFlow to create and utilize the tunnels required in this kind of network virtualization. It is not unreasonable to think of these overlay networks as stepping stones toward a more complete SDN solution which includes SDN and OpenFlow for addressing both the virtual as well as the physical needs of the network. In Chapter 6 we will delve more deeply into the SDN alternatives introduced in this chapter as well as other alternatives not yet discussed. First, though, in the next chapter we provide a detailed overview of the OpenFlow specification.

DISCUSSION QUESTION:

Of the three different types of SDN we've introduced—Open SDN, SDN via APIs, and SDN via Overlays—which do you think will be most popular in the near future, and why? Or longer term?

REFERENCES

[1] Shenker S. The future of networking, and the past of protocols. Open Networking Summit, October 2011, Stanford University.

[2] Industry Leaders Collaborate on OpenDaylight Project, Donate Key Technologies to Accelerate Software-Defined Networking. OpenDaylight, April 2013. Retrieved from: http://www.opendaylight.org/announcements/2013/04/industry-leaders-collaborate-opendaylight-project-donate-key-technologies.

[3] Kerner S. Brocade Doubles Down on Open SDN. Enterprise Networking Planet, May 23, 2014. Retrieved from: http://www.enterprisenetworkingplanet.com/netsysm/brocade-doubles-down-on-open-sdn.html.

[4] NetFlow traffic analyzer. Solarwinds. Retrieved from: http://www.solarwinds.com/netflow-traffic-analyzer.aspx.

[5] Production quality, multilayer open virtual switch. Open vSwitch. Retrieved from: http://openvswitch.org.

[6] Open thin switching, open for business. Big Switch Networks, June 27, 2013. Retrieved from: http://www.bigswitch.com/topics/introduction-of-indigo-virtual-switch-and-switch-light-beta.

[7] Kerner S. New Cisco CEO's Message? Faster, Faster, Faster. Enterprise Networking Planet, October 7, 2015. Retrieved from: http://www.enterprisenetworkingplanet.com/netsysm/new-cisco-ceos-message-faster-faster-faster.html.

[8] OpenFlow Management and Configuration Protocol (OF-Config 1.1.1). Open Networking Foundation, March 23, 2013. Retrieved from: https://www.opennetworking.org/sdn-resources/onf-specifications.

[9] Pfaff B, Davie B. The Open vSwitch Database Management Protocol. Internet Draft, Internet Engineering Task Force, October 2013.

[10] Wang K. Floodlight documentation. Project Floodlight, December 2013. Retrieved from: http://docs.projectfloodlight.org/display/floodlightcontroller.
[11] Learn REST: A RESTful Tutorial. Retrieved from: http://www.restapitutorial.com.
[12] Lawson S. Network heavy hitters to pool SDN efforts in OpenDaylight project. Network World, April 8, 2013. Retrieved from: http://www.networkworld.com/news/2013/040813-network-heavy-hitters-to-pool-268479.html.
[13] Open daylight technical overview. Retrieved from: http://www.opendaylight.org/project/technical-overview.
[14] Level 3's Waters on integrating tw telecom, SDN and deepening the metro network footprint. FierceTelecom, November 12, 2014. Retrieved from: http://www.fiercetelecom.com/special-reports/level-3s-waters-integrating-tw-telecom-sdn-and-deepening-metro-network-foot.
[15] Nox. Retrieved from: http://www.noxrepo.org.
[16] Erikson D. OpenFlow @ Stanford, February 2013. Beacon. Retrieved from: https://openflow.stanford.edu/display/Beacon/Home.
[17] Ruby: a programmer's best friend. Retrieved from: http://www.ruby-lang.org.
[18] Trema: full-stack OpenFlow framework in Ruby and C. Retrieved from: http://trema.github.io/trema.
[19] ONOS project partners with Linux Foundation. Retrieved from: http://onosproject.org/.
[20] Goldstein P. AT&T's Donovan: Traffic from 'millions' of wireless customers now running on SDN architecture. FierceWireless, October 8, 2015. Retrieved from: http://www.fiercewireless.com/story/atts-donovan-traffic-millions-wireless-customers-now-running-sdn-architectu/2015-10-08.
[21] Baldwin H. Are carriers really leading the charge to SDN? Infoworld, November 19, 2014. Retrieved from: http://www.infoworld.com/article/2850034/are-carriers-really-leading-the-charge-to-sdn.html.
[22] Project BOULDER: Intent Northbound Interface (NBI). Retrieved from: http://opensourcesdn.org/ossdn-projects/.
[23] Atlas A, Halpern J, et al. An architecture for the interface to the routing system. Internet Draft, Internet Engineering Task Force, February 20, 2016.
[24] Mahalingam M, Dutt D, Duda K, Agarwal P, Kreeger L, Sridhar T, et al. VXLAN: a framework for overlaying virtualized layer 2 networks over layer 3 networks. Internet Draft, Internet Engineering Task Force, August 26, 2011.
[25] Sridharan M, et al. NVGRE: network virtualization using generic routing encapsulation. Internet Draft, Internet Engineering Task Force, September 2011.
[26] Davie B, Gross J. STT: a stateless transport tunneling protocol for network virtualization (STT). Internet Draft, Internet Engineering Task Force, March 2012.

CHAPTER

THE OpenFlow SPECIFICATION

5

Casual followers of networking news can experience confusion as to whether SDN and OpenFlow are one and the same thing. Indeed, they are not. OpenFlow is definitely a distinct subset of the technologies included under the big tent of SDN. An important tool and catalyst for innovation, OpenFlow defines both the communications protocol between the SDN data plane and the SDN control plane, as well as part of the behavior of the data plane. It does not describe the behavior of the controller itself. While there are other approaches to SDN, OpenFlow is today the only nonproprietary, general-purpose protocol for programming the forwarding plane of SDN switches. As such, this chapter will focus only on the protocol and behaviors dictated by OpenFlow, as there are no directly competing alternatives.

An OpenFlow system consists of an *OpenFlow controller* that communicates to one or more *OpenFlow Switches*. The OpenFlow protocol defines the specific messages and message formats exchanged between controller (control plane) and device (data plane). The OpenFlow behavior specifies how the device should react in various situations, and how it should respond to commands from the controller. There have been various versions of OpenFlow, which we examine in detail in this chapter. We will also consider some of the potential drawbacks and limitations of the OpenFlow specification.

Note that our goal in this chapter is not to provide an alternative to a detailed reading of the specification itself. We hope to provide the reader with a rudimentary understanding of the elements of OpenFlow and how they interoperate to provide basic switch and routing functions, as well as an experimental platform that allows each version to serve as a springboard for innovations that may appear in future versions of OpenFlow. In particular, while we provide tables that list the different ports, messages, instructions and actions that make up OpenFlow, we do not attempt to explain each in detail, as this would essentially require that we replicate the specification here. We will use the most important of these in clear examples so that the reader is left with a basic understanding of OpenFlow operation. We provide a reference to each of the four versions of OpenFlow in the section where it is introduced.

5.1 CHAPTER-SPECIFIC TERMINOLOGY

The following new operations are introduced for this chapter.

To *pop* an item is to remove it from a *Last-In-First-Out* (LIFO) ordered list of like items.

To *push* an item is to add it to a LIFO ordered list of like items.

These terms are frequently used in conjunction with the term *stack* which is a computer science term for a LIFO ordered list. If one pictures a stack of items such as books, and three books are added to that stack, the order in which those books are normally retrieved from the pile is a LIFO order. That is, the first book pulled off the top of the pile is the last one that was added to it. If we alter slightly the perception of the stack of books to imagine that the stack exists in a spring-loaded container such

Software Defined Networks. http://dx.doi.org/10.1016/B978-0-12-804555-8.00005-3

that the top-most book is level with the top of the container, the use of the terms pop and push become apparent. When we add a book to such a stack we *push* it onto the top and later *pop* it off to retrieve it. The OpenFlow specification includes several definitions that use these concepts. In this context, the push involves adding a new header element to the existing packet header, such as an MPLS label. Since multiple such labels may be pushed on before any are popped off, and they are popped off in LIFO order, the stack analogy is apt.

5.2 OpenFlow OVERVIEW

The OpenFlow specification has been evolving for a number of years. The nonprofit Internet organization, *openflow.org* was created in 2008 as a mooring to promote and support OpenFlow. While openflow.org existed formally on the Internet, in the early years the physical organization was really just a group of people that met informally at Stanford University. From its inception OpenFlow was intended to belong to the research community to serve as a platform for open network switching experimentation, with an eye on commercial use through commercial implementations of this public specification. The first release Version 1.0.0 appeared in December 31, 2009, though numerous point prereleases existed before then that were made available for experimental purposes as the specification evolved. At this point and continuing up through release 1.1.0, development and management of the specification was performed under the auspices of openflow.org. On March 21, 2011 the *Open Network Foundation* (ONF) was created for the express purpose of accelerating the delivery and commercialization of SDN. As we will explain in Chapter 6, there are a number of proponents of SDN that offer SDN solutions that are not based on OpenFlow. For the ONF, however, OpenFlow remains at the core of their SDN vision for the future. For this reason the ONF has become the responsible entity for the evolving OpenFlow specification. Starting after the release of V.1.1, revisions to the OpenFlow specification are released and managed by the ONF.

One may get the impression from the fanfare surrounding OpenFlow that the advent of this technology has been accompanied by concomitant innovation in switching hardware. The reality is a bit more complicated. The OpenFlow designers realized a number of years ago that many switches were really built around ASICs controlled by rules encoded in tables that could be programmed. Over time, fewer home-grown versions of these switching chips were being developed, and there was greater consolidation in the semiconductor industry. More manufacturers' switches were based on ever-consolidating switching architecture and programmability, with ever-increasing use of programmable switching chips from a relatively small number of merchant silicon vendors. OpenFlow is an attempt to allow the programming, in a generic way, of the different implementations of switches that conform to this new paradigm. OpenFlow attempts to exploit the table-driven design extant in many of the current silicon solutions. As the number of silicon vendors consolidates, there should be greater possibility for alignment with future OpenFlow versions.

It is worth pausing here to remark on the fact that we are talking a lot about ASICs for a technology called *Software* Defined Networking. Hardware must be part of the discussion since it is necessary to use this specialized silicon in order to switch packets at high line rates. We explained in Chapter 4 that while pure software SDN implementations exist, they cannot switch packets at sufficiently high rates to keep up with high-speed interfaces. What is really meant by the word *software* in the name SDN, then, is that the SDN devices are fully programmable, not that everything is done using software running on a traditional CPU.

In the sections that follow we will introduce the formal terminology used by OpenFlow and provide basic background that will allow us to explore the details of the different versions of the OpenFlow specification that have been released up to the time of the writing of this book.

5.2.1 THE OpenFlow SWITCH

Fig. 5.1 depicts the basic functions of an OpenFlow V.1.0 switch and its relationship to a controller. As would be expected in a packet switch, we see that the core function is to take packets that arrive on one port (path *X* on port 2 in the figure) and forward it through another port (port *N* in the figure), making any necessary packet modifications along the way. A unique aspect of the OpenFlow switch is embodied in the *packet matching function* shown in Fig. 5.1. The adjacent table is a *Flow Table* and we will give separate treatment to this below in Section 5.3.2. The wide, grey, double arrow in Fig. 5.1 starts in the decision logic, shows a match with a particular entry in that table, and directs the now-matched packet to an *action box* on the right. This action box has three fundamental options for the disposition of this arriving packet:

- *A*: Forward the packet out a local port, possibly modifying certain header fields first.
- *B*: Drop the packet.
- *C*: Pass the packet to the controller.

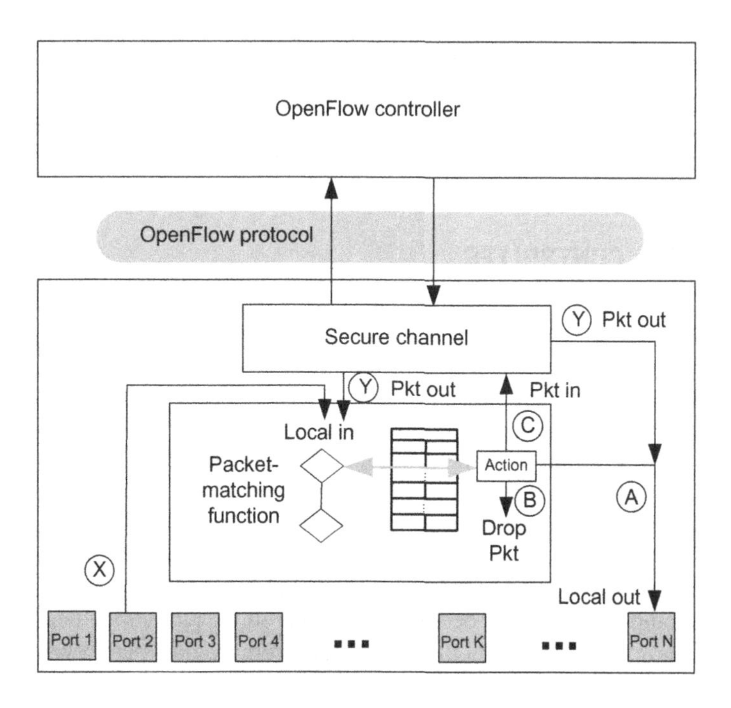

FIG. 5.1

OpenFlow V.1.0 switch.

These three fundamental packet paths are illustrated in Fig. 5.1. In the case of path *C*, the packet is passed to the controller over the *secure channel* shown in the figure. If the controller has either a control message or a data packet to give to the switch, the controller uses this same secure channel in the reverse direction. When the controller has a data packet to forward out through the switch, it uses the OpenFlow PACKET_OUT message. We see in Fig. 5.1 that such a data packet coming from the controller may take two different paths through the OpenFlow logic, both denoted *Y*. In the rightmost case, the controller directly specifies the output port and the packet is passed to that port *N* in the example. In the leftmost path *Y* case, the controller indicates that it wishes to defer the forwarding decision to the packet matching logic. We will see in Section 5.3.4 that the controller dictates this by stipulating the virtual port TABLE as the output port.

A given OpenFlow switch implementation is either *OpenFlow-only* or *OpenFlow-hybrid.* An OpenFlow-only switch is one that forwards packets *only* according to the OpenFlow logic described above. An OpenFlow hybrid is a switch that can also switch packets in its legacy mode as an Ethernet Switch or IP router. One can view the hybrid case as an OpenFlow switch residing next to a completely independent traditional switch. Such a hybrid switch requires a preprocessing classification mechanism that directs packets to either OpenFlow processing or the traditional packet processing. It is probable that hybrid switches will be the norm during the migration to pure OpenFlow implementations.

Note that we use the term OpenFlow *switch* in this chapter instead of the term OpenFlow *device* we customarily use. This is because switch is the term used in the OpenFlow specification. In general, though, we opt to use the term device since there are already nonswitch devices being controlled by OpenFlow controllers, such as wireless access points.

DISCUSSION QUESTION:

Why are there two PKT_OUT options emerging from the Secure Channel box in Fig. 5.1?

5.2.2 THE OpenFlow CONTROLLER

Modern Internet switches make millions of decisions per second about whether or not to forward an incoming packet, to what set of output ports it should be forwarded, and what header fields in the packet may need to be modified, added, or removed. This is a very complex task. The fact that this can be carried out at line rates on multigigabit media is a technological wonder. The switching industry has long understood that not all functions on the switching datapath can be carried out at line rates, so there has long been the notion of splitting the pure data plane from the control plane. The data plane matches headers, modifies packets and forwards them based on a set of forwarding tables and associated logic, and it does this very, very fast. The rate of decisions being made as packets stream into a switch through a 100Gbs interface is astoundingly high. The control plane runs routing and switching protocols and other logic to determine what the forwarding tables and logic in the data plane should be. This process is very complex and cannot be done at line rates as the packets are being processed, and it is for this reason we have seen the control plane separated from the data plane even in legacy network switches.

The OpenFlow control plane differs from the legacy control plane in three key ways. First, it can program different data plane elements with a common, standard language, OpenFlow. Second, it exists on a separate hardware device than the forwarding plane, unlike traditional switches where the control plane and data plane are instantiated in the same physical box. This separation is made possible because the controller can program the data plane elements remotely over the Internet. Third, the controller can program multiple data plane elements from a single control plane instance.

The OpenFlow controller is responsible for programming all of the packet matching and forwarding *rules* in the switch. Whereas a traditional router would run routing algorithms to determine how to program its forwarding table, that function or *an equivalent replacement* to it is now performed by the controller. Any changes that result in recomputing routes will be programmed onto the switch by the controller.

5.2.3 THE OpenFlow PROTOCOL

As shown in Fig. 5.1 the OpenFlow protocol defines the communication between an OpenFlow controller and an OpenFlow switch. This protocol is what most uniquely identifies OpenFlow technology. At its essence, the protocol consists of a set of messages that are sent from the controller to the switch and a corresponding set of messages that are sent in the opposite direction. The messages, collectively, allow the controller to program the switch so as to allow fine-grained control over the switching of user traffic. The most basic programming defines, modifies and deletes flows. Recall that in Chapter 4 we defined a flow as *a set of packets transferred from one network endpoint (or set of endpoints) to another endpoint (or set of endpoints). The endpoints may be defined as IP address-TCP/UDP port pairs, VLAN endpoints, layer three tunnel endpoints, and input port among other things.* One set of rules describes the forwarding actions that the device should take for all packets belonging to that flow. When the controller defines a flow, it is providing the switch with the information it needs to know how to treat incoming packets that match that flow. The possibilities for treatment have grown more complex as the OpenFlow protocol has evolved, but the most basic prescriptions for treatment of an incoming packet are denoted by paths *A*, *B*, and *C* in Fig. 5.1. These three options are to forward the packet out one or more output ports, drop the packet, or pass the packet to the controller for exception handling.

The OpenFlow protocol has evolved significantly with each version of OpenFlow, so we will cover the detailed messages of the protocol in the version-specific sections that follow. The specification has evolved from development point release 0.2.0 on March 28, 2008 through release V.1.5.0, released in 2014. Numerous point releases over the intervening years have addressed problems with earlier releases and added incremental functionality. OpenFlow was viewed primarily as an experimental platform in its early years. As such, there was little concern on the part of the development community advancing this standard to provide for interoperability between releases. As OpenFlow began to see more widespread commercial deployment, backwards compatibility has become an increasingly important issue. There are many features, however, that were introduced in earlier versions of OpenFlow that are no longer present in the current version. Since the goal of this chapter is to provide a roadmap to understanding OpenFlow as it exists today, we will take a hybrid approach of covering the major releases that have occurred since V.1.0. We focus on those key components of each release that became the basis for the advances in subsequent releases and do not focus on functionality in earlier releases that has been subsumed by new features in subsequent releases.

5.2.4 THE CONTROLLER-SWITCH SECURE CHANNEL

The secure channel is the path used for communications between the OpenFlow controller and the OpenFlow device. Generally, this communication is secured by TLS-based asymmetrical encryption, though unencrypted TCP connections are allowed. These connections may be *in-band* or *out-of-band*. Fig. 5.2 depicts these two variants of the secure channel. In the out-of-band example, we see in the figure that the secure channel connection enters the switch via port Z, which is *not* switched by the OpenFlow data plane. Some legacy network stack will deliver the OpenFlow messages via the secure channel to the secure channel process in the switch, where all OpenFlow messages are parsed and handled. Thus, the out-of-band secure channel is only relevant in the case of a OpenFlow-hybrid switch.

In the in-band example, we see the OpenFlow messages from the controller arriving via port *K*, which is part of the OpenFlow data plane. In this case these packets will be handled by the OpenFlow packet matching logic shown in the figure. The flow tables will have been constructed so that this OpenFlow traffic is forwarded to the LOCAL virtual port, which results in the messages being passed to the secure channel process. We discuss the LOCAL virtual port in Section 5.3.4.

Note that when the controller and all the switches it controls are located entirely within a tightly controlled environment such as a data center, it may be wise to consider not using TLS-based encryption to secure the channel. This is because there is a performance overhead incurred by using this type of security and if it is not necessary it is preferable to not pay this performance penalty. Another argument against using TLS-based encryption is that it may be used incorrectly. To put TLS in practice one must

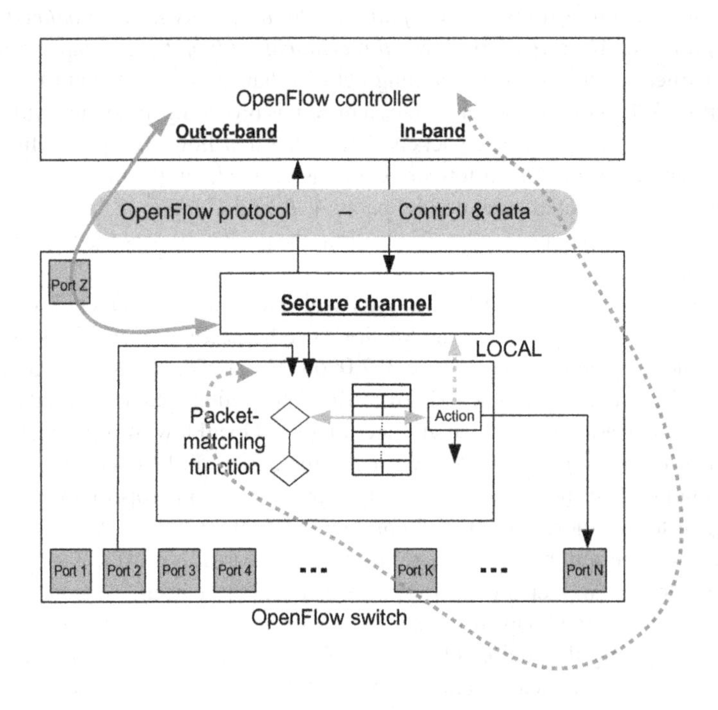

FIG. 5.2

OpenFlow controller-switch secure channel.

obtain and configure security certificates for each device, which can be time-consuming and error-prone for someone not familiar with these methods.

5.3 OpenFlow 1.0 AND OpenFlow BASICS

OpenFlow 1.0 [1] was released on December 31, 2009. For the purposes of this work, we will treat OpenFlow 1.0 as the initial release of OpenFlow. Indeed, years of work and multiple point releases preceded the OpenFlow 1.0 release, but we will subsume all of this incremental progress into the single initial release of 1.0 as if it had occurred atomically. In this section we will describe in detail the basic components of this initial OpenFlow implementation.

5.3.1 PORTS AND PORT QUEUES

The OpenFlow specification defines the concept of an *OpenFlow Port*. An OpenFlow V.1.0 port corresponds to a physical port. This concept is expanded in subsequent releases of OpenFlow. For many years, sophisticated switches have supported multiple *queues* per physical port. These queues are generally served by scheduling algorithms that allow the provisioning of different *Quality of Service* (QoS) levels for different types of packets. OpenFlow embraces this concept and permits a flow to be mapped to an already-defined queue at an output port. Thus, if we look back to Fig. 5.1 the output of a packet on port *N* may include specifying onto which queue on port *N* the packet should be placed. Hence, if we zoom in on option *A* in Fig. 5.1 we reveal what we now see in Fig. 5.3. In our zoomed-in figure we can see that the *actions* box specifically enqueued the packet being processed to queue 1 in port *N*.

Note that the support for QoS is very basic in V.1.0. QoS support in OpenFlow was expanded considerably in later versions (see Section 5.6.3).

5.3.2 FLOW TABLE

The *flow table* lies at the core of the definition of an OpenFlow switch. We depict a generic flow table in Fig. 5.4. A flow table consists of *flow entries*, one of which is shown in Fig. 5.5. A flow entry consists of *header fields*, *counters*, and *actions* associated with that entry. The header fields are used as match criteria to determine if an incoming packet matches this entry. If a match exists, then the packet belongs to this flow. The counters are used to track statistics relative to this flow, such as how many packets have been forwarded or dropped for this flow. The actions fields prescribe what the switch should do with a packet matching this entry. We describe this process of packet matching and actions in Section 5.3.3.

5.3.3 PACKET MATCHING

When a packet arrives at the OpenFlow switch from an input port (or, in some cases, from the controller), it is matched against the flow table to determine if there is a matching flow entry. The following match fields associated with the incoming packet may be used for matching against flow entries:

- switch input port
- VLAN ID

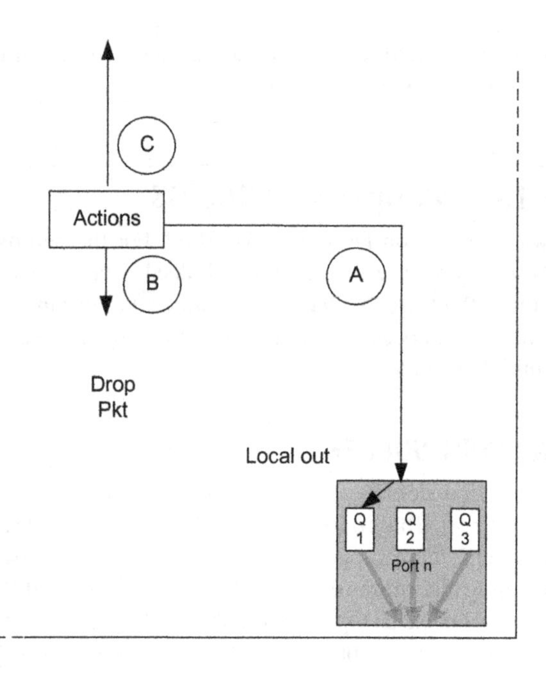

FIG. 5.3

OpenFlow support for multiple queues per port.

Flow entry 0		Flow entry 1			Flow entry F			Flow entry M	
Header fields	Inport 12 192.32.10.1, Port 1012	Header fields	Inport * 209.*.*.*, Port *		Header fields	Inport 2 192.32.20.1, Port 995		Header fields	Inport 2 192.32.30.1, Port 995
Counters	val	Counters	val	...	Counters	val	...	Counters	val
Actions	val	Actions	val		Actions	val		Actions	val

FIG. 5.4

OpenFlow V.1.0 flow table.

- VLAN priority
- Ethernet source address
- Ethernet destination address
- Ethernet frame type
- IP source address
- IP destination address
- IP protocol

Header fields	Field value
Counters	Field value
Actions	Field value

FIG. 5.5

Basic flow entry.

- IP *Type of Service* (ToS) bits
- TCP/UDP source port
- TCP/UDP destination port

These twelve match fields are collectively referred as the basic twelve-tuple of match fields. The flow entry's match fields may be wild-carded using a bit mask, meaning that any value that matches on the unmasked bits in the incoming packet's match fields will be a match. Flow entries are processed in order, and, once a match is found, no further match attempts are made against that flow table. (We will see in subsequent versions of OpenFlow that there may be *additional* flow tables against which packet matching may continue.) Because of this, it is possible for there to be multiple matching flow entries for a packet to be present in a flow table. Only the first flow entry to match is meaningful—the others will not be found as packet matching stops upon the first match.

The V.1.0 specification is silent about which of these twelve match fields are required versus optional. The ONF has clarified this confusion by defining three different types of conformance in their V.1.0 conformance testing program. The three levels are: *full conformance*, meaning all twelve match fields are supported, *layer two conformance*, when only the layer two header field matching is supported, and, finally, *layer three conformance*, when only layer three header field matching is supported.

If the end of the flow table is reached without finding a match, this is called a *table-miss*. In the event of a table-miss in V.1.0, the packet is forwarded to the controller. (Note that this is no longer strictly true in later versions.) If a matching flow entry is found, the actions associated with that flow entry determine how the packet is handled. The most basic action prescribed by an OpenFlow switch entry is how to forward this packet. We discuss this in the following section.

It is important to note that this V.1.0 packet matching function was designed as an abstraction of the way that real-life switching hardware works today. Early versions of OpenFlow were designed to specify the forwarding behavior of existing commercial switches via this abstraction. A good abstraction hides the details of the thing being abstracted while still permitting sufficiently fine-grained control to accomplish the needed tasks. As richer functionality is added in later versions of the OpenFlow protocol, we will see that the specification outpaces the reality of today's hardware. At that point, it is no longer providing a clean abstraction for current implementations but specifying behavior for switching hardware that has yet to be built.

5.3.4 ACTIONS AND PACKET FORWARDING

The required actions that must be supported by a flow entry are to either *output (forward)* or *drop* the matched packet. The most common case is that the output action specifies a physical port on which the packet should be forwarded. There are, however, five special virtual ports defined in V.1.0 that have special significance for the output action. They are *LOCAL*, *ALL*, *CONTROLLER*, *IN_PORT*, and *TABLE*. We depict the packet transfers resulting from these various virtual port designations in Fig. 5.6. In the figure, the notations in brackets next to the wide shaded arrows represent one of the virtual port types discussed in this section.

LOCAL dictates that the packet should be forwarded to the switch's local OpenFlow control software, circumventing further OpenFlow pipeline processing. LOCAL is used when OpenFlow messages from the controller are received on a port that is receiving packets switched by the OpenFlow data plane. LOCAL indicates that the packet needs to be processed by the local OpenFlow control software.

ALL is used to flood a packet out all ports on the switch except the input port. This provides rudimentary broadcast capability to the OpenFlow switch.

CONTROLLER indicates that the switch should forward this packet to the OpenFlow controller.

IN_PORT instructs the switch to forward the packet back out of the port on which it arrived. Effectively, IN_PORT normally creates a loopback situation, which could be useful for certain scenarios. One scenario where this is useful is the case of an 802.11 wireless port. In this case, it is quite normal to receive a packet from that port from one host and to forward it to the receiving host via the same port. This needs to be done very carefully to not create unintended loopback situations. Thus the protocol requires explicit stipulation of this intent via this special virtual port.

Another instance when IN_PORT is required is *Edge Virtual Bridging* (EVB) [2]. Edge Virtual Bridging defines a reflective relay service between a physical switch in a data center and a lightweight virtual switch within the server known as a *Virtual Edge Port Aggregator* (VEPA). The standard IEEE 802.1Q bridge at the edge of the network will reflect packets back out the port on which they arrive to

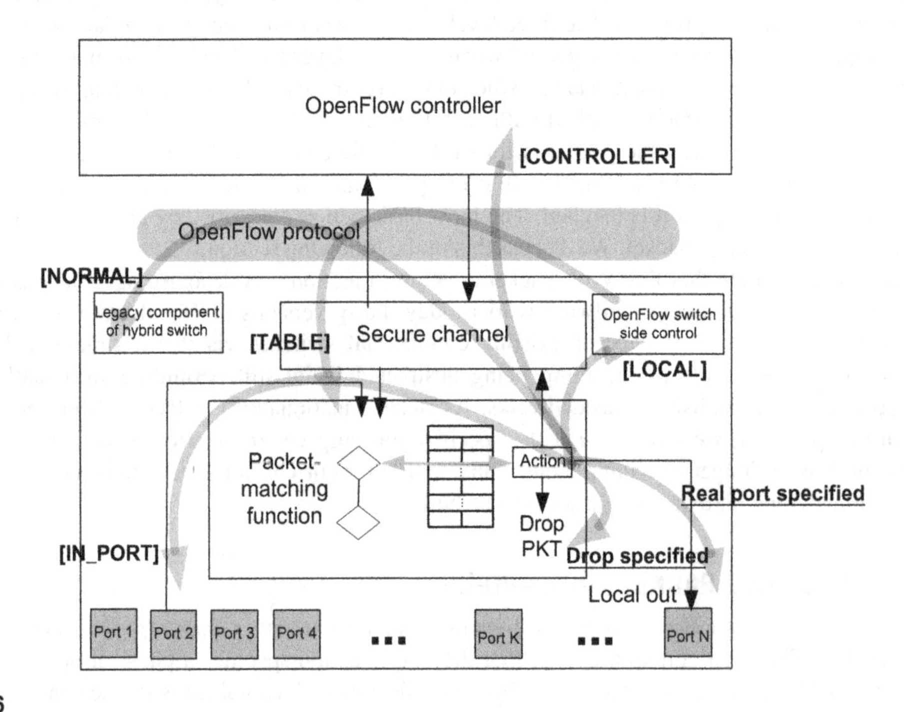

FIG. 5.6

Packet paths corresponding to virtual ports.

allow two virtual machines on the same server to talk to one another. This reflective relay service can be supported by the IN_PORT destination in OpenFlow rules.

Finally, there is the TABLE virtual port, which only applies to packets that the controller sends to the switch. Such packets arrive as part of the PACKET_OUT message from the controller, which includes an *action-list*. This action-list will generally contain an output action, which will specify a port number. The controller may wish to directly specify the output port for this data packet, or, if it wishes the output port to be determined by the normal OpenFlow packet processing pipeline, it may do so by stipulating TABLE as the output port. These two options are depicted in the two *Y* paths shown in Fig. 5.1.

There are two additional virtual ports, but support for these is optional in V.1.0. The first is the NORMAL virtual port. When the output action forwards a packet to the NORMAL virtual port, it sends the packet to the legacy forwarding logic of the switch. We contrast this with the LOCAL virtual port, which designates that the packet be passed to the local OpenFlow control processing. Conversely, packets whose matching rule indicates NORMAL as the output port will remain in the ASIC to be looked up in other forwarding tables which are populated by the local (nonOpenFlow) control plane. Use of NORMAL only makes sense in the case of a hybrid switch.

The remaining virtual port is FLOOD. In this case, the switch sends a copy of the packet out all ports except the ingress port.

Note that we have excluded the ALL and FLOOD representations from Fig. 5.6 as the number of arrows would have overly congested the portrayal. The reader should understand that ALL and FLOOD would show arrows to all ports on the OpenFlow switch except the ingress port. The reader should also note that we depict in the figure the more normal *nonvirtual* cases of *real port specified* and *drop specified* underlined without brackets. When the V.1.0 switch passes a packet to the controller as a result of finding no table matches, the path taken in Fig. 5.6 is the same as that denoted for the virtual port CONTROLLER.

There are two optional actions in V.1.0, *enqueue* and *modify-field*. The enqueue action selects a specific queue belonging to a particular port. This would be used in conjunction with the output action and is used to achieve desired QoS levels via the use of multiple priority queues on a port. Lastly, the modify-field action informs the switch how to modify certain header fields. The specification contains a lengthy list of fields that may be modified via this action. In particular, VLAN headers, Ethernet source and destination address, IPv4 source and destination address and TTL field may be modified. Modify-field is essential for the most basic routing functions. In order to route layer three packets, a router must decrement the TTL field before forwarding the packet out the designated output port.

When there are multiple actions associated with a flow entry, they appear in an action-list, just like the PACKET_OUT message described above. The switch must execute actions in the order in which they appear on the action-list. We will walk through a detailed example of V.1.0 packet forwarding in Section 5.3.7.

DISCUSSION QUESTION:

Looking back there are seven packet paths emerging from the Action box in Fig. 5.6. Five of these are for virtual ports. Which of these seven paths corresponds to the typical case of a packet arriving on one port of the switch and being forwarded out another port with minimal transformation? Which path will be followed in the case of an arriving control plane packet such as an OSPF packet destined for this switch?

5.3.5 MESSAGING BETWEEN CONTROLLER AND SWITCH

The messaging between the controller and switch is transmitted over a secure channel. This secure channel is implemented via an initial TLS connection over TCP. (Subsequent versions of OpenFlow allow for multiple connections within one secure channel.) If the switch knows the IP address of the controller, then the switch will initiate this connection. Each message between controller and switch starts with the OpenFlow header. This header specifies the OpenFlow version number, the message type, the length of the message and the transaction ID of the message. The various message types in V.1.0 are listed in Table 5.1. The messages fall into three general categories: *Symmetric*, *Controller-Switch*, and *Async*. We will explain the categories of messages shown in Table 5.1 in the following paragraphs. We suggest that the reader refer to Fig. 5.7 as we explain each of these messages below. Fig. 5.7 shows the most important of these messages in a normal context and illustrates whether it normally is used during the *initialization*, *operational*, or *monitoring* phases of the controller-switch dialogue. The operational and monitoring phases generally overlap, but for the sake of clarity we show them as disjoint in the figure. In the interest of brevity, we have truncated the OFPT_ prefix from the message names shown in Table 5.1. We will follow this convention whenever referring to these message names in the balance of the book, however they appear in the index with their full name, including their OFPT_ prefix.

Table 5.1 OFPT Message Types in OpenFlow 1.0

Message Type	Category	Subcategory
HELLO	Symmetric	Immutable
ECHO_REQUEST	Symmetric	Immutable
ECHO_REPLY	Symmetric	Immutable
VENDOR	Symmetric	Immutable
FEATURES_REQUEST	Controller-Switch	Switch Configuration
FEATURES_REPLY	Controller-Switch	Switch Configuration
GET_CONFIG_REQUEST	Controller-Switch	Switch Configuration
GET_CONFIG_REPLY	Controller-Switch	Switch Configuration
SET_CONFIG	Controller-Switch	Switch Configuration
PACKET_IN	Async	NA
FLOW_REMOVED	Async	NA
PORT_STATUS	Async	NA
ERROR	Async	NA
PACKET_OUT	Controller-Switch	Cmd from controller
FLOW_MOD	Controller-Switch	Cmd from controller
PORT_MOD	Controller-Switch	Cmd from controller
STATS_REQUEST	Controller-Switch	Statistics
STATS_REPLY	Controller-Switch	Statistics
BARRIER_REQUEST	Controller-Switch	Barrier
BARRIER_REPLY	Controller-Switch	Barrier
QUEUE_GET_CONFIG_REQUEST	Controller-Switch	Queue configuration
QUEUE_GET_CONFIG_REPLY	Controller-Switch	Queue configuration

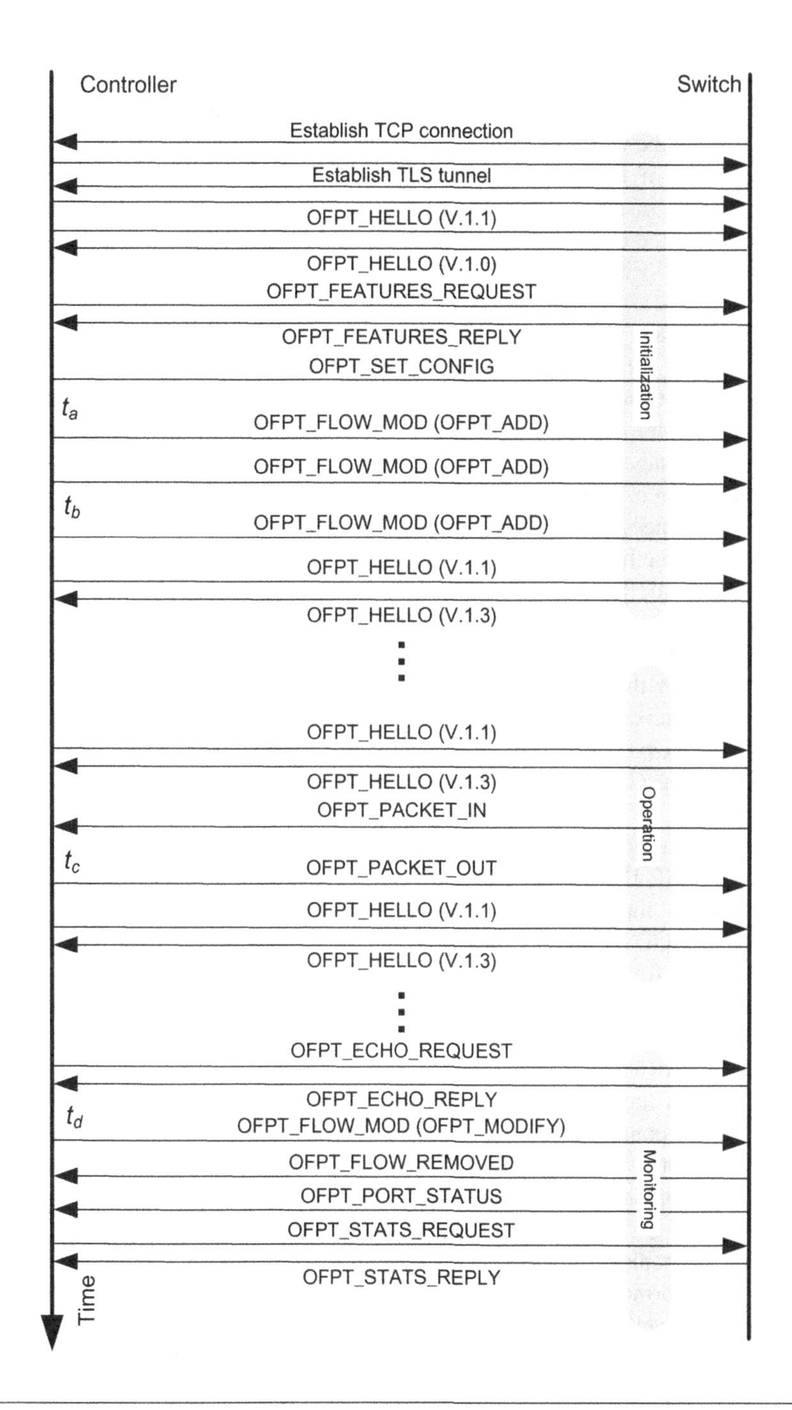

FIG. 5.7

Controller-Switch Protocol Session.

Symmetric messages may be sent by either the controller or the switch, without having been solicited by the other. The HELLO messages are exchanged after the secure channel has been established to determine the highest OpenFlow version number supported by the peers. The protocol specifies that the lower of the two versions is to be used for Controller-Switch communication over this secure channel instance. ECHO messages are used by either side during the life of the channel to ascertain that the connection is still alive and to measure the current latency or bandwidth of the connection. The VENDOR messages are available for vendor-specific experimentation or enhancements.

Async messages are sent from the switch to the controller without having been solicited by the controller. The PACKET_IN message is how the switch passes data packets back to the controller for exception handling. Control plane traffic will usually be sent back to the controller via this message. The switch can inform the controller that a flow entry is removed from the flow table via the FLOW_REMOVED message. PORT_STATUS is used to communicate changes in port status whether by direct user intervention or by a physical change in the communications medium itself. Finally, the switch uses the ERROR message to notify the controller of problems.

Controller-Switch is the broadest category of OpenFlow messages. In fact, as shown in Table 5.1, they can be divided into five subcategories: *Switch Configuration*, *Command From Controller*, *Statistics*, *Queue Configuration*, and *Barrier*. The Switch Configuration messages consist of a unidirectional configuration message and two request-reply message pairs. The unidirectional message, SET_CONFIG, is used by the controller to set configuration parameters in the switch. In Fig. 5.7 we see the SET_CONFIG message sent during the initialization phase of the controller-switch dialogue. The FEATURES message pair is used by the controller to interrogate the switch about which features it supports. Similarly, the GET_CONFIG message pair is used to retrieve a switch's configuration settings.

There are three messages comprising the *Command From Controller* category. PACKET_OUT is the analog of the PACKET_IN mentioned above. It is used by the controller to send data packets to the switch for forwarding out through the data plane. The controller modifies existing flow entries in the switch via the FLOW_MOD message. PORT_MOD is used to modify the status of an OpenFlow port.

Statistics are obtained from the switch by the controller via the STATS message pair. The BARRIER message pair is used by the controller to ensure that a particular OpenFlow command from the controller has finished executing on the switch. The switch must complete execution of all commands received prior to the BARRIER_REQUEST before executing any commands received after it, and notifies the controller of having completed such preceding commands via the BARRIER_REPLY message sent back to the controller.

The Queue Configuration message pair is somewhat of a misnomer in that actual queue configuration is beyond the scope of the OpenFlow specification and is expected to be done by an unspecified out-of-band mechanism. The QUEUE_GET_CONFIG_REQUEST and QUEUE_GET_CONFIG_REPLY message pair is the mechanism by which the controller learns from the switch how a given queue is configured. With this information, the controller can intelligently map certain flows to specific queues to achieve desired QoS levels.

Note that *immutable* in this context means that the message types will not be changed in future releases of OpenFlow. *Author's note: It is somewhat remarkable that the immutable characteristic is cited as distinct in OpenFlow. In many more mature communications protocols, maintaining backwards compatibility is of paramount importance. This is not the case with OpenFlow where support for some message formats in earlier versions is dropped in later versions. This situation is handled within the OpenFlow environment by the fact that two OpenFlow implementations coordinate their version numbers via the HELLO protocol, and presumably the higher-versioned implementation reverts to the older version of the protocol for the sake of interoperability with the other device.*

In the event that the HELLO protocol detects a loss of the connection between controller and switch, the V.1.0 specification prescribes that the switch should enter *emergency mode* and reset the TCP connection. All flows are to be deleted at this time except special flows that are marked as being part of the *emergency flow cache*. The only packet matching that is allowed in this mode is against those flows in that emergency flow cache. Which flows should be in this cache was not prescribed in the specification, and subsequent versions of OpenFlow have addressed this area differently and more thoroughly.

DISCUSSION QUESTION:

Fig. 5.7 depicts three groupings of controller-switch communications. OFPT_FLOW_MOD messages appear in both the initialization and monitoring groupings. What is the difference between the messages? Do you think that the type of OFPT_FLOW_MOD messages shown in the initialization phase could be interspersed with messages in operation and monitoring?

5.3.6 EXAMPLE: CONTROLLER PROGRAMMING FLOW TABLE

In Fig. 5.8 two simple OpenFlow V.1.0 flow table modifications are performed by the controller. We see in the figure that the initial flow table has three flows. In the quiescent state before time t_a, we see an exploded version of flow entry zero. It shows that the flow entry specifies that all Ethernet frames entering the switch on input port K with a destination Ethernet address of 0x000CF15698AD should be output on output port N. All other match fields have been wildcarded, indicated by the asterisks in their respective match fields in Fig. 5.8. At time t_a, the controller sends a FLOW_MOD(ADD) command to the switch adding a flow for packets entering the switch on any port, with source IP addresses 192.168.1.1 and destination IP address 209.1.2.1, source TCP port 20 and destination port 20. All other match fields have been wildcarded. The outport port is specified as P. We see that after this controller command is received and processed by the switch, the flow table contains a new flow entry F corresponding to that ADD message. At time t_b, the controller sends a FLOW_MOD(MODIFY) command for flow entry zero. The controller seeks to modify the corresponding flow entry such that there is a one hour (3600 second) idle time on that flow. The figure shows that after the switch has processed this command, the original flow entry has been modified to reflect that new idle time. Note that idle time for a flow entry means that after that number of seconds of inactivity on that flow, the flow should be deleted by the switch. Looking back at Fig. 5.7, we see an example of such a flow expiration just after time t_d. The FLOW_REMOVED message we see there indicates that the flow programmed at time t_b in our examples in Figs. 5.7 and 5.8 has expired. This controller had requested to be notified of such expiration when the flow was configured and the FLOW_REMOVED message serves this purpose.

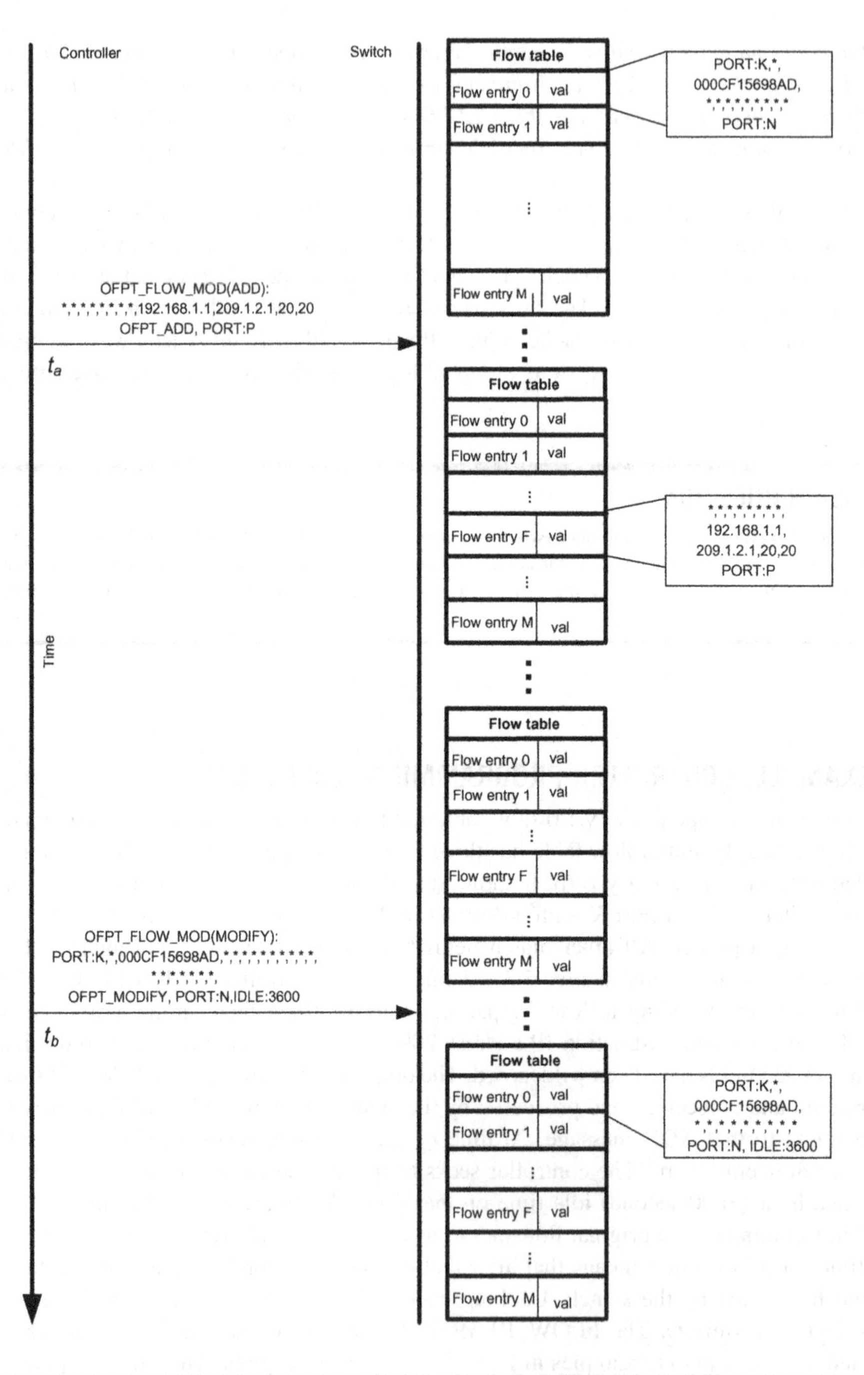

FIG. 5.8

Controller programming flow entries in V.1.0.

5.3.7 EXAMPLE: BASIC PACKET FORWARDING

We illustrate the most basic case of OpenFlow V.1.0 packet forwarding in Fig. 5.9. The figure depicts a packet arriving at the switch through port 2 with source IPv4 address of 192.168.1.1 and destination IPv4 address of 209.1.2.1. The packet matching function scans the flow table starting at flow entry 0 and finds a match in flow entry *F*. Flow entry *F* stipulates that a matching packet should be forwarded out port *P*. The switch does this, completing this simple forwarding example.

An OpenFlow switch forwards packets based on the header fields it matches. The network programmer designates layer three switch behavior by programming the flow entries to try to match layer three headers such as IPv4. If it is a layer two switch, the flow entries will dictate matching on layer two headers. The semantics of the flow entries allow matching for a wide variety of protocol headers but a given switch will be programmed only for those that correspond to the role it plays in packet forwarding. Whenever there is overlap in potential matches of flows, the priority assigned the flow entry by the controller will determine which match takes precedence. For example, if a switch was both a layer two and a layer three switch, placing layer three header matching flow entries at a higher priority would ensure that if possible, layer three switching is done on that packet.

5.3.8 EXAMPLE: SWITCH FORWARDING PACKET TO CONTROLLER

We showed in the last section an example of an OpenFlow switch forwarding an incoming data packet to a specified destination port. Another fundamental action of the OpenFlow V.1.0 switch is to forward packets to the controller for exception handling. The two reasons for which the switch may forward a packet to the controller are OFPR_NO_MATCH and OFPR_ACTION. Obviously OFPR_NO_MATCH is used when no matching flow entry is found. OpenFlow retains the ability to specify that a particular matching flow entry should always be forwarded to the controller. In this case OFPR_ACTION is specified as the reason. An example of this would be a control packet such as a routing protocol packet that always needs to be processed by the controller. In Fig. 5.10 we show an example of an incoming data packet that is an OSPF routing packet. There is a matching table entry for this packet which specifies that the packet should be forwarded to the controller. We see in the figure that a PACKET_IN

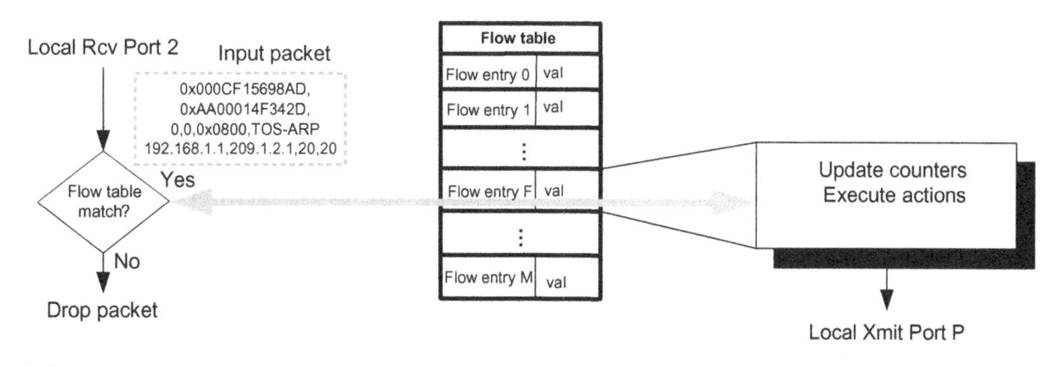

FIG. 5.9

Packet matching function—basic packet forwarding V.1.0.

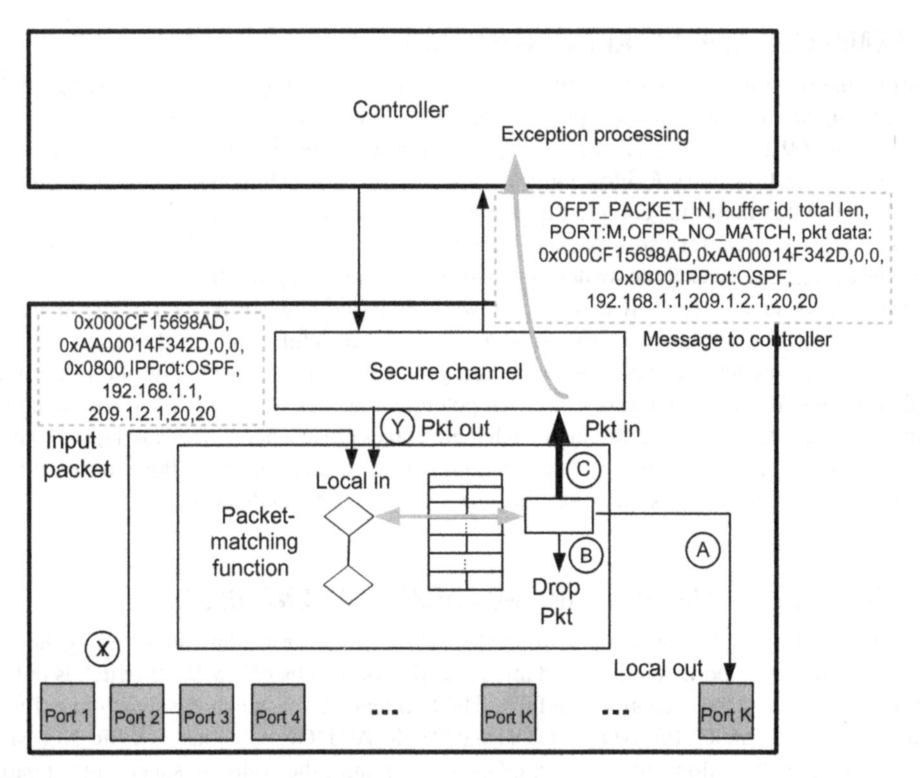

FIG. 5.10

Switch forwarding incoming packet to controller.

message is sent via the secure channel to the controller, handing off this routing protocol update to the controller for exception processing. The processing that would likely take place on the controller is that the OSPF routing protocol would be run, potentially resulting in a change to the forwarding tables in the switch. The controller could then modify the forwarding tables via the brute force approach of sending FLOW_MOD commands to the switch modifying the output port for each flow in the switch affected by this routing table change. (We will see in Section 5.4.2 that there is a more efficient way for the controller to program the output port for flows in a layer three switch where multiple flows share the same next hop IP address.)

At a minimum, the controller needs access to the packet header fields to determine its disposition of the packet. In many cases, though not all, it may need access to the entire packet. This would in fact be true in the case of the OSPF routing packet in this example. In the interest of efficiency, OpenFlow allows the optional buffering of the full packet by the switch. In the event of a large number of packets being forwarded from the switch to the controller for which the controller only needs to examine the packet header, significant bandwidth efficiency gains are achieved by buffering the full packet in the switch and only forwarding the header fields. Since the controller will sometimes need to see the balance of the packet, a buffer ID is communicated with the PACKET_IN message. This buffer

ID may be used by the controller to subsequently retrieve the full packet from the switch. The switch has the ability to age out old buffers that have not been retrieved by the controller.

There are other fundamental actions that the switch may take on an incoming packet: (1) to flood the packet out all ports except the port on which it arrived or, (2) to drop the packet. We trust that the reader's understanding of the drop and flood functions will follow naturally from the two examples just provided, and we'll now move on to discuss the extensions to the basic OpenFlow functions provided in V.1.1.

DISCUSSION QUESTION:

Explain how the controller would use the BUFFER ID field in the message sent to it shown in Fig. 5.10.

5.4 OpenFlow 1.1 ADDITIONS

OpenFlow 1.1 [3] was released on February 28, 2011. Table 5.2 lists the major new features added in this release. Multiple flow tables and group table support are the two most prominent features added to OpenFlow in release V.1.1. We highlight these new features in Fig. 5.11. From a practical standpoint, V.1.1 had little impact other than as a stepping stone to V.1.2. This was because it was released just before the ONF was created and the SDN community waited for the first version following the ONF transition (i.e., V.1.2) before creating implementations. It is important that we cover it here, though, as the following versions of OpenFlow are built upon some of the V.1.1 feature set.

5.4.1 MULTIPLE FLOW TABLES

V.1.1 significantly augments the sophistication of packet processing in OpenFlow. The most salient shift is due to the addition of *multiple flow tables*. The concept of a single flow table remains much like it was in V.1.0, however, in V.1.1 it is now possible to defer further packet processing to subsequent matching in other flow tables. For this reason, it was necessary to break the execution of actions from its direct association with a flow entry. With V.1.1 the new *instruction* protocol object is associated with a flow entry. The processing pipeline of V.1.1 offered much greater flexibility than what was available in V.1.0. This improvement derives from the fact that flow entries can be *chained* by an instruction

Table 5.2 Major New Features Added in OpenFlow 1.1

Feature Name	Description
Multiple Flow Tables	See Section 5.4.1
Groups	See Section 5.4.2
MPLS and VLAN Tag Support	See Section 5.4.3
Virtual Ports	See Section 5.4.4
Controller Connection Failure	See Section 5.4.5

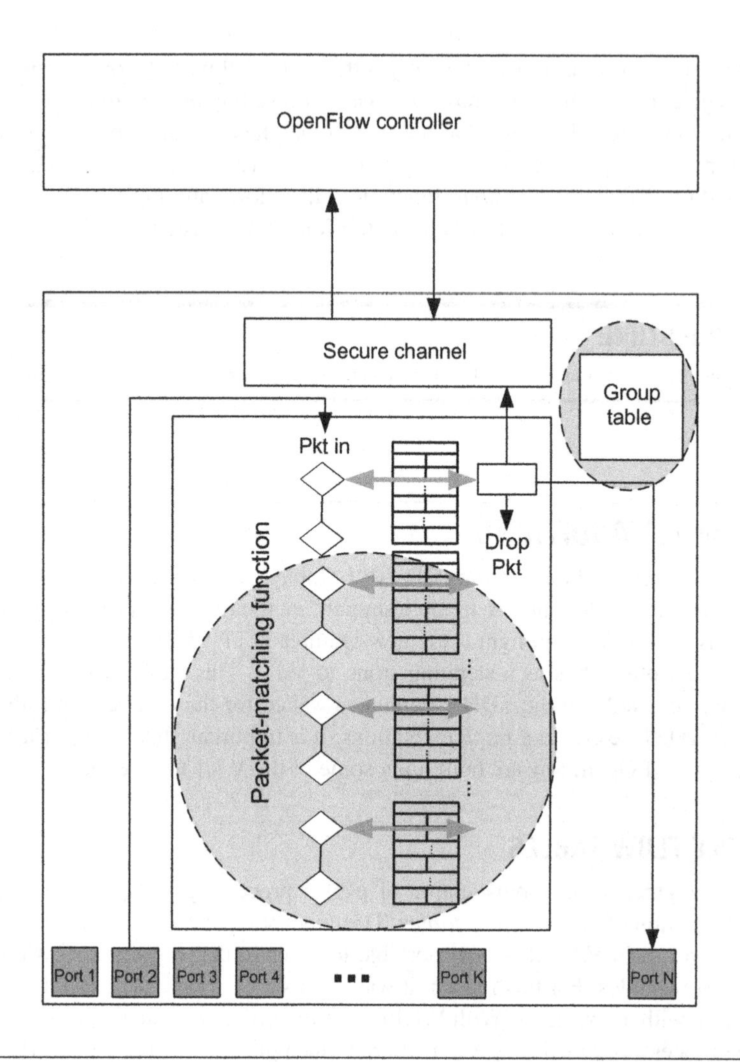

FIG. 5.11

OpenFlow V.1.1 switch with expanded packet processing.

in one flow entry pointing to another flow table. This is called a *GOTO* instruction. When such an instruction is executed, the packet matching function depicted earlier in Figs. 5.9 and 5.10 is invoked again, this time starting the match process with the first flow entry of the new flow table. This new pipeline is reflected in the expanded packet matching function shown in Fig. 5.11. The pipeline allows all the power of an OpenFlow V.1.1 instruction to test and *modify* the contents of a packet to be applied multiple times, with different conditions used in the matching process in each flow table. This allows dramatic increases in both the complexity of the matching logic as well as the degree and nature of the packet modifications that may take place as the packet transits the OpenFlow V.1.1 switch. V.1.1

packet processing, with different instructions chained through complex logic as different flow entries are matched in a sequence of flow tables, constitutes a robust packet processing *pipeline*.[1]

In Section 5.3.4 we described the actions supported in V.1.0 and the order in which they were executed. The V.1.1 instructions form the conduit that controls which actions are taken and in what order. They can do this in two ways. In the first, they add actions to an *action-set*. The action set is initialized and modified by all the instructions executed during a given pass through the pipeline. Instructions delete actions from the action-set or merge new actions into the action-set. When the pipeline ends, the actions in the action-set are executed in the following order:

1. copy TTL inwards
2. pop
3. push
4. copy TTL outwards
5. decrement TTL
6. set: apply all set_field actions to the packet
7. qos: apply all QoS actions to the packet, such as set_queue
8. group: if a group action is specified, apply the actions to the relevant action buckets (which can result in the packet being forwarded out the ports corresponding to those buckets)
9. output: if no group action was specified, forward the packet out the specified port

The output action, if such an instruction exists (and it normally does), is executed last. This is logical because other actions often manipulate the contents of the packet and this clearly must occur before the packet is transmitted. If there is neither a group nor output action, the packet is dropped. We will explain the most important of the actions listed above in discussion and examples in the following sections.

The second way that instructions can invoke actions is via the *Apply-Actions* instruction. This is used to execute certain actions immediately between flow tables, while the pipeline is still active, rather than waiting for the pipeline to reach its end, which is the normal case. Note that the action list that is part of the Apply-Actions has exactly the same semantics as the action list in the PACKET_OUT message, and both are the same as the action list semantics of V.1.0.

In the case of the PACKET_OUT message being processed through the pipeline, rather than the action list being executed upon a flow entry match, its component actions on the action list are merged into the action set being constructed for this incoming packet. If the current flow entry includes a GOTO instruction to a higher-numbered flow table, then further action lists may be merged into the action set before it is ultimately executed.

When a matched flow entry does not specify a GOTO flow table, the pipeline processing completes, and whatever actions have been recorded in the action set are then executed. In the normal case, the final action executed is to forward the packet to an output port, to the controller or to a *group* table, which we describe below.

[1] This increased robustness comes at a price. It has proven challenging to adapt existing hardware switches to support multiple flow tables. We discuss this further in Section 5.6. In Sections 14.9.3 and 15.3.7 we discuss attempts to design hardware specifically designed with this capability in mind.

5.4.2 GROUPS

V.1.1 offers the *group* abstraction as a richer extension to the FLOOD option. In V.1.1 there is the group table, which first appeared in Fig. 5.11 and is shown in detail in Fig. 5.12. The group table consists of *group entries*, each entry consisting of one or more *action buckets*. The buckets in a group have actions associated with them that get applied before the packet is forwarded to the port defined by that bucket. Refinements on flooding, such as multicast, can be achieved in V.1.1 by defining groups as specific sets of ports. Sometimes a group may be used when there is only a single output port, as illustrated in the case of group 2 in Fig. 5.12. A use case for this would be when many flows all should be directed to the same next hop switch and it is desirable to retain the ability to change that next hop with a single configuration change. This can be achieved in V.1.1 by having all the designated flows pointing to a single group entry that forwards to the single port connected to that next hop. If the controller wishes to change that next hop due to a change in IP routing tables in the controller, all the flows can

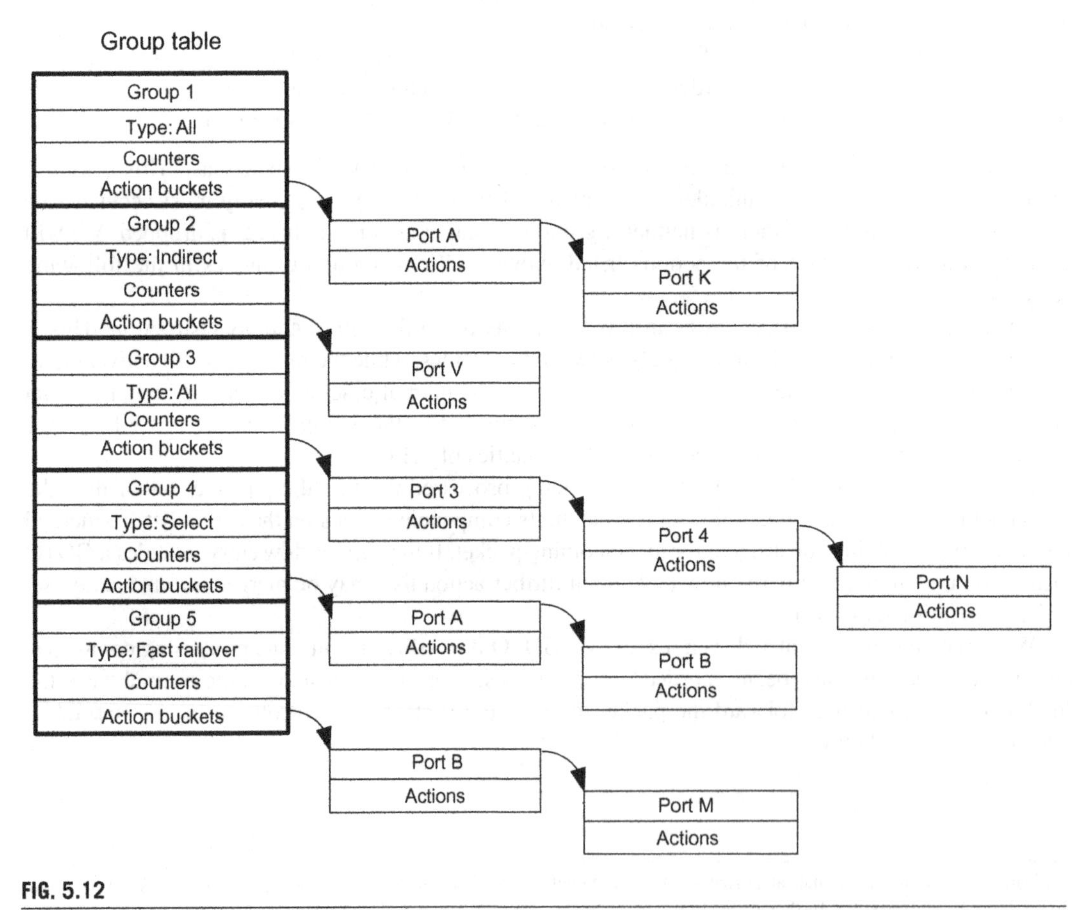

FIG. 5.12

OpenFlow V.1.1 group table.

be rerouted by simply reprogramming the single group entry. This provides a more efficient way of handling the routing change from the OSPF update example presented earlier in Section 5.3.8. Clearly, changing a single group entry's action bucket is faster than updating the potentially large number of flow entries whose next hop changed as a result of a single routing update. Note that this is a fairly common scenario. Whenever a link or neighbor fails or is taken out of operation, *all flows* traversing that failed element need to be rerouted by the switch that detects that failure.

Note that one group's buckets may forward to other groups, providing the capability to chain groups together.

5.4.3 MPLS AND VLAN TAG SUPPORT

V.1.1 is the first OpenFlow version to provide full VLAN support. Since providing complete support for multiple levels of VLAN tags required robust support for the *popping* and *pushing* of multiple levels of tags, support for MPLS tagging followed naturally and is also part of V.1.1. For the interested reader unfamiliar with MPLS technology, we provide a good reference in [4]. Both the new PUSH and POP actions, as well as the chaining of flow tables, were necessary to provide this generic support for VLANs and MPLS. When a PUSH action is executed, a new header of the specified type is inserted in front of the current outermost header. The field contents of this new header are initially copied from the corresponding existing fields in the current outermost header, should one exist. If it does not exist, they are initialized to zero. New header values are then assigned to this new outermost header via subsequent SET actions. While the PUSH is used to add a new tag, the POP is used to remove the current outermost tag. In addition to PUSH and POP actions, V.1.1 also permits modification of the current outermost tag, whether it be a VLAN tag or an MPLS shim header.

While V.1.1 correctly claims full MPLS and VLAN tag support, that support requires specification of very complex matching logic that has to be implemented when such tags are encountered in the incoming packet. The *extensible match support* that will be introduced in Section 5.5.1 provides the more generalized semantics to the controller such that this complex matching logic in the switch can be disposed of. Since this V.1.1 logic is replaced by the more general solution in Section 5.5.1, we will not confuse the reader by providing details about it here.

5.4.4 VIRTUAL PORTS

In V.1.0, the concept of an output port mapped directly to a physical port, with some limited use of *virtual ports*. While the TABLE and other virtual ports did exist in earlier versions of OpenFlow, the concept of virtual ports has been augmented in V.1.1. A V.1.1 switch groups ports into the categories of *standard ports* and *reserved virtual ports*. Standard ports consist of:

- Physical Ports
- Switch-Defined Virtual Ports: In V.1.1 it is now possible for the controller to forward packets to an abstraction called Switch-Defined Virtual Ports. Such ports are used when more complex processing will be required on the packet than simple header field manipulation. One example of this is when the packet should be forwarded via a tunnel. Another use case for a virtual port is *Link Aggregation* (LAG). For more information about LAGs, we refer the reader to [5].

Reserved virtual ports consist of:

- ALL: This is the straightforward mechanism to flood packets out all standard ports except the port on which the packet arrived. This is similar to the effect provided by the optional FLOOD reserved virtual port below.
- CONTROLLER: Forwards the packet to the controller in an OpenFlow message.
- TABLE: Processes the packet through the normal OpenFlow pipeline processing. This only applies to a packet that is being sent from the controller (via a PACKET_OUT message). We explained the use of the TABLE virtual port in Section 5.3.4.
- IN_PORT: This provides a loopback function. The packet is sent back out on the port on which it arrived.
- LOCAL (optional): This optional port provides a mechanism where the packet is forwarded to the switch's local OpenFlow control software. As a LOCAL port may be used both as an output port and an ingress port, this can be used to implement an in-band controller connection, obviating the need for a separate control network for the controller-switch connection. We refer the reader to our earlier discussion on LOCAL in Section 5.3.3 for an example of how LOCAL may be used.
- NORMAL (optional): This directs the packet to the normal nonOpenFlow pipeline of the switch. This differs from the LOCAL reserved virtual port in that it may only be used as an output port.
- FLOOD (optional): The general use of this port is to send the packet out all standard ports except the port on which it arrived. The reader should refer to the OpenFlow V.1.1 specification for the specific nuances of this reserved virtual port.

5.4.5 CONTROLLER CONNECTION FAILURE

Loss of connectivity between the switch and controller is a serious, and real possibility, and the OpenFlow specification needs to specify how it should be handled. The *emergency flow cache* was included in V.1.0 in order to handle such a situation, but support for this was dropped in V.1.1 and replaced with two new mechanisms, *fail secure mode* and *fail stand-alone mode*.[2] The V.1.1 switch immediately enters one of these two modes upon loss of connection to the controller. Which of these two modes is entered will depend upon which is supported by the switch or, if both are supported, by user configuration. In the case of fail secure mode, the switch continues to operate as a normal V.1.1 switch, except that all messages destined for the controller are dropped. In the case of fail stand-alone mode, the switch additionally ceases its OpenFlow pipeline processing and continues to operate in its native, underlying switch or router mode. When the connection to the controller is restored, the switch resumes its normal operation mode. The controller, having detected the loss and restoration of the connection, may choose to delete existing flow entries and begin to configure the switch anew. *(Author's note: This is another example of the aforementioned lack of backwards compatibility in the OpenFlow specifications.)*

[2]A common synonym for fail secure mode is *fail open mode*. Similarly, a synonym for fail stand-alone mode is *fail closed mode*.

5.4.6 EXAMPLE: FORWARDING WITH MULTIPLE FLOW TABLES

Fig. 5.13 expands on our earlier example of V.1.0 packet forwarding that was presented in Fig. 5.9. Assuming that the incoming packet is the same as in Fig. 5.9, we see in Fig. 5.13 that there is a match in the second flow entry in flow table 0. Unlike V.1.0, the pipeline does not immediately execute actions associated with that flow entry. Its counters are updated, and the newly initialized action-set is updated with those actions programmed in that flow entry's instructions. One of those instructions is to continue processing at table *K*. We see this via the jump to processing at the letter *A*, where we resume the pipeline process matching the packet against table *K*. In this case, a matching flow entry is found in flow entry *F*. Once again, this match results in the flow entry's counters being updated and its instructions being executed. These instructions may apply further modifications to the action set that has been carried forward from table 0. In our example, an additional action *A*3 is merged into the action set. Since there is no GOTO instruction present this time, this represents the end of the pipeline processing and the actions in the current action set are performed in the order specified by OpenFlow V.1.1. (The difference between action set and action-list was explained earlier in Section 5.4.1.) The OUTPUT action, if it is present, is the last action performed before the pipeline ends. We see that the OUTPUT action was present in our example as we see the packet forwarded out port *P* in Fig. 5.13.

5.4.7 EXAMPLE: MULTICAST USING V.1.1 GROUPS

We direct the reader to Fig. 5.12 discussed previously. In this figure we see that group 3 has group type ALL and has three action buckets. This group configuration prescribes that a packet sent for processing by this group entry should have a copy sent out each of the ports in the three action buckets. Fig. 5.14 shows that the packet matching function directs the incoming packet to group 3 in the group table for processing. Assuming that the group 3 in Fig. 5.14 is as is shown in Fig. 5.12, the packet is multicast out ports 3, 4, and *N* as indicated by the dashed arrows in the figure.

DISCUSSION QUESTION:

The group table depicted in Fig. 5.14 is the one shown earlier in Fig. 5.12. Explain why, in Fig. 5.14, there appear three copies of the incoming packet being forwarded out ports 3, 4, and N, respectively.

5.5 OpenFlow 1.2 ADDITIONS

OpenFlow 1.2 [6] was released on December 5, 2011. Table 5.3 lists the major new features added in this release. We discuss the most important of these in the following sections.

5.5.1 EXTENSIBLE MATCH SUPPORT

Due to the relatively narrow semantics in the packet matching prior to V.1.2, it was necessary to define complex flow diagrams that described the logic of how to perform packet parsing. The packet matching capability provided in V.1.2 provides sufficient richness in the packet matching descriptors that the

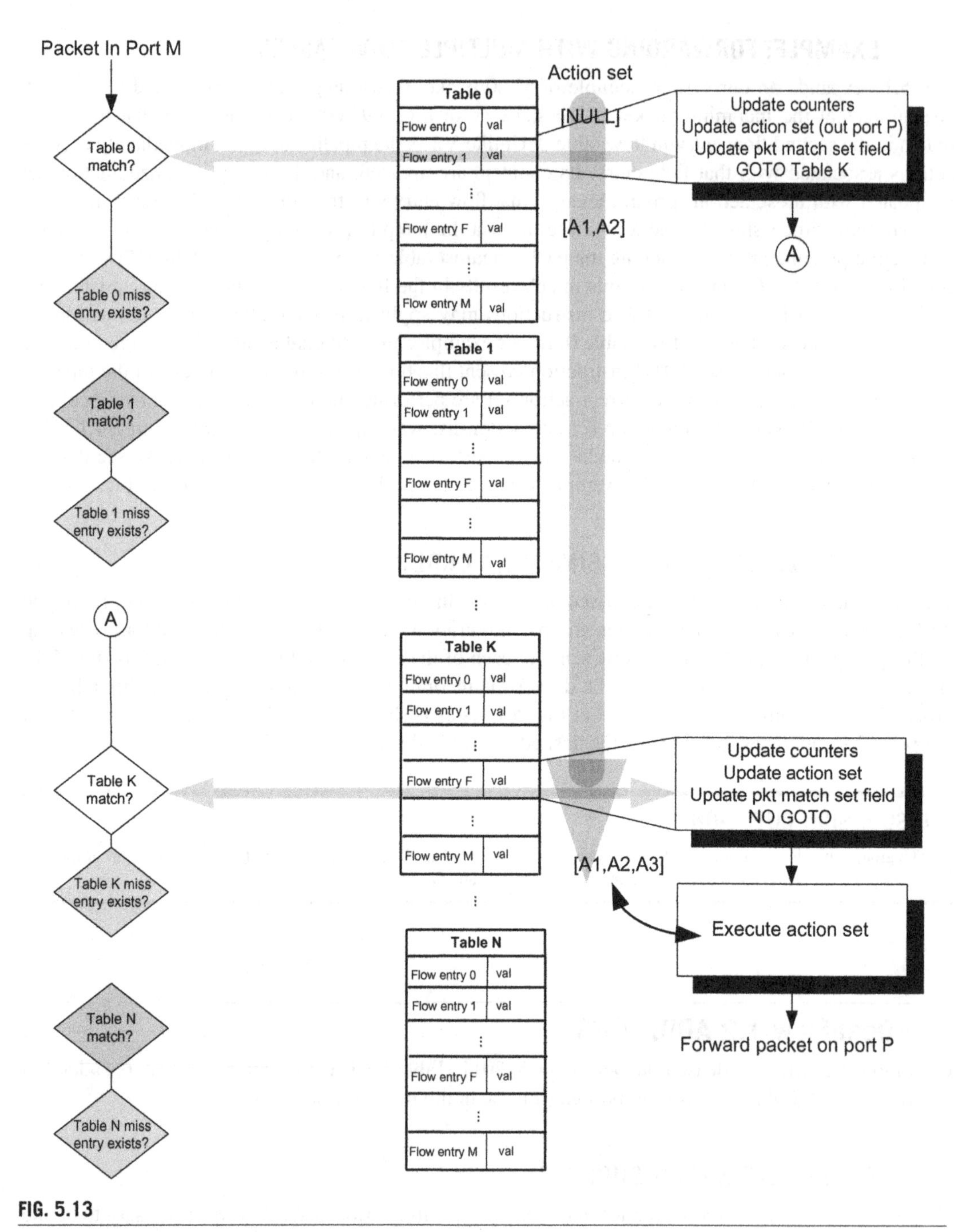

FIG. 5.13

Packet matching function—basic packet forwarding V.1.1.

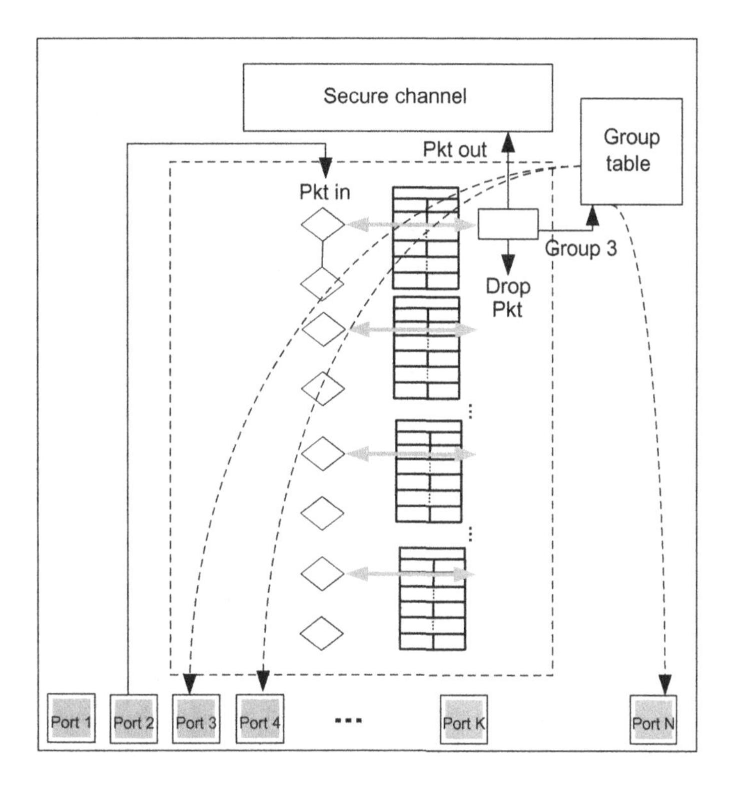

FIG. 5.14

Multicast using group table in V.1.1.

Table 5.3 Major New Features Added in OpenFlow 1.2

Feature Name	Description
Extensible Match Support	See Section 5.5.1
Extensible set_field Packet Rewriting Support	See Section 5.5.2
Extensible Context Expression in "packet-in"	See Section 5.5.3
Extensible Error Messages via Experimenter Error Type	See [6]
IPv6 Support	See Section 5.5.2
Simplified Behavior of Flow-Mod Request	See [6]
Removed Packet Parsing Specification	See Section 5.5.1
Multiple Controller Enhancements	See Section 5.5.4

controller can encode the desired logic in the rules themselves. This obviates the earlier versions' requirement that the behavior be hard-coded into the switch logic.

A generic and extensible packet matching capability has been added in V.1.2 via the *OpenFlow Extensible Match* (OXM) descriptors. OXM defines a set of *type-length-value* (TLV) pairs that can describe or define virtually any of the header fields an OpenFlow switch would need to use for matching. This list is too long to enumerate here, but in general any of the header fields that are used in matching for Ethernet, VLAN, MPLS, IPv4, and IPv6 switching and routing may be selected and a value (with bitmask wildcard capability) provided. Prior versions of OpenFlow had a much more static match descriptor which limited flexibility in matches and made adding future extensions more difficult. An example of this is that in order to accommodate the earlier fixed structure, the single match field TCP Port was overloaded to also mean UDP Port or ICMP code depending on the context. This confusion has been eliminated in V.1.2 via OXM. Because the V.1.2 controller can encode more descriptive parsing into the pipeline of flow tables, complex packet parsing does not need to be specified inside the V.1.2 switch.

This ability to match on any combination of header fields is provided within the OPENFLOW_BASIC *match class*. V.1.2 expands the possibilities for match fields by allowing for multiple match classes. Specifically, the EXPERIMENTER match class is defined, opening up the opportunity for matching on fields in the packet payload, providing a near limitless horizon for new definitions of flows. The syntax and semantics of EXPERIMENTER are left open so that the number and nature of different fields is subject to the experimental implementation.

5.5.2 EXTENSIBLE SET_FIELD PACKET REWRITING SUPPORT

The V.1.2 switch continues to support the action types we saw in previous versions. A major enhancement provides for the ability to set the value of *any* field in the packet header that may be used for matching. This is due to the fact that the same OXM encoding described above for enhanced packet matching is made available in V.1.2 for generalized setting of fields in the packet header. For example, IPv6 support falls out naturally from the fact that any of the fields that may be described by an OXM TLV may also be set using the set_field action. The ability to match an incoming IPv6 header against a flow entry and, when necessary, to set an IPv6 address to a new value, together provide support for this major feature of V.1.2. CRC recalculation is performed automatically when a set-field action changes the packet's contents.

Since the EXPERIMENTER action type can be a modification or extension to any of the basic action types, or even something totally novel, this allows for packet fields not part of the standard OXM header fields to be modified. In general, if a packet field may be used for matching in V.1.2, then it may also be modified.

5.5.3 EXTENSIBLE CONTEXT EXPRESSION IN PACKET_IN

The OXM encoding is also used to extend the PACKET_IN message sent from the switch to the controller. In previous versions this message included the packet headers used in matching that resulted in the switch deciding to forward the packet to the controller. In addition to the packet contents, the packet matching decision is influenced by *context* information. Formerly, this consisted of the input port identifier. In V.1.2 this context information is expanded to include the input virtual port, the input physical port, and *metadata* that has been built up during packet matching pipeline processing. Metadata semantics is not prescribed by the OpenFlow specification, other than that the OXM metadata TLV may be initialized, modified or tested at any stage of the pipeline processing. Since the OXM encoding described above contains TLV definitions for all the context fields, as well as all matchable packet headers, the OXM format provides a convenient vehicle to communicate the packet matching state when the switch decides to forward the packet to the controller. The switch may forward a packet to the controller because:

- there was no matching flow or
- an instruction was executed in the pipeline prescribing that a matching packet be forwarded to the controller or
- the packet had an invalid TTL.

5.5.4 MULTIPLE CONTROLLERS

In previous versions of OpenFlow, there was very limited support for backup controllers. In the event that communication with the current controller was lost, the switch entered either fail secure mode or fail stand-alone mode. The notion that the switch would attempt to contact previously configured backup controllers was encouraged by the specification, but not explicitly described. In V.1.2 the switch may be configured to maintain simultaneous connections to multiple controllers. The switch must ensure that it only sends messages to a controller pertaining to a command sent by that controller. In the event that a switch message pertains to multiple controllers, it is duplicated and a copy sent to each controller. A controller may assume one of three different roles relative to a switch:

- Equal
- Slave
- Master

These terms relate to the extent to which the controller has the ability to change the switch configuration. The least powerful is obviously slave mode, wherein the controller may only request data from the switch, such as statistics, but may make no modifications. Both equal and master modes allow the controller the full ability to program the switch, but in the case of master mode the switch enforces that only one controller be in master mode and all others are in slave mode. The multiple controllers feature is part of how OpenFlow addresses the *high availability* (HA) requirement.

5.5.5 EXAMPLE: BRIDGING VLANS THROUGH SP NETWORKS

V.1.2 is the first release to provide convenient support for bridging customer VLANs through service provider (SP) networks. This feature is known by a number of different names, including *provider bridging*, *Q-in-Q* and *Stacked VLANs*. These all refer to the tunneling of an edge VLAN through a provider-based VLAN. There are a number of variants of this basic technology, but they are most often based on the nesting of multiple IEEE 802.1Q tags [7]. In this example we will explain how this can be achieved by stacking one VLAN tag inside another, which involves the PUSHing and POPping of VLAN tags that was introduced in Section 5.4.3. We deferred this example until our discussion of V.1.2 since the V.1.1 packet matching logic to support this was very complex and not easily extended. With the extensible match support in V.1.2, the logic required to provide VLAN stacking in the OpenFlow pipeline is much more straightforward.

This feature is implemented in OpenFlow V.1.2 by encoding a flow entry to match a VLAN tag from a particular customer's layer two network. In our example, this match occurs at an OpenFlow switch serving as a service provider access switch. This access switch has been programmed to match VLAN 10 from site A of customer X and to tunnel this across the service provider core network to the same VLAN 10 at site B of this same customer X. The OpenFlow pipeline will use the extensible match support to identify frames from this VLAN. The associated action will PUSH service provider VLAN tag 40 onto the frame, and the frame will be forwarded out a port connected to the provider backbone network. When this doubly tagged frame emerges from the backbone, the OpenFlow-enabled service provider access switch matches VLAN tag 40 from that backbone interface, and the action associated with the matching flow entry POPs off the outer tag and the singly tagged frame is forwarded out a port connected the site B of customer X.

Note that while our example uses VLAN tags, similar OpenFlow processing can be used to tunnel customer MPLS connections through MPLS backbone connections. In that case, an outer MPLS label is PUSHed onto the packet and then POPped off upon exiting the backbone.

5.6 OpenFlow 1.3 ADDITIONS

OpenFlow V.1.3 [8] was released on April 13, 2012. This release was a major milestone. Many implementations are being based on V.1.3 in order to stabilize controllers around a single version. This is also true about ASICs. Since V.1.3 represents a major new leap in functionality and has not been followed quickly by another release, this provides an opportunity for ASIC designers to develop hardware support for many of the V.1.3 features with the hope of a more stable market into which to sell their new chips. This ASIC opportunity notwithstanding, there are post V.1.0 features that are very challenging to implement in hardware. For example, the iterative matching actions we described in Section 5.4.1 are difficult to implement in hardware at line rates. It is likely that the real-life chips that support V.1.3 will have to limit the number of flow tables to a manageable number.[3] Such limitations are not imposed on software-only implementations of OpenFlow, and, indeed, such implementations will provide full V.1.3 support. Table 5.4 lists the major new features added in this release. A discussion of the most prominent of these new features is found in the following sections.

5.6.1 REFACTOR CAPABILITIES NEGOTIATION

There is a new MULTIPART_REQUEST/MULTIPART_REPLY message pair in V.1.3. This replaces the READ_STATE messaging in prior versions that used STATS_REQUEST/STATS_REPLY to get this information. Rather than having this capability information embedded as if it were a table statistic, the new message request-reply pair is utilized and the information is conveyed using a standard *type-length-value* (TLV) format. Some of the capabilities that can be reported in this new manner include *next-table*, *table-miss flow entry*, and *experimenter*. This new MULTIPART_REQUEST/REPLY message pair subsumes the older STATS_REQUEST/REPLY pair and is now used for both reporting of statistics as well as capability information. The data formats for the capabilities are the TLV format and this capability information has been removed from the table statistics structure.

[3]We discuss the possibility of ASICs that would allow more general support of these features in Section 14.9.3 and Section 15.3.7.

Table 5.4 Major New Features Added in OpenFlow 1.3

Feature Name	Description
Refactor Capabilities Negotiation	See Section 5.6.1
More Flexible Table Miss Support	See Section 5.6.2
IPv6 Extension Header Handling Support	See [8]
Per Flow Meters	See Section 5.6.3
Per Connection Event Filtering	See Section 5.6.4
Auxiliary Connections	See Section 5.6.5
MPLS BoS Matching	Bottom of Stack bit (BoS) from the MPLS header may now be used as part of match criteria
Provider Backbone Bridging Tagging	See Section 5.6.7
Rework Tag Order	Instruction execution order now determines tag order rather than it being statically specified
Tunnel-ID Metadata	Allows for support of multiple tunnel encapsulations
Cookies in PACKET_IN	See Section 5.6.6
Duration for Stats	The new *duration* field allows more accurate computation of packet and byte rate from the counters included in those statistics.
On Demand Flow Counters	Disable/enable packet and byte counters on a per-flow basis

5.6.2 MORE FLEXIBLE TABLE-MISS SUPPORT

We have explained that a table-miss is when a packet does not match any flow entries in the current flow table in the pipeline. Formerly there were three configurable options for handling such a table-miss. This included dropping the packet, forwarding it to the controller or continuing packet matching at the next flow table. V.1.3 expands upon this limited handling capability via the introduction of the *table-miss flow entry*. The controller programs a table-miss flow entry into a switch in much the same way it would program a normal flow entry. The table-miss flow entry is distinct in that it is by definition of lowest priority (zero) and all match fields are wildcards. The zero-priority characteristic guarantees that it is the last flow entry that can be matched in the table. The fact that all match fields are wild cards means that *any* packet being matched against the table-miss will be a match. The upshot of these characteristics is that the table-miss entry serves as a kind of backstop for any packets that would otherwise have found no matches. The advantage of this approach is that the full semantics of the V.1.3 flow entry, including instructions and actions, may be applied to the case of a table-miss. It should be obvious to the reader that the table-miss base cases of dropping the packet, forwarding it to the controller, or continuing at next flow table are easily implemented using the flow entry instructions and actions. Interestingly, though, by using the generic flow entry semantics, it is now conceivable to treat table-misses in more sophisticated ways, including passing the packet that misses to a higher-numbered flow table for further processing. This capability adds more freedom to the matching "language" embodied by the flow tables, entries and associated instructions and actions.

5.6.3 PER FLOW METERS

V.1.3 introduces a flexible meter framework. Meters are defined on a per-flow basis and reside in a *meter table*. Fig. 5.15 shows the basic structure of a meter and how it is related to a specific flow. A given meter is identified by its *meter ID*. V.1.3 instructions may direct packets to a meter identified by its meter ID. The framework is designed to be extensible to support the definition of complex meters in future OpenFlow versions. Such future support may include color-aware meters that will provide DiffServ QoS capabilities. V.1.3 meters are only rate-limiting meters. As we see in Fig. 5.15 there may be multiple *meter bands* attached to a given meter. Meter 3 in the example in the figure has three meter bands. Each meter band has a configured bandwidth rate and a type. Note that the units of bandwidth are not specified by OpenFlow. The type determines the action to take when that meter band is processed. When a packet is processed by a meter, at most one band is used. This band is selected based on the highest bandwidth rate band that is lower than the current measured bandwidth for that flow. If the current measured rate is lower than all bands, no band is selected and no action is taken. If a band is selected, the action taken is that prescribed by the band's type field. There are no required types in V.1.3. The *optional* types described in V.1.3 consist of DROP and *DSCP remark*. DSCP remark indicates that the DiffServ drop precedence field of the *Differentiated Services Code Point* (DSCP)

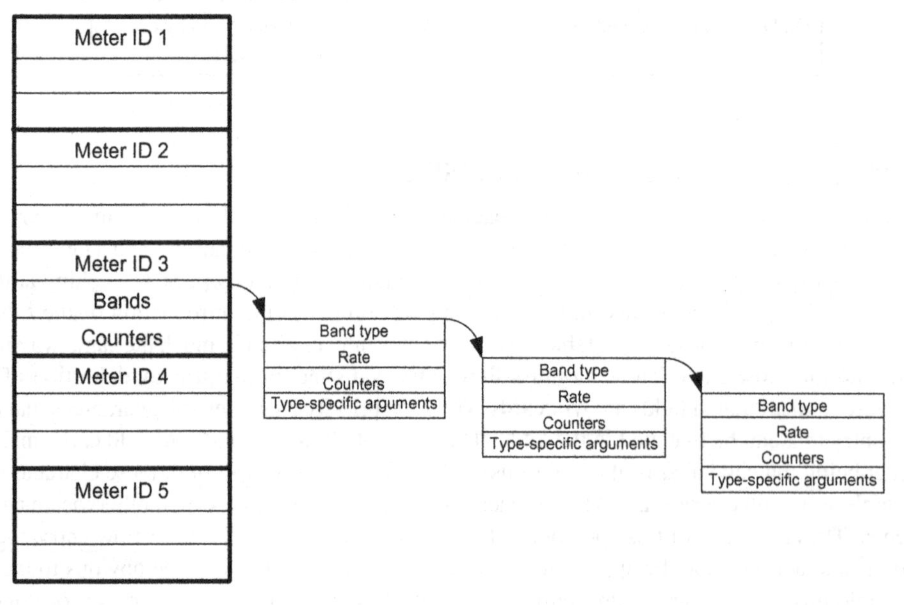

FIG. 5.15

OpenFlow V.1.3 meter table.

field should be decremented, increasing the likelihood that this packet will be dropped in the event of queue congestion. Well known and simple rate-limiting algorithms, such as the *leaky bucket* [9], may be implemented in a straightforward manner under this framework. This feature provides direct QoS control at the flow level to the OpenFlow controller by careful programming of the meters.

As seen in Fig. 5.15 there are meter-level counters and meter-band-level counters. The meter counters are updated for all packets processed by the meter and the per-meter-band counters are updated only when that particular band is selected. The dual level of counters is key in that presumably the majority of the packets processed by the meter suffer no enforcement, but their passage through the meter must be recorded in order to track the current measured rate. The reader should understand now that when a packet is processed by a band, it has exceeded a bandwidth threshold, which results in some kind of enforcement (e.g., dropping of the packet or marking it as *drop eligible* via DSCP remark).

5.6.4 PER CONNECTION EVENT FILTERING

We saw in V.1.2 the introduction of the notion of multiple controllers. Multiple controllers, able to take on different roles, achieve an improved level of fault tolerance and load balancing. The earlier discussion explained that the switches would only provide replies to the controller that initiated the request that elicits that reply. This, however, did not prescribe how to damp down the communication of asynchronous messages from the switch to its family of controllers. The idea that the different controllers have asymmetrical roles was not as effective as it could have been if all controllers must receive the same kind and quantity of asynchronous notifications from the switches. For example, a slave controller may not want to receive all types of asynchronous notifications from the switch. V.1.3 introduces a SET_ASYNC message that allows the controller to specify which sorts of async messages it is willing to receive from a switch. It additionally allows the controller to filter out certain reason codes that it does not wish to receive. A controller may use two different filters, one for the master/equal role and another for the slave role. Note that this filter capability exists in addition to the ability to enable or disable asynchronous messages on a per-flow basis. This new capability is controller-oriented rather than flow-oriented.

5.6.5 AUXILIARY CONNECTIONS

We have already seen that earlier releases of OpenFlow allowed some parallelism in the switch-controller channel via the use of multiple controllers. In this way, multiple parallel communication *channels* existed from a single switch to multiple controllers. V.1.3 introduces an additional layer of parallelism by allowing multiple *connections* per communications channel. That is, between a single controller and switch, multiple connections may exist. We depict these multiple connections between the switch and the MASTER controller in Fig. 5.16. The figure shows that there are also multiple channels connecting the switch to different controllers. The advantage provided by the additional connections on a channel lies in achieving greater overall throughput between the switch and the controller. Because of the flow-control characteristics of a TCP connection, it is possible that the connection be forced to quiesce (due to TCP window closing or packet loss) when there is actually bandwidth available on the physical path(s) between the switch and controller. Allowing multiple parallel connections allows the switch to take advantage of that. The first connection in the channel is specified to be a TCP connection. The specification indicates that other, nonreliable connection types,

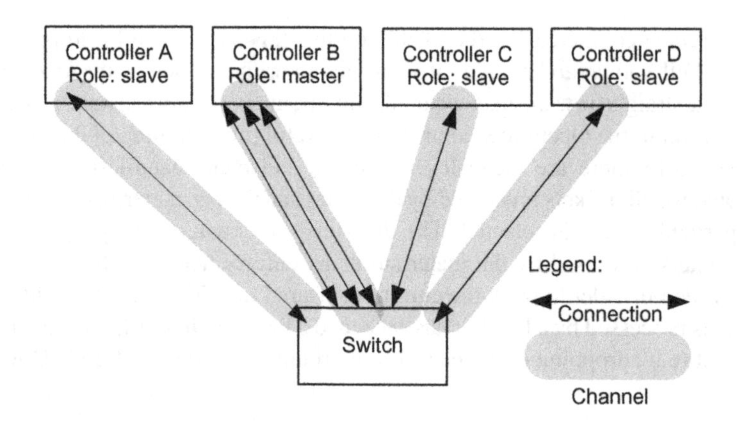

FIG. 5.16

Multiple connections and multiple channels from a switch.

such as UDP, may be used for the secondary connections, but the specification is fairly clear that the problems of potential data loss and sequencing errors that can arise from the use of such alternative transport layers is beyond the scope of the specification.

If bandwidth between the controller and switch becomes a constraint, it is most likely due to a preponderance of data packets. Some loss of data packets can be tolerated. Thus, the primary intended use of the auxiliary connections is to transmit and receive data packets between the switch and controller. This presumes that any control messages would be sent over the reliable, primary connection.

One example of the utility of auxiliary connections is that when a switch has many PACKET_IN messages to send to the controller, this can create a bottleneck due to congestion. The delay due to this bottleneck could prevent important OpenFlow control messages, such as a BARRIER_REPLY message, from reaching the controller. Sending PACKET_IN data messages on the UDP auxiliary connection will obviate this situation. Another possible extension to this idea is that the OpenFlow pipeline could send UDP-based PACKET_IN messages directly from the ASIC to the controller without burdening the control code on the switch CPU.

5.6.6 COOKIES IN PACKET-IN

The straightforward handling of a PACKET_IN message by the controller entails performing a complete packet matching function to determine what existing flow this packet relates to, if any. As the size of commercial deployments of OpenFlow grows, performance considerations have become increasingly important. Accordingly, the evolving specification includes some features that are merely designed to increase performance in high bandwidth situations. Multiple connections per channel, discussed above, is such an example. In the case of PACKET_IN messages, it is somewhat wasteful to require the controller to perform a complete packet match for every PACKET_IN message, considering that this look-up has already just occurred in the switch and, indeed, is the very reason the PACKET_IN message is being sent to the controller (that is, either a specific instruction to forward this packet to the controller was encountered or a table-miss resulted in the packet being handed off to the controller).

This is particularly true if this is likely to happen over and over for the same flow. In order to render this situation more efficient, V.1.3 allows the switch to pass a *cookie* with the PACKET_IN message. This cookie allows the switch to cache the flow entry pointed to by this cookie and circumvent the full packet matching logic. Such a cookie would not provide any efficiency gain the first time it is sent by the switch for this flow, but once the controller cached the cookie and pointer to the related flow entry, considerable performance boost can be achieved. The actual savings accrued would be measured by comparing the computational cost of full packet header matching versus performing a hash on the cookie. As of this writing, we are not aware of any study that has quantified the potential performance gain.

The switch maintains the cookie in a new field in the flow entry. Indeed, the flow entries in V.1.3 have expanded considerably as compared to the basic flow entry we presented in Fig. 5.5. The V.1.3.0 flow entry is depicted in Fig. 5.17. The header, counter and actions fields were present in Fig. 5.5 and were covered earlier in Section 5.3.2. The priority field determines where this particular flow entry is placed in the table. A higher priority places the entry lower in the table such that it will be matched before a lower priority entry. The timeouts field can be used to maintain two timeout clocks for this flow. Both an idle timer and a hard timeout may be set. If the controller wishes for the flow to be removed after a specified amount of time without matching any packets, then the idle timer is used. If the controller wishes for the flow to exist only for a certain amount of time regardless of the amount of traffic, the hard timeout is used and the switch will delete this flow automatically when the timeout expires. We provide an example of such a timeout condition in Section 5.3.6. We covered the application of the cookie field earlier in this section.

5.6.7 PROVIDER BACKBONE BRIDGING TAGGING

Earlier versions of OpenFlow supported both VLAN and MPLS tagging. Another WAN/LAN technology, known as *Provider Backbone Bridging* (PBB) is also based on tags similar to VLAN and MPLS. PBB allows for user LANs to be layer two bridged across provider domains, allowing complete domain separation between the user layer two domain and the provider domain via the use of *MAC-in-MAC* encapsulation. For the reader interested in learning more about PBB, we provide a reference

Header fields	Field value
Priority	Field value
Counters	Field value
Actions/instructions	Field value
Timeouts	Field value
Cookie	Field value

FIG. 5.17

V.1.3 flow entry.

to the related standard in [10]. This is a useful technology already supported by high-end commercial switches, and support for this in OpenFlow falls out easily as an extension to the support already provided for VLAN tags and MPLS shim headers.

5.6.8 EXAMPLE: ENFORCING QOS VIA METER BANDS

Fig. 5.18 depicts the new V.1.3 meter table as well as the group table that first appeared in V.1.1. We see in the figure that the incoming packet is matched against the second flow entry in the first flow table, and that the associated instructions direct the packet to Meter 3. The two dashed lines emanating from the meter table show that, based on the current measured bandwidth of the port selected as the output port (port *N* in this case), the packet may be dropped as exceeding the bandwidth limits, or it may be forwarded to port *N* if the meter does not indicate that bandwidth enforcement is necessary. As we have explained earlier, if the packet passes the meter control without being dropped or marked as dropped-eligible, it is *not* automatically forwarded to the output port, but may undergo any further processing indicated by the instructions in its packet processing pipeline. This could entail being matched against a higher-numbered flow table with its own instructions and related actions.

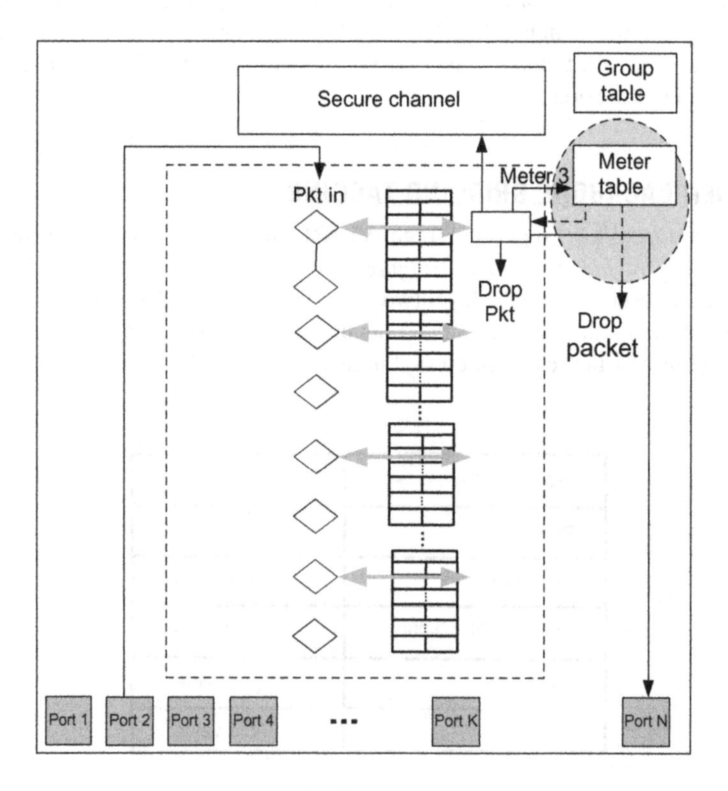

FIG. 5.18

Use of meters for QoS control in V.1.3.

DISCUSSION QUESTION:

Explain why, in Fig. 5.18, if the meter table does not drop the packet, the alternative dashed line indicates that the packet is not sent directly to the output port but rather forwarded back through the packet processing logic.

5.7 OpenFlow 1.4 ADDITIONS

OpenFlow V.1.4.0 [11] was released on October 14, 2013. Table 5.5 lists the major new features added in this release. In the case of the more significant features, the table description references one of the following sections where we provide more explanation of the feature. For other, more minor features, such as new error codes or clarification of the protocol, the table refers the interested reader to [11]. As of this writing, one switch manufacturer, Pica8, has already announced support for OpenFlow 1.4 [12]. The primary drivers behind this decision were the OpenFlow 1.4 support for bundles, eviction and vacancy events, and improved support for multiple controllers. Another notable feature of OpenFlow 1.4 was the new port descriptor which contained fields to support optical ports. We will provide further details on these more important features of OpenFlow 1.4 in the following sections.

5.7.1 BUNDLES

The bundle feature described in [11] provides enhanced transactional capabilities to the OpenFlow controller. In earlier versions, related, but more primitive control came from using a BARRIER_REQUEST

Table 5.5 New Features Added in OpenFlow 1.4

Feature Name	Description
More Extensible Wire Protocol	Standardize on TLV format: See [11]
More Descriptive Reasons for PACKET-IN	Clarification: See [11]
Optical Port Properties	See Section 5.7.4
Flow-removed reasons for meter delete	Clarification: See [11]
Flow monitoring	See Section 5.7.3
Role status events	See Section 5.7.3
Eviction	See Section 5.7.2
Vacancy Events	See Section 5.7.2
Bundles	See Section 5.7.1
Synchronized Tables	See Section 5.7.5
Group and Meter change notifications	See Section 5.7.3
Error code for bad priority	New Error Codes: See [11]
Error code for Set-async-config	New Error Codes: See [11]
PBB UCA header field	See [11]
Error code for duplicate instruction	New Error Codes: See [11]
Error code for multipart timeout	New Error Codes: See [11]
Change default TCP port to 6653	Official IANA port number assigned. See [11]

message whereby the switch must complete execution of all commands received prior to the BARRIER_REQUEST before executing any commands received after it. The OpenFlow 1.4 concept of a bundle is a set of OpenFlow messages that are to be executed as a single transaction. No component message should be acted upon by the switch until the entire set of commands in the bundle has been received and *committed*. The controller uses the new BUNDLE_CONTROL message to create, destroy or commit bundles. OpenFlow 1.4 also defines the two new commands BUNDLE_ADD_MESSAGE and BUNDLE_FAILED to add messages to a bundle and to report the failure of a bundle operation, respectively.

5.7.2 EVICTION AND VACANCY EVENTS

On physical switches, flow tables have very finite capacity and running out of room in a flow table on a switch can cause significant problems for the controller managing that switch. OpenFlow 1.4 contemplates a notion of *vacancy* whereby the switch can inform the controller that a critical level of table capacity has been reached. The motivation behind this is that by being given advance notice that the table is reaching capacity, the controller may be able to take logical steps to prune less important table entries and thus allow continued smooth operation of the switch. OpenFlow 1.4 allows the controller to act on this in a pro-active fashion by denoting certain flows as candidates for automatic eviction by the switch in the event that a vacancy threshold has been reached. Passing a vacancy threshold is communicated to the controller via a TABLE_STATUS message that includes reason codes supporting the vacancy feature. Parameters are provided to set the vacancy threshold itself as a means to quiesce flapping vacancy messages due to too-frequent passing of the threshold. Similarly, the eviction mechanism described in [11] allows configuration of various levels of importance to a flow entry to provide an orderly hierarchy of auto-eviction decisions taken by the switch.

DISCUSSION QUESTION:

Why are the eviction and vacancy events features more relevant to a ASIC-based switch than to a purely software-based switch?

5.7.3 ENHANCED SUPPORT FOR MULTIPLE CONTROLLERS

While OpenFlow 1.2 already included the concept of master and slave controllers, when one controller set itself to master, the previous master controller was not informed of this. In [11], a role status message is defined that allows the switch to inform a controller of a change to its master-slave role.

In addition, OpenFlow 1.4 permits definition of a set of monitors each of which may specify a subset of the flow tables on that switch. When a flow entry is added, deleted or changed within that subset, a notification is sent to the particular controller associated with that monitor. This capability is generically described as *flow monitoring* in [11].

Similarly, V.1.4 supports *group and meter change notifications* whereby the other controllers managing a switch are notified of group and meter changes effected by one controller.

5.7.4 OPTICAL PORT SUPPORT

The port descriptor is augmented in [11] to permit configuration and reporting on a number of optical-specific properties. These include fields for the transmit power of a laser, its transmit and receive frequencies or wavelengths, whether the transmitter and receiver are tunable, and whether the transmit power is configurable. These fields permit the configuration and monitoring of both Ethernet optical ports or the ports on optical circuit switches. Note that while these OpenFlow 1.4 extensions are a start for providing full OpenFlow support for optical ports, the ONF's *Optical Transport Working Group* (OTWG) has continued work in this area. The first specification emanating from this working group is discussed below in Section 5.10.

5.7.5 FLOW TABLE SYNCHRONIZATION

When two flow tables are synchronized, this signifies that when one table is changed the other table will be automatically updated. This is more complicated than it may first appear since the modification to one table may differ from that made to the second table. For example, if the first table matches on the source Ethernet address, the synchronization with the second table may transpose this to the destination Ethernet address. This is precisely the case for the example provided in [11] for flow table synchronization, which is a MAC learning bridge. In this example, when a new Ethernet address is learned on a port, an entry needs to be created in the first table (the learning table) with an idle timer so that inactive addresses can be later pruned. A *forwarding* table could be synchronized with the learning table and the newly learned (timed-out) addresses would be added (deleted) in that second table. The exact details of how the synchronization would occur are not specified in [11].

5.8 OpenFlow 1.5 ADDITIONS

OpenFlow V.1.5.0 [13] was released on December 19, 2014. Table 5.6 lists the major new features added in this release. In the case of the more significant features, the table description references one of the following sections where we provide more explanation of the feature. For other, more minor features related to consistency, flexibility or clarification of the protocol, the table refers the interested reader to [13]. A cursory inspection of the table of contents of the V.1.5 specification leads to the conclusion that there are far more additions to V.1.4 than is in fact the case. This is due to the fact that the document has been reorganized and section labels have been introduced to improved readability. There are, however several important new features introduced which we cover in the sections that follow.

5.8.1 ENHANCED L4–L7 SUPPORT

Even prior to OpenFlow 1.5, directing a packet to a logical port could be used to insert a network service into the flow. With the addition of *port recirculation* in V.1.5, it is now possible for the packet to be directed back to the OpenFlow packet processor after completing the logical port processing. This can enable *service chaining*, which is the sequencing of L4–L7 services such as firewalls or load balancers. This would occur by effectively chaining multiple logical ports such that a packet would be processed by each in sequence in the same switch.

Table 5.6 New Features Added in OpenFlow 1.5

Feature Name	Description
Egress Tables	See Section 5.8.3
Packet Type Aware Pipeline	See Section 5.8.4
Extensible Flow Entry Statistics	See Section 5.8.8
Flow Entry Statistics Trigger	See Section 5.8.8
Copy-Field action to copy between two OXM fields	See Section 5.8.2
Packet Register pipeline fields	See Section 5.8.2
TCP flags matching	See Section 5.8.1
Group Command for selective bucket operation	Flexibility: See [13]
Allow set-field action to set metadata field	See Section 5.8.1
Allow wildcard to be used in set-field action	See Section 5.8.1
Scheduled Bundles	See Section 5.8.5
Controller Connection Status	See Section 5.8.6
Meter Action	Flexibility: See [13]
Enable setting of all pipeline fields in PACKET_OUT	See Section 5.8.7
Port properties of all pipeline fields	See Section 5.8.7
Port Property for recirculation	See Section 5.8.1
Clarify and improve BARRIER	Consistency and clarification: See [13]
Always generate port status on port config change	See Section 5.8.6
Make all Experimenter OXM-IDs 64 bits	Consistency and clarification: See [13]
Unified requests for group, port and queue multiparts	Consistency and clarification: See [13]
Rename some types for consistency	Consistency and clarification: See [13]
Specification Reorganization	Consistency and clarification: See [13]

V.1.5 also offers the capability to maintain flow state via the storing and accessing flow metadata that persists for the lifetime of the flow, not just the current packet. Another new feature of V.1.5 that enhances L4 support is the ability to detect the start and end of TCP connections. This capability comes from the newly added capability of matching on the flag bits in the TCP header.

5.8.2 PIPELINE PROCESSING ENHANCEMENTS

The new *copy-field* action permits the copying of one header or pipeline field into another header or pipeline field. Prior to V.1.5, such fields could only be set to statically programmed values via the *set-field* action. *Packet registers*, new in V.1.5, provide scratchpad space that may be used to pass nonstandard information through the pipeline steps. These V.1.5 features, along with other, more minor ones appearing in Table 5.6, provide new robustness to the transformations that may occur to a packet as it passes through an OpenFlow pipeline.

5.8.3 EGRESS TABLES

V.1.5 introduces the notion of *egress* tables. In earlier versions, a flow table was consulted as part of the matching process on an incoming packet. This traditional role of the flow table is now referred to as an *ingress* table. Ingress table processing is still a mandatory part of the OpenFlow pipeline, whereas egress tables may or may not exist. In a switch where such two-stage pipeline processing is configured, matching occurs against the egress tables once the output port(s) have been identified in the ingress processing. An example would be where the ingress processing consults a group table which results in the packet being forwarded out three output ports. By including egress tables, three distinct packet transformations may thus be performed in the separate contexts of each of the three output ports. This allows much greater flexibility in encapsulation than was formerly possible.

5.8.4 FITNESS FOR CARRIER USE

While a number of the V.1.5 features mentioned above improve OpenFlow's suitability in a carrier environment, most salient is the support of other data plane technologies beyond Ethernet. V.1.5 includes a new OXM pipeline field that identifies the packet type. The additional packet types supported in V.1.5 are IPV4 (no header in front), IPV6 (no header in front), no packet (e.g., circuit switch), and EXPERIMENTER.

5.8.5 BUNDLE ENHANCEMENTS

We introduced the concept of bundles in Section 5.7.1. V.1.5 now allows for an execution time to be specified with the bundle so that the bundle may be delivered to the switch in advance of the time that the switch should begin acting on the messages contained therein. This new version of OpenFlow also allows a controller to ask the switch to report its bundle capabilities, so that the controller will know what set of bundle features it may invoke on that switch.

5.8.6 ENHANCED SUPPORT FOR MULTIPLE CONTROLLERS

Some new features in V.1.5 are specifically targeted at facilitating more than one controller managing the same switch. One of these is *controller connection status*. This allows one controller to ask the switch the status of its connections to *any* controller, thereby affording one controller knowledge of other controllers managing the same switch. Another weakness in earlier versions was that when a port status was changed via OpenFlow, no port status change message was sent to the controller. In a single controller environment this is perfectly reasonable since the one controller would be aware that it itself effected the port status change. In a multicontroller environment, however, the remaining controllers would not be aware of the port status change which is potentially problematic. In V.1.5 a port status change message will always be sent to all controllers connected to the switch where the port was altered.

5.8.7 ENHANCED SUPPORT FOR TUNNELS

Starting with V.1.5, a logical port's properties allow specification of which OXM pipeline fields should be provided for a packet being sent to that port as well as which OXM fields should be included when a packet is transmitted from that logical port. This can provide enhanced tunnel support since the pipeline field *tunnel_id* can be used to hold the metadata associated for the encapsulation used by a tunnel. In support of this, V.1.5 also includes the ability to set *all* pipeline fields in the PACKET_OUT message, whereas earlier versions only supported the setting of the *in_port* field. In particular, should a logical port's V.1.5 properties denote that it *consumes* (i.e., requires) the *tunnel_id* field, the packet forwarding logic sending to that logical port would be able set the *tunnel_id* field thus passing information relevant to the aforementioned tunnel encapsulation.

5.8.8 ENHANCEMENTS TO FLOW ENTRY STATISTICS

Flow entry statistics are represented in a more extensible format starting in V.1.5. Statistic fields supported in earlier versions such as flow duration, and counters for flows, packets and bytes are now expressed in standard TLV format. Flow idle time is now captured as a standard statistic. In addition, support for arbitrary experimenter-based statistics is included. Another important performance enhancement pertaining to flow entry statistics is the addition of a trigger capability. This eliminates the overhead of the controller having to poll the switch for statistics and enables the switch to proactively send statistics to the controller on the basis of a trigger. This trigger may be based on either a time period or on the basis of a statistic reaching a certain threshold value.

5.8.9 SUMMARY

While OpenFlow V.1.3 provided the flexibility to handle most challenges posed by the data center, it did not meet the much greater flexibility required by network operators in the WAN [14]. The changes encompassed by V.1.4 and V.1.5 discussed above have provided many features needed for such WAN deployments. Commercial adoption of V.1.4 and V.1.5 has so far, however, been weak. Part of this stems from persistent problems achieving interoperability between mixes of controllers and switches from different manufacturers that all claim compliance with the standards. We discuss some of the ONF's approaches to address this issue in the next section.

5.9 IMPROVING OpenFlow INTEROPERABILITY

The move away from the single flow table of OpenFlow 1.0 to multiple flow tables, specified in OpenFlow 1.1, and implemented by most vendors in OpenFlow 1.3 was instigated in order to open up the rich potential of SDN to an ever increasing spectrum of applications. The single, very flexible flow table did not map well to the hardware pipelines available in advanced switching silicon [15]. While the chaining of multiple flow tables into a pipeline does indeed more closely approximate real hardware implementations, differences between these implementations required that any application wishing to use these detailed features needed to be developed with intimate knowledge of those hardware details. One reason for the difference in these implementations is that support for many OpenFlow 1.3 features is considered optional, further confusing the task for the application developer. The situation was such that applications intended for use on hardware switches needed to be developed specifically for those switches. While it was possible to implement uniform OpenFlow support on software-based switches, limiting an application to such switches usually reduces its usefulness.

In order resolve this interoperability problem, the ONF has espoused two new abstractions that are intended to facilitate interoperation between OpenFlow 1.3 switches. These two abstractions are *Table Type Patterns* (TTPs) and *Flow Objectives.* We discuss these in the following sections.

5.9.1 TABLE TYPE PATTERNS

In 2014, the ONF published the first TTP specification [16]. A TTP is a formal and detailed description of a complete set of logical switch behavior [15]. Many *OpenFlow Logical Switches* (OFLSs) may coexist inside a given physical switch. From an application developer's perspective, an OFLS possesses the switching capabilities required by a particular application. From the switch vendor's perspective, the behavior of an OFLS is described succinctly in the OpenFlow specification terminology of the TTP. The fact that a particular TTP is succinct and limited to one application or class of applications makes it feasible for the switch vendor to confirm support or nonsupport for that particular OFLS. As the OFLS is described by a TTP, the TTP becomes the language that the controller and switch can use to confirm or deny the existence of a needed set of capabilities.

The formal language to describe a TTP is described in [15] and is based on *JavaScript Object Notation* (JSON). A very useful sample TTP describing an OFLS implementing a simple L2 and L3 forwarding switch is provided in [15].

5.9.2 FLOW OBJECTIVES

While TTPs provide a uniform way of describing a particular pipeline needed in a multi-table OpenFlow-enabled switch, the language of TTPs still requires detailed knowledge of the OpenFlow details that implement that pipeline. Flow objectives are intended to abstract multitable OpenFlow-level actions into generic application *objectives* [17]. These services-oriented objectives describe relatively high-level switching functions needed by an application but implemented via different pipelines on different physical switches. In Fig. 5.19 we show where flow objectives are defined within the *Open Network Operating System* (ONOS) embedded within the *Atrium* open source solution provided by the ONF [18]. Examples of these services-oriented objectives would include a filtering objective, a forwarding objective, and a next-hop objective.

The goal of exposing the flow objective abstraction is to permit application developers to build applications that benefit from scalable, multitable OpenFlow-based switching platforms without being aware of the details of the OpenFlow pipelines that would implement the needed services. As shown in Fig. 5.19, the pipeline-specific drivers architecturally below the flow objectives mask the differences between the different pipelines allowing the application developer to specify services at a more abstract level than using OpenFlow terminology.

DISCUSSION QUESTION:

Both TTPs and flow objectives are intended to enhance interoperability between OpenFlow implementations and as a result increase overall adoption of OpenFlow. In what ways do these two concepts differ in their approach to accomplishing these goals? In what way are they similar?

5.10 OPTICAL TRANSPORT PROTOCOL EXTENSIONS

The OpenFlow *Optical Transport Protocol Extensions* [19] was released on March 15, 2015. This specification builds upon the *port attribute extensibility* defined in OpenFlow 1.4. While that V.1.4 extension does facilitate the configuration and reporting of optical-specific port parameters, there are a number of areas that remain to be addressed to provide for robust OpenFlow support of practical optical networking. We discuss these below.

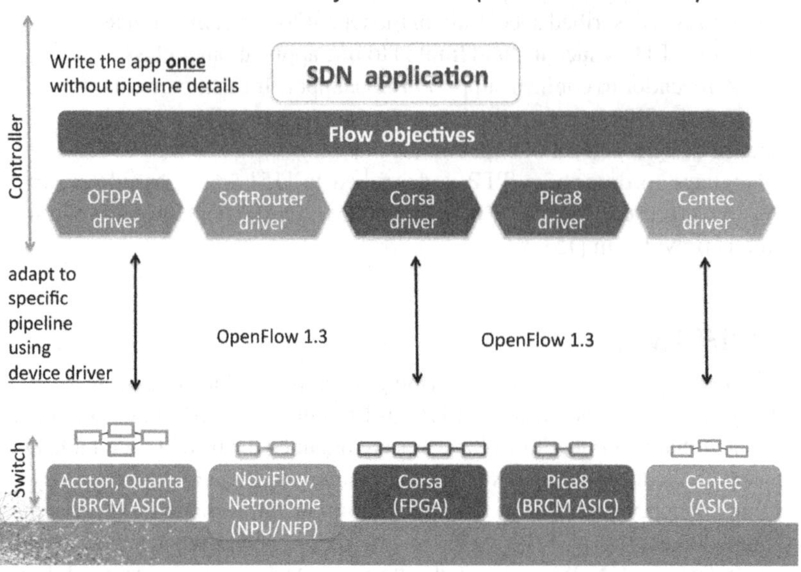

FIG. 5.19

Flow objectives abstraction.

Reproduced with permission from Saurav D. ATRIUM Open SDN Distribution, SDN Solutions Showcase, ONS 2015, Dusseldorf. Retrieved from: https://www.opennetworking.org/images/stories/news-and-events/sdn-solutions-showase/Atrium-ONS-Live.pdf.

5.10.1 MATCH/ACTION SUPPORT

In packet switching, much of the control functionality can be achieved by the control plane injecting control packets into the data plane. For example, knowledge about the adjacencies that together form the topology of a network can be collected by the exchange of such control packets between neighboring control planes. This method does not work with optical switching where the switches cannot simply inject packets in-line.

Similarly, optical switches are not packet switches, they are circuit switches. Implicit in this distinction, an OpenFlow-enabled circuit switch will not typically inspect packet headers to use as match criteria for flow tables, and thus not use that information to take actions such as forward or drop. Instead, the circuit switch will base these match and action decisions for a signal on the basis of an *Optical Channel* (OCh) characteristic such as wavelength for L0 switching or on the basis of an *Optical Data Unit* (ODU) characteristic like ODU tributary slot for L1 switching. The ODU tributary slot would identify the circuit by its position in time whereas individual physical optical channels are distinguished by transmission on different wavelengths. In either case, identification of the circuit permits a flow table-based set of rules to prescribe what action the switch should take for this circuit. Note that in this *Optical Transport Network* (OTN) context, we use the terms signal, channel, and circuit interchangeably.

5.10.2 OPTICAL PORT ATTRIBUTE SUPPORT

In [19] the OTWG defines *OTN port description extensions*. These extensions are used to identify and define the port types for L0 and L1 circuit switching control. More specifically, the extensions are used to specify and report on the characteristics of *Optical Transmission Section* (OTS), *Optical Physical Section* (OPS) and *Optical Channel Transport Unit* (OTU) ports. Examples of a characteristic that would be specified would be the specific layer class of an OTN signal. Examples of these layer classes are:

- OCH (OCh layer signal type)
- ODU (ODU layer signal type)
- ODUCLT (ODU Client layer signal type).

Within each of these layer classes, there are many possibilities for the specific signal type. This specific signal type is also specified in the OTN extensions. There are numerous other OTN-specific fields defined in these extensions and it is beyond the scope of this book to delve into the details of optical networking. We refer the reader to [19] for further details.

In [19] two methods are provided by which these OTN port description extensions may be passed between the controller and the switch, one based on OpenFlow 1.3 and one based on OpenFlow 1.4. One of the changes in OpenFlow 1.4 was extension of the basic port descriptor to provide support for a broader range of port types. While this expanded port descriptor provided a richer set of parameters for describing a port, it still lacks the full richness required to fully describe an OTN port. Thus, in OpenFlow 1.4 and beyond, the OTN extensions are captured by a combination of the extended port descriptor plus a set of OTN extensions defined in [19]. For OpenFlow 1.3, Ref. [19] specifies an experimenter multipart message whose body will contain the OpenFlow 1.4 port descriptor followed by those OTN extensions. (Note that these OTN extensions are sometimes referred to as the OTWG OpenFlow 1.5 OTN extensions.)

5.10.3 ADJACENCY DISCOVERY

Since optical switches cannot inject control packets in-line into the data plane to pass control and topology information through the network as packet switches currently do, other mechanisms are necessary to provide important basic features such as adjacency discovery. One proposed mechanism is based on using the control information known as the *Trail Trace Identifier* (TTI). Current ITU-sponsored standards provide that TTI be passed in-line within the data plane. In the optical world, this information is available to the control plane. The extensions contemplated in [19] allow for this TTI information to be made available via OpenFlow to the controller. The controller, in turn, can use this TTI information reported from different switches to synthesize adjacencies and thus network topology in much the way it does for packet switching. This represents a feasible mechanism for adjacency discovery for OpenFlow enabled optical circuit switches.

5.10.4 FUTURE WORK NEEDED

A number of areas that were identified as important for OpenFlow support of OTN require further specification than that provided in [19]. These include:

- Support for carrier reliability mechanisms for the data plane, in particular support for OAM monitoring of OTN network links as well as support for rapid protection switching functions that guarantee recovery from link failure.

- Support for network elements handling multiple technology layers according to transport layering models.
- Support for a hierarchy of controllers. In such a hierarchy, OpenFlow messages are exchanged between parent and child controllers in a *Control Virtual Network Interface* (CVNI) context. This is important as the scale required to support carrier-grade optical networks will require large numbers of OpenFlow controllers arranged in such a hierarchy.

As well as the future work identified in [19], the Calient corporation has identified areas that are urgently needed to foster the use of OpenFlow in optical circuit switches. These additional proposed extensions are described in [20].

DISCUSSION QUESTION:

Could OpenFlow-enabled switches support optical ports without the extensions described in Section 5.10? What major new class of OpenFlow-enabled switches is contemplated via these extensions?

5.11 OpenFlow LIMITATIONS

As OpenFlow remains in a state of rapid evolution, it is difficult to pin down precise limitations as these may be addressed in subsequent releases. One limitation is that the currently defined match fields are limited to the packet header. Thus, *Deep Packet Inspection*, (DPI) where fields in the packet's payload may be used to distinguish flows, is not supported in standard OpenFlow. Nonetheless, the EXPERIMENTER modes that are permitted within OpenFlow do open the way for such application-layer flow definition in the future. Secondly, some OpenFlow abstractions may be too complex to implement directly in today's silicon. This is unlikely to remain an insurmountable obstacle for long, however, as the tremendous momentum behind SDN is likely to give birth to switching chips that are designed explicitly to implement even OpenFlow's most complicated features. Another limitation is the possible processing delay if there is no matching entry in the OpenFlow switch's flow table, in which case processing the packet mandates that it be sent to the controller. While this is just a default behavior and the switch can be easily programmed to handle the packet explicitly, this potential delay is inherent to the OpenFlow paradigm. There are also a number of possible security vulnerabilities introduced by OpenFlow, such as (1) *Distributed Denial of Service* (DDoS) risks that can make the controller unavailable, and (2) failure to implement switch authentication in the controller. A brief discussion on these and other potential OpenFlow security vulnerabilities can be found in [21].

If we cast a wider net and consider limitations of all of Open SDN, there are a number of other areas to consider. We discuss these in Section 6.1.

5.12 CONCLUSION

In this chapter we have attempted to provide the reader with a high-level understanding of the general OpenFlow framework. This covered the protocol that an OpenFlow controller uses to configure and control an OpenFlow switch. We have also presented the switch-based OpenFlow abstractions that must be implemented on a physical switch as an interface to the actual hardware tables and forwarding engine of the switch in order for it to behave as an OpenFlow switch. For the reader interested in a

Table 5.7 OpenFlow Protocol Constant Classes

Prefix	Description
OFPT	OpenFlow Message Type
OFPPC	Port Configuration Flags
OFPPS	Port State
OFPP	Port Numbers
OFPPF	Port Features
OFPQT	Queue Properties
OFPMT	Match Type
OFPXMC	OXM Class Identifiers
OFPXMT	OXM Flow Match Field Types for Basic Class
OFPIT	Instruction Types
OFPAT	Action Types
OFPC	Datapath Capabilities and Configuration
OFPTT	Flow Table Numbering
OFPFC	Flow Modification Command Type
OFPFF	Flow Modification Flags
OFPGC	Group Modification Command Type
OFPGT	Group Type Identifier
OFPM	Meter Numbering
OFPMC	Meter Commands
OFPMF	Meter Configuration Flags
OFPMBT	Meter Band Type
OFPMP	Multipart Message Types
OFPTFPT	Flow Table Feature Property Type
OFPGFC	Group Capabilities Flags

deeper understanding of these issues, whether out of sheer academic interest or a need to implement one of the versions of OpenFlow, there is no substitute for reading the OpenFlow specifications cited in this chapter, especially [13], as it is generally an updated superset of its predecessors. These specifications are, however, very detailed and are not especially easy for the OpenFlow novice to follow. We believe that this chapter will serve as an excellent starting point and guide for that reader who needs to delve into the lengthy specifications that comprise the OpenFlow releases covered here. In particular, these specifications contain a plethora of structure and constant names. To aid in reading the specifications, we draw your attention to Table 5.7 in this chapter as a quick reference and aid to understanding to what part of the protocol a specific reference pertains.

REFERENCES

[1] OpenFlow Switch Specification, Version 1.0.0 (Wire Protocol 0x01). Open Networking Foundation, December 31, 2009. Retrieved from: https://www.opennetworking.org/sdn-resources/onf-specifications.

[2] IEEE Standard for Local and Metropolitan Area Networks—Media Access Control (MAC) Bridges and Virtual Bridged Local Area Networks—Amendment 21: Edge Virtual Bridging. IEEE 802.1Qbg, July 2012. New York: IEEE; 2012.

[3] OpenFlow Switch Specification, Version 1.1.0 (Wire Protocol 0x02). Open Networking Foundation, December 31, 2009. Retrieved from: https://www.opennetworking.org/sdn-resources/onf-specifications.
[4] Davie B, Doolan P, Rekhter Y. Switching in IP networks. San Francisco: Morgan Kaufmann; 1998.
[5] IEEE Standard for Local and Metropolitan Area Networks–Link Aggregation. IEEE 802.1AX, November 3, 2008. USA: IEEE Computer Society; 2008.
[6] OpenFlow Switch Specification, Version 1.2.0 (Wire Protocol 0x03). Open Networking Foundation, December 5, 2011. Retrieved from: https://www.opennetworking.org/sdn-resources/onf-specifications.
[7] IEEE Standard for Local and Metropolitan Area Networks—Media Access Control (MAC) Bridges and Virtual Bridged Local Area Networks. IEEE 802.1Q, August 2011. New York: IEEE; 2011.
[8] OpenFlow Switch Specification, Version 1.3.0 (Wire Protocol 0x04). Open Networking Foundation, June 25, 2012. Retrieved from: https://www.opennetworking.org/sdn-resources/onf-specifications.
[9] Shenker S, Partridge C, Guerin R. Specification of guaranteed quality of service. RFC 2212. Internet Engineering Task Force; 1997.
[10] Draft Standard for Local and Metropolitan Area Networks—Virtual Bridged Local Area Networks—Amendment 6: Provider Backbone Bridges. IEEE P802.1ah/D4.2, March 2008. New York: IEEE; 2008.
[11] OpenFlow Switch Specification, Version 1.4.0 (Wire Protocol 0x05). Open Networking Foundation, October 14, 2013. Retrieved from: https://www.opennetworking.org/images/stories/downloads/sdn-resources/onf-specifications/openflow/openflow-switch-v1.4.0.noipr.pdf.
[12] Oliver B. Pica8: First to adopt openflow 1.4; why isn't anyone else? Tom's ITPRO, May 2, 2014. Retrieved from: http://www.tomsitpro.com/articles/pica8-openflow-1.4-sdn-switches,1-1927.html.
[13] OpenFlow Switch Specification, Version 1.5.0 (Wire Protocol 0x06). Open Networking Foundation, December 19, 2014. Retrieved from: https://www.opennetworking.org/images/stories/downloads/sdn-resources/onf-specifications/openflow/openflow-switch-v1.5.0.noipr.pdf.
[14] Hunt G. The Quest for Dominance: OpenFlow or NETCONF for Networks Outside the Data Center? Network Matter, February 6, 2015. Retrieved from: http://networkmatter.com/2015/02/06/the-quest-for-dominance-openflow-or-netconf-for-networks-outside-the-data-center/.
[15] Dixon C. OpenFlow & Table Type Patterns in OpenDaylight's Next Release. SDX Central, August 11, 2014. Retrieved from: https://www.sdxcentral.com/articles/contributed/openflow-table-type-patterns-opendaylight-next-release-colin-dixon/2014/08/.
[16] OpenFlow Table Type Patterns, Version 1.0 (ONF TS-017). Open Networking Foundation, August 15, 2014. Retrieved from: https://www.opennetworking.org/images/stories/downloads/sdn-resources/onf-specifications/openflow/OpenFlowTableTypePatternsv1.0.pdf.
[17] Pitt D. Paving the way For SDN interoperability. InformationWeek Network Computing, October 9, 2015. Retrieved from: http://www.networkcomputing.com/networking/paving-the-way-for-sdn-interoperability-/a/d-id/1322562.
[18] Saurav D. ATRIUM Open SDN Distribution. SDN Solutions Showcase, ONS 2015, Dusseldorf. Retrieved from: https://www.opennetworking.org/images/stories/news-and-events/sdn-solutions-showcase/Atrium-ONS-Live.pdf.
[19] Optical Transport Protocol Extensions, Version 1.0 (ONF TS-022). Open Networking Foundation, March 15, 2015. Retrieved from: https://www.opennetworking.org/images/stories/downloads/sdn-resources/onf-specifications/openflow/Optical_Transport_Protocol_Extensions_V1.0.pdf.
[20] Optical Circuit Switch OpenFlow Protocol Extensions. Calient, June 2, 2015. Retrieved from: https://wiki.opendaylight.org/images/1/1d/OCS_OF_Protocol_Extensions_Rev._0.4.pdf.
[21] Benton K, Camp L, Small C. OpenFlow Vulnerability Assessment. HotSDN, SIGCOMM 2013, August 2013, Hong Kong. Retrieved from: http://conferences.sigcomm.org/sigcomm/2013/papers/hotsdn/p151.pdf.

CHAPTER

ALTERNATIVE DEFINITIONS OF SDN

6

There are likely to be networking professionals who dismiss Open SDN as much ado about nothing. To be sure, some of the criticism is a natural reaction against change. Whether you are a network equipment manufacturer or a customer and any part of the vast industry gluing those endpoints together, the expectation of disruptive change such as that associated with SDN is unsettling. There are indeed valid drawbacks of the Open SDN concept. In the first part of this chapter, we inventory the drawbacks most commonly cited about Open SDN. For each, we attempt to assess whether there are reasonable solutions to these drawbacks proposed by the Open SDN community.

In some cases, the disadvantages of Open SDN have been addressed by proposed alternative SDN solutions that do not exhibit these flaws. In Chapter 4 we introduced a number of alternative SDN technologies that propose SDN solutions that differ from Open SDN. In a field with so much revenue-generating opportunity, there may be a tendency by some vendors to jump on the SDN bandwagon and claim that their product is an SDN solution. It is therefore important that we have a way of determining whether or not we should consider a particular alternative a true SDN solution. Recall that in Section 4.1 we said that we characterize an SDN solution as possessing all of the five following traits: *plane separation*, *a simplified device*, *centralized control*, *network automation and virtualization*, and *openness*. As we discuss each of the alternative SDN technologies in the latter part of this chapter, we will evaluate it against these five criteria. Admittedly, this remains a bit subjective, as our litmus test is not absolute. Some of the alternatives do exhibit some of our defined SDN traits, but not all. In addition, we will evaluate whether or not each alternative suffers from the same drawbacks noted early in the chapter for Open SDN. Without providing any absolute judgment about winners or losers, our hope is that this SDN report card will help others make decisions about which, if any, SDN technology is appropriate for their needs.

6.1 POTENTIAL DRAWBACKS OF OPEN SDN

While the advent of the idea of Software Defined Networking has been met with enthusiasm by many, it is not without critics. They point to a number of shortcomings related to both the technology and its implementation. Some of the most vocal critics are the network vendors themselves. This is not surprising, since they are the most threatened by such a radical change in the landscape of networking. Some critics are skeptics who do not believe that this new technology will be successful, for various reasons, which we examine in this chapter.

Software Defined Networks. http://dx.doi.org/10.1016/B978-0-12-804555-8.00006-5

6.1.1 TOO MUCH CHANGE, TOO QUICKLY

In Chapter 3 we introduced the creation of the *Clean Slate* project at Stanford, where researchers were encouraged to consider what networking might be like if we started anew without the baggage of our legacy networks. Out of Clean Slate came the creation of OpenFlow and the basic concepts of Open SDN as defined in this book. Some network professionals feel that this approach is not practical since real-world customers do not have the luxury of being able to discard old technology wholesale and start fresh.

The following reasons are often cited as to why such a comprehensive change would negatively impact network equipment customers:

- **The Expense of New Equipment.** A complete change to the way networking is performed is too expensive because it requires a *forklift* change to the network. Large quantities of existing network equipment must be discarded, as new equipment is brought en masse into customer networking environments. *Counterargument: If this is a greenfield deployment then this point is moot as new equipment is being purchased anyway. If this is an existing environment with equipment being continually upgraded, it is possible to upgrade your network to SDN incrementally with a sound migration plan.*
- **Too Risky.** A complete change to the networking process is too risky because the technology is new, possibly untested, and is not backed by years of troubleshooting and debugging experience. *Counterargument: Continued use of legacy protocols also incurs risk due to the increased costs of administering them, the inability to easily automate the network, and the inability to scale virtual networks as needed. It is unreasonable to consider only the risks of migrating to SDN without also considering the risk of maintaining the status quo.*
- **Too Revolutionary.** There are tens of thousands of network engineers and IT staff personnel who have had extensive training in the networking technology that exists today. While it may be true that Open SDN has the potential to yield benefits in terms of lowered operational expenses, in the short term these expenses will be higher. This increased expense will in large part be due to the necessary training and education for network engineers and IT personnel. *Counterargument: The scale and growth of data centers has itself been revolutionary and it is not unreasonable to think that a solution to such a problem may need to be revolutionary.*

The risk associated with too much change too quickly can be mitigated by deploying Open SDN in accordance with a carefully considered migration plan. The network can be converted to Open SDN in stages, targeting specific network areas for conversion and rolling out the changes incrementally. Fig. 6.1 shows a simple example of such phasing. First, the SDN technology is tested in a lab

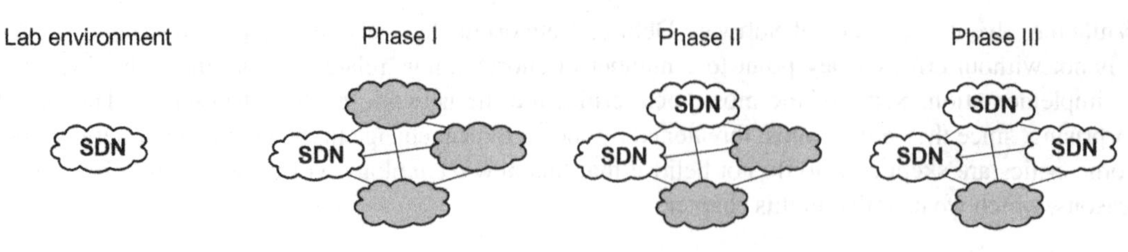

FIG. 6.1

Phased deployment of SDN.

environment. It is subsequently rolled out to various subnets or domains incrementally, based on timing, cost and sensitivity to change. This phased approach can help mitigate the cost issue as well as the risk. It also provides time for the re-training of skilled networking personnel. Nevertheless, for those areas where Open SDN is newly deployed, the change will be radical. If this radical change brings about radical improvements in an environment previously careening out of control, then this is a positive outcome. Indeed, sometimes radical change is the only escape from an out-of-control situation.

6.1.2 SINGLE POINT OF FAILURE

One of the advantages of distributed control is that there is no single point of failure. In the current model, a distributed, resilient network is able to tolerate the loss of a single node in the infrastructure and reconfigure itself to work around the failed module. However, Open SDN is often depicted as a single controller responsible for overseeing the operation of the entire network. Since the control plane has been moved to the controller, while the SDN switch can forward packets matching currently defined flows, no changes to the flow tables are possible without that single controller. A network in this condition cannot adapt to any change. In these cases, the controller can indeed become a single point of failure.

Fig. 6.2 shows a single switch failing in a traditional network. Prior to the failure, we see that two flows, $F1$ and $F2$, were routed via the switch that is about to fail. We see in the figure that upon detecting the failed node the network automatically uses its distributed intelligence to reconfigure itself to overcome the single point of failure. Both flows are routed via the same alternate route.

Fig. 6.3 shows a recovery from the failure of a single node in an SDN network. In this case, it is the intelligence in the SDN controller that reconfigures the network to circumvent the failed node. One of the premises of SDN is that a central controller will be able to deterministically reconfigure the network in an optimal and efficient manner, based on global knowledge of the network. Because of other information about the overall network that the SDN controller has at its disposal, we see in the figure that it reroutes $F1$ and $F2$ over different paths. One scenario that would cause this would be that the two flows have specific QoS guarantees that cannot be met if they are both placed on the same alternate path that was chosen in Fig. 6.2. Finding optimal paths for different flows is relatively straightforward within

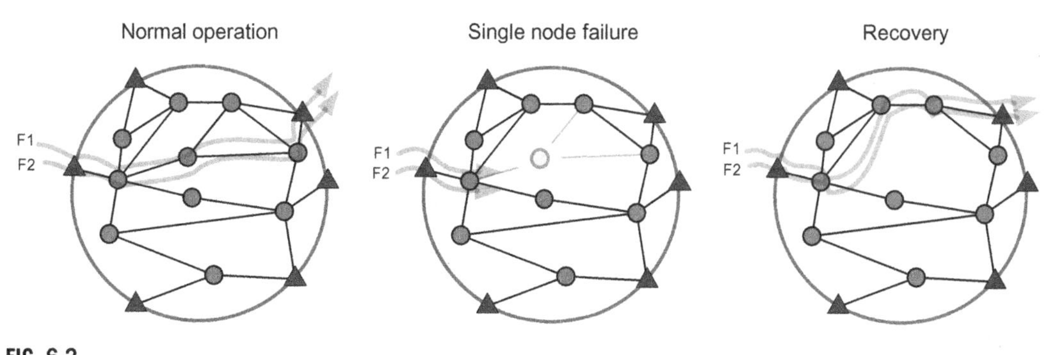

FIG. 6.2

Traditional network failure recovery.

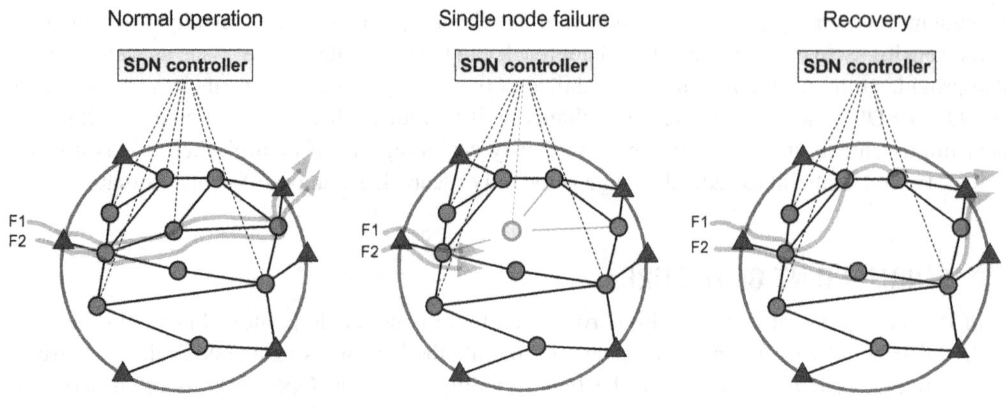

FIG. 6.3

SDN network failure recovery.

the Open SDN model, yet is very challenging to implement in traditional networking with autonomous devices. We revisit this topic via a specific use case in Section 9.1.2.

If, however, the single point of failure is the controller itself, the network is vulnerable, as shown in Fig. 6.4. As long as there is no change in the network requiring a flow table modification, the network can continue to operate without the controller. However, if there is any change to the topology, the network is unable to adapt. The loss of the controller leaves the network in a state of reduced functionality, unable to adapt to failure of other components or even normal operational changes. So, failure to the centralized controller could potentially pose a risk to the entire network. The SDN

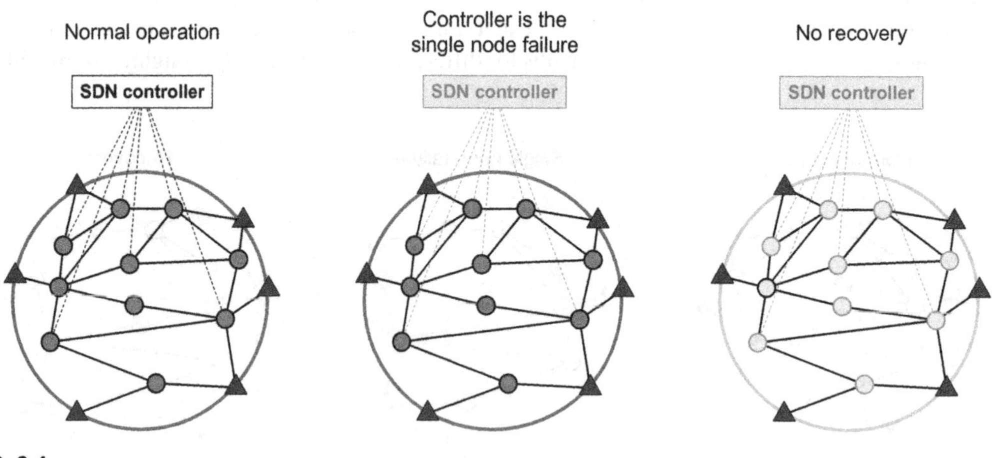

FIG. 6.4

SDN controller as single point of failure.

controller is vulnerable to both hardware and software failures, as well as to malicious attacks. Note that in addition to such negative stimuli such as failures or attacks, a sudden surge in flow entry modifications due to a sudden increase in network traffic could also cause a bottleneck at the controller, even if that sudden surge is attributable to an entirely positive circumstance such as a flash mob!

The SDN controller runs on a physical compute node of some sort, probably a standard off-the-shelf server. These servers, like any other hardware components, are physical entities, frequently with moving parts, and, as such, they are subject to failure. Software is subject to failure due to poor design, bugs, becoming overloaded due to scale, and other typical software issues which can cause the failure of the controller. A single SDN controller is an attractive target for malicious individuals wishing to attack an organization's networking infrastructure. *Denial of Service* (DoS) attacks, worms, bots, and similar malware can wreak havoc on an SDN controller which has been compromised, just as they have wreaked havoc in legacy networks for years.

Critics of SDN point to these as evidence of the susceptibility of SDN to catastrophic failure. If indeed the SDN controller is allowed to become that single point of failure, this does represent a structural weakness. Clearly, SDN designers did not intend that operational SDN networks be vulnerable to the failure of a single controller. In fact, the Open SDN pioneers described the concept of a *logically centralized controller*. In Open SDN, one has the flexibility to decide how many control nodes (controllers) the network has. This can range from a totally centralized situation with a single control node to one control node per device. The latter extreme may still theoretically fit the Open SDN paradigm, but it is not practical. The disparate control nodes will face an unnecessary challenge in trying to maintain a consistent global view of the network. The other extreme of the single controller is also not realistic. A reasonable level of controller redundancy should be used in any production network.

High availability controller with hardened links

SDN controllers must use *High Availability* (HA) techniques and/or redundancy. SDN controllers are not the only systems in the broad computer industry which must be highly available. Wireless LAN controllers, critical servers, storage units, even networking devices themselves, have for some time relied on HA and redundancy to ensure that there is no suspension of service should there be a failure of a hardware or software component. Fig. 6.5 shows an example of a well-designed controller configuration with redundant controllers and redundant controller databases. Redundancy techniques ranges from $N + 1$ approaches where a single hot-standby stands ready to assume the load of any of the N active controllers, to the more brute-force approach of having a hot-standby available for each controller. A version of hot-standby for each controller is to use HA designs internal to the controller hardware such that all critical components are redundant within that server (e.g., mirrored disks, redundant power supplies). Redundancy is also reflected by allowing the devices themselves to switch to a backup controller, as we discussed in Section 5.5.4.

Detecting complete failure can be relatively simple compared to ascertaining the correct operation of a functional component. However, instrumentation exists to monitor and evaluate systems to make sure they are working as intended and to notify network administrators if they begin to deviate from the norm. Thus, SDN controllers must have monitoring and instrumentation in place to ensure correct operation. Fig. 6.5 depicts a monitor which is constantly alert for changes in the behavior and responsiveness of the SDN controllers under its purview. When a problem arises, the monitor is able to take corrective action.

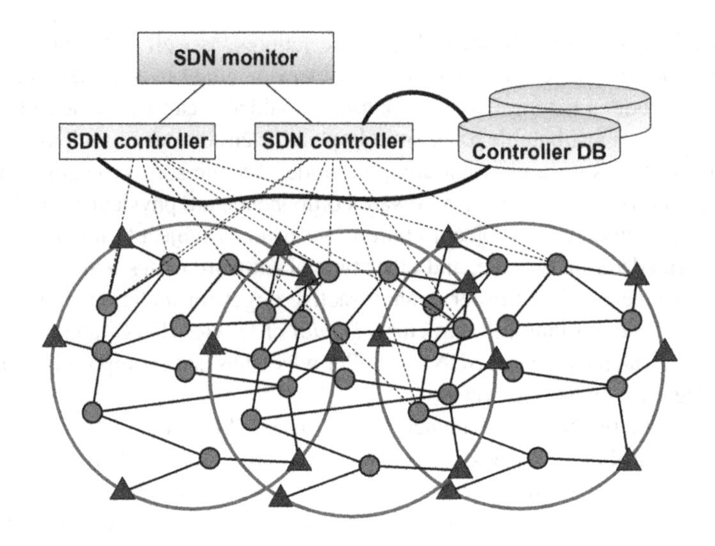

FIG. 6.5

Controller high availability and monitoring.

With respect to malicious attacks, our intuition leads us to the conclusion that by centralizing the control plane the network now has a vulnerable central point of control that must be protected against attack or the entire network is vulnerable. This is in fact true, and the strongest security measures available must be used to defend against such attacks. In practice, both the controller and the links to it would be hardened to attack. For example, it is common practice to use out-of-band links or high priority tunnels for the links between the devices and the controller. In any case, it is naive to assume that the distributed control system such as that used in legacy networks is immune to attack. In reality, distributed systems can present a large attack surface to a malicious agent. If one device is hacked, the network can self-propagate the poison throughout the entire system. In this regard, *a single SDN Controller presents a smaller attack surface.* If we consider the case of a large data center, there are thousands or tens of thousands of control nodes available to attack. This is because each network device has a locally resident control plane. In an equivalent-sized SDN network there might be ten to twenty SDN controllers. Assuming that the compromising of any control node can bring down the entire network, that means that the SDN network presents two orders of magnitude fewer control nodes susceptible for attack as compared to its traditional equivalent. Through hardening to attack both the controllers and the secure channels, network designers can focus extra protection around these few islands of intelligence in order to keep malicious attacks at bay.

In summary, while the introduction of a single point of failure is indeed a vulnerability, with the right architecture and safeguards in place an Open SDN environment may actually be a safer and more secure network than the earlier, distributed model.

Resilience in SDN is an active research area. In addition to dealing with controller availability, this research crosses a set of disciplines, including survivability, dependability, traffic tolerance, and security. Security encompasses not only protecting against DDoS attacks, but also issues of *confidentiality* and *integrity*. There are also research efforts dealing specifically with the consistency between the network controller and redundant backups, and others investigating placement strategies of the network controller. This last item is known as *the controller placement problem.* The goal of this research is to find optimal places in the topology to place a controller in order to minimize connectivity issues in case a network partition occurs. A comprehensive discussion of this work on resilience in SDN can be found in [1].

6.1.3 PERFORMANCE AND SCALE

As the saying goes, "Many hands make light work." Applied to networking, this would imply that the more the networking intelligence is spread across multiple devices, the easier it is to handle the overall load. In most cases this is certainly true. Having a single entity responsible for monitoring as well as switching and routing decisions for the entire network can create a processing bottleneck. The massive amount of information pertaining to all end-nodes in a large network such as a data center can indeed become an issue. In spite of the fact that the controller software is running on a high-speed server with large storage capabilities, there is a limit at which even the capacity of such a server becomes strained and performance could suffer. An example of this performance sensitivity is the potential for request congestion on the controller. The greater the quantity of network devices which are dependent on the SDN controller for decisions regarding packet processing, the greater the danger of overloading the input queues of the controller such that it is unable to process incoming requests in a timely manner, again causing delays and impacting network performance.

We have dedicated a considerable part of this book to promoting the practice of moving the control plane off of the switch and onto a centralized controller. The larger the network under the management of the SDN controller, the higher the probability of network delays due to physical separation or network traffic volume. As network latency increases, it becomes more difficult for the SDN controller to receive requests and respond in a timely manner.

Issues related to performance and scalability of Open SDN are real and should not be dismissed. It is worth noting, however, that while there can be performance issues related to scale in a network design involving a centralized controller, the same is true when the network intelligence is distributed. One such issue is that those distributed entities will often need to coordinate and share their decisions with one another; as the network grows, more of these distributed entities need to coordinate and share information, which obviously impacts performance and path convergence times. Indeed, there is evidence that for Open SDN the convergence times in the face of topology changes should be *as good or better* than with traditional distributed routing protocols [2]. Additionally, these distributed environments have their own frailties regarding size and scale, as we have pointed out in earlier chapters, in areas such as MAC address table size and VLAN exhaustion. Also, in large data center networks, with VMs and virtual networks constantly being ignited and extinguished, path convergence may never occur with traditional networks.

A further issue with autonomous, distributed architectures is that in real-life networks the protocols that comprise the control plane are implemented by different groups of people at different times, usually working at multiple NEMs. While we have standards for these protocols, the actual implementations on the devices are different from one version to another and from one manufacturer to another. In addition to the vagaries that result from different groups of humans trying to do anything, even *identical* implementations will behave differently from one device to another due to different CPU speeds, differing buffering constraints and the like. These differences can create instability amongst the group of devices. In many legacy situations there is no way to scale the computation needed to manage the topology effectively. The centralized design has fewer differences in the implementation of the control plane because there are fewer implementations running it and it *can* be scaled, as we have explained above, using traditional compute scaling approaches, as well as by simply buying a bigger server. The upgrade process to higher performing control planes is easier to address in this centralized environment. Conversely, in the embedded, distributed world of traditional networking, we are stuck with low-powered CPUs and a nonupgradable memory footprint that came with the devices when they were purchased.

There are ways to improve Open SDN's performance and scalability through appropriate controller design. In particular, it is important to employ network design where more than one controller is used to distribute the load. There are two primary ways of doing this. The first is to deploy multiple controllers in a *cluster*. As network size grows, there will be a need for multiple coordinated controllers. Note that the logistics required to coordinate a number of these controllers does not approach the complexity of logistics and coordination required with thousands of network devices, so the problem is easier to solve. In this type of deployment, the controllers are able to share the load of the large set of network devices under their control. Fig. 6.6 shows three controllers spreading the load of the management of a large number of devices. While the details of how to implement clusters of controllers is currently out of the scope of the OpenFlow specification, the specification does state that a multiple controller scenario is likely and some aspects of the protocol have been extended to describe messaging to multiple controllers. This notion of clusters of controllers has been in widespread use for some time with wireless controllers. Some NEMs have published guidelines for how multiple wireless controllers for the same network should interact, notably Cisco in [3] and HP in [4].

A second means of utilizing multiple controllers is to organize the controllers into a *hierarchy of controllers*. As network size grows, it may be necessary to create such a hierarchy. This model allows for the off-loading of some of the burden from the *leaf* controllers to be handled by the *root* controller. The concept of controller hierarchies for SDN networks is explored in detail in [5]. Fig. 6.7 shows a two-level hierarchy of controllers, with the root controller managing the coordination between the

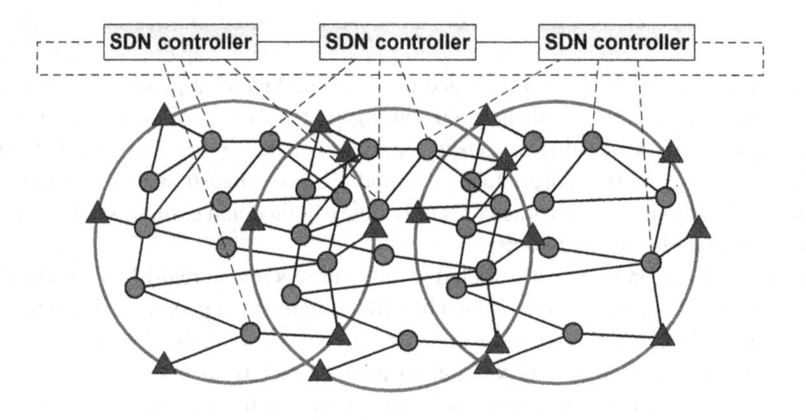

FIG. 6.6

Controller cluster.

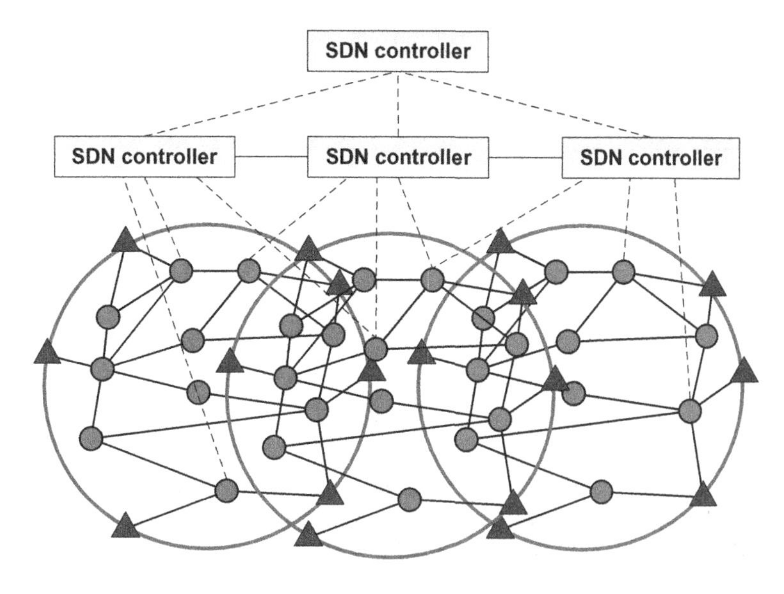

FIG. 6.7

Controller hierarchy.

leaf controllers' peers below. In this type of deployment there can be a separation of responsibilities as well; the highest-level controller can have global responsibility for certain aspects of the network, while lower-level responsibilities that require more geographical proximity may be handled by the controllers at the lower tier. This geographical proximity of the leaf controllers can ameliorate the problems of increased latency that come with increased network size.

As mentioned in Section 5.10.4, controller hierarchies are not yet well defined within the OpenFlow family of specifications. This topic has been identified as one urgently needing to be addressed.

To summarize, performance issues can arise in large Open SDN networks, just as they can in non-SDN networks. Most of these issues are due to the logically centralized controller and the possible path latency between the switch and the controller. Engineering solutions are feasible with the scale of today's Open SDN networks but as these scale additional work is needed. There are a number of research efforts focused on performance issues in Open SDN. One of these is *DevoFlow* [6], whose goal it is to reduce the amount of communication between the control and data planes. Another is found in [7], where the authors study the scalability of the SDN controller. In [8], the authors examine compression methods to reduce the amount of redundant rules in order to expedite traffic management. Finally, the performance of the *Northbound API* (NBI) is studied in [9].

6.1.4 DEEP PACKET INSPECTION

There are a number of network applications which may have requirements that exceed the capabilities of SDN or, more specifically, of OpenFlow, as it is defined today. In particular, there are applications which need to see more of the incoming packet in order to make matches and take actions.

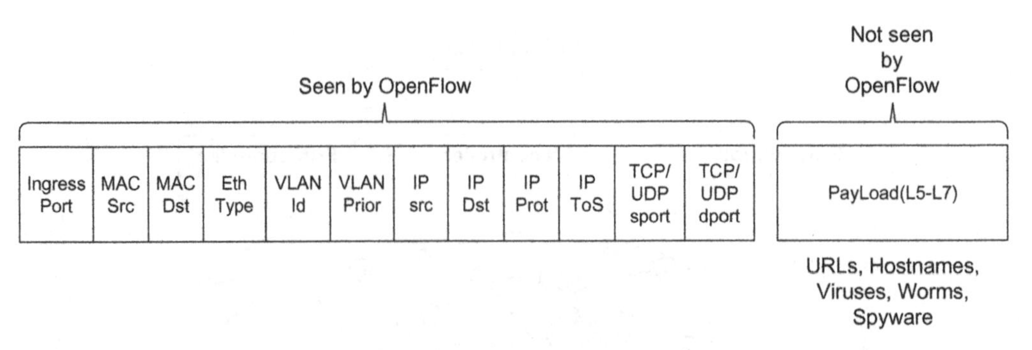

FIG. 6.8

Deep packet inspection.

One type of application that requires *Deep Packet Inspection* (DPI) is a firewall. Simple firewalls may make decisions based on standard fields available in OpenFlow, such as destination IP address or TCP/UDP port. More advanced firewalls may need to examine and act on fields that are not available to OpenFlow matches and actions. If it is impossible to set flows which match on appropriate fields, then the device will be unable to perform local firewall enforcement.

Another application of this sort is a load balancer. Simple load balancers, like firewalls, make decisions based on standard fields available in OpenFlow, such as source or destination IP address, or source or destination TCP/UDP port. This is sufficient for many cases, but more advanced load balancers may want to forward traffic based on fields not available in OpenFlow today, such as the specific destination URL. Fig. 6.8 shows the fields that OpenFlow 1.0 devices are able to match. As the figure illustrates, OpenFlow is not able to match against fields in the payload. The payload data includes many important items, such as URLs. It is possible that even simple load balancing decisions might be made on criteria in the payload. Also, detecting viruses, worms, and other spyware may require examining and processing such data. Since OpenFlow devices are not able to match data in the packet payload, they cannot make enforcement decisions locally.

There are two distinct challenges brought to light in the scenario described above. The first is to detect that a packet might need special treatment. If the criteria for that decision lie within the packet payload, an OpenFlow device cannot perform that test. Secondly, even if the device can determine that a packet needs to be examined for further scrutiny, the OpenFlow device will not have the processing horsepower to perform the packet analysis that many *Intrusion Detection Systems* (IDS) and *Intrusion Prevention Systems* (IPS) perform. In this case, there is a solution that dovetails well with OpenFlow capabilities. When the OpenFlow pipeline determines that a packet does not need deeper inspection, perhaps because of packet size or TCP/UDP port number, it can forward the packet around the *bump in the wire* that the in-band IDS/IPS normally constitutes. This has the advantage of allowing the vast majority of the packets transiting the network to completely bypass the IDS/IPS, helping to obviate the bottlenecks that these systems frequently become.

Rather than directing selected packets around an in-line appliance, OpenFlow can sometimes be used to shunt them to a deep-packet inspection device and allow normal packets to proceed unfettered. Such a device could actually reside inside the same switch in a separate network processor card. This is the idea behind the *Split SDN Data Plane Architecture* presented in [10].

In any event, as of this writing, deep packet inspection is still absent from the latest version of the OpenFlow standard. As this is a recognized limitation of OpenFlow, it is possible that a future version of the standard will provide support for matching on fields in the packet payload.

Limitations on matching resulting from security mechanisms

Note that the crux of the problem of deep packet inspection is the inability to make forwarding decisions based on fields that are not normally visible to the traditional packet switch. What we have presented thus far refers to fields that are in the packet payload, but the same problem can exist for certain packet header fields in the event that those fields are encrypted by some security mechanism. An example of this would be encrypting the entire payload of a layer two frame. Obviously, this obscures the layer three headers fields from inspection. This is not a problem for the traditional layer two switch which will make its forwarding decisions solely on the basis of the layer two header fields which are still in the clear. An Open SDN switch, on the other hand, would lose its ability to make fine-grained flow-based forwarding decisions using all of the twelve-tuple of packet header fields normally available as criteria. In this sense, such encryption renders normal header fields unavailable to the OpenFlow pipeline much like fields deep within the packet payload.

6.1.5 STATEFUL FLOW AWARENESS

An OpenFlow 1.3 device examines every incoming packet independently, without consideration of packets arriving earlier which may have affected the state of either the device or the transaction taking place between the two communicating nodes. We describe this characteristic as a lack of *stateful flow awareness* or *statefulness*. An OpenFlow 1.3 device attempting to perform a function that requires statefulness will not be adequate for the task. Examples of such functions include stateful firewalls or more sophisticated IPS/IDS solutions. As we noted in Section 6.1.4, in some cases OpenFlow can help the network scale by shunting packets that are candidates for stateful tracking to a device capable of such processing, diverting much of the network flows away from those processing-intensive appliances.

Most network applications use standard and well-known TCP or UDP ports. For example, HTTP traffic uses TCP port 80, *Domain Name System* (DNS) uses TCP and UDP port 53, *Dynamic Host Configuration Protocol* (DHCP) uses UDP ports 67 and 68, and so on. These are easy to track and to establish appropriate OpenFlow rules. However, there are certain protocols, for example FTP and *Voice over IP* (VoIP), which start on a standard and fixed port (twenty-one for FTP, 5060 for the VoIP *Session Initiation Protocol* (SIP) protocol) for the initial phase of the conversation, but which then set up an entirely separate connection using dynamically allocated ports for the bulk of their data transfer. Without stateful awareness, it is impossible for an OpenFlow device to have flow rules which match these dynamic and variable ports. Fig. 6.9 shows an example of stateful awareness where an FTP request is made to the FTP service on TCP port 21, which creates a new FTP instance that communicates with the FTP client on an entirely new TCP port chosen from the user space. An ability to track this state of the FTP request and subsequent user port would be helpful for some applications that wish to deal with FTP traffic in a particular manner.

Some solutions, such as load balancers, require the ability to follow transactions between two end-nodes in order to apply policy for the duration of the transaction. Without stateful awareness of the transaction, an OpenFlow device is unable to independently make decisions based on the state of ongoing transactions, and to take action based on a change in that state. Fig. 6.10 shows a sequence

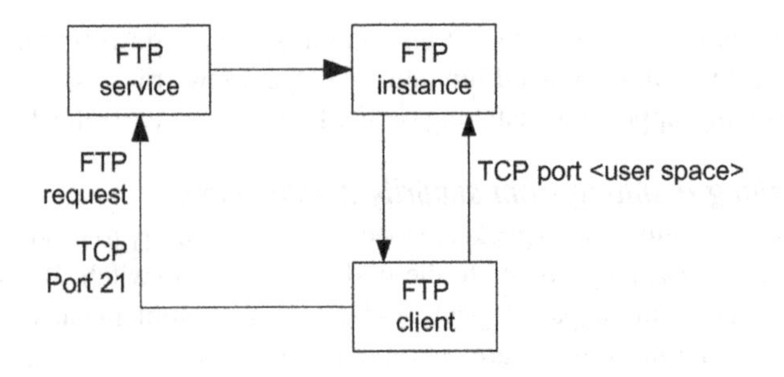

FIG. 6.9

Stateful awareness for special protocols.

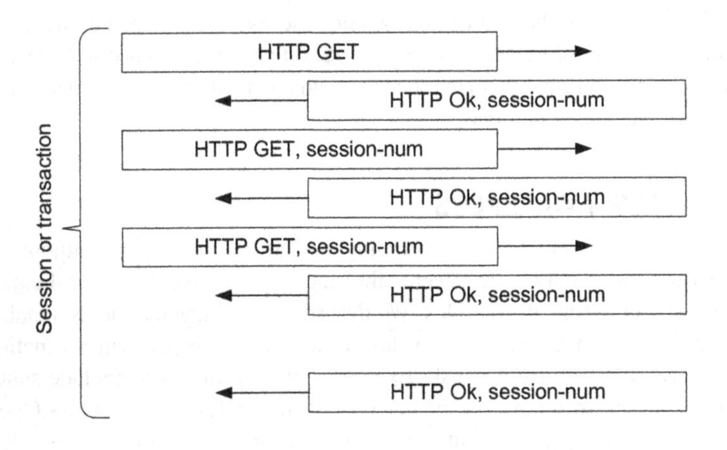

FIG. 6.10

Stateful awareness of application-level transactions.

of HTTP requests which together constitute an entire transaction. The transaction is identified by the session number. Stateful awareness would allow the transaction to be dealt with as an integral unit.

It should be noted that contemporary Internet switches are generally stateless. Higher-function specialized appliances such as load balancers and firewalls do implement stateful intelligence, however. Since Open SDN is often touted as a technology which could replace certain purpose-built network appliances such as these, the lack of stateful awareness is a genuine drawback. Some vendors of these appliances will justifiably advertise their capability for deep packet inspection as well as their ability to track state and provide features based on these two capabilities.

As with deep packet inspection, the lack of stateful awareness can be a genuine limitation for certain applications.

For this reason, OpenFlow 1.5 contemplates the possibility of maintaining flow state across packets. We provided a brief review of this new feature in Section 5.8.1. As of this writing, there are few details available about implementations based on this feature, so for the time being, in those instances where such stateful flow treatment is required, the user will likely have to continue to rely on specialized hardware or software rather than Open SDN.

6.1.6 SUMMARY

In this section we have reviewed a number of often-cited shortcomings of Open SDN. In some cases, there are documented Open SDN strategies for dealing with the issue. In other cases, workarounds outside the scope of Open SDN need to be used to address the specific application. In the balance of this chapter we will examine alternative SDN technologies in greater depth than we have previously. For each, we will determine whether the Open SDN drawbacks discussed in this section apply to that alternative SDN technology as well. This is important, since these alternatives are often promoted specifically as a response to the perceived limitations of Open SDN. We will also assess how well that alternative meets our criteria to be considered an SDN technology.

DISCUSSION QUESTION:

We have discussed potential drawbacks and limitations of SDN. Were the arguments convincing, or do you feel that these SDN limitations are debilitating for the adoption of SDN? What other drawbacks might there be that we have not mentioned?

6.2 SDN VIA APIs

One of the alternative SDN approaches that we introduced in Chapter 4 was *SDN via Existing APIs*.[1] Superficially, one might be tempted to dismiss SDN via APIs as an attempt by incumbent network vendors to protect their turf against encroachment by radical new technologies. But we would do it a disservice if we did not consider the positive elements of this solution. One such positive element is the movement toward controller-driven networks.

This brings us to the first of two artificial constraints we will use in our definition of SDN via APIs. First, we acknowledge that there are some proponents of controller-less SDN via APIs. Juniper's *JunOS SDK* [11] provides a rich set of network device programmability and was initially released before SDN became known to the industry. It is not controller-based, as the JunOS SDK applications actually reside in the devices themselves. Through the SDK APIs, these applications may wield control over a wide variety of device functions, including packet processing in the data plane. The details of the individual platforms' data plane hardware are abstracted so that a network programmer can affect packet processing without worrying about the vagaries of one hardware implementation versus another. Nonetheless, we consider such controller-less approaches to be outliers, and we will exclude them from consideration in this category. Looking back at our introduction to SDN via APIs in Chapter 4, Figs. 4.8 and 4.9 portrayed a controller accessing APIs residing in the devices. This model will guide our definition of the SDN via APIs approach.

The second constraint is on the very definition of API. We wish to delineate the Open SDN approach from SDN via APIs. Nonetheless, OpenFlow itself is clearly the preeminent southbound API to the networking devices in Open SDN. Since we are trying to bring clarity to a confusing situation, when we say SDN via APIs we refer to APIs that program or configure a control plane that is active in the device. Our introduction to SDN via APIs in Chapter 4 used the term *SDN via Existing APIs*. While

[1] For the sake of brevity, we will use the term *SDN via APIs* in the balance of this book.

that suited the context of that chapter, the word *existing* is not sufficiently precise. Legacy APIs all programmed the control plane that was resident on the device. If a brand-spanking-new API continues to program a control plane residing in the device, it will still fall into this category. We contrast this with OpenFlow. As we explained in Chapter 5, OpenFlow is used to directly control the data plane. This distinction will be very important as we now review both legacy and new APIs that are part of the SDN via APIs paradigm.

We have spent significant time in this book discussing the history behind the distributed-intelligence model of networking and the stagnation that has occurred due to the resulting closed system. SDN via APIs actually does open up the environment to a certain degree, as there are opportunities for software developed for the SDN controller to use those APIs in new and better ways, and, hence, to advance networking technology.

This movement toward networking control residing in software, in a controller, is definitely a step forward in the evolution of networking. Another major step forward is the improvement in APIs on the devices themselves, which we discuss in the next section.

6.2.1 LEGACY APIS IN NETWORK DEVICES

Currently, networking devices have APIs which allow for a certain amount of configuration. There are a number of network device APIs in current use:

- The *Simple Network Management Protocol* (SNMP) first was standardized in 1988 and it has been the standard for management interactions with networking devices since that time. The general mechanism involved a simple protocol allowing the controlling function to GET and SET objects on a device, which have been defined in a *Management Information Base* (MIB). The functionality offered by SNMP can be extended by the definition of new and sometimes proprietary MIBs, allowing access to more information and attributes on a device. While SNMP does provide both GET and SET primitives, it is primarily used for monitoring, using GET, rather than for configuration which uses SET.
- The *Command Line Interface* (CLI) is the most fundamental way of accessing a device, and is designed with a human end-user in mind. Often the networking device functionality placed into a device is only accessible via CLI (not SNMP). Consequently, management software attempting to access this data must emulate a user logging into the CLI and executing CLI commands and attempting to read the results. Even when the CLIs are enhanced to support scripting, this merely automates an interface designed for static configuration changes.
- *Remote Authentication Dial In User Service* (RADIUS), which we introduced in Section 3.2.2, has been widely used in residential broadband and has been used to push policy, ACLs and VLANs onto devices.
- The *Transaction Language 1* (TL1) was developed for telecommunications equipment in the 1980s and is similar in some respects to SNMP in that its intent is to provide a language for communication between machines.
- The *TR-069 CPE WAN Management Protocol* [12] is intended to control the communication between *Customer Premises Equipment* (CPE) and an *Auto-Configuration Server* (ACS). In our context, the CPE is the network device. The ACS is a novel contribution of TR-069. The ACS provides secure auto-configuration of the network device, and also incorporates other device management functions into a unified framework.

These APIs have been used for years primarily by network management applications. While we consider these to be legacy APIs, this should not imply that they are only for configuring old features. They have been extended to provide support for cutting-edge features, such as Cisco's PBR [13]. However, these extended APIs have not scaled well into the current world of networking, especially in data centers. When there is a fully distributed control plane on hundreds or thousands of devices, this means that the configuration of the control plane on each device is very complex.

Three examples of APIs that seem to scale well as SDN APIs are NETCONF, the BGP-LS/PCE-P plugin, and the *Representational State Transfer* (REST) API. We discuss these in the following sections.

6.2.2 NETCONF/YANG

In Chapter 7 we will delve into the details of the emerging trend of using the *Network Configuration Protocol* (NETCONF) as the southbound API for SDN applications and how network management-style SDN is gaining favor as an evolutionary path toward a more complete SDN solution. In this section we explore the advantages NETCONF provides over SNMP as a protocol for implementing SDN-style behavior in a network. NETCONF was developed in the last decade to be a successor to protocols such as SNMP, since SNMP was mainly being used to monitor networks, not to configure them. The configuration payload utilizes an *Extensible Markup Language* (XML) format for passing data and commands to and from the networking devices. There are a number of strengths that NETCONF exhibits versus SNMP as a protocol for both network management and for SDN:

- **Security**: NETCONF has operated over a secure channel from its beginning, while SNMP had to evolve in order to support rudimentary secure exchanges of data with the management server.
- **Organization**: NETCONF separates *configuration* and *operational* data, making organization and control over different parts of the data much simpler.
- **Announcement of capabilities**: The first exchange between a NETCONF server on the device, and the client on the management station, is the announcement by the device of all the YANG models that it supports. YANG, which we describe in Section 7.2.2, provides a formal language for expressing the device's capabilities. Thus there is no ambiguity regarding what functionality the device supports.
- **Operations**: NETCONF *operations* are *Remote Procedure Calls* (RPCs). This capability allows the SDN controller to instruct the device to take a particular action, passing a set of parameters to the RPC. This enables the controller to more easily manipulate forwarding behavior on the device.

As an SDN API, NETCONF has enjoyed particular popularity amongst large service providers. Service providers have long been lobbying networking vendors to support NETCONF on their large routers. Vendors such as Juniper and Cisco have thus been compelled to implement NETCONF on devices targeted at these customers. This has created a groundswell of support for NETCONF in these environments.

One drawback is that NETCONF suffers from a paucity of standard YANG models. Consider SNMP, which from an early period had MIB definitions for many critical pieces of data on switches and routers. In comparison, the industry still struggles to create standard YANG models for routing, switching, interfaces, ACLs, and other fundamental network components. This hinders SDN application developers, who must write device-specific code in order to control the many different YANG models that exist for the same functionality. We revisit this YANG-related hurdle in Section 7.2.2. Nevertheless, the evidence is that NETCONF will be one of the primary protocols used by future SDN application developers.

6.2.3 BGP-LS/PCE-P

In this section we discuss how the BGP-LS/PCE-P plugin is used to create a more functional and extensive SDN solution, using existing protocols. In Sections 7.2.1, 7.2.4 and 7.2.5 we will delve into greater detail about how this plugin works and its role in an emerging trend in SDN solutions, particularly on the *OpenDaylight* (ODL) controller. At a broad-brush level, we divide the tasks performed by this plugin into two categories, *gathering topology information* and *setting paths and routes*.

Via this plugin, the SDN controller uses the *Border Gateway Protocol* (BGP) to gather exterior routing topology, the *BGP Link State* (BGP-LS) protocol to retrieve interior routing topology, and the *Path Computation Element Protocol* (PCE-P) to acquire topology information related to MPLS. The controller gathers this information to provide various topology views about the network to requesters, thus allowing participating applications to make intelligent decisions about routes and paths.

If so requested by an application, the BGP-LS/PCE-P plugin can program routes in the network by emulating a BGP node and using the BGP protocol to insert routes into the network devices. Similarly, the plugin uses the PCE-P protocol to set up and tear down MPLS *Label Switched Paths* (LSPs) per the application's requests.

We contrast this ODL plugin with other API-based solutions where the API is provided directly on the device or on a controller which then interprets those requests into the appropriate protocol to talk directly to devices. BGP-LS/PCE-P operates at a higher level of abstraction, providing topology, path and routing APIs to applications. This genre of plugin is relatively new but is part of the general momentum toward more abstract APIs, relieving the SDN application of the burden of dealing directly with a large collection of devices.

6.2.4 REST

A new technology being successfully used as an SDN API is REST, which uses HTTP or HTTPS. APIs based on this technology are called *RESTful* interfaces. REST technology was introduced a number of years ago and has been used primarily for access to information through a web service. The movement toward REST as one of the means for manipulating SDN networks is based on a number of its perceived advantages:

- **Simplicity**. The web-based REST mechanism utilizes the simple HTTP GET, PUT, POST, and DELETE commands. Access to data involves referencing *resources* on the target device using normal and well-understood URL encoding.
- **Flexibility**. Requesting entities can access the defined configuration components on a device using REST resources which are represented as separate URLs. There is no need for complicated schemas or MIB definitions to make the data accessible.
- **Extensibility**. Support for new resources does not require recompilation of schemas or MIBs, but only requires that the calling application access the appropriate URL.

- **Security**. It is very straightforward to secure REST communications. Simply by running this web-based protocol through HTTPS adequately addresses security concerns. This has the advantage of easily penetrating firewalls, a characteristic not shared by all network security approaches.

A potential risk in using a protocol as flexible and extensible as REST is that it lacks the strict formalism and type-checking of some other methods.

A related use of REST is in the RESTCONF protocol. While more often used as a northbound API to the controller, RESTCONF can also be used as a device API. RESTCONF is an implementation of NETCONF/YANG using JSON or XML over HTTP rather than XML RPCs over SSH.

It is important to point out that thus far we have been discussing APIs for devices such as switches. This is the context in which we have referred to REST as well as the earlier API technologies. The reader should note that REST is also frequently used as a northbound API to the controller. Such northbound APIs are used at different levels of the controller and also may exist as high as the management layer that oversees a cluster of controllers. Since this kind of northbound API plays an important role in how network programmers actually program an SDN network, they certainly are SDN-appropriate APIs. Nonetheless, the reader should distinguish these from the device APIs that we refer to when discussing SDN via APIs.

6.2.5 EXAMPLES OF SDN VIA APIs

Cisco

Cisco originally had its own proprietary SDN program called *Cisco onePK* [14] that consists of a broad set of APIs in the devices. These APIs provide SDN functionality on legacy Cisco switches. Cisco onePK, currently de-emphasized by Cisco, does not rely on traditional switch APIs such as the CLI, but extends the switch software itself to support a richer API to support user application control of routing and traffic steering, network automation, policy control and security features of the network, among others. The more recent *Cisco Extensible Network Controller* (XNC) provided the original controller foundation for the ODL controller, discussed below.

Cisco has two other SDN-via-API controllers, the *Application Policy Infrastructure Controller—Enterprise Module* (APIC-EM) and the *Application Policy Infrastructure Controller—Data Center* (APIC-DC). The APIC-EM solution is targeted at enterprise customers who wish to control access and prioritize traffic between end users and resources such as servers. APIC-EM is unique in that its southbound protocol is actually the CLI. The controller takes higher-level policy information about users, resources, and access rights, and converts those into CLI settings on the switches and routers in the enterprise network. The APIC-DC solution is targeted at data centers, and it is built using an entirely new protocol called *OpFlex*. The OpFlex protocol is policy-based, meaning that the controller exchanges policy-level information with the devices in the network. The devices interpret the policy instructions and make the appropriate configuration modifications.

OpenDaylight

A hybrid approach is the ODL open source project [15]. ODL supports OpenFlow as one of its family of southbound APIs for controlling network devices. Used in this way, ODL clearly can fit into the Open SDN camp. The ODL controller also supports southbound APIs that program the legacy control plane on network devices, using protocol plugins such as NETCONF and BGP-LS/PCE-P, presented in Sections 6.2.2 and 6.2.3, respectively. When used with those legacy southbound APIs, it falls into the controller-based SDN via APIs category. Due to this chameleon-like role, we will not categorize it as exclusively either SDN via APIs or Open SDN. We discuss the ODL program in more detail in Chapters 7 and 13.

Juniper

Juniper has created an API-based controller called *Contrail*. This controller uses NETCONF and *Extensible Messaging and Presence Protocol* (XMPP) to create a data center solution which has parallels in the service provider space, using MPLS-like features to route traffic through the network. As Juniper has historically been a significant force in service provider markets, this provides a convenient segue for them into data centers. Juniper has open-sourced the base Contrail controller, calling this version *OpenContrail*, encouraging customers and vendors to contribute code to the project. At the time of this writing, there seems to be a smaller community of open source developers committed to OpenContrail than with ODL.

Arista

Arista is another commercial example of a company that asserts that SDN is not about the separation of the control plane from the data plane and is instead about scaling the existing control and data plane with useful APIs [16]. Arista is basing its company's messaging squarely around this API-centric definition of SDN. Arista claims that they support both centralized controllers and an alternative of distributed controllers that may reside on devices themselves.

6.2.6 RANKING SDN VIA APIs

In Table 6.1 we provide a ranking of SDN via APIs versus the other technologies discussed in this chapter. As we said in the introduction to this chapter, despite our best efforts, there is surely some subjectivity in this ranking. Rather than assign strict letter grades for each criterion, we simply provide a high, medium or low ranking as an assessment of how these alternatives stack up against one another. We said earlier that for the sake of clarity we have tightly constrained our definitions of each category which may exclude an implementation that an individual reader may consider important. We also acknowledge that there may be criteria that are extremely important for some network professionals or users that are entirely absent from our list. For this, we apologize and hope that this attempt at ranking at least serves to provoke productive discussions.

The first five criteria are those that define how well a technology suits our definition of an SDN technology. Of these five, one criterion where SDN via APIs seems to exemplify a true SDN technology is in the area of improved network automation and virtualization. Networking devices which are only changed in that they have better APIs are certainly not simplified devices. Such devices retain the ability to operate in an autonomous and independent manner, which means that the device itself will be no simpler and no less costly than it is today. Indeed, by our very definition, these APIs still program the locally resident control plane on the device. As we stated at the start of Section 6.2, our discussion of SDN via APIs presumes use of a centralized controller model. For this reason, we give it a grade of high for centralized controller. The APIs defined by vendors are generally unique to their own devices, and, hence, usually proprietary.[2] The end result is that SDN applications written to one vendor's set of APIs will often not work with network devices from another vendor. While it may be true that the Cisco APIs have been mimicked by its competitors in order to make their devices interoperable with Cisco-based management tools, this is not true interoperability. Cisco continually creates new APIs so as to differentiate themselves from competitors, which puts its competitors in constant catch-up mode. This is not the case with NETCONF and BGP-LS/PCE-P, however, as both of these are open specifications. While these two specifications are open, the lack of standardization of YANG models limits this openness somewhat. Thus, the grade for openness for SDN via APIs will be a dual ranking of [*Low*, *Medium*] depending on whether the API is a proprietary or open protocol. Finally, the control plane remains in place in the device and only a small part of forwarding control has been moved to a centralized controller.

When ranked against the drawbacks of Open SDN, however, this technology fares somewhat better. Since the network is still composed of the same autonomous legacy switches with distributed control, there is neither too much change nor a single point of failure. We rank it medium with respect to performance and scale, as legacy networks have proven that they can handle sizable networks but are showing inability to scale to the size of mega-data centers. SDN via APIs solutions support neither deep packet inspection nor stateful flow awareness. Nothing in these solutions directly addresses the problems of MAC forwarding table or VLAN ID exhaustion.

So, in summary, SDN via APIs provides a mechanism for moving toward a controller-based networking model. Additionally, these APIs can help to alleviate some of the issues raised by SDN around the need to have a better means of automating the changing of network configurations in order to attempt to keep pace with what is possible in virtualized server environments.

DISCUSSION QUESTION:

SDN via APIs help customers to protect existing investments in their legacy equipment, but they can also help vendors protect their profit margins. Which do you think may have been the primary motivation? In what ways does SDN via APIs limit or even hinder technological progress in the networking domain?

6.3 SDN VIA HYPERVISOR-BASED OVERLAYS

In Chapter 4 we introduced SDN via Hypervisor-Based Overlay Networks. This hypervisor-based overlay technology creates a completely new virtual network infrastructure that runs independently on top of the underlying physical network, as was depicted in Fig. 4.12. In that diagram we saw that the overlay networks exist in an abstract form *above* the actual physical network below. The overlay networks can be created without requiring reconfiguration of the underlying physical network, which is independent of the overlay virtual topology.

With these overlays it is possible to create networks which are separate and independent of each other, even when VMs are running on the same server. As was shown in Fig. 4.12, a single physical server hosting multiple VMs can have each VM be a member of a separate virtual network. This VM

[2]An important exception relevant to SDN is Nicira's *Open vSwitch Database Management Protocol* (OVSDB). OVSDB is an open interface used for the configuration of OVS virtual switches.

can communicate with other VMs in its virtual network, but usually does not cross boundaries and talk to VMs which are part of a different virtual network. When it is desirable to have VMs in different networks communicate, they do so by using a virtual router between them.

It is important to notice in the diagram that traffic moves from VM to VM without any dependence on the underlying physical network, other than its basic operation. The physical network can use any technology and can be either a layer two or a layer three network. The only matter of importance is that the virtual switches associated with the hypervisor at each endpoint can talk to each other.

What is actually happening is that the virtual switches establish communication tunnels amongst themselves using general IP addressing. As packets cross the physical infrastructure, as far as the virtual network is concerned, they are passed directly from one virtual switch to another via virtual links. These virtual links are the tunnels just described. The virtual ports of these virtual switches correspond to the *Virtual Tunnel Endpoints* (VTEPs) defined in Section 4.6.3. The actual payload of the packets between these vSwitches is the original layer two frame being sent between the VMs. In Chapter 4 we defined this as *MAC-in-IP* encapsulation, which was depicted graphically in Fig. 4.13. We will provide a detailed discussion about MAC-in-IP encapsulation in Chapter 8.

Much as we did for SDN via APIs, we restrict our definition of SDN via Hypervisor-based Overlays to those solutions that utilize a centralized controller. We acknowledge that there are some SDN overlay solutions that do not use a centralized controller (e.g., MidoNet, which has MidoNet agents located in individual hypervisors), but we consider those to be exceptions that cloud the argument that this alternative is, generally speaking, a controller-based approach.

6.3.1 OVERLAY CONTROLLER

By our definition, SDN via Hypervisor-based Overlays utilizes a central controller, as do the other SDN alternatives discussed thus far. The central controller keeps track of hosts and edge switches. The overlay controller has knowledge of all hosts in its domain, along with networking information about each of those hosts, specifically their IP and MAC addresses. These hosts will likely be virtual machines in data center networks. The controller must also be aware of the edge switches that are responsible for each host. Thus, there will be a mapping between each host and its adjacent edge switch. These edge switches will most often be a virtual switch associated with a hypervisor resident in a physical server and attaching the physical server's virtual machines to the network.

In an overlay environment, these edge switches act as the endpoints for the tunnels which serve to carry the traffic across the top of the physical network. These endpoints are the VTEPs mentioned above. The hosts themselves are unaware that anything other than normal layer two Ethernet forwarding is being used to transport packets throughout the network. They are unaware of the tunnels and operate just as if there were no overlays involved whatsoever. It is the responsibility of the overlay controller to keep track of all hosts and their connecting VTEPs so that the hosts need not concern themselves with network details.

6.3.2 OVERLAY OPERATION

The general sequence of events involved with sending a packet from Host A to a remote Host B in an Overlay network is as follows:

1. Host A wants to send payload P to Host B.
2. Host A discovers the IP address and MAC address for Host B using normal methods (DNS and ARP). While Host A uses its ARP broadcast mechanism to resolve Host B's MAC address, the

means by which this layer two broadcast is translated to the tunneling system varies. The original VXLAN tries to map layer two broadcasts directly to the overlay model by using IP multicast to flood the layer two broadcasts. This allows the use of Ethernet-style *MAC address learning* to map between virtual network MAC addresses and the virtual network IP addresses. There are also proprietary control plane implementations where the tunnel endpoints exchange the VM-MAC address to VTEP-IP address mapping information. Another approach is to use MP-BGP to pass MPLS VPN membership information between controllers where MPLS is used as the tunneling mechanism. In general, this learning of the virtual MAC addresses across the virtual network is the area where virtual network overlay solutions differ most. We provide this sampling of some alternative approaches only to highlight these differences, and do not attempt an authoritative review of the different approaches.

3. Host A constructs the appropriate packet, including MAC and IP addresses, and passes it upstream to the local switch for forwarding.
4. The local switch takes the incoming packet from Host A, looks up the destination VTEP to which Host B is attached, and constructs the encapsulating packet as follows:
    - The outer destination IP address is the destination VTEP and the outer source IP address is the local VTEP.
    - The entire layer two frame that originated from Host A is encapsulated within the IP payload of the new packet.
    - The encapsulating packet is sent to the destination VTEP.
5. The destination VTEP receives the packet, strips off the outer encapsulation information, and forwards the original frame to Host B.

Fig. 6.11 shows these five steps. From the perspective of the two hosts, the frame transmitted from one to the other is the original frame constructed by the originating host. However, as the packet traverses the physical network, it has been encapsulated by the VTEP and passed directly from one VTEP to the other, where it is finally decapsulated and presented to the destination host.

6.3.3 EXAMPLES OF SDN VIA HYPERVISOR-BASED OVERLAYS

Prior to Nicira's acquisition by VMware in 2012, Nicira's *Network Virtualization Platform* (NVP) was a very popular SDN via Hypervisor-Based Overlays offering in the market. NVP has been widely deployed in large data centers. It permits virtual networks and services to be created independently of the physical network infrastructure below. Since the acquisition, NVP is now bundled with VMware's *vCloud Network and Security* (vCNS) and is marketed as VMware NSX [17],[3] where it continues to enjoy considerable market success. The NVP system includes an open source virtual switch, *Open vSwitch*, (OVS) that works with the hypervisors in the NVP architecture. Open vSwitch has become the most popular open source virtual switch implementation even outside of NVP implementations. Interestingly, NVP uses OpenFlow as the southbound interface from its controller to the OVS switches that form its overlay network. Thus, it is both an SDN via Hypervisor-Based Overlays *and* an Open SDN implementation.

[3]Since we focus on the NVP component in this work, we will usually refer to that component by its original, preacquisition name, rather than NSX.

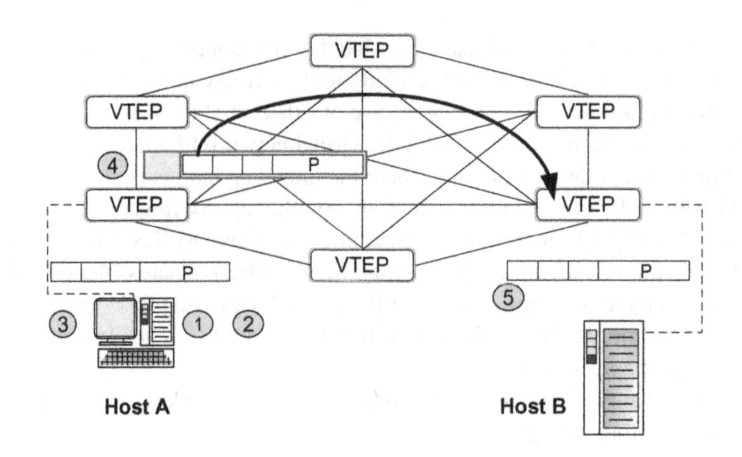

FIG. 6.11

Overlay operation.

Juniper's JunosV Contrail [18], however, does not use OpenFlow. We introduced Contrail in Section 6.2.5 as an example of the use of NETCONF as SDN via APIs. While that is indeed the case, it is also true that Contrail is a controller-based virtual overlay approach to network virtualization. It uses XMPP to program the control plane of the virtual switches in its overlay networks. Because the southbound APIs are NETCONF and XMPP, it is indeed a standards-based approach. Since the control plane still resides on those virtual switches and Contrail does not directly program the data plane, we do not consider it Open SDN. It is clearly an SDN via Hypervisor-Based Overlays approach, though, as well as a NETCONF-based SDN via APIs approach.

Another important SDN via Hypervisor-Based Overlays offering is IBM's *Software Defined Network for Virtual Environments* (SDN VE) [19]. SDN VE is based on IBM's established overlay technology, *Distributed Overlay Virtual Ethernet* (DOVE).

6.3.4 RANKING SDN VIA HYPERVISOR-BASED OVERLAYS

In Table 6.1 we include our ranking of SDN via Hypervisor-Based Overlays versus the alternatives. SDN via Hypervisor-Based Overlays is founded on the concepts of network automation and virtualization and, thus, scores high in those categories. With respect to openness, this depends very much on the particular implementation. NVP, for example, uses OpenFlow, OVSDB, and OVS, all open technologies. Other vendors use proprietary controller-based strategies. There are both closed and open subcategories within this alternative, so we have to rate it with a dual ranking [*Medium, High*] against this criterion, depending on whether or not the overlay is implemented using Open SDN. As we restricted our definition of SDN via Hypervisor-Based Overlays to controller-based strategies, it ranks high in this category. With respect to device simplification, the overlay strategy does allow the physical network to be implemented by relatively simple IP layer three switches, so we will give this a ranking of medium with respect to device simplification.

Table 6.1 SDN Technologies Report Card

Criterion	Open SDN	SDN via APIs	SDN via Hypervisor-Based Overlays
Plane Separation	High	Low	Medium
Simplified Device	High	Low	Medium
Centralized Control	High	High	High
Network Automation and Virtualization	High	High	High
Openness	High	[Low, Medium][b]	[Medium, High][c]
Too Much Change	Low	High	Medium
Single Point of Failure	Low	Medium	N/A[d]
Performance and Scale	Medium	Medium	Medium
Deep Packet Inspection	Low	Low	Medium
Stateful Flow Awareness	Low[a]	Low	Medium
MAC Forwarding Table Overflow	High	Low	High
VLAN Exhaustion	High	Low	High

[a] *While OpenFlow V.1.5 describes the possibility of maintaining flow state across packets, there are few details and no commercial implementations that we are aware of so we leave this ranked low. (see Section 6.1.5).*
[b] *When the API is a proprietary protocol, we do not consider this open, hence it gets a low ranking. As we point out in Section 6.2.6, some APIs such as NETCONF are somewhat more open, and in those cases this deserves a medium ranking.*
[c] *Either medium or high depending on the openness of the particular implementation (see Section 6.3.4).*
[d] *Since SDN via Hypervisor-Based Overlays may or may not be based on the centralized controller concept we feel that ranking is not applicable (see Section 6.3.4).*

A sidebar is called for here. The SDN via Hypervisor-Based Overlays is often touted as delivering network virtualization on top of the existing physical network. Taken at face value, this does not force any device simplification, it merely allows it. A customer may be very pleased to be able to keep his very complex switches if the alternative means discarding that investment in order to migrate to simple switches. This actually holds true for OpenFlow as well. Many vendors are adding the OpenFlow protocol to existing, complex switches, resulting in hybrid switches. Thus, OpenFlow itself does not mandate simple devices either, but it does allow them.

Plane separation is more complicated. In the sense that the topology of the physical network has been abstracted away and the virtual network simply layers tunnels on top of this abstraction, the control plane for the virtual network has indeed been separated. The physical switches themselves, however, still implement locally their traditional physical network control plane, so we give this a score of medium.

The criterion of too much change is also difficult to rank here. On one hand, the same physical network infrastructure can usually remain in use, so no forklift upgrade of equipment is necessarily required. On the other hand, converting the network administrators' mindset to thinking of virtual networks is a major transformation and it would be wrong to underestimate the amount of change this represents. Thus, we rank this medium.

Since SDN via Hypervisor-Based Overlays may or may not be based on the centralized controller concept, it gets a N/A ranking in the single point of failure category. Similarly, we feel that it deserves the same medium ranking in performance and scale as Open SDN. It fares better than Open SDN with respect to deep packet inspection and stateful flow awareness because the virtual switches are free to implement this under the overlay paradigm. They may map different flows between the same hosts to different tunnels. This freedom derives from the fact that these virtual switches may implement any propriety feature, such as deep packet inspection, and are not restricted by a standard like OpenFlow. Since these features may be supported by proprietary overlay solutions but are not currently supported by some of the largest overlay incumbents, we give these both a medium ranking.

SDN via Hypervisor-Based Overlays receives the highest marks in the final two categories. The MAC forwarding table size problem is solved since the physical network device only deals with the MACs of the VTEPs, which is a much smaller number than if it had to deal with all the VM MAC addresses. Similarly, since virtualization can be achieved by tunnels in this technology, the system does not need to overrely on VLANs for virtualization.

To summarize, as SDN via Hypervisor-Based Overlay Networks was specifically designed for the data center, it was not tailored to other environments, such as the Campus, Service Provider, Carrier, and Transport Networks. For this reason, while it directly addresses some major needs in the data center, it has fewer applications in other networking environments. In the overlay alternative the underlying physical network does not fundamentally change. As we examine SDN use cases outside the data center in Chapter 9, we will see that in many cases the solution depends on changing the underlying physical network. In such cases, the overlay approach is clearly not appropriate. While SDN via Hypervisor-Based Overlays is an important solution gaining considerable traction in data center environments, it does not necessarily offer the device simplification or openness of Open SDN.

DISCUSSION QUESTION:

SDN via Overlays run across existing hardware and utilize a software-only design. In what ways is this solution is superior to Open SDN and SDN via APIs? Do you think that future networking devices and solutions will attempt to create environments that are software-only?

6.4 SDN VIA OPENING UP THE DEVICE

There is a another approach to Software Defined Networking which has received considerably less attention than the other alternatives. This approach attempts to provide SDN characteristics by opening up the networking device itself for software development. We examine this alternative in the sections which follow.

We have stated that traditional networking devices themselves are closed environments. Only Cisco engineers can write networking software for Cisco devices, only Juniper engineers can write software for Juniper devices, and so on. The ramifications of this are that the evolution of networking devices has been stunted because so few people are enabled to innovate and develop new networking technologies.

The approaches we have discussed thus far, Open SDN, SDN via APIs, and SDN via Hypervisor-Based Overlays, generally push networking intelligence off the switch onto a controller, which sees the entire network and on which it is possible to write networking applications. To varying degrees, these applications may be written by developers from outside the switch manufacturers. We have explained the advantages of this in our discussion of the SDN trait of openness. A fundamentally different approach to SDN is to provide networking devices which *are themselves open*, capable of running software which has been written by the open source community.

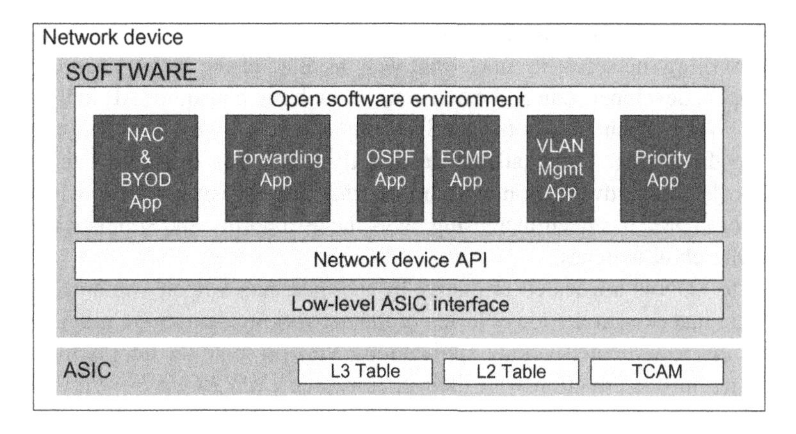

FIG. 6.12

Opening up the device.

This *Open Device* concept is represented in Fig. 6.12. The device itself, at the lower levels, is the same as in any other device, with L2 and L3 forwarding tables, and TCAMs for the more complex matching and actions. However, above this layer is an API which allows networking software applications to be written and installed into the device. Under the Open Device approach, the current distributed, independent and autonomous device model remains unchanged. The network intelligence is still distributed throughout the network on the devices, and there is no centralized controller. The nature of the networking devices changes from being closed and proprietary to being open and standardized, such that any outside party can write software to run on these open devices. In order to be successful, there needs to be some level of standardization of the APIs which will be used by the networking software running on these devices. This is similar to the experience with Linux, where the operating system of the networking device provides some APIs which are relatively universal, so that software can be written once and run on multiple open device platforms. It is important to recognize that the Linux experience was only possible because of the prevalence of a standardized PC hardware platform, which is a result of the Wintel de facto partnership.

This situation does not yet exist with enterprise class switches, but may begin to occur as more *Original Device Manufacturers* (ODMs) build enterprise class switches all based on the same reference design from the same merchant silicon vendors such as Broadcom or Intel. Since the switching chip manufacturers are few in number, the number of different reference designs will be small, and we may see de facto standardization of these switching platforms. Interestingly, this is already occurring with consumer wireless router hardware, which provides a concrete example of how *opening up the device* can work.

OpenWRT [20] is an open source implementation of a consumer wireless router. This is a Linux-based distribution that has been extended to include the IEEE 802.11, routing and switching software necessary to support the home AP functionality. OpenWRT has enjoyed considerable success due to the fact that the consumer AP ODMs have largely followed the model we described above: a very large number of devices are manufactured from the same reference design and, thus, are compatible from a software point of view. This is evidenced by the very long list of supported hardware documented in [20]. This hardware compatibility allows the OpenWRT to be *reflashed* onto the hardware, overwriting the software image that was installed at the factory. Since the source code for OpenWRT is open, developers can add extensions to the basic consumer AP, and bring that hybrid product to market. While OpenWRT is not an SDN implementation, the model it uses is very much that of opening up the device. The challenge in directly translating this model to SDN is that the consumer wireless device hardware is much simpler than an enterprise class switch. Nonetheless, the sophistication of ODMs has been increasing, as is the availability and sophistication of reference designs for enterprise class switches.

This approach to SDN is advocated primarily by the merchant silicon vendors who develop the ASICs and hardware that exist at the lower levels of the networking device. By providing an open API to anybody who wants to write networking applications, they promote the adoption of their products. Intel has taken a step in this direction with their open source *Data Plane Development Kit* (DPDK) [21,22]. Despite this momentum, such a standardized hardware platform for enterprise class switches does not yet exist. The success of SDN via opening up the device is predicated on this becoming a reality.

DISCUSSION QUESTION:

There is a push towards creating networking devices that are open, or run Linux inside, such that software can be written to extend or enhance the capabilities of devices. What are the ways in which this is a good idea? And are there ways in which this is a bad idea?

6.5 NETWORK FUNCTIONS VIRTUALIZATION

We cannot complete our chapter on SDN alternatives without a mention of *Network Functions Virtualization* (NFV), as there is confusion in the industry as to whether or not NFV is a form of SDN [23]. NFV is the implementation of the functionality of network appliances and devices in software. This means implementing a generalized component of your network such as a load balancer or IDS in software. It can even be extended to the implementation of routers and switches. This has been made possible by advancements in general purpose server hardware. Multicore processors and network interface controllers that offload network-specific tasks from the CPUs are the norm in data centers. While these hardware advances have been deployed initially to support the large numbers of VMs and virtual switches required to realize network virtualization, they have the advantage that they are fully programmable devices and can thus be reprogrammed to be many different types of network devices. This is particularly important in the service provider data center, where the flexibility to dynamically reconfigure a firewall into an IDS [24] as a function of tenant demand is of enormous value.

The reason for the confusion about whether or not NFV is a form of SDN is that the two ideas are complementary and overlapping. SDN may be used very naturally to implement certain parts of NFV. NFV may be implemented by non-SDN technologies, however, just as SDN may be used for many purposes not related to NFV. While NFV is not just another approach to SDN, the relationship between the two is sufficiently intimate that we dedicate the entire Chapter 10 to NFV.

6.6 ALTERNATIVES OVERLAP AND RANKING

While we have tried to define a clear framework and delineated four distinct SDN alternatives, the lines between them are sometimes blurred. For example:

- NVP uses OpenFlow as the southbound interface from its controller to the OVS switches that form its overlay network, so it is an SDN via Hypervisor-Based Overlays *and* an Open SDN implementation.
- The OpenDaylight controller uses several different southbound APIs to control the network devices. One of those southbound APIs is OpenFlow. Thus, depending on the exact configuration, OpenDaylight may be considered to implement SDN via APIs or Open SDN.
- As explained in Section 6.3.3, Juniper's Contrail is both an SDN via Hypervisor-Based Overlays approach as well as a NETCONF-based SDN via APIs approach.

In Fig. 6.13 we attempt to depict the possible overlap between our four major SDN approaches. The category SDN via Opening Up the Device is a difficult one for comparison. As Fig. 6.13 indicates, this category actually subsumes all the other categories and includes alternatives not yet contemplated. After all, if the device is truly opened up, nothing restricts that implementation to stick with any current or future paradigm. For example, Big Switch's Indigo provides OpenFlow switch software freely to device vendors who wish to use it. This should encourage ODMs to build standard hardware and software so

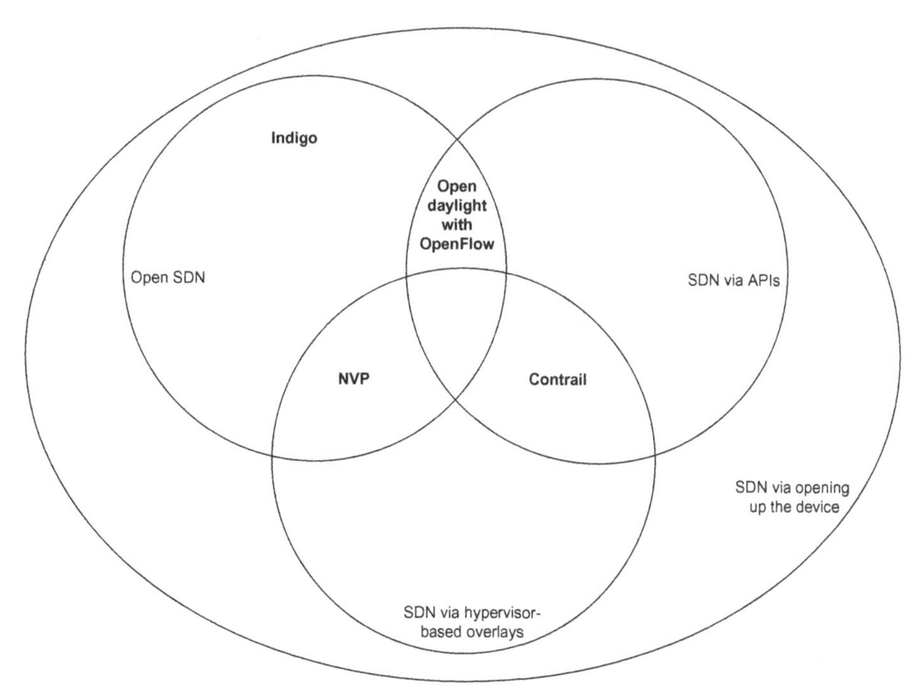

FIG. 6.13

SDN alternatives overlap.

that they can use software like Indigo to OpenFlow-enable their switch without developing the switch-side OpenFlow software. So, one could argue that Indigo is an example of SDN via opening up the device. But Indigo is clearly an example of Open SDN as it is an implementation of OpenFlow. For this reason, we exclude SDN via Opening Up the Device from Table 6.1, as the comparisons against something that can mean all things to all people are not reasonable.

Before examining Table 6.1 in detail, we remind the reader to recall certain constraints on our definitions of the alternatives:

- We restrict our definition of both SDN via APIs and SDN via Hypervisor-based Overlays to controller-based approaches.
- We restrict our definition of SDN via APIs to those approaches where the APIs reside on the devices and are used to program a locally resident control plane on those devices
- It is possible to implement SDN via Hypervisor-based Overlays using Open SDN. We thus define two subcategories for this alternative. The first subcategory is not based on Open SDN, and the second is using Open SDN and, when appropriate, the ranking will reflect those two subcategories (i.e., [*notOpenSDN*, *OpenSDN*]).

DISCUSSION QUESTION:

We have described different alternatives to SDN. Which SDN alternative do you think is the most promising for the future of networking? The most revolutionary? The most pragmatic? The most likely to succeed in the long run?

6.7 CONCLUSION

In this chapter we have examined some well-publicized drawbacks of Open SDN and looked at three alternative definitions of SDN. We have examined the ways in which each of these approaches to SDN meets our definition of what it means to be SDN, as well as to what degree they suffer from those same well-publicized drawbacks. Each of the alternatives, SDN via APIs, SDN via Hypervisor-Based Overlays, and SDN via Opening Up the Device, has strengths and areas where they address real SDN-related issues. Conversely, none of the alternatives matches our criteria for what constitutes an SDN system as well as does Open SDN. Thus, our primary focus as we study different use cases in Chapters 8 and 9 will be how Open SDN can be used to address these real life problems. We will consider these alternative definitions where they have proven to be particularly effective at addressing a given use case. Before that, in the next chapter we survey new models of SDN protocols, controllers and applications that have emerged since this book was first published.

REFERENCES

[1] Silva AS, Smith P, Mauthe A, Schaeffer-Filho A. Resilience support in software-defined networking: a survey. Comput Netw 2015;92(1):189–207.

[2] Casado M. Scalability and reliability of logically centralized controller. Stanford CIO Summit, June 2010.

[3] Wireless LAN Controller (WLC) Mobility Groups FAQ. Cisco Systems. Retrieved from: http://www.cisco.com/en/US/products/ps6366.
[4] HP MSM7xx Controllers Configuration Guide. Hewlett-Packard, March 2013.
[5] Yeganeh S, Ganjali Y. Kandoo: a framework for efficient and scalable offloading of control applications. In: Hot Topics in Software Defined Networking (HotSDN), ACM SIGCOMM, August 2012, Helsinki; 2012.
[6] Curtis AR, Mogul JC, Tourrilhes J, Yalagandula P, Sharma P, Banerjee S. Devoflow: scaling flow management for high-performance networks. In: Proceedings of the ACM SIGCOMM 2011 conference, SIGCOMM'11. New York, NY, USA: ACM; 2011. p. 254–65.
[7] Veisllari R, Stol N, Bjornstad S, Raffaelli C. Scalability analysis of SDN-controlled optical ring MAN with hybrid traffic. In: IEEE international conference on communications (ICC); 2014. p. 3283–8. http://dx.doi.org/10.1109/ICC.2014.6883827.
[8] Zhang Y, Natarajan S, Huang X, Beheshti N, Manghirmalani R. A compressive method for maintaining forwarding states in SDN controller. In: Proceedings of the third workshop on hot topics in software defined networking, HotSDN'14. New York, NY, USA: ACM; 2014. p. 139–44. http://dx.doi.org/10.1145/2620728.2620759.
[9] Zhou W, Li L, Chou W. SDN northbound REST API with efficient caches. In: IEEE international conference on web services (ICWS). 2014. p. 257–64. http://dx.doi.org/10.1109/ICWS.2014.46.
[10] Narayanan R, Kotha S, Lin G, Khan A, Rizvi S, Javed W, et al. Macroflows and microflows: enabling rapid network innovation through a split SDN data plane. In: European Workshop on Software Defined Networking (EWSDN), October 25–26, Darmstadt; 2012.
[11] Creating innovative embedded applications in the network with the Junos SDK. Juniper Networks, 2011. Retrieved from: http://www.juniper.net/us/en/local/pdf/whitepapers/2000378-en.pdf.
[12] TR-069 CPE WAN Management Protocol. Issue 1, Amendment 4, Protocol Version 1.3, July 2011. The Broadband Forum—Technical Report; 2011.
[13] Configuring Policy-Based Routing. Cisco Systems. Retrieved from: http://www.cisco.com/en/US/docs/switches/lan/catalyst4500/12.2/20ew/configuration/guide/pbroute.html.
[14] Cisco onePK Developer Program. Cisco Systems. Retrieved from: http://developer.cisco.com/web/onepk-developer.
[15] Open Daylight Technical Overview. Retrieved from: http://www.opendaylight.org/project/technical-overview.
[16] Bringing SDN to reality. Arista Networks, March 2013. Retrieved from: http://www.aristanetworks.com.
[17] Nsx. VMware. Retrieved from: https://www.vmware.com/products/nsx/.
[18] McGillicuddy S. Contrail: the Juniper SDN controller for virtual overlay network. Techtarget, May 9, 2013. Retrieved from: http://searchsdn.techtarget.com/news/2240183812/Contrail-The-Juniper-SDN-controller-for-virtual-overlay-network.
[19] Chua R. IBM DOVE Takes Flight with New SDN Overlay, Fixes VXLAN Scaling Issues. SDN Central, March 26, 2013. Retrieved from: http://www.sdncentral.com/news/ibm-dove-sdn-ve-vxlan-overlay/2013/03.
[20] OpenWRT—Wireless Freedom. Retrieved from: https://openwrt.org/.
[21] Intel DPDK: Data Plane Development Kit. Retrieved from: http://www.dpdk.org.
[22] McGillicuddy S. Intel DPDK, switch and server ref designs push SDN ecosystem forward. Techtarget, April 23, 2013. Retrieved from: http://searchsdn.techtarget.com/news/2240182264/Intel-DPDK-switch-and-server-ref-designs-push-SDN-ecosystem-forward.
[23] Pate P. NFV and SDN: what's the difference? SDN Central, March 30, 2013. Retrieved from: http://www.sdncentral.com/technology/nfv-and-sdn-whats-the-difference/2013/03.
[24] Jacobs D. Network functions virtualization primer: software devices take over. Techtarget, March 8, 2013. Retrieved from: http://searchsdn.techtarget.com/tip/Network-functions-virtualization-primer-Software-devices-take-over.

CHAPTER

EMERGING PROTOCOL, CONTROLLER, AND APPLICATION MODELS

7

The landscape of SDN technologies is changing on a day-to-day basis. This makes keeping a book about SDN up to date more challenging than would be the case for a more established technology. In this chapter we explore new and emerging trends in SDN.

7.1 EXPANDED DEFINITIONS OF SDN

When the first edition of this book was published, the commercial SDN landscape was fluid as different vendors and startups sought to develop SDN solutions to customer issues. However, from a technological standpoint, SDN was fairly stable. SDN had emerged from universities in the years following the *Clean Slate* program we reviewed in Section 3.2.6. Most vendors planned to use OpenFlow in their SDN solutions. In the past few years, as major networking incumbents entered the SDN marketplace, we have begun to see changes. In the next sections we examine the impact these vendors have brought to bear on the SDN industry.

7.1.1 IMPACT OF MAJOR NEMS IN THE SDN ARENA

In the earlier days of SDN, the networking vendors most involved were those more interested in industry disruption. These were the vendors who would benefit most from a change to the status quo. They wanted to establish a beachhead in the various markets that were dominated by the larger vendors such as Cisco. Therefore, companies such as Hewlett-Packard, IBM, and NEC, as well as a number of startups, developed solutions oriented around Open SDN and the OpenFlow protocol.

When SDN began to take root and major vendors such as Cisco and Juniper took notice, this landscape and even the very definition of SDN began to shift. In Sections 4.6.1 and 6.2 we discussed the definition of *SDN via APIs*, which relied not on the new OpenFlow-based models of implementing SDN, but on more traditional models based on existing protocols. This contest has continued since the first edition of this book, with momentum swinging away from Open SDN toward API-based SDN solutions involving existing protocols.

For established vendors this achieved different goals, the priority of which depend on one's perspective. Two of these goals were:

- **Customer Protection**: One argument in favor of using traditional protocols is that it protects customer investments in current devices and technologies. Customers generally do not want to replace their networking equipment wholesale, and even a gradual evolution to a new paradigm carries potential risk and must be carefully evaluated.

Software Defined Networks. http://dx.doi.org/10.1016/B978-0-12-804555-8.00007-7

- **Vendor Protection**: Another argument is that using existing protocols and devices helps established vendors continue to dominate markets in which they are already the leader.

The emerging trends toward legacy protocols such as *Network Configuration Protocol* (NETCONF) and *Border Gateway Protocol* (BGP), and toward the SDN controllers that support these protocols are discussed in the sections that follow.

DISCUSSION QUESTION:

Discuss whether you believe that the influence on SDN by dominant networking vendors is a positive or negative influence on the advancement of networking technology.

7.1.2 NETWORK MANAGEMENT VERSUS SDN

One of the protocols that is being used for the application of SDN-based policies is NETCONF, which we examine in detail in the sections that follow. But the use of such a protocol, which was developed specifically as a means of improving the effectiveness of network management, raises the issue of where network management ends and SDN begins. Is this type of solution just an improved network management or is it really software defined networking?

In Chapter 6 we compared and contrasted three classes of SDN solutions: *Open SDN*, *SDN via APIs*, and *SDN via Overlays*. As that chapter illustrated, it is difficult to precisely circumscribe SDN. For the purposes of this discussion, since network management in general shares many of the same attributes as SDN (i.e., centralized control, network-wide views, network-wide policies), we consider such network management-based solutions to also fall under the larger SDN umbrella.

Fig. 7.1 shows the spectrum of solutions being promoted as SDN. On the right hand side of the picture is *reactive* OpenFlow, which involves packets getting forwarded to the controller via PACKET_IN messages. This type of SDN solution is the most dramatically different from traditional networking technologies, as highlighted in the figure. At the other end of the spectrum is network management, which is the most similar to what we see in traditional networks today. Between those extremes reside NETCONF, *Border Gateway Protocol Link State* (BGP-LS) and *Path Computation Element Protocol* (PCE-P), which we describe later in this chapter. Note that the term PCEP is frequently used as a substitute for PCE-P.

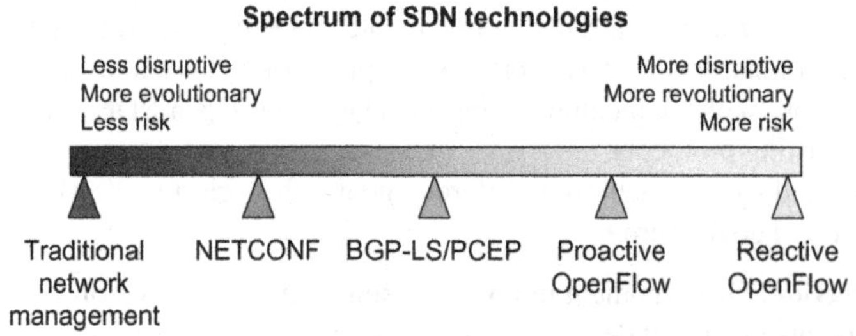

FIG. 7.1

SDN spectrum.

These SDN solutions all are considered valid approaches to SDN, based on the current and common use of the term, although they differ in their approach, their implementation and their suitability for various customer and domain needs. Each has benefits and limitations, which we examine next. The reader may wish to occasionally refer to Fig. 7.1 as our discussion below will generally follow the figure moving from left to right.

7.1.3 BENEFITS OF NETWORK MANAGEMENT-BASED SDN

We list here some benefits of SDN based on an evolved version of network management:

- **Least Disruptive**: This type of SDN is least disruptive to the current operation of the network because it operates with the current network infrastructure and capabilities of the networking staff.
- **Least Costly**: This type of SDN does not require new equipment, nor does it require a great deal of new training of IT personnel.
- **Least Risky**: This type of SDN introduces the least amount of risk into the network, since it runs on the same hardware and device software, with changes to networking behavior limited to what the SDN application is able to do through more traditional network management channels.

These advantages make network management-based SDN attractive to customers who want to eventually reach an SDN-based future through a gradual, more evolutionary path. Indeed, such solutions do move networking technology toward an SDN-based future.

7.1.4 LIMITATIONS OF NETWORK MANAGEMENT-BASED SDN

While there are benefits of network management-based SDN, there are limitations as well:

- **Limited Improvement**: Because it is restricted to using current devices, often without the capabilities for controlling forwarding and shaping of traffic, this type of SDN is limited in how much improvement it can provide in a network.
- **Limited Innovation**: Because this type of SDN is restricted to using currently configurable functionality in devices, it limits opportunities for truly innovative networking.
- **Limited Business Opportunity**: From an entrepreneurial standpoint, this type of SDN may not provide the opportunity to create disruptive and revolutionary new players in the networking device industry.

Customers with established networking environments may decide that the benefits of network management-based SDN outweigh the limitations, while others may opt for the more radical change offered by protocols such as OpenFlow.

DISCUSSION QUESTION:

Network management has been around for some time, and was never considered any type of "software defined networking." Now it is. Discuss whether you think that network management-based SDN should be considered "real" SDN or not, and defend your point of view.

7.2 ADDITIONAL SDN PROTOCOL MODELS

This book has predominantly focused on OpenFlow as the original and the most prominent SDN protocol being used in research, entrepreneurial efforts, and even in some established commercial environments (e.g., Google). However, current trends indicate that much SDN development today focuses on other protocols. We examine these protocols, controllers, and application development trends in the following sections.

7.2.1 USING EXISTING PROTOCOLS TO CREATE SDN SOLUTIONS

Assuming that one of the goals of emerging SDN solutions is to reduce risk and to make use of existing customer and vendor equipment, it is consistent that these solutions utilize existing protocols when feasible. Established protocols such as NETCONF, BGP, and *Multiprotocol Label Switching* (MPLS) are all potentially relevant here. These are mature protocols that are used in massively scaled production networks. Utilizing these protocols to implement SDN solutions makes sense for those interested in directing their SDN efforts along an evolutionary path utilizing existing technologies.

Fig. 7.2 shows a real-life example of existing protocols being used to create an SDN solution. The figure depicts some of the main components involved in the *OpenDaylight* (ODL) controller's BGP-LS/PCE-P plugin. Starting at the top of the figure, we see that there are three distinct protocols involved:

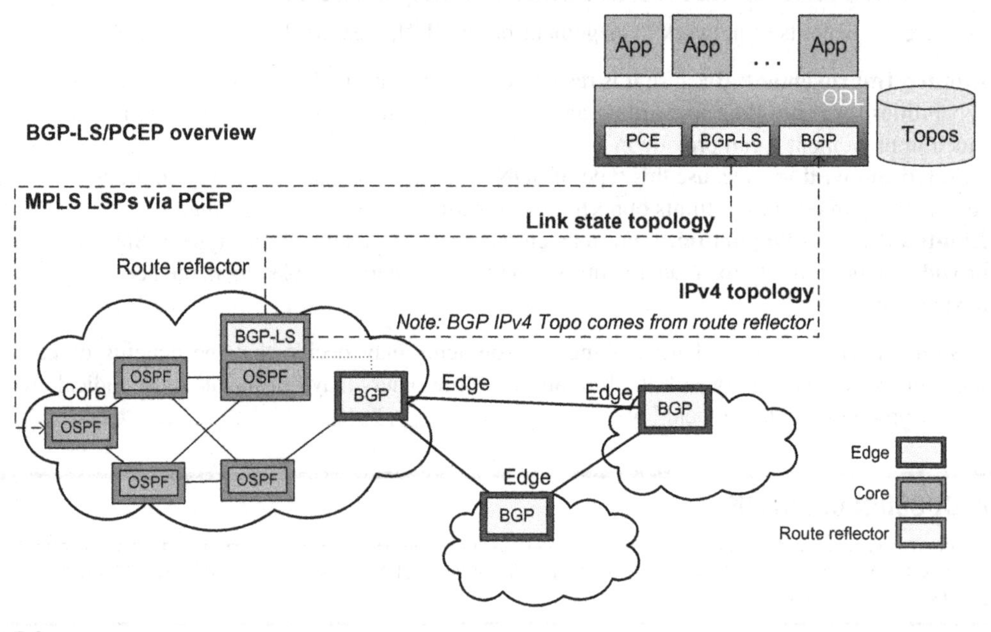

FIG. 7.2

BGP-LS/PCE-P overview.

- **BGP-LS**: The BGP-LS protocol is used by ODL to gather link state topology information from the routing protocols running in the clouds in the figure. This topology reflects routers and interconnecting links within the *Open Shortest Path First* (OSPF) or *Intermediate System to Intermediate System* (IS-IS) domains.
- **BGP**: The BGP protocol is used by ODL to gather IP *Exterior Gateway Protocol* (EGP) topology from the BGP routers connecting the clouds (domains) in the picture.
- **PCE-P**: The PCE-P protocol is used by ODL to configure MPLS *Label Switched Paths* (LSPs) for forwarding traffic across those networks.

The SDN solution in Fig. 7.2 will be discussed further in the following sections. The interested reader can find detailed information in [1]. In the next sections we will examine these protocols as well as NETCONF in order to understand their roles in SDN.

As we consider this use of existing protocols, a helpful perspective may be to look at the different control points shown in Fig. 7.3 that are managed and configured by the SDN application. These control points are *Config*, where general configuration is done, *Routing Information Base* (RIB), where routes (e.g., prefixes and next-hops) are set, and *Forwarding Information Base* (FIB), which is lower level and can be considered *flows*, where packet headers are matched and actions are taken.

The association between these control points and existing protocols is shown in Table 7.1. The table shows control points in general terms. NETCONF's role is for setting configuration parameters. BGP is

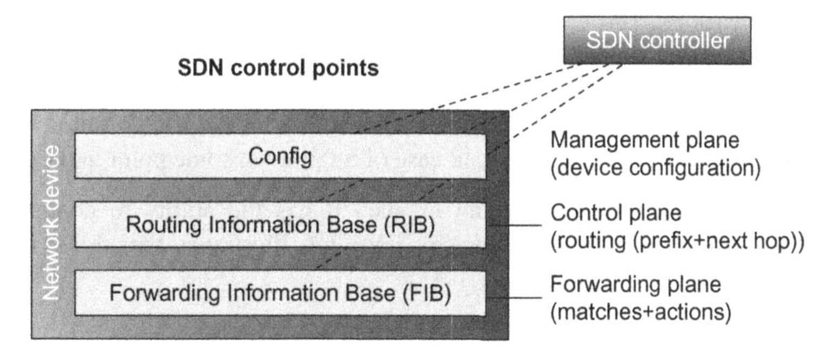

FIG. 7.3

SDN control points.

Table 7.1 Comparison of Existing Protocols for SDN

Protocol	Control Point	Details
NETCONF	Config	Interfaces, ACLs, Static routes
BGP-LS	-	Topology discovery is used to pass link-state IGP information about topology to ODL.
BGP	RIB	Topology discovery and setting RIB
PCE-P	MPLS	PCE to set MPLS LSPs. Used to transmit routing information from the PCE Server to the PCE Clients in the network.
BGP-FS	Flows	BGP-FlowSpec to set matches and actions

involved in setting RIB entries, and PCE-P is used for setting MPLS paths through the network. BGP-LS is used to gather topology information from the RIB. *BGP-FlowSpec* (BGP-FS) is employed to set matches and actions, similar to what is done with OpenFlow, using instead the BGP-FS functionality of the router. BGP-FS leverages the BGP *Route Reflection* infrastructure and can use BGP *Route Targets* to define which routers get which routes. Unlike OpenFlow, BGP-FS does not support layer 2 matches but only layer 3 and above.

7.2.2 USING THE NETCONF PROTOCOL FOR SDN

NETCONF is a protocol developed in an *Internet Engineering Task Force* (IETF) working group and became a standard in 2006, published in *Request for Comments* (RFC) 4741 [2] and later revised in 2011 and published in RFC 6241 [3]. The protocol was developed as a successor to the *Simple Network Management Protocol* (SNMP) and attempted to address some of SNMP's shortcomings. Some key attributes of NETCONF are:

- **Separation of configuration and state (operational) data.** Configuration data is set on the device to cause it to operate in a particular way. State (operational) data is set by the device as a result of dynamic changes on the device due to network events and activities.
- **Support for *Remote Procedure Call* (RPC)-like functionality.** Such functionality was not available in SNMP. With NETCONF, it is possible to invoke an operation on a device, passing parameters and receiving returned results, much like RPC calls in the programming paradigm.
- **Support for Notifications.** This capability is a general event mechanism, whereby the managed device can notify the management station of significant events. Within SNMP this concept is called a trap.
- **Support for transaction-based configurations.** This allows for the configuration of multiple devices to be initiated, but then rolled back in case of a failure at some point in the process.

NETCONF is a management protocol and as such it has the ability to configure only those capabilities which are exposed by the device. Fig. 7.4 illustrates the difference between a

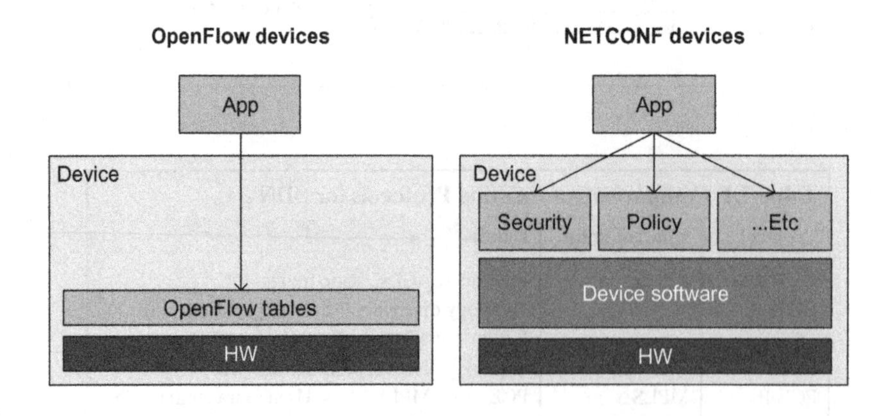

FIG. 7.4

NETCONF versus OpenFlow.

NETCONF-controlled and an OpenFlow-controlled device. We see in the figure that OpenFlow configures the lower levels of the networking device, that is, the ASIC containing the TCAM. This entails setting the matches and actions of the FIB.

NETCONF, on the other hand, performs traditional configuration of the device but via the NETCONF protocol rather than via the CLI or SNMP. The SDN application developer is limited by what the device exposes as configurable. If the device exposes NETCONF data models that allow for the configuration of *Access Control Lists* (ACLs), then the application can configure those. Similarly, if the device has data models for configuring *Quality of Service* (QoS) or static routes, then those will be possible as well.

The NETCONF programmer can learn which capabilities are exposed via the *Yet Another Next Generation data modeling language* (YANG) data models supported by the device.

NETCONF and YANG

NETCONF itself is a network management protocol, just as SNMP is a network management protocol. Such protocols become useful via the data models which convey information and requests to and from the device. With SNMP, the data models took the form of a *Management Information Base* (MIB), which we defined using *Structure of Management Information* (SMI). Contrasting the SNMP and NETCONF paradigms, SMI is analogous to YANG and the MIB is analogous to the YANG data model.

YANG provides a standardized way for devices to support and advertise their capabilities. One of the first operations that takes place between a NETCONF client on the controller and a NETCONF server running on the device is for the device to inform the client which data models are supported. This allows the SDN application running on the controller to know which operations are possible on each device. This granularity is key, since different devices will often vary in their capabilities. One of the current drawbacks of NETCONF and YANG is that different vendors often support different YANG models. This sometimes even occurs within different product families from the same vendor. Unlike the case of SNMP and standard MIBs (e.g., MIB-II, the Interfaces MIB, and the RMON MIB), there is currently no consistent set of YANG data models supported across the industry. It is currently necessary for applications to request and set data on different devices.

YANG models are still relatively new, and it is likely that standardized models will be defined in the near future. This will be facilitated by the fact that modern networking devices have better internal configuration schemas than in the early days of SNMP, so for most vendors it is relatively easy to auto-generate YANG data models that map onto those schemas. Note that in our discussions about NETCONF throughout this book we assume that YANG is used as its data modeling language.

NETCONF and RESTCONF

NETCONF uses the *Extensible Markup Language* (XML) to communicate with devices, which can be cumbersome. Fortunately, SDN controllers often support *REST-based NETCONF* (RESTCONF) as a mechanism for communicating between the controller and devices. RESTCONF works like a general REST API in that the application will use HTTP methods such as GET, POST, PUT, and DELETE in order to make NETCONF requests to the device.

As with normal REST communication, the URL specifies the specific resource that is being referenced in the request. The payload used for the request can be carried in either XML or *JavaScript Object Notation* (JSON).

Software developers with web programming backgrounds often find RESTCONF easier to work with than traditional use of NETCONF. This is due to the fact that REST APIs and JSON are generally more familiar to web developers than XML RPCs. Hence, using RESTCONF makes communication between the SDN controller and devices much simpler than it might be otherwise.

7.2.3 USING THE BGP PROTOCOL FOR SDN

Another protocol being promoted as a mechanism for SDN solutions is BGP. As we explained in Section 1.5.2, BGP is the EGP routing protocol used in the Internet. In addition to this traditional role, it is also used internally in some data centers. Consequently, the prospect of configuring BGP routes dynamically in a software defined manner is appealing. There are two major aspects of the BGP functionality currently used in ODL. These are:

- **IPv4 Topology**: The BGP plugin running inside ODL is implementing an actual BGP node, and as such it has access to topological information via the *Route Reflector* (RR). This information provides the topology between devices implementing the EGP, often referred to as the IPv4 topology. Fig. 7.5 shows an EGP network with routers supporting BGP, and an RR communicating topology information to the BGP node running inside the ODL controller. This information helps to provide the network-wide views characteristic of SDN solutions, and it can be used to dynamically configure intelligent routing paths throughout the network, via RIB configuration. This network-wide view is seen in Fig. 7.5 in the network topology to the right of the ODL controller. Note that while we specifically cite the IPv4 topology here, other topologies, such as the IPv6 topology, can be reported by the BGP plugin.

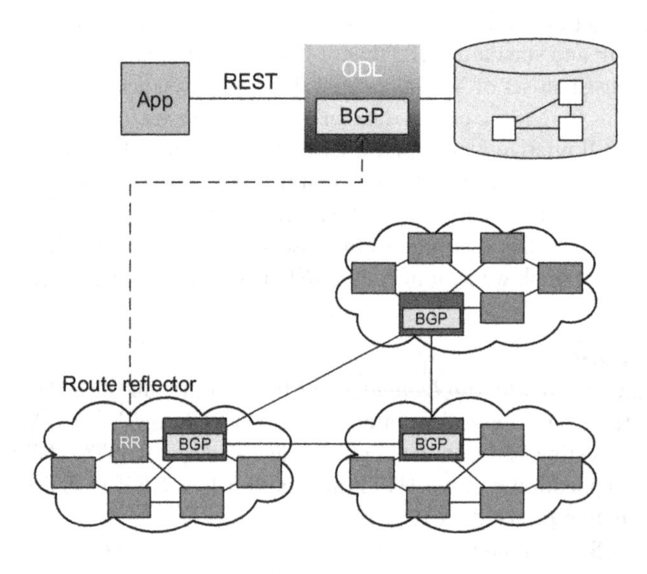

FIG. 7.5

SDN BGP topology.

- **RIB Configuration**: Within ODL there are APIs for creating a RIB application which can be used to inject routes into the network, based on the topology information, traffic statistics, congestion points to be avoided, as well as other possible relevant data.

A key aspect of this technique is that the ODL's controller's BGP plugin appears to the network as a normal BGP node. Note that the plugin does not advertise itself as a next hop for any of the routes. It will, however, effect changes on the adjacent nodes that will in turn propagate routing information throughout the network. In this way, an SDN application running on ODL can force RIB changes throughout the network, thereby achieving SDN-like agility and automatic network reconfiguration.

7.2.4 USING THE BGP-LS PROTOCOL FOR SDN

Figs. 7.6 and 7.7 depict the operation of the BGP-LS/PCE-P plugin on ODL.

- **BGP-LS** is used to pass link-state (OSPF or IS-IS) *Interior Gateway Protocol* (IGP) information about topology to ODL.
- **PCE-P** is used to transmit routing information from the PCE Server to the PCE clients in the network. A PCE client is also more simply known as a *Path Computation Client* (PCC).
- **MPLS** will be used to forward packets throughout the network, using the Label Switched Paths (LSPs) transmitted to head-end nodes via PCE-P.

Fig. 7.6 illustrates an IGP network of OSPF-supporting routers, sharing topology information with ODL. At a high level, BGP-LS is running on one of the OSPF (or another IGP) nodes in the network, and the IGP shares topology information with BGP-LS running on that node. That BGP-LS node in turn shares the topology information with the BGP-LS plugin running in ODL. That topology information is made available to the SDN application running on ODL, which can combine that knowledge with

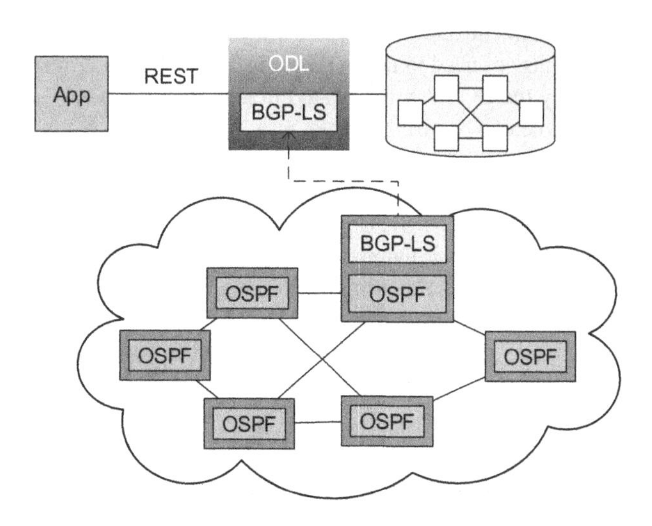

FIG. 7.6

SDN BGP-LS topology.

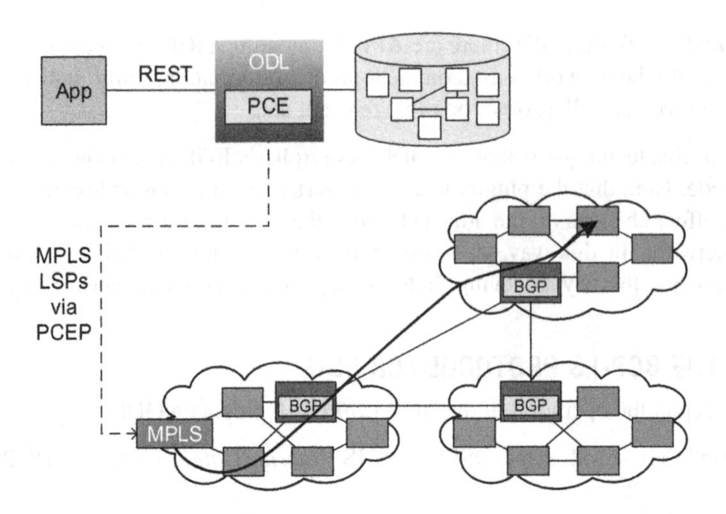

FIG. 7.7

SDN PCE-P and MPLS.

other knowledge about congestion, traffic, bandwidth, prioritization policies, and the like. This can be combined by the SDN application, which will determine optimal routing paths, and will communicate those MPLS paths to the PCCs in the network.

7.2.5 USING THE PCE-P PROTOCOL FOR SDN

PCE and its associated protocol PCE-P, have been in existence since roughly 2006 [4] and address the need to compute complex paths through IGP networks, as well as across *Autonomous System* (AS) boundaries via BGP. These paths are used in networks that support *MPLS Traffic Engineering* (MPLS-TE). The computation done by the PCE can be located in any compute node—in an MPLS head-end router, in the cloud, or on a dedicated server.

In Fig. 7.7, ODL (driven by an SDN application) sets MPLS *Label Switched Paths* (LSPs) using PCE-P. Communication is between the PCE server in ODL and the PCC on the MPLS router.

The PCC runs on the *head-end* of each LSP. Using these LSPs, the router is able to route traffic using MPLS through the network in an optimal manner. Using PCE-P in this fashion has advantages over a pre-SDN counterpart called *Constrained Shortest Path First* (CSPF). Like our PCE-P model described above, CSPF computes the LSPs but is limited to the topology of the IGP domains to which it belongs. Conversely, PCE-P can run across multiple IGP domains. Another advantage of PCE-P is that it can perform global optimization contrary to the CSPF model where each head-end router performs local optimization only.

7.2.6 USING THE MPLS PROTOCOL FOR SDN

MPLS will be used to forward packets throughout the network using the LSPs transmitted to head-end nodes via PCE-P. In the SDN solution described in the previous section the role of MPLS is to forward traffic according to the paths configured by the SDN application running on ODL. Configuration takes place on the MPLS head-end router shown in Fig. 7.7. This router will receive the matching packets at the edge of the MPLS network, and packets will then be forwarded on the LSP that has been configured by PCE-P.

This solution does not require a new protocol such as OpenFlow in order to control the network's forwarding behavior, but rather uses these existing protocols (i.e., BGP, BGP-LS, PCE-P, and MPLS) to achieve intelligent routing of packets based on paths that have been configured by the centralized controller. This emerging solution holds promise for customers looking to utilize their existing infrastructure in a new, SDN-based manner.

DISCUSSION QUESTION:

Both the NETCONF and the BGP-LS/PCE-P southbound protocol plugins attempt to provide SDN capabilities using existing protocols. Which of these two seems to be more "SDN" and why? And what are some potential dangers in using BGP to set RIBs in the SDN controller?

7.3 ADDITIONAL SDN CONTROLLER MODELS

In this section we turn our attention to SDN controllers and controller technologies that have gained recent prominence in SDN. Hot topics in recent SDN controller innovation have included southbound protocol plugins, internal architectures, service provider solutions, scalability, and northbound interfaces to SDN applications.

As the dust begins to settle on SDN, two commercially viable controllers stand out: ODL and *Open Network Operating System* (ONOS). In the discussion that follows we are interested in the technologies used by these controllers. There are also business ramifications of the dominance that these controllers have asserted, but we defer our treatment of those until Chapter 14. Table 7.2 lists the primary areas of emphasis of each of these two controllers.

We caution the reader that the emphases denoted in Table 7.2 are just that—*emphases*. One should not infer that the other controller ignores these issues. For example, ONOS has projects for alternative southbound protocols like NETCONF and PCE-P, and ODL has projects and functionality related to service providers, scalability, and *intents*. We will see in the following sections that both controllers implement functionality in all of these areas.

7.3.1 CONTROLLERS WITH MULTIPLE SOUTHBOUND PLUGINS

Prior to the advent of ODL, almost every general-purpose SDN controller used the OpenFlow protocol as the sole southbound protocol. Other controllers were either not targeted at the open application development community or did not garner a sizeable community of developers. To gain traction,

Table 7.2 ODL and ONOS

Controller	Organization	Emphases
ODL	Linux Foundation	Multiple southbound, MD-SAL
ONOS	ON.Lab	Service providers, scalability, *intents*

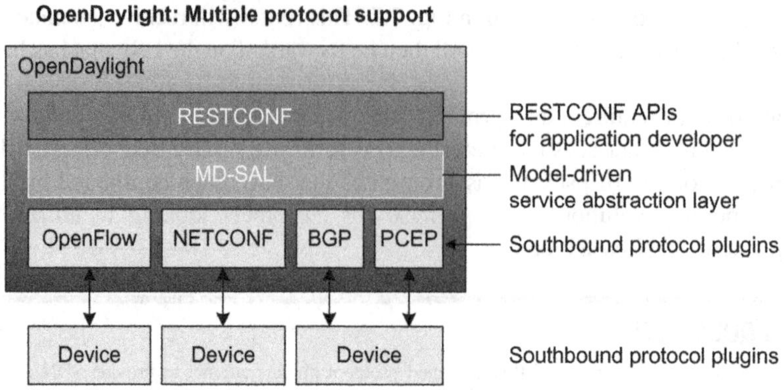

FIG. 7.8

ODL southbound protocol plugins.

an open controller effort requires a cadre of developers, some willing to write applications for that controller and others contributing to the controller's code base itself. OpenDaylight has emerged at the forefront of open SDN controller solutions.

OpenDaylight has enjoyed this success due to the support of the Linux Foundation and Cisco. In fact, the core of the ODL controller initially was Cisco's own XNC product. Another catalyst of its meteoric rise is the fact that ODL proposed and delivered on the promise of providing an SDN controller that supports not only OpenFlow, but also other suitable southbound protocol plugins. Consequently protocols such as NETCONF, SNMP, BGP, and PCE-P have gained traction as potential southbound protocols for SDN using the ODL controller.

Fig. 7.8 provides a high-level architectural view of ODL and its support for multiple southbound plugins. This affords customers an SDN controller that works with their legacy devices. The limitations of the protocol corresponding to the plugin are still present, but the architecture in Fig. 7.8 benefits from the SDN model of making networks software-centric and centrally programmed.

7.3.2 CONTROLLERS WITH MODEL-DRIVEN INTERNAL DESIGN

For some time, controllers and network management platforms have utilized *model-based* technologies for communicating with devices, particularly since object-oriented programming became prevalent in the industry. ODL, with significant implementation commencing in the Helium release and continuing with the Lithium release, builds upon this theme by making all internal ODL applications (i.e., modules) adhere to their *Model-Driven Service Abstraction Layer* (MD-SAL) architecture. The goals of MD-SAL are as follows:

- **Abstraction**: Earlier versions of ODL resulted in every southbound protocol having its own protocol-specific set of APIs. This clearly is not desirable for application developers, having to develop against varying APIs for multiple different protocols. MD-SAL permits communication to devices only through models, thus yielding an abstraction that provides protocol-independent APIs for application developers to use.

- **Standardized Communication**: When an MD-SAL module is created, APIs for the module are automatically generated by tools included within the MD-SAL environment. Thus, these APIs (both RESTful and internal) are more standardized than would otherwise be true if they were created ad hoc by human developers. Significantly, these APIs are generated *automatically* by the build framework, and thus require no extra design or implementation effort by the API developer. This facilitates easier intermodule and interapplication communication.
- **Microservices**: Microservices are a software architecture style in which complex applications are composed of small, independent processes communicating with each other using standardized APIs. MD-SAL facilitates such an environment in ODL through the use of YANG models to define every service. Hence every MD-SAL application is a service with its own auto-generated APIs.

7.3.3 CONTROLLERS TARGETING SERVICE PROVIDER SOLUTIONS

Since the early days of SDN, service providers have played a major role in defining standards and helping to make the new technology suitable for the largest of networks. Companies such as AT&T, Deutsche Telekom, and NTT were all key players in the *Open Networking Foundation*, (ONF) as well as delivering featured keynotes at various SDN conferences. The two dominant emerging controller technologies, ODL and ONOS have both targeted the service provider market and have developed their products accordingly.

ODL and service providers

The first release of ODL, Hydrogen, had a specific distribution of the controller geared toward service providers. Since that time, vendors such as Brocade and Cisco have openly courted service providers with their branded versions of ODL, the Brocade SDN Controller and the Cisco Open SDN Controller, respectively.

Although service providers have participated in the definition of OpenFlow, they also have massive networks of existing equipment upon which vast amounts of traffic and numerous organizations depend. Converting this established environment to OpenFlow, no matter how promising the potential benefits, is not a strategy these service providers will embrace overnight. Rather, most of these large organizations will pursue a more evolutionary approach, utilizing NETCONF configuring existing devices. Consider that Deutsche Telekom announced in 2013 [5] that it was selecting Tail-f as its SDN technology of choice (Tail-f provides solutions primarily involving NETCONF). Likewise, AT&T's Domain 2.0 project [6] uses NETCONF with ODL in order to reach their stated goal of being 75% SDN-based by 2020 [7].

Service providers also have very large deployments of IPv4 and IPv6 BGP. One way ODL supports those service providers is through the implementation of *BGP Monitoring Protocol* (BMP). Through this mechanism ODL can learn about best performing routes in the network, which can be useful for determining optimal paths and balancing traffic across the network.

ONOS and service providers

The ONOS controller primarily uses OpenFlow, but is built specifically for service providers. This service provider orientation is not related to the protocol used, but rather the controller's architecture. The ONOS controller is designed natively with a distributed core, meaning that ONOS scales to support massive numbers of network devices [8]. We look more closely at scalability and the architectures of both ODL and ONOS in the next section.

7.3.4 CONTROLLERS BUILT FOR SCALABILITY

For service providers as well as for the majority of networks today, scalability is of paramount importance, especially as the demands on networking and bandwidth continue to grow. This is self-evident for service providers and data centers. Even enterprise networks, with their growing need for speed and dependence on the Internet, require a level of performance and scalability not seen before. Hence, any SDN solution will need to cope with rapidly expanding workloads. Networking environments today rely on three interrelated attributes, *Scalability*, *Performance*, and *High-Availability* (HA):

- **Scalability**: Technologies such as the cloud and streaming video, combined with the advent of the *Internet of Things* (IoT) demand that SDN controllers scale to handle the growth in traffic as well as the exploding number of network devices and hosts.
- **Performance**: This increased load requires commensurate improvement in the performance, speed, and efficiency of network devices and from the controllers that manage them. Therefore network solutions must be able to adapt dynamically to the frequent changes in network traffic loads and congestion. Overloaded or sluggish controllers will only exacerbate the issues faced by the network in such situations.
- **HA**: The critical role played by SDN controllers creates increasing dependence on the uninterrupted availability of the SDN controller. While certain SDN solutions place higher demands for HA, virtually every SDN solution requires some amount of resiliency. Whether it is achieved via clustering, teaming, or some other form of active-active or active-passive strategy, the imperative is that production SDN solutions must provide constant controller availability.

Different controllers employ varied strategies for scalability, performance, and HA. The specific methodologies are in flux as limitations are confronted and new technologies evolve. In the balance of this section, we look at the strategies employed by ODL and ONOS. For more detail, we refer the interested reader to [9] for ODL and [10] for ONOS.

ODL and scalability, performance, and high-availability

The very first release of ODL, Hydrogen, was more of a proof-of-concept and, hence, did not feature robust scalability, performance, and HA. With Helium, and later with the Lithium release, ODL adopted an Akka-based [11] solution for HA. Akka was developed originally for Erlang [12] and was adopted for use in Scala and Java environments. It is an open-source toolkit with the goal of simplifying the creation of applications that run in a distributed environment. Much of Akka involves improvements on concurrency, an area whose challenges are normally exacerbated when running on multiple systems.

Some of the tenets of Akka are: (a) concurrency is message-based and asynchronous; (b) components interact with each other in the same manner, regardless of whether they are running on the same or different hosts; and (c) components adhere to a hierarchical model, allowing supervision of child nodes to help prevent or resolve failures.

Another feature of ODL's controller clustering solution involves the use of *shards* [13], which are slices of functionality that are deployed in a distributed manner in order to balance the load across the systems in a cluster. As of this writing, shards are organized based on components, and so

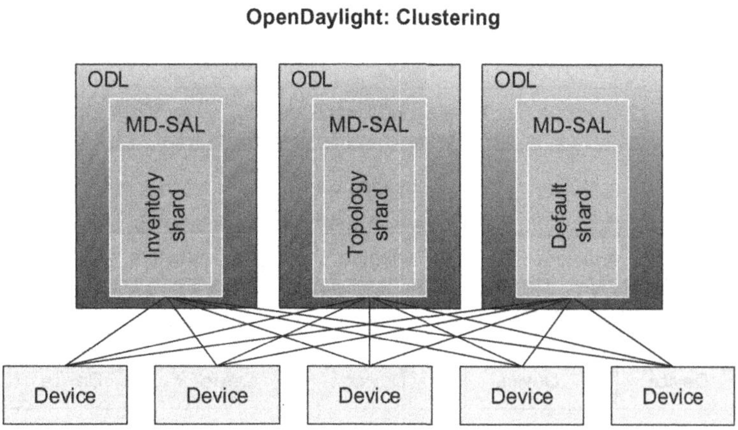

OpenDaylight: Clustering

ODL
MD-SAL
Inventory shard
ODL
MD-SAL
Topology shard
ODL
MD-SAL
Default shard
Device
Device
Device
Device
Device

FIG. 7.9

ODL clustering.

components of the same functionality are typically co-located for performance, which is advantageous in the nonfailed state, but which requires some recovery time if a host in the cluster goes down. It is also possible to design your cluster such that shards are split across the nodes in the cluster, such that the functionality is distributed, with a designated leader, thus resulting in faster recovery times, but potentially slower performance.

Fig. 7.9 illustrates an example of ODL clustering, with shards implementing inventory and topology, and a default shard for other running services. The different components may interact with any of the controlled devices. ODL's clustering strategy has expanded over the course of successive ODL releases as demands from service providers and others help to ensure that their needs regarding performance, scalability, and high-availability are met.

ONOS and scalability, performance, and high-availability

The ONOS controller was conceived and built for scalability and performance. Consequently, its initial design was distributed, allowing it to control massively scaled environments. The controller is built around the distributed core shown in Fig. 7.10. According to the diagram, ONOS distributes its *adapter* (protocol) functionality onto all instances of the controller, but has a single distributed core functionality shared across all hosts running ONOS. Fig. 7.10 also depicts a northbound API which is based on intents, which is described in the next section.

DISCUSSION QUESTION:

Compare and contrast OpenDaylight and ONOS. Which do you believe is more "SDN"? Which do you believe has the greatest chance of being successful as the dominant controller solution?

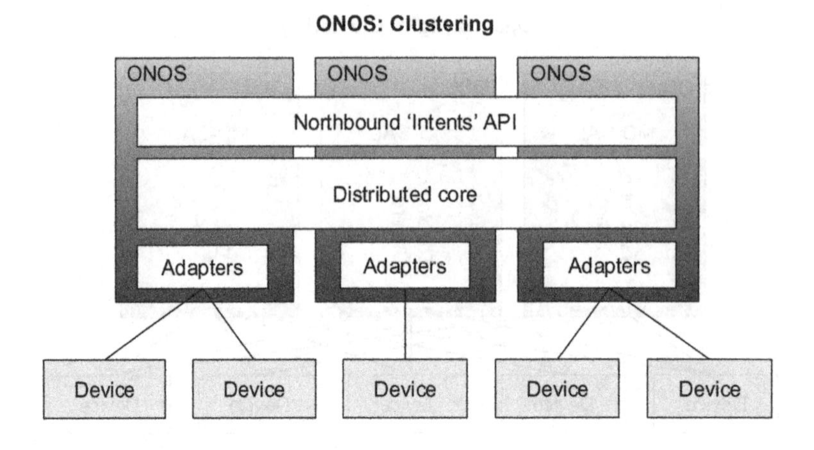

FIG. 7.10

ONOS clustering.

7.3.5 CONTROLLERS BUILT FOR INTENTS-BASED APPLICATIONS

In the early days of SDN, controllers were built with the sole purpose of creating a Java or REST API to allow SDN applications to program the OpenFlow tables resident in networking devices. This led to issues related to the fact that each SDN controller created its own API for setting OpenFlow rules. Each SDN application had to be developed as a function of the controller vendor.

This led to an attempt at standardization by the ONF. The ONF started a *Northbound Interface* (NBI) working group tasked with addressing this issue. Rather than define yet another northbound API, this working group defined an API at a level of abstraction above the base controller APIs, providing interface capabilities based on intents rather than on raw programming of OpenFlow tables.

As an example, an application wishing to provide some level of connectivity between two hosts A and B, using existing base controller APIs would program flows on switches all across the topology, from the switch connecting A to the network, to the switch connecting B to the network, including all switches along the path between the hosts. This clearly increases the complexity of the SDN application.

On the other hand, an intents-based API would allow the application to specify something more abstract, such as *the desire to allow host A to communicate with host B*. Invoking that API would result in the SDN controller performing flow programming of the switches on the path from host A to host B, assuming that OpenFlow was the SDN protocol in use. If some other type of protocol was being used, this would not perturb the application, since the controller would be responsible for making that connectivity happen with whatever southbound API was appropriate. Consider three features of this concept:

- **Abstraction**: The goal of an SDN controller, as with operating systems in general, is to *abstract* the details of the hardware below from the application running above.
- **Declarative**: Specifying *what* to do, rather than *how* to do it, is a characteristic of *declarative* systems.
- **Protocol-agnostic**: An abstract declarative interface hides details of how the network programming occurs, allowing different protocols to be used in different situations.

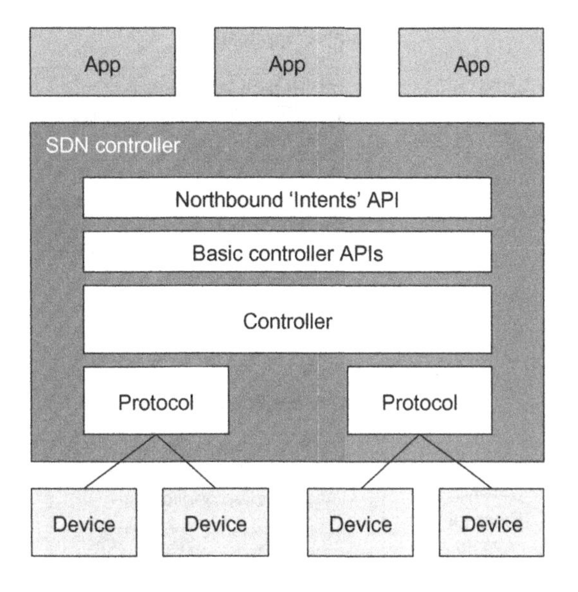

FIG. 7.11

Northbound intents-based interface.

Fig. 7.11 shows a controller with an intents-based API superimposed between the applications and the controller APIs. In [14] the reader will find a description of a functioning example of this concept. The intents-based API described in [14] actually runs on both ODL and ONOS. This is a step toward creating an API which frees SDN applications from having to know and understand the details of the devices in the physical network.

DISCUSSION QUESTION:

We discussed a number of important emerging attributes of SDN controllers. Comparing them to each other, which do you feel is the most important for SDN going forward? Which of them might be less important than the others?

7.4 ADDITIONAL APPLICATION MODELS

In addition to the introduction of new protocols for SDN, and evolving controller trends for SDN, there have also been shifts regarding SDN application models. We look at these in the next sections.

7.4.1 PROACTIVE APPLICATION FOCUS

Two fundamental classes of applications have emerged since the inception of SDN. Fig. 7.12 shows the two general designs of SDN applications, *proactive* and *reactive*. The figure reveals a number of differences, but for this discussion the most important difference between the two is that reactive

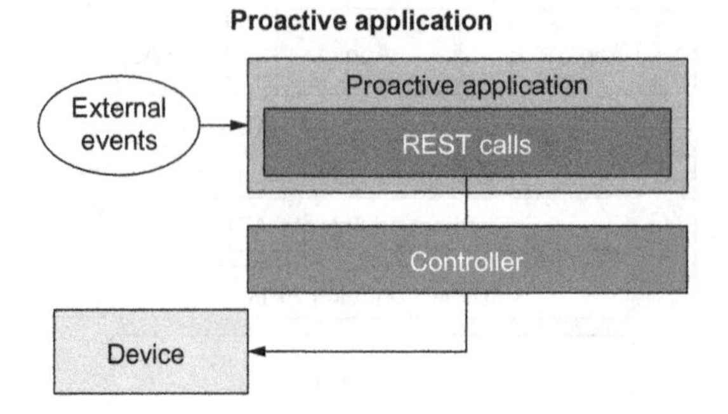

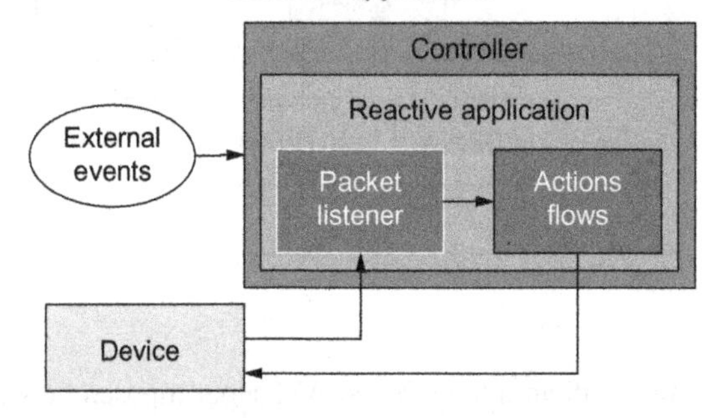

FIG. 7.12

SDN application types.

applications sometimes receive packets forwarded by the networking device. This is shown in the figure by the arrow that goes from the device to the *Packet Listener* component in the Reactive Application. We elaborate on the differences between these two applications classes in Chapter 12.

The idea that forwarding rules might not be programmed until related packets arrived was one of the more disruptive aspects of early SDN research. This concept fostered a wave of new innovation. However, reactive applications also engender concerns about scalability, reliability, and susceptibility to *Denial of Service* (DoS) attacks.

For these reasons, current emphasis on production applications focuses on proactive application designs. We defer detailed discussion of both classes of applications to Chapter 12. For this discussion, the main differentiators of a proactive application are: (1) setting of forwarding behavior a priori, rather than waiting for packets to be forwarded to the controller for processing, (2) some dynamism is achieved by responding to external events, but far less frequently than reactive applications, and (3) the ability to write applications that reside externally from the controller, which can be written in any programming language, since they often utilize the REST APIs of the controller.

The advantages of proactive applications involve simplicity, programming language flexibility, deployment flexibility, and security. We will delve more deeply into these advantages in Chapter 12.

7.4.2 DECLARATIVE APPLICATION FOCUS

As we discussed in Section 7.3.5, early SDN controller designs provided low-level APIs. Such a system, where the application developer dictates to each network device exactly how it was supposed to behave, is called an *imperative* system. In current programming and in systems development, the trend is toward *declarative* APIs and solutions. We distinguish imperative from declarative systems as follows:

- An **imperative** system requires that it be told exactly *how* to do something. The net result of all the *hows* is that the system will achieve the *what* that was intended in the first place.
- A **declarative** system needs only be told *what* needs to be done; the system itself figures out *how* to accomplish the goal.

An intents-based API clearly is more declarative in nature than the base APIs offered by controllers in the past. An internal architecture such as that provided by MD-SAL establishes the building blocks for creating declarative solutions. In fact, Brocade's SDN solution [15] has produced such an interface using MD-SAL, called the *Flow Manager* [16]. This MD-SAL application builds an intents-based API on top of ODL, yielding a more declarative API for application developers.

In addition to ODL and ONOS, other commercial SDN solutions such as Cisco's *Application Policy Infrastructure Controller—Data Center* (APIC-DC) and *Application Policy Infrastructure Controller—Enterprise Module* (APIC-EM), provide policy-level APIs for customers who want to write applications for those controllers. These APIs are oriented toward allowing developers to specify the *what*, allowing the system to decide *how* to make that happen.

7.4.3 EXTERNAL APPLICATION FOCUS

There is also an increased focus on creating external rather than internal SDN applications. The two types of applications are shown in Fig. 7.13. In most controllers, internal applications run inside the *Open Services Gateway initiative* (OSGi) [17] Java container in which the controller is running. Applications running internally must adhere to the standards defined by the internal controller environment. In general, there are more constraints when creating an internal application. External applications, however, have much more freedom:

- External SDN applications can be written in many different programming languages including Python, Ruby, PERL, and even scripting languages.
- External SDN applications can run on the same system as the controller, on a different but geographically local system, on a remote system, or even in the cloud.
- External SDN applications generally have a reduced impact on the operation of the controller (e.g., internal applications can cause failures or bottlenecks on the controller).

This relative ease of use and increased safety have been factors accelerating the migration toward external applications in SDN.

One drawback of external applications is the fact that REST APIs by themselves do not provide a mechanism for asynchronous notification back to the calling application. While this is true, some controller implementations do provide some mechanisms to achieve this goal. This is often done using some socket-based functionality, wherein the application opens a socket and listens for notifications, after *registering* itself with the controller. In this manner, the controller issues notifications to registered

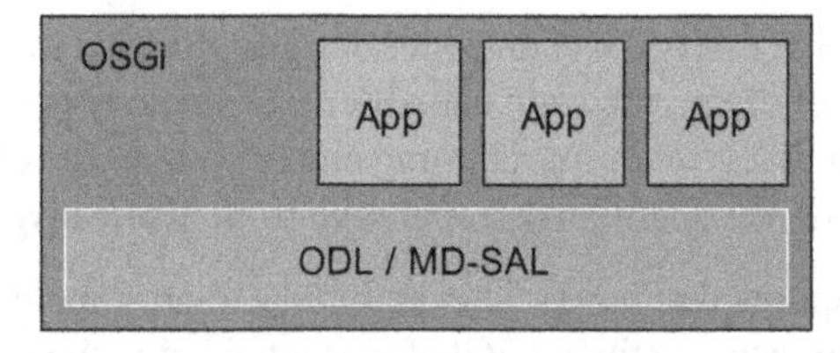

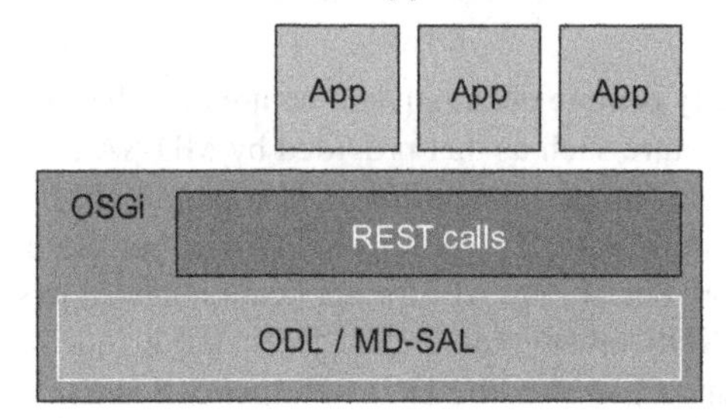

FIG. 7.13

External vs. internal SDN applications.

listeners for subscribed events (e.g., topology change notifications). These notifications are not as fast as would be the case with an internal application, so care must be taken regarding the design of the application, such that this does not become a bottleneck.

DISCUSSION QUESTION:

We have looked at a number of trends related to SDN application development. Which application type, proactive or reactive, will be the most important in the long term regarding SDN? Which in the short term? Do you believe there is a place for internal SDN applications in the future of SDN?

7.5 NEW APPROACHES TO SDN SECURITY

This chapter has described trends toward using existing protocols and using proactive external applications in SDN. These trends have ramifications on SDN security as well. Fortunately, these approaches have facilitated the task of securing an SDN environment. In the following sections we explore the impact of these trends on SDN security.

7.5.1 SECURITY ASPECTS OF REACTIVE APPLICATIONS

The model of reactive applications presumes that certain packets received by the OpenFlow device will be forwarded to the controller, exposing a number of security vulnerabilities. One such vulnerability is the possibility of a DoS attack. In this type of attack, a rogue device connects to the controller and forwards massive numbers of packets to the controller, overwhelming it and its applications, thus bringing down the network. Admittedly, there are safeguards in place with OpenFlow to make it difficult for a rogue device to become trusted by the controller and, hence, some level of security in this regard is maintained; but the threat still exists. With the rise of proactive applications, the risk of this type of threat is diminished, as devices do not forward any user traffic to the controller. Even if this were to occur, in the proactive paradigm the SDN controller and applications will ignore and discard such packets.

7.5.2 SECURITY FOR NETWORK MANAGEMENT APPLICATIONS

We look now at the security aspects of implementing SDN using existing network management protocols, such as NETCONF. An SDN application making use of these protocols is essentially performing an enhanced style of network management. Such network management protocols and applications have already weathered many years of production use and have already been hardened to remove security vulnerabilities. Thus, SDN applications based on network management are already using existing, secure technologies. Since they are at the core of a network management application, they intervene with the device far less then a reactive, intensely dynamic SDN application. Consequently, networking configurations and policy changes are made far less frequently and are not as deep as those of an OpenFlow-based reactive application. We can illuminate what we mean by deep by looking back at the control points shown in Fig. 7.3. In that diagram we see OpenFlow applications set matches and actions directly at the lower level of devices, labeled FIB in the diagram. Conversely, network management-style SDN applications access the configuration layer of the device. That configuration layer is less *deep* in the sense that it only exposes configuration control points that ensure that the device cannot be compromised, rendering the device more secure.

7.5.3 SECURITY BENEFITS OF EXTERNAL APPLICATIONS

One final aspect of security is related to the trend toward developing external SDN applications. An application running externally has limited access to the internals of the SDN controller as it can only interact with the controller via RESTful APIs. Internal applications, on the other hand, are normally operating within the Java runtime environment (OSGi) and, hence, have access to the internals of the controller. While there may be safeguards in place to prevent the loading of malicious applications onto SDN controllers, should those safeguards be compromised, the danger for the SDN controller is very real.

To a certain extent, maintaining constant availability and high performance of the controller are also part of keeping an SDN system secure. As described in Section 7.4.3, an external application can fail or crash with limited effect on the SDN controller. However, an internal application that crashes can cause side effects which negatively impact the operation of the controller. If the internal application fails while it holds important resources, like shared threads or data locks, the impact on the rest of the controller can be significant, if not catastrophic. Similarly, a poorly performing external application will have limited effect on the controller, whereas a poorly performing internal application can have dreadful performance consequences, especially when using shared resources such as threads and data.

Since external applications are less likely to introduce the issues described here, their use contributes to a more secure and reliable SDN environment.

Before we leave this topic of security, we should point out to the reader that there are more aspects of SDN security than we can reasonably cover in this book. Two examples would be the issues of *integrity* and *confidentiality* between SDN components. We refer the interested reader to [18] for more information on these topics.

DISCUSSION QUESTION:

From our discussion and your own experience, do you believe that security is a major threat to the success of SDN? Do you believe that the emerging trends presented in this chapter are addressing those threats? Are there other threats that were not mentioned in the text?

7.6 THE P4 PROGRAMMING LANGUAGE

A novel SDN idea that is beginning to gain traction is a programming language specifically designed to tell networking devices how to handle incoming packets. P4 [19] is an example of such a purpose-built language. Some of the basic attributes of P4 are the following:

- **Language**: The language itself is declarative, and is syntactically similar to C.
- **Matches**: Matching tables define against what the incoming packets will be compared.
- **Actions**: Action tables define what should be done to the packet after a match has occurred.
- **Compile-time vs. Run-time**: The language is intended to be compiled into run-time format for efficient execution.

Today, the matches supported by P4 are those common for policy-based routing (e.g., ACLs, forwarding). This entails matching on the header fields specified by OpenFlow and other configuration protocols, and ultimately implemented on the device in TCAMs. The actions are also familiar from OpenFlow, such as *forwarding out a specific set of ports* or *modifying fields*. P4 represents an intriguing new frontier for SDN and it will be interesting to observe its impact in the coming years.

There are other academic research efforts aiming to provide high-level languages and abstractions for programming software defined networks. These include the work done in [20] using the *Pyretic* language and that using the *Procera* language in [21].

7.7 CONCLUSION

The SDN world is a fast-changing place. In this chapter, we have looked at trends toward using existing protocols for SDN applications, in particular as an alternative to OpenFlow. Our discussion presented two controllers that now dominate most of the SDN dialogue. We explained the growing emphasis on proactive and external applications. We now forge into the phalanx of data center racks where the need for SDN first emerged.

REFERENCES

[1] BGP LS PCEP:Main. OpenDaylight Wiki. Retrieved from: https://wiki.opendaylight.org/view/BGP_LS_PCEP:Main.

[2] NETCONF Configuration Protocol. IETF Proposed Standard, December 2006.

[3] Network Configuration Protocol (NETCONF). IETF Proposed Standard, June 2011.

[4] Path Computation Element (PCE) Communication Protocol (PCEP). IETF Proposed Standard, March 2009.

[5] Deutsche Telekom selects Tail-f as provider of SDN. Tail-f press release. Retrieved from: http://www.tail-f.com/deutsche-telekom-selects-tail-f-as-provider-of-software-defined-networking-sdn-in-terastream-project/.

[6] AT&T Domain 2.0 Vision White Paper. November 13, 2013. Retrieved from: https://www.att.com/Common/about_us/pdf/AT&TDomain 2.0 Vision White Paper.pdf.

[7] AT&T to virtualize 75% of its network by 2020. Wall Street Journal, December 16, 2014. Retrieved from: http://blogs.wsj.com/cio/2014/12/16/att-to-virtualize-75-of-its-network-by-2020/.

[8] Overview of ONOS architecture. ONOS Wiki. Retrieved from: https://wiki.onosproject.org/display/ONOS/Overview+of+ONOS+architecture.

[9] OpenDaylight Wiki. Retrieved from: https://wiki.opendaylight.org.

[10] ONOS Wiki. Retrieved from: https://wiki.onosproject.org/display/ONOS/Wiki+Home.

[11] akka. Retrieved from: http://akka.io.

[12] Erlang. Retrieved from: http://erlang.org.

[13] OpenDaylight Controller:MD-SAL:Architecture:Clusteringi. Retrieved from: https://wiki.opendaylight.org/view/OpenDaylight_Controller:MD-SAL:Architecture:Clustering.

[14] Intent framework. ONOS Architecture Guide. Retrieved from: https://wiki.onosproject.org/display/ONOS/Intent+Framework.

[15] Brocade SDN controller. Retrieved from: http://www.brocade.com/en/products-services/software-networking/sdn-controllers-applications/sdn-controller.html.

[16] Brocade flow manager. Retrieved from: http://www.brocade.com/content/brocade/en/products-services/software-networking/sdn-controllers-applications/flow-manager.html.

[17] OSGi: the dynamic module system for Java. Retrieved from: https://www.osgi.org.

[18] Silva AS, Smith P, Mauthe A, Schaeffer-Filho A. Resilience support in software-defined networking: a survey. Comput Netw 2015;92(1):189–207.

[19] Matsumoto C. P4 SDN language aims to take SDN beyond OpenFlow. SDXCentral, May 30, 2015. Retrieved from: https://www.sdxcentral.com/articles/news/p4-language-aims-to-take-sdn-beyond-openflow/2015/05/.

[20] Monsanto C, Reich J, Foster N, Rexford J, Walker D. Composing software-defined networks. In: Proceedings of the 10th USENIX conference on Networked systems design and implementation (NSDI'13). Berkeley, CA, USA: SENIX Association; 2013.

[21] Voellmy A, Kim H, Feamster N. Procera: a language for high-level reactive network control. In: Proceedings of the first workshop on Hot topics in software defined networks (HotSDN '12). New York, NY, USA: ACM; 2012.

CHAPTER

SDN IN THE DATA CENTER

8

More than most mainstream deployments, the modern data center has stretched traditional networking to the breaking point. Previous chapters in this book have highlighted the importance of data center technology in accelerating the urgency and relevance of SDN. In its various incarnations discussed thus far, SDN has evolved specifically to address the limitations of traditional networking that have presented themselves most dramatically in today's mega-data center. In this chapter we delve more deeply into data center networking technology, both as it exists today and as it will adapt in the future through the benefit of SDN. Specifically, in this chapter we examine:

- **Data Center Needs**. What are the specific shortcomings that exist in data center networks today?
- **Data Center Technologies**. What technologies are employed in the data center, both now and with the advent of SDN?
- **Data Center Use Cases**. What are some specific applications of SDN in the data center?

In answering these questions, we hope to illuminate the close relationship that exists between SDN's technical strengths and the modern data center.

8.1 DATA CENTER DEFINITION

Let us define what exactly is a data center in order to better understand its needs and the technologies that have arisen to address those needs. The idea of a data center is not new. Densely packed racks of high-powered computers and storage have been around for decades, originally providing environments for mainframes. These eventually gave way to minicomputers and then to the servers that we know and deal with today. Over time the density of those servers and the arrays of storage that served them has made it possible to host a tremendous amount of compute power in a single room.

Today's data centers hold thousands, even tens of thousands, of physical servers. These data centers can be segregated into the following three categories:

- **Private Single-Tenant**. Individual organizations which maintain their own data centers belong in this category. The data center is for the private use of the organization and there is only the one organization or *tenant* using the data center.
- **Private Multitenant**. Organizations which provide *specialized* data center services on behalf of other client organizations belong in this category. IBM and EDS (now HP) are examples of companies which host such data centers. They are built and maintained by the organization providing the service and there are multiple clients whose data is stored there, suggesting

Software Defined Networks. http://dx.doi.org/10.1016/B978-0-12-804555-8.00008-9

the term *multitenant*. These data centers are private because they offer their services contractually to specific clients.

- **Public Multitenant**. Organizations which provide *generalized* data center services to any individual or organization belong in this category. Examples of companies which provide these services include Google and Amazon. These data centers offer their services to the public. Anybody who wishes to use them may access them via the world-wide web, be they individuals or large organizations.

In the past, these data centers were often hosted such that they could be reached only through private communication channels. However, in recent years, these data centers have begun to be designed to be accessible through the Internet. These types of data centers are often referred to as residing in *the cloud*. Three subcategories of clouds are in common use, *public* cloud, *private* cloud, and *hybrid* cloud. In a public cloud, a service provider or other large entity makes services available to the general public over the Internet. Such services may include a wide variety of applications or storage services, among other things. Examples of a public cloud offering include *Microsoft Azure Services Platform* and *Amazon Elastic Compute Cloud*. In a private cloud, a set of server and network resources is assigned to one tenant exclusively and protected behind a firewall specific to that tenant. The physical resources of the cloud are owned and maintained by the cloud provider, which may be a distinct entity from the tenant. The physical infrastructure may be managed using a product such as VMware's vCloud. Amazon Web Services is an example of how a private cloud may also be hosted by a third party (i.e., Amazon). An increasingly popular model is the hybrid cloud, where part of the cloud runs on resources dedicated to a single tenant but other parts reside on resources that are shared with other tenants. This is particularly beneficial when the shared resources may be acquired and released dynamically as demand grows and declines. Verizon's *cloud bursting* that we present in Section 9.2.3 provides an example of how a hybrid cloud might work. Note that the ability to dynamically assign resources between tenants and to the general public is a major driver behind the interest in network virtualization and SDN.

Combining the increasing density of servers and storage and the rise in networking speed and bandwidth, the trend has been to host more and more information in ever-larger data centers. Add to that the desire of enterprises to reduce operational costs and we find that merging many data centers into a larger single data center is the natural result.

Concomitant with this increased physical density is a movement toward virtualization. Physical servers now commonly run virtualization software which allows a single server to actually run a number of virtual machines. This move toward virtualization in the compute space reaps a number of benefits, from decreasing power requirements to more efficient use of hardware and the ability to quickly create, remove, and grow or shrink applications and services.

One of the side effects of the migration to cloud computing is that it is usually associated with a usage-sensitive payment system. The three main business models of cloud computing are *Infrastructure as a Service* (IaaS), *Platform as a Service* (PaaS), and *Software as a Service* (SaaS). These three are all provided and sold on an on-demand, usage-sensitive basis, shifting CAPEX costs to OPEX. The popularity of this new business model has been a major driver for the explosion in the size and number of data centers. We discuss the business ramifications of this new model on SDN in Chapter 14.

Thus, the three trends toward cloud, increased density and virtualization have placed demands on data center networking that did not exist before. We examine those needs in the next section.

8.2 DATA CENTER DEMANDS

As a result of the evolutionary trends identified in the previous section, there have arisen a number of critical networking needs that must be met in order for data center technology to thrive. We have mentioned some of these needs in previous chapters. We examine them in greater detail here.

8.2.1 OVERCOMING CURRENT NETWORK LIMITATIONS

The potential to easily ignite and tear down VMs is a radical departure from the physical world that network managers have traditionally faced. Networking protocols were coupled with physical ports which is *not* the case moving forward. The dynamic nature and sheer number of VMs in the data center have placed demands on the capacity of network components that were earlier thought to be safe from such pressures. In particular, these areas include *MAC Address Table Size*, *Number of VLANs*, and *Spanning Tree*.

MAC address explosion

In switches and routers, the MAC Address Table is used by the device to quickly determine the port or interface out of which the device should forward the packet. For speed, this table is implemented in hardware. As such, it has a physical limit to its size. Naturally, device vendors will determine the maximum number of entries to hold in the MAC table based on controlling their own costs, while at the same time providing an adequate number of entries for the demands of the network. If the table is too large, the vendor will have spent too much money on creating the device. If the table is too small, then the customer will experience the problems we describe below.

Networks in the past had manageable limits on the maximum number of MAC addresses that would need to be in the MAC Address Table at any given time. This was partly attributed to physical limitations of data centers and the number of servers that would be part of a single layer two network. It is also affected by the physical layout of the network. In the past, physically separate layer two networks remained logically separate. Now, with network virtualization and the use of Ethernet technology across WANs, the layer two networks are being stretched geographically as never before. With server virtualization, the number of servers possible in a single layer two network has increased dramatically. With numerous virtual NICs on each virtual server, this problem of a skyrocketing number of MAC addresses is exacerbated. Layer two switches are designed to handle the case of a MAC table miss by flooding the frame out all ports except the one on which it arrived, as shown in Fig. 8.1. This has the benefit of ensuring that the frame reaches its destination, if that destination exists on the layer two network. Under normal circumstances, when the destination receives that initial frame, this will prompt a response from the destination. Upon receipt of the response, the switch is able to learn the port on which that MAC address is located and populates its MAC table accordingly. This works well unless the MAC table is full, in which case it cannot learn the address. Thus, frames sent to that destination continue to be flooded. This is a very inefficient use of bandwidth and can have significant negative performance impact. This problem is exacerbated in the core of the data center network. When the traditional nonvirtualized network design is used, where all the MAC addresses are visible throughout the topology, the pressure on the MAC address tables is intense.

Since VLANs are used extensively in modern layer two networks, let us take a closer VLAN-centric look at Fig. 8.1. When a VLAN-tagged packet fails its match it is flooded out only to all ports on that

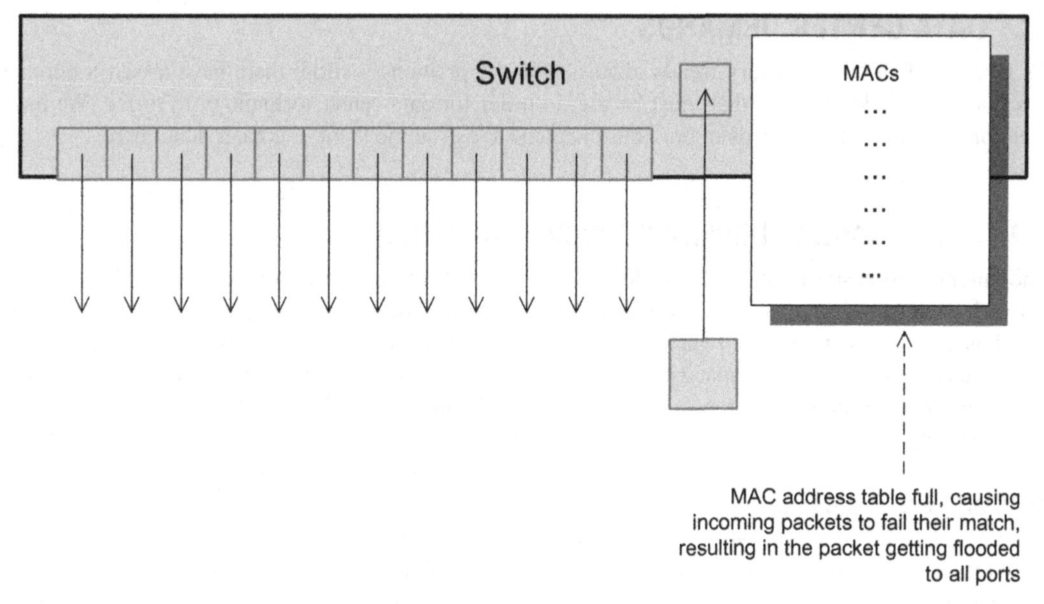
Switch
MACs
...
...
...
...
...
...
MAC address table full, causing incoming packets to fail their match, resulting in the packet getting flooded to all ports

FIG. 8.1

MAC address table overflow.

VLAN, somewhat mitigating the flooding inefficiency. Conversely, a host may be a member of multiple VLANs, in which case it will occupy one MAC address table entry per VLAN, further exacerbating the problem of MAC address table overflow.

Number of VLANs

When the IEEE 802.1 working group created the 802.1Q extension to the definition of local area networks, they did not anticipate that there would be a need for more than twelve bits to hold potential VLAN IDs. The IEEE 802.1Q Tag for VLANs is shown in Fig. 8.2. The tag depicted in Fig. 8.2 supports $2^{12} - 2$ (4094) VLANs (the subtraction of two is due to the fact that all zeros and all ones are reserved). When these tags were introduced in the mid-to-late 1990s, networks were smaller and there was very limited need for multiple virtual domains within a single physical network.

The introduction of data centers created the need to segregate traffic so as to ensure security and separation of the various tenant networks' traffic. The technology responsible for providing this separation is VLANs. As long as data centers remained single-tenant networks, the maximum number of 4094 VLANs seemed more than sufficient. To have expanded this twelve-bit field unnecessarily was not wise, as different tables in memory and in the ASICs had to be large enough to accommodate the maximum number. Thus, to have made the maximum larger than 4094 had a definite downside at the time the design work was performed.

When data centers continued to expand, however, especially with multitenancy and server virtualization, that number of VLANs required could easily exceed 4094. When there are no more available VLANs, the ability to share the physical resources amongst the tenants quickly becomes complex.

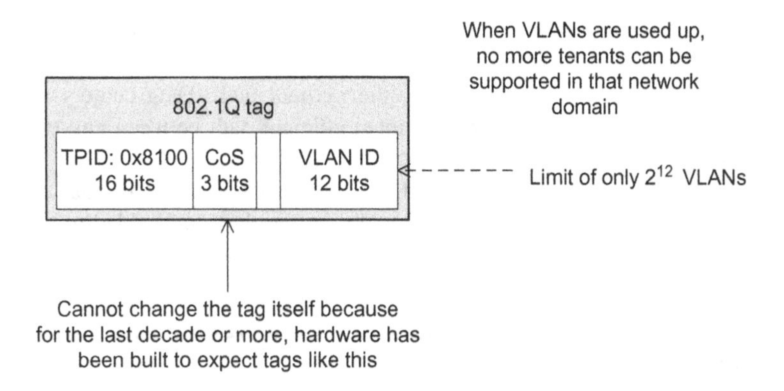

FIG. 8.2

VLAN exhaustion.

Since the number of bits allocated to hold the VLAN is fixed in size, and hardware has been built for years which depends on that specific size, it is nontrivial to expand the size of this field to accommodate more VLANs. Thus, some other solution is needed to overcome this limitation.

An upshot of the limit of 4094 VLANs has been an increase in the use of MPLS. MPLS does not suffer the same limitation in the number of MPLS tags as exists with VLAN IDs, and they need not be isolated to a single layer two network. Precisely because it is a layer three technology, with the correspondingly more complex control plane, MPLS has primarily been deployed in the WAN. It is likely, though, that MPLS will see more use in the data center [1]. One example of this is Juniper's Contrail product, which we presented in Section 6.2.5. We explore the use case of MPLS with SDN in the WAN in Section 9.1.1.

Spanning tree

In the early years of Ethernet, bridges were built as *transparent* devices capable of forwarding packets from one segment to another without explicit configuration of forwarding tables by an operator. Forwarding tables were learned by the bridges by observing the traffic being forwarded through them. Distributed intelligence in the devices was capable of collectively determining if there were loops in the network and truncating those loops so as to prohibit issues such as broadcast storms from bringing down the network.

This was accomplished by the bridges or switches collectively communicating with one another to create a *spanning tree* which enforces a loop-free hierarchical structure on the network in situations where physical loops do exist. This spanning tree was then calculated using the *Spanning Tree Protocol* (STP) we briefly reviewed in Section 1.5.1. The process of determining this spanning tree is called convergence, and in the early days it would take some time (dozens of seconds) for the spanning tree to converge after a physical change to the networking devices and their links had taken place.

Over time, through improvements to STP, the process of converging the spanning tree increased in speed. For example, with the improvements in recent versions of the standard, convergence times for conventionally large networks have decreased dramatically.[1] Despite these improvements, STP still leaves perfectly functional links unused and with frames being forwarded to the root of the spanning tree, which is certainly not universally the optimal path. Data centers need more *cross-sectional* bandwidth. By this we mean using the most efficient path between any two nodes without imposing an artificial hierarchy in the traffic patterns.

The fact that the fluidity of data center virtualization has increased the frequency of changes and disruptions, thus requiring re-convergence to occur more often, has only added to the inefficiency of STP in the data center. STP was simply not built for the modern data center world of virtual switches and ephemeral MAC addresses.

DISCUSSION QUESTION:

We've listed three major network limitations in the data center. Can you think of others that would cause data center administrators to look for answers via SDN or other technologies?

8.2.2 ADDING, MOVING, AND DELETING RESOURCES

Today's virtualized data centers are able to make changes much quicker than in the past. Networks need to adapt in order to keep pace with the virtualization capabilities of servers and storage. Speed and automation are of critical importance when it comes to handling the rapid changes demanded by virtualized servers and storage.

It is clearly important that changes to networking infrastructure take place at the same rate as is possible with servers and storage. Legacy networks require days or weeks for significant changes to VLANs and other network plumbing needs. A large part of the reason for this is that the repercussions of a mistaken network change can easily impact all the data center resources, including not only networking but also its compute and storage components. Coordinating changes with all of these disparate departments is unavoidably complex. These changes need to have the ability to be automated so that changes that must happen immediately can take place without human intervention. Legacy protocols were designed to react *after* the newly ignited service comes online. With SDN one can use the foreknowledge that a new service is about to be initiated and proactively allocate the network capacity it will require.

8.2.3 FAILURE RECOVERY

The size and scale of data centers today makes recovery from failure a complex task, and the ramifications of poor recovery decisions are only magnified as scale grows. Determinism, predictability, and optimal reconfiguration are among the most important considerations related to this area.

With the distributed intelligence of today's networks, recovery from failures may result in unpredictable behavior. It is desirable that the network move to a known state given a particular failure.

[1] In practice, including link failure detection time, convergence times have been reduced to the order of hundreds of milliseconds.

Distributed protocols render this difficult. It requires a complete view of the network in order to make the recovery process yield the best result.

8.2.4 MULTITENANCY

Data center consolidation has resulted in more and more clients occupying the same set of servers, storage and network. The challenge is to keep those individual clients separated and insulated from each other and to utilize network bandwidth efficiently.

In a large multitenant environment, it is necessary to keep separate the resources belonging to each client. For servers this may mean not mixing clients' virtual machines on the same physical server. For networks, this may mean segregation of traffic using a technology which ensures that packets from two distinct tenants remain insulated from one another. This is needed not only for the obvious security reasons, but also for QoS and other service guarantees.

With a small number of tenants in a data center, it was possible to arrange traffic and bandwidth allocation using scripts or some other rudimentary techniques. For example, if it was known that backups would be run at a certain time, arrangements could be made to route traffic appropriately around that heavy load. However, in an environment with hundreds (large data center) or thousands (cloud) of tenants with varying needs and schedules, a more fluid and robust mechanism is required to manage traffic loads. This topic is examined in more detail in the next section.

8.2.5 TRAFFIC ENGINEERING AND PATH EFFICIENCY

As data centers have grown, so has the need to make better use of the resources being shared by all of the applications and processes using the physical infrastructure. In the past this was less of an issue because overprovisioning could make up for inefficiencies. But with the current scale of data centers, it is imperative to optimally utilize the capacity of the network. This entails proper use of monitoring and measurement tools for more informed calculation of the paths that network traffic takes as it makes its way through the physical network.

In order to understand traffic loads and take the appropriate action, the traffic data must be monitored and measured. Gathering traffic intelligence has often been a luxury for networks. But with the need to make the most efficient use of the links and bandwidth available, this task has become an imperative.

State-of-the-art methods of calculating routes use link-state technology, which provides the network topology for the surrounding network devices which are part of the same Autonomous System (AS). This allows path computation to determine the *shortest path*. However, the shortest path as defined by traditional technology is not always the optimal path, since it does not take into consideration dynamic factors such as traffic load.

One of the reasons for the increasing attention on traffic engineering and path efficiency in the data center has been the rise of *East-West* traffic relative to *North-South* traffic. In Section 1.3 we defined these traffic types as follows: *East-West traffic is composed of packets sent by one host in a data center to another host in that same data center. Analogously,* North-South *traffic is traffic entering (leaving) the data center from (to) the outside world.*

Facebook provides a good example of what is driving the growth in East-West traffic. When you bring up your newsfeed page on Facebook, it has to pull in a multitude of data about different people and events in order to build the webpage and send it to you. That data resides throughout the data center.

The server that is building and sending you your Facebook newsfeed page has to pull data from servers in other parts of the data center. That East-West traffic is at least as large as the North-South data (i.e., the final page it sends you) and in fact can be larger if you account for the fact that they are moving data around even when no one is requesting it (i.e., replication, staging data in different data centers, etc.).

In the older data center model, when most traffic was North-South, traffic engineering inside the data center was mostly a question of ensuring that congestion and loss would not occur as the data moved from the servers back toward the outside world. With so much East-West traffic, traffic engineering in the data center has become much more complex. These new traffic patterns are a large part of the reason for the interest in the *Ethernet fabrics* we discuss in Section 8.5. An example of this problem is how to distribute East-West elephant flows across multiple paths in the data center. For instance, if internal servers have 10Gbps uplinks and the data center has a 40Gbps core with multiple equal cost paths, it is very likely that without explicit traffic engineering the full-speed 10Gbps flows will experience congestion at some point(s) in the core.

We have now described some of the most critical needs that exist in data center networks as a result of server virtualization and the massive scale that occurs due to data center consolidation and the advent of the cloud. In response to these needs, the networking industry has developed new technologies to mitigate some of these issues. Some of these technologies are directly related to SDN and some are not. We examine a number of these technologies in the sections that follow.

DISCUSSION QUESTION:

We have listed adds, moves, deletes, failure recovery, multitenancy, and traffic engineering as issues in data centers today. Which of these do you think has been magnified by the emergence of the cloud? Which of these is critical because of virtualization of servers and storage in data centers? Which of these is critical because of applications such as video and voice?

8.3 TUNNELING TECHNOLOGIES FOR THE DATA CENTER

Previously in this chapter we have noted the impact of server virtualization. One of the responses to server virtualization has been the birth of the concept of *network virtualization*. We defined network virtualization in Section 3.7. We also discussed one of the alternative definitions of SDN, which we referred to as *SDN via Hypervisor-Based Overlays*. At that time we showed how hypervisor-based tunneling technologies are employed to achieve this virtualization of the network. There are a number of tunneling technologies used in the data center, and some of these data center-oriented tunneling technologies predate SDN. There are three main technologies being used today to address many of the data center needs we presented in Section 8.2. These tunneling methods are *Virtual eXtensible Local Area Network* (VXLAN) [2], *Network Virtualization using Generic Routing Encapsulation* (NVGRE) [3], and *Stateless Transport Tunneling* (STT) [4].

All of these tunneling protocols are based on the notion of encapsulating an entire layer two MAC frame inside an IP packet. This is known as *MAC-in-IP tunneling*. The hosts involved in communicating with each other are unaware that there is anything other than a traditional physical network between them. The hosts construct and send packets in exactly the same manner as they would had there been no network virtualization involved. In this way, network virtualization resembles server virtualization, where hosts are unaware that they are actually running in a virtual machine environment. Admittedly, the fact that the packets are encapsulated as they transit the physical network does reduce the *maximum transmission unit* (MTU) size, and the host software must be reconfigured to account for that detail.

All three of these tunneling methods may be used to form the tunnels that knit together the *Virtual Tunnel End Points* (VTEPs) in an SDN via Hypervisor-Based Overlays system.[2] Given this background and common behavior of these tunneling technologies, the next sections take a deeper look at these technologies and examine how they differ from one another. These three technologies compete for market share, and while adoption rate statistics are elusive, we presume that:

- VXLAN will benefit from its considerable support by networking heavyweight Cisco.
- NVGRE will be adopted widely in Microsoft's popular Azure data centers.
- STT may struggle for adoption due to its relative newness, and the fact that VMware was a primary contributor and promoter of VXLAN, and hence has a significant installed base.

8.3.1 VIRTUAL EXTENSIBLE LOCAL AREA NETWORK

The VXLAN technology was developed primarily by VMware and Cisco, as a means to mitigate the inflexibility and limitations of networking technology. Some of the main characteristics of VXLAN are:

- VXLAN utilizes MAC-in-IP tunneling.
- Each virtual network or *overlay* is called a VXLAN segment.
- VXLAN segments are identified by a 24 bit segment ID, allowing for up to 2^{24} (approximately 16 million) segments.
- VXLAN tunnels are stateless.
- VXLAN segment end points are the switches that perform the encapsulation and are called VTEPs.
- VXLAN packets are unicast between the two VTEPs and use UDP over IP packet formats.
- It is UDP-based. The UDP port number for VXLAN is 4789.

Fig. 8.3 illustrates the format of a VXLAN packet. As you can see in the diagram, the outer header contains the MAC and IP addresses appropriate for sending a unicast packet to the destination switch, acting as a virtual tunnel end point. The VXLAN header follows the outer header and contains a VXLAN Network Identifier of 24 bits in length, sufficient for about 16 million networks.

The proponents of VXLAN technology argue that it assists load balancing within the network. Much of the load balancing within the data center network is based on the values of various packet fields including the destination and source port numbers. The fact that the source port of the UDP session is derived by hashing other fields of the flow results in a smooth distribution function for load balancing with the network.

[2]These are sometimes called *VXLAN Tunnel Endpoints*, which essentially means the same thing as the more generic and commonly used *Virtual Tunnel End Point*.

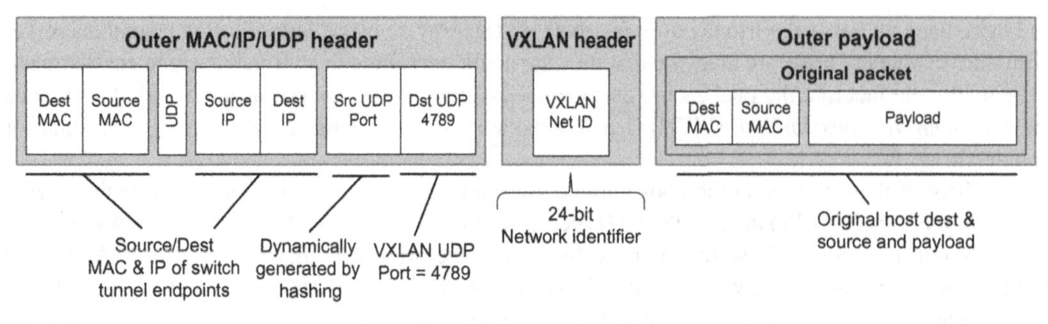

FIG. 8.3

VXLAN packet format.

8.3.2 NETWORK VIRTUALIZATION USING GRE

The NVGRE technology was developed primarily by Microsoft, with other contributors including Intel, Dell, and Hewlett-Packard. Some of the main characteristics of NVGRE are:

- NVGRE utilizes MAC-in-IP tunneling.
- Each virtual network or *overlay* is called a virtual layer two network.
- NVGRE virtual networks are identified by a 24 bit Virtual Subnet Identifier, allowing for up to 2^{24} (16 million) networks.
- NVGRE tunnels, like GRE tunnels, are stateless.
- NVGRE packets are unicast between the two NVGRE end points, each running on a switch. NVGRE utilizes the header format specified by the GRE standard [5,6].

Fig. 8.4 shows the format of an NVGRE packet. The outer header contains the MAC and IP addresses appropriate for sending a unicast packet to the destination switch, acting as a virtual tunnel end point, just like VXLAN. Recall that for VXLAN the IP protocol value was UDP. For NVGRE, the IP protocol value is 0x2F, which means GRE. GRE is a separate and independent IP protocol in

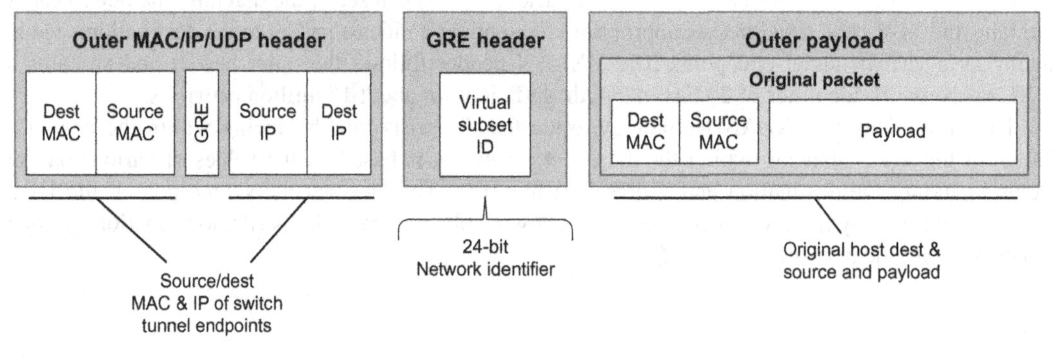

FIG. 8.4

NVGRE packet format.

the same class as TCP or UDP. Consequently, as you can see in the diagram, there are no source and destination TCP or UDP ports. The NVGRE header follows the outer header and contains an NVGRE Subnet Identifier of 24 bits in length, sufficient for about 16 million networks.

8.3.3 STATELESS TRANSPORT TUNNELING

Stateless Transport Tunneling (STT) is a recent entry into the field of tunneling technologies used for network virtualization. Its major sponsor was originally Nicira. Some of the main characteristics of STT are:

- STT utilizes MAC-in-IP tunneling.
- The general idea of a virtual network exists in STT, but is enclosed in a more general identifier called a Context ID.
- STT context IDs are 64 bits, allowing for a much larger number of virtual networks and a broader range of service models.
- STT attempts to achieve performance gains over NVGRE and VXLAN by leveraging the *TCP Segmentation Offload* (TSO) found in the *Network Interface Cards* (NICs) of many servers. TSO is a mechanism implemented on server NICs which allows large packets of data to be sent from the server to the NIC in a single send request, thus reducing the overhead associated with multiple smaller requests.
- STT, as the name implies, is stateless.
- STT packets are unicast between tunnel end points, utilizing TCP in the stateless manner associated with TSO. This means that it does not use the typical TCP windowing scheme, which requires state for TCP synchronization and flow control.

Fig. 8.5 shows the format of an STT packet. Like the other tunneling protocols discussed here, the outer header contains the MAC and IP addresses appropriate for sending a unicast packet to the destination switch, acting as a VTEP. For VXLAN, the IP protocol value was UDP and for NVGRE the IP protocol value was GRE. For STT, the IP protocol is TCP. The TCP port for STT has yet to be ratified, but, tentatively, the value 7471 has been used. The STT header follows the outer header and

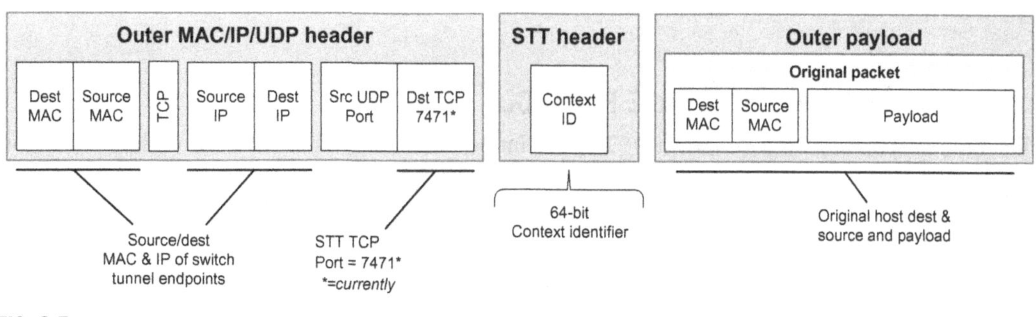

FIG. 8.5

STT packet format.

contains an STT Context Identifier of 64 bits in length, which can be subdivided and used for multiple purposes; however that is done, there is ample space for as many virtual networks as required.

8.4 PATH TECHNOLOGIES IN THE DATA CENTER

One of the critical issues in current data centers is the efficient use of the physical network links that connect network devices forming the data center's network infrastructure. With the size and demands of data centers, it is imperative that all links be utilized to their full capacity.

As we explained in Section 8.2.1, one of the shortcomings of layer two network technologies such as spanning tree is that it will intentionally block certain links in order to ensure a hierarchical network without loops. Indeed, in the data center, this drawback has become more important than the problem of slow convergence times which formed the bulk of the early critiques of STP.

Layer three networks also require intelligent routing of packets as they traverse the physical network. This section explores some path-related technologies that are intended to provide some of the intelligence required to make the most efficient use of the network and its interconnecting links.

8.4.1 GENERAL MULTIPATH ROUTING ISSUES

In a typical network of layer three subnets, there will be multiple routers connected in some manner as to provide redundant links for failover. These multiple links have the potential to be used for load balancing. Multipath routing is the general category for techniques which make use of multiple routes in order to balance traffic across a number of potential paths. Issues which must be considered in any multipath scheme are:

- The potential for *out-of-order delivery* (OOOD) of packets, which take different paths and must be reordered by the recipient, and
- The potential for maximum packet size differences on links within the different paths, causing issues for certain protocols, such as TCP and its path MTU discovery.

While these are general multipath issues, they are largely avoidable in a properly configured modern data center. Provided that the network performs per-flow load balancing and the core MTUs are much higher than the access MTUs these problems should be avoidable.

8.4.2 MULTIPLE SPANNING TREE PROTOCOL

The *Multiple Spanning Tree Protocol* (MSTP) was introduced to achieve better network link utilization with spanning tree technology when there are multiple VLANs present. Each VLAN would operate under its own spanning tree. The improved use of the links was to have one VLAN's spanning tree use unused links from another VLANs when reasonable to do so. MSTP was originally introduced as IEEE 802.1s. It is now part of IEEE 802.1Q-2005 [7]. It was necessary to have a large number of VLANs in order to achieve a well distributed utilization level across the network links as the only distribution was at VLAN-granularity. MSTP predates the Shortest Path Bridging protocol discussed below, which does not suffer this limitation.

8.4.3 SHORTEST PATH BRIDGING

IEEE 802.1aq is the *Shortest Path Bridging* (SPB) standard and its goal is to enable the use of multiple paths within a layer two network. Thus, SPB allows all links in the layer two domain to be active.

SPB is a link state protocol, which means that devices have awareness of the topology around them and are able to make forwarding decisions by calculating the best path to the destination. It uses the *Intermediate System to Intermediate System* (IS-IS) routing protocol [8] to discover and advertise network topology and to determine the paths to all other bridges in its domain.

SPB accomplishes its goals by using encapsulation at the edge of the network. This encapsulation can be either MAC-in-MAC (IEEE 802.1ah) or Q-in-Q (IEEE 802.1ad). We discussed Q-in-Q encapsulation briefly in Section 5.5.5.

Fig. 8.6 shows the frame format for Q-in-Q. As can be seen in the figure, Q-in-Q means pushing VLAN tags into the packet such that the inner tag becomes the original VLAN, also called the *Customer VLAN ID* (C-VLAN) or *C-VID*. The outer tag has multiple names, depending on the context of its use. They are known as *Metro Tag* for its use in Metro Ethernet backbones, and *Provider Edge VLAN* (PE-VLAN), *Service Provider VLAN ID* (S-VLAN) or *S-VID* for its use in provider VLANs. The *C-VID* represents the original customer VLAN tag.

When the packet arrives at the edge of the provider's interior network, a new VLAN tag is pushed into the packet identifying the VLAN on which the packet will travel when it is inside the provider's network. When it exits, that VLAN tag is removed.

Fig. 8.7 shows the frame format for MAC-in-MAC. The new MAC header includes the backbone's source (SA) and destination (DA) addresses. Also part of the header is the VLAN identifier for the backbone provider's network (B-VID), as well as the backbone's service instance ID (I-SID) which allows the provider to organize customers into specific services and provide appropriate levels of QoS and throughput to each service. When a packet arrives at the provider edge, the new header gets attached to the beginning of the packet and is sent through the provider's network. When it leaves the provider's domain, the header is stripped.

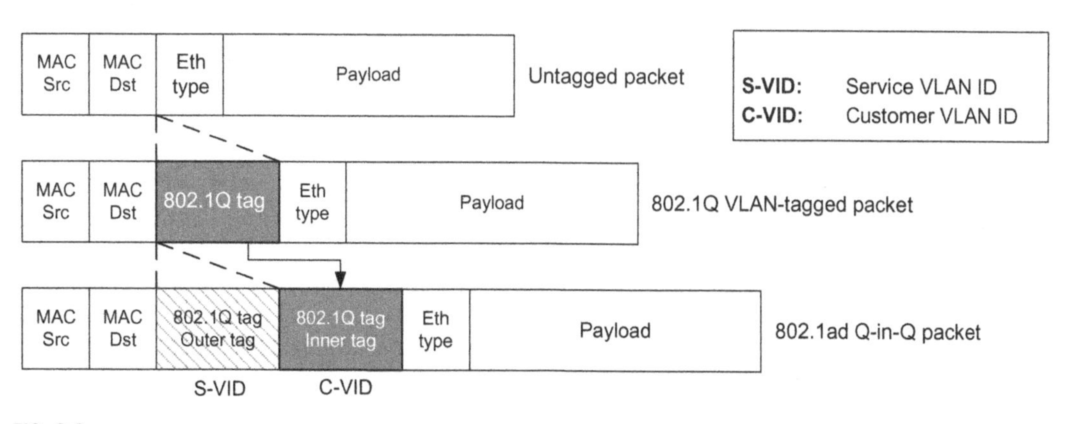

FIG. 8.6

Shortest path bridging—Q-in-Q (VLAN).

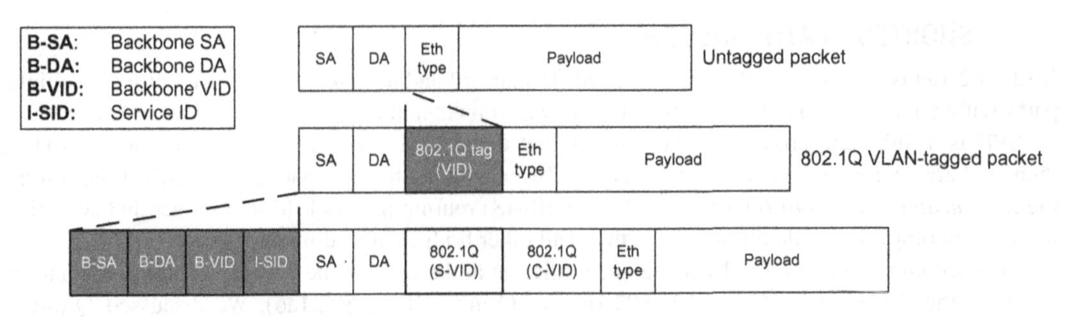

FIG. 8.7

Shortest path bridging—MAC-in-MAC.

8.4.4 EQUAL-COST MULTIPATH

One of the most important operations that must be undertaken in large networks is optimal path computation. Merely getting a packet to its destination is far simpler than getting it there in the most efficient way possible. In Section 1.5.2 we introduced OSPF and IS-IS as modern link-state protocols used for calculating routes in complex layer three networks. We just explained in Section 8.4.3 that IS-IS is used as the basis for calculating routes in the layer two SPB protocol. There is a general routing strategy called *Equal-cost multipath* (ECMP) that is applicable in the data center. Multipath routing is a feature that is explicitly allowed in both OSPF and IS-IS. The notion is that when more than one equal-cost path exists to a destination, these multiple paths can be computed by a shortest path algorithm and exposed to the packet forwarding logic. At that point some load balancing scheme must be used to choose between the multiple available paths. As several routing protocols can derive the multiple paths and there are many ways in which to load balance across the multiple paths, ECMP is more of a routing strategy than a specific technology. A discussion of some of the issues with ECMP can be found in [9]. An analysis of a specific ECMP algorithm is found in [10].

Another sophisticated path computation technology is the *Path Computation Element (PCE)-based Architecture*. As this has been used primarily in the WAN and not seen much use in the data center, we will defer deeper treatment of this topic to Section 9.1.1. It is likely that PCE will be applied in the data center for directing elephant flows.

8.4.5 SDN AND SHORTEST-PATH COMPLEXITY

Throughout this chapter we have made frequent reference to the importance of calculating optimal paths. We have implied that a centralized controller can perform this task better than the traditional distributed algorithm approach because: (1) it has a more stable, global view of the network, (2) it can take more factors into consideration than current distributed approaches do, including current bandwidth loads, and (3) the computation can be performed on the higher capacity memory and processor of a server rather than the more limited memory and processor available on the network devices. While all this is true, we should point out that the fundamental computational complexity of shortest-path calculation is not changed by any of these factors. Shortest-path remains a fundamentally hard problem which grows harder very fast as the number of switches and links scales. The well-known

Dijkstra's algorithm [11] for shortest path remains in wide use in *Interior Gateway Protocols* (IGPs) (it is used in both OSPF and IS-IS) and SDN is unlikely to alter that. Having a faster machine and a more stable data base of the network graph helps compute optimal paths more quickly; they do not affect the complexity of the fundamental problem.

It is worthwhile noting that some *Massively Scaled Data Centers* (MSDCs) are now abandoning an IGP altogether and instead use the *External Border Gateway Protocol* (EBGP) for all routing. This implies the use of a path-vector routing protocol instead of one based on the Dijkstra algorithm. This should scale better computationally as the network grows while maintaining awareness of distant link failures since a failed link will result in paths being removed from the routing table.

8.5 ETHERNET FABRICS IN THE DATA CENTER

Traffic engineering in the data center is challenging using traditional Ethernet switches arranged in a typical hierarchy of increasingly powerful switches as one moves up the hierarchy closer to the core. While the interconnecting links increase in bandwidth as one moves closer to the core, these links are normally heavily oversubscribed and blocking can easily occur. By oversubscription we mean that the aggregate potential bandwidth entering one tier of the hierarchy is greater than the aggregate bandwidth going to the next tier.

An alternative to a network of these traditional switches is the Ethernet fabric [12]. Ethernet fabrics are based on a nonblocking architecture where every tier is connected to the next tier with equal or higher aggregate bandwidth. This is facilitated by a topology referred to as *fat tree* topology. We depict such a topology in Fig. 8.8. In a fat tree topology, the number of links entering one switch in a given tier of the topology is equal to the number of links leaving that switch toward the next tier. This implies that as we approach the core, the number of links entering the switches is much greater than those toward the leaves of the tree. The root of the tree will have more links than any switch lower than it. This topology is also often referred to as a *Clos* architecture. Clos switching architecture dates from the early days of crossbar telephony and is an effective nonblocking switching architecture. Note that the Clos architecture is not only used in constructing the network topology but may also be used in the internal architecture of the Ethernet switches themselves.

For Ethernet fabrics the key characteristic is that the aggregate bandwidth not decrease from one tier to the next as we approach the core. Whether this is achieved by a larger number of low bandwidth links or a lower number of high bandwidth links does not fundamentally alter the fat tree premise.

As we approach the core of the data center network, these interconnecting links are increasingly built from 100Gb links and it is costly to simply try to overprovision the network with extra links. Thus, part of the Ethernet fabric solution is to combine it with ECMP technology so that *all* of the potential bandwidth connecting one tier to another is available to the fabric.

DISCUSSION QUESTION:

We have discussed a number of tunneling and path technologies as potential solutions to issues faced by networking administrators in the data center today. Of these, which do you believe is best suited to solving these issues, and why? Might there be solutions consisting of a combination of these technologies?

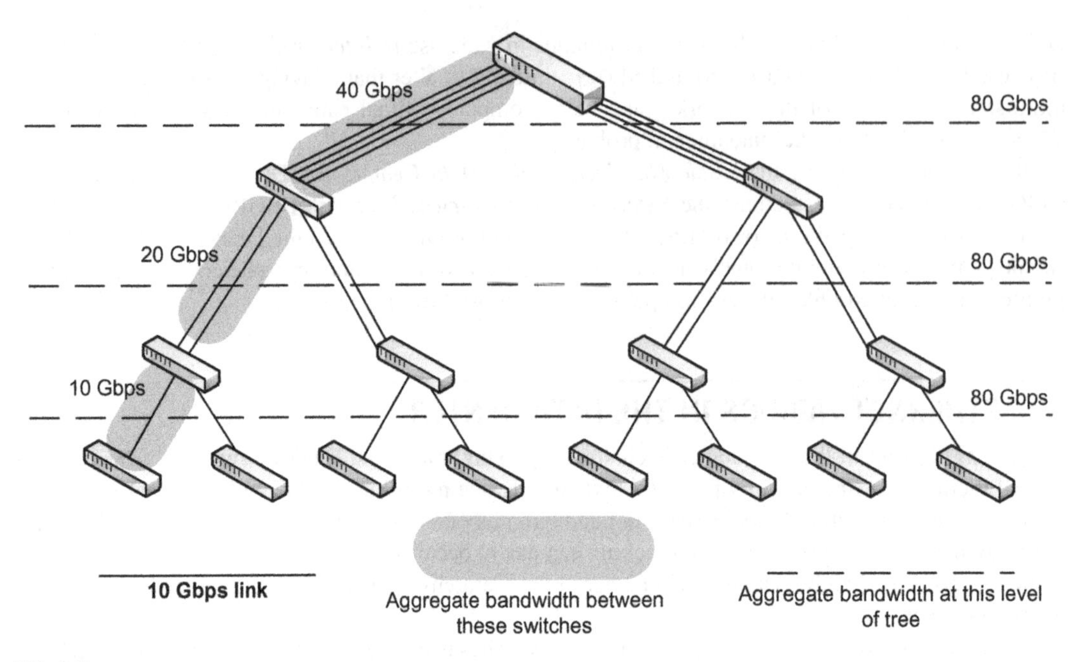

FIG. 8.8

Fat tree topology.

8.6 SDN USE CASES IN THE DATA CENTER

We have described several definitions of software defined networking: the original SDN which we call Open SDN, SDN via APIs, and SDN via Hypervisor-Based Overlays. We additionally described SDN via Opening Up the Device, but here we will focus on the first three, as they are the only ones to have gained significant commercial traction as of this writing. How do these versions of SDN address the needs of today's data center as we have described them in this chapter? The following sections answer that question.

For each data center requirement we presented in Section 8.2, we consider each of the three SDN technologies. We present them in the order SDN via Overlays first, followed by Open SDN, and, lastly, SDN via APIs. The reader should note that in this book when we use the expression SDN via Overlays we mean SDN via *Hypervisor-Based* Overlays. This distinction is important as there are other overlay mechanisms that are not hypervisor-based. The reason for this ordering is because SDN via Overlays is a technology specifically addressing the data center and so one would expect the Overlay solution to directly address most of these issues.

Note that Table 8.1 briefly summarizes the ability of the various SDN definitions to address the data center needs that we have mentioned in this chapter. We discuss aspects of this table in the sections that follow.

Table 8.1 SDN Alternatives' Support of Data Center Needs

Need	SDN via Overlays	Open SDN	SDN via APIs
Overcoming Network Limitations	Inherent	Feasible	Implementation-dependent
Adds, Moves, Deletes	Inherent	Inherent	Implementation-dependent
Failure Recovery	Unaddressed	Inherent	Implementation-dependent
Multitenancy	Inherent	Inherent	Implementation-dependent
Traffic Engineering and Path Efficiency	Unaddressed	Inherent	Implementation-dependent

8.6.1 OVERCOMING CURRENT NETWORK LIMITATIONS

In this section we examine how SDN via Overlays, Open SDN and SDN via APIs address the limitations of MAC address table size and maximum number of VLANs, and other aspects related to Table 8.1.

SDN via Overlays

The simplest and most readily available solution to these problems involves tunneling, so *SDN via Overlays* is an obvious choice here. The only MAC addresses visible through the physical network are the MAC addresses of the tunnel end points, which are at the hypervisors. As a trivial example, if there are eight VMs per hypervisor, then you have reduced the total number of MAC addresses by a factor of eight. If the tunnel end points are further upstream or the number of VMs per hypervisor is higher, the MAC address savings are even greater.

For the issue of VLAN exhaustion, that is, exceeding the limit of 4094, this solution is superior, as the new mechanism for multitenancy is tunnels, not VLANs. As we explained in Section 8.3, the number of tunneled networks or *segments* can be 16 million or greater using VXLAN, NVGRE, or STT tunneling technologies.

As for the issue of spanning tree convergence and using all links in the physical infrastructure, SDN via Overlays does not address issues related to the physical infrastructure. The network designer would have to use current non-SDN technologies to mitigate these types of physical limitations.

When we say that SDN via Overlays leaves unaddressed the issues of Failure Recovery and Traffic Engineering, it is important for the reader to understand that this technology does not prevent solutions to these issues to be implemented in the underlay network. We simply mean that SDN via Overlays does not *itself* provide the solution.

Open SDN

Open SDN has the capability of addressing these network limitations as well. It does not, however, *inherently* resolve these limitations in the same way as does SDN via Overlays due to that alternative's basic nature of using tunnels. Moving control functionality off of the device to a centralized controller does not directly address limitations such as MAC address table size and the maximum number of VLANs.

However, Open SDN is well-suited to creating a solution that is an instance of SDN via Overlays. The SDN controller can create tunnels as required at what will become the tunnel end points, and then OpenFlow rules are used to push traffic from hosts into the appropriate tunnel. Since hardware exists that has built-in tunneling support, SDN devices can be built that derive these benefits from tunneling, but with the performance gain of hardware. Thus, Open SDN can solve these network limitations similarly to the Overlay alternative.

SDN via APIs

Adding SDN APIs to networking devices does not directly address network limitations as we have discussed them in this chapter. However, some data center SDN via APIs solutions have been created using existing or new protocols. In particular:

- Cisco's *APIC-Data Center* (APIC-DC) utilizes a new protocol called *OpFlex*. This protocol is designed to carry policy-level information from an the APIC-DC controller, to intelligent devices capable of rendering the policies received.
- Juniper's *Contrail* and its open-source variant *OpenContrail*, utilize existing protocols such as NETCONF, MPLS, and XMPP combined with APIs existing in devices today to create a virtualized network targeted at the data center.

So while it is true that SDN via APIs inherently does not address network limitations, these creative uses of existing protocols and APIs can yield improvements in networking behavior.

8.6.2 ADDING, MOVING, AND CHANGING RESOURCES

Agility, automation, fast and efficient adds, moves, and changes—this type of functionality is critical in data centers in order to keep pace with speeds brought about by automation of servers and storage. We now examine the ways in which SDN technology can address these issues.

SDN via Overlays

The primary attribute of SDN via Overlays as it addresses adds, moves, and changes is that the technology revolves around virtualization. It does not deal with the physical infrastructure at all. The networking devices that it manipulates are most often the virtual switches that run in the hypervisors. Furthermore, the network changes required to accomplish the task are simple and confined to the construction and deletion of virtual networks, which are carried within tunnels that are created expressly for that purpose. These are easily manipulated via software.

Consequently, the task of making adds, moves, deletes, and changes in an overlay network is quite straightforward and is easily automated. Because the task is isolated and constrained to tunnels, problems of complexity are less prevalent compared to what would be the case if the changes needed to be applied and replicated on all the physical devices in the network. Thus, many would argue that overlays are the simplest way to provide the automation and agility required to support frequent adds, moves, deletes, and changes.

A downside of this agility is that, since the virtual networks are not tightly coupled with the physical network, it is easy to make these adds, moves, and changes without being certain that the underlying physical network has the capacity to handle them. The obvious solution is to overengineer the physical network with great excess capacity, but this is not an efficient solution to the problem.

Open SDN

As we discussed in the previous section, if Open SDN is being used to create tunnels and virtual networks, then it is straightforward for Open SDN to achieve the same results as discussed earlier. The task is to create the overlay tunnels as required and to use OpenFlow rules to push packets into the appropriate tunnels.

In addition to the advantages of virtual networks via tunnels, Open SDN offers the ability to change the configuration and operation of the physical network below—what some refer to as the *underlay*. The real advantage of this capability is described in Section 8.6.5.

SDN via APIs

APIs provide a programmatic framework for automating tasks that would otherwise require manual intervention. Having a controller aware of server virtualization changes that can respond with changes to the network is a definite advantage. Good examples include complete API-based solutions in the data center such as Cisco's APIC-DC and Juniper's Contrail. Both of these solutions use existing or new proprietary APIs provided on their respective devices in order to foster the agility required in data centers. A possible limitation of the API solution is that, if using legacy protocols, the controller may be constrained by the limitations of the devices themselves (a limitation which is not true for Open SDN or Overlays). However, remember that the main facet of supporting adds, moves, and changes, is the ability to do so in an automated and agile manner. From this perspective, with a centralized controller managing the network, SDN via APIs should provide significant improvement over current legacy networks.

8.6.3 FAILURE RECOVERY

Data centers today have mechanisms for achieving high availability, redundancy, and recovery in the case of failure. However, as mentioned earlier, these mechanisms have deficiencies in the predictability and optimization of those recovery methods. Consequently, we would expect SDN to provide some assistance in meeting those specific needs.

Note that failures on the compute side, i.e., physical servers or virtual machines, are really just a subset of the previous section on *adds, moves, and changes* and have been addressed in that context.

SDN via Overlays

Because it does not deal at all with the physical network below it, overlay technology offers little in terms of improving the failure recovery methods in the data center. If there are failures in the physical infrastructure, those must be dealt with via the mechanisms already in place, apart from overlays. In addition, the interplay between the virtual and physical topologies can sometimes be difficult to diagnose when there are problems.

Open SDN

One of the stated benefits of Open SDN is that with a centralized controller the whole network topology is known and routing (or, in this case, rerouting) decisions can be made which are consistent and predictable. Furthermore, those decisions can incorporate other sources of data related to the routing decisions, such as traffic loads, time of day, even scheduled or observed loads over time. Creating a solution such as this is not trivial, but the SDN application responsible for such failure recovery functionality can leverage existing and emerging technologies, such as IS-IS and PCE.

SDN via APIs

Versions of SDN via APIs such as Cisco's APIC-DC and Juniper's Contrail can utilize knowledge about the network in a similar manner to Open SDN, and can make use of those existing legacy APIs on devices to achieve superior availability and recovery from failures. This is in fact the direction that API-based solutions have been headed since 2013.

8.6.4 MULTITENANCY

With large data centers and cloud computing, it is frequently the case that more and more tenants are passing traffic along the same physical network infrastructure and therefore sharing the same physical communications links. The traditional way of achieving the separation required by this sharing of network bandwidth has been through the use of VLANs. We have shown how the maximum number of VLANs is no longer adequate for today's data centers.

SDN via Overlays

Overlay technology resolves the multitenancy issue by its very nature through the creation of virtual networks that run on top of the physical network. These substitute for VLANs as the means of providing traffic separation and isolation. In overlay technologies, VLANs are only relevant within a single tenant. For each tenant, there is still the 4094 VLAN limit, but that seems to currently suffice for a single tenants's traffic.

Open SDN

Open SDN can implement network virtualization using layer three tunnel-based overlays in a manner very similar to SDN via Overlays. Open SDN offers other alternatives as well, however. Other types of encapsulation (e.g., MAC-in-MAC, Q-in-Q) can also be employed to provide layer two tunneling, which can provide multiplicative increases in the number of tenants. For example, using Q-in-Q can theoretically result in 4094 times 4094 VLANs, or roughly 16 million, the same value as is possible with the Overlay solutions utilizing VXLAN or NVGRE.

SDN via APIs

Many legacy networks depend on VLANs to provide network virtualization, and those VLANs are established and configured via APIs. As such, SDN via APIs can play a role in addressing multitenancy issues. Moreover, as technologies such as VXLAN become more prevalent in network devices, it becomes possible to utilize these capabilities using APIs on devices. While traditionally used in service provider networks, protocols such as MPLS are now being applied in the data center (e.g., Contrail), using these legacy APIs to provide a level of multitenancy well beyond what was possible with VLANs. For these reasons, SDN via APIs can enhance multitenancy support in data center environments.

8.6.5 TRAFFIC ENGINEERING AND PATH EFFICIENCY

In the past, networking hardware and software vendors created applications which measured traffic loads on various links in the network infrastructure. Some even recommended changes to the infrastructure in order to make more efficient use of network bandwidth. As good as these applications may have been, the harsh reality was that the easiest way to ensure optimal response times and minimal latency was to overprovision the network. In other words, the simplest and cheapest solution was almost always to purchase more and newer networking gear to make certain that links were uncongested and throughput was maximized.

However, with the massive expansion of data centers due to consolidation and virtualization, it has become imperative to make the most efficient use of networking bandwidth. *Service level agreements* (SLAs) with clients and tenants must be met, while at the same time costs need to be cut in order for the data center service to be competitive. This has placed an added importance on traffic load measurements and using that information to make critical path decisions for packets as they traverse the network.

SDN via Overlays

In a similar manner to the issue of failure recovery SDN via Overlays does not have much to contribute in this area because of the fact that it does not attempt to affect the physical network infrastructure. Traffic loads as they pass from link to link across the network are not a part of the discussion concerning traffic which is tunneled across the top of the physical devices and interfaces. Thus, SDN via Overlays is dependent on existing underlying network technology to address these types of issues.

Open SDN

Open SDN has major advantages here in the areas of centralized control and having complete control of the network devices. Open SDN has centralized control with a view of the full network and can make decisions predictably and optimally. It also controls the individual forwarding tables in the devices and, as such, maintains direct control over routing and forwarding decisions.

Two other major ways that Open SDN can impact the traffic moving through the network are: (1) path selection and (2) bandwidth management on a per-flow basis. Open SDN can make optimal path decisions for each flow. It can also prioritize traffic down to the flow level, allowing for granular control of queuing, using methods such as IEEE 802.1 *Class of Service* (CoS), IP-level *Type of Service* (ToS), and *Differentiated Services Code Point* (DSCP).

Fig. 8.9 provides an illustration of the difference between shortest path via the current node-by-node technology and optimal path, which is possible with an SDN controller which can take into consideration more network parameters such as traffic data, in making its decision.

Open SDN is explicitly designed to be able to directly control network traffic on the switching hardware down to the flow level. We contrast this with the overlay alternative where network traffic does exist at the flow level in the virtual links, but not natively in the switching hardware. Therefore, Open SDN is particularly well-suited to addressing traffic engineering and path efficiency needs.

SDN via APIs

The core feature of SDN via APIs is to set configuration of networking devices in a simpler and more efficient manner. Since this is done from a centralized controller, a solution using APIs would enhance automation, and will yield a greater degree of agility. While legacy APIs were generally not designed to handle the time-critical, highly granular needs inherent in traffic engineering, *extended* APIs can improve traffic engineering even in legacy devices with local control planes. This is rendered feasible through the use of SDN applications running on an API-based controller such as OpenDaylight,

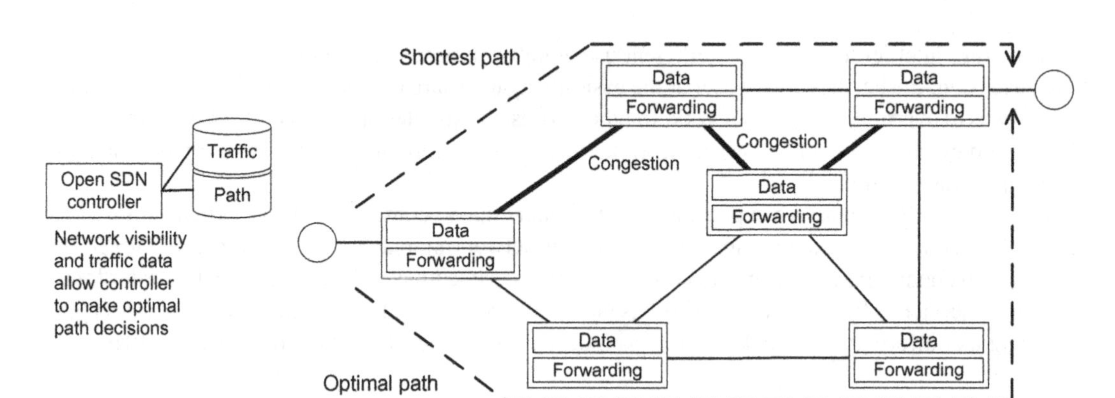

FIG. 8.9

Optimal path selection by SDN controller.

Contrail, or APIC-DC. In fact, though Fig. 8.9 illustrates the capabilities of OpenFlow for optimal path selection, a similar solution is provided via OpenDaylight's BGP-LS/PCE-P plugin.

It is worth noting that *policy-based routing* (PBR) has the ability to direct packets across paths at a fairly granular level. In theory one could combine current traffic monitoring tools with PBR and use current SNMP or CLI APIs to accomplish the sort of traffic engineering we discuss here. RSVP and MPLS-TE are examples of traffic engineering protocols that may be configured via API calls to the device's control plane.

However, those APIs and PBR did not traditionally enable frequent and dynamic changes to paths and cannot react as quickly in the face of change as can OpenFlow. However, with new solutions such as APIC-DC and Contrail, utilizing policy-based and path-based protocols, vendors are thrusting SDN APIs into a more agile SDN-based future.

8.7 COMPARISON OF OPEN SDN, OVERLAYS, AND APIs

We see in Table 8.1 a comparison of the different types of SDN and their contribution to alleviating problems with network in the data center. We take a closer look in the following sections.

SDN via Overlays

- The technology is designed for solving data center issues.
- The technology typically utilizes contemporary data center technologies, specifically VXLAN and NVGRE, and, thus, the technology is not entirely new.
- The technology creates a virtual overlay network and thus does a very good job with overcoming networking limitations and improving agility.
- The technology does not address any issues related to improving the behavior of the physical infrastructure. This has the advantage, though, of incurring little or no costs related to new networking equipment.

- It is sometimes difficult to diagnose problems and determine if they relate to the virtual network or the physical one.

Open SDN

- The technology is designed for solving networking issues in a wide range of domains, not just data centers.
- Open SDN can be implemented using networking devices designed for Open SDN. Legacy switching hardware can often be enhanced by adding OpenFlow to function as Open SDN devices.
- The technology, used in conjunction with virtual networks (e.g., VXLAN and NVGRE), does a very good job addressing all the networking issues we have discussed.
- The technology is broader in scope; hence, it is more disruptive and may pose more transitional issues than does SDN via Overlays.
- Although both Open SDN and SDN via Overlays allow significant innovation, the Open SDN use cases discussed in Chapter 9 will show how implementers have *radically* redefined the network. In so doing, they abandon legacy constructs and structure the network to behave in exact alignment with application requirements.

SDN via APIs

- The technology can utilize existing network devices with minimal upgrades.
- Some solutions (e.g., APIC-DC and OpFlex) are natively policy-based solutions, whereas others are added onto an existing solution.
- Other solutions translate existing successful architectures from one domain into the data center domain. This is the case for Contrail which uses MPLS technology in the data center, though MPLS has traditionally been used in the WAN. The effectiveness of this translation has yet to be proven, but it certainly holds promise.
- There is major vendor support from Juniper and Cisco, each for their own data center solution, which can prove to be significant in determining the success or failure of a new technology.

These points indicate that SDN via Overlays is a good solution and may be an easier transitional phase in the near term, while Open SDN holds the most promise in a broader and more comprehensive sense. While Table 8.1 illustrated that SDN via APIs may not inherently be an optimal solution in the data center due to dealing with legacy issues, we have pointed out that a number of data center problems may be addressed by particular SDN via APIs solutions. The more simple SDN via APIs solutions may in fact represent the easiest transition plan for network programmers in the short term as less radical change is needed than that required moving to either the Overlay or Open SDN approach.

DISCUSSION QUESTION:

Which of the types of SDN (Open, APIs, Overlays) is your favorite for resolving immediate issues in the data center? Which do you consider to be the best long-term strategy? Do you believe it is wise to favor near-term pragmatism, or long-term viability, for solving these issues?

8.8 REAL-WORLD DATA CENTER IMPLEMENTATIONS

In this chapter we have spent time defining the needs of modern data centers, and we have investigated the ways that each SDN technology has addressed those needs. One may wonder if anybody is actually implementing these SDN technologies in the data center. The answer is that, although it is still a work in progress, there are organizations which are running SDN pilots and in some cases placing it into production. In Table 8.2 we list a number of well-known enterprises and their early adopter implementations of SDN in their data centers. The table shows a mix of Open SDN and SDN via Overlays, but no examples of SDN via APIs. The bulk of the API-oriented efforts are not really what we have defined as SDN. They use technologies similar to what we described in Sections 3.2.3 and 3.2.4, that is, orchestration tools and VMware plugins to manipulate networks using SNMP and CLI. Some experimentation with newer SDN APIs is presumably underway, but as of this writing production data center SDN networks tend more to be hypervisor-based overlays with a few examples using Open SDN. It is worth pointing out that Juniper's Contrail is somewhat of an outlier here. We have defined Contrail as an example of both SDN via APIs (Section 6.2.5) *and* an SDN via Overlays technology (Section 6.3.3). Contrail has been deployed in a commercial data center environment [13] and thus might be considered as an example of a production SDN via APIs deployment. In any event, between Open SDN and the overlay approach, our observation is that the extremely large greenfield data center examples gravitate toward Open SDN, whereas hypervisor-based overlays have seen greater adoption in the more typical-scale data centers.

8.9 CONCLUSION

This chapter has described the many significant and urgent needs that have caused data center networking to become a stumbling block to technological advancement in that domain. We examined

Table 8.2 Data Center SDN Implementations

Enterprise	SDN Type	Description
Google	Open SDN	Has implemented lightweight OpenFlow switches, an OpenFlow Controller and SDN Applications for managing the WAN connections between their data centers. They are moving that Open SDN technology into the data centers.
Microsoft Azure	Overlays	Implementing Overlay technology with NVGRE in vSwitches, communicating via enhanced OpenFlow, creating tens of thousands of virtual networks.
Ebay	Overlays	Creating public cloud virtual networks, using VMware's Nicira solution.
Rackspace	Overlays	Creating large multitenant public cloud using VMware's Nicira solution.

new technologies and standards that attempt to address those needs. Then we presented how the three main SDN alternatives made use of those technologies, as well as new ideas introduced by SDN to address those needs. Lastly, we listed how some major enterprises are using SDN today in their data center environments.

In Chapter 9 we will consider other use cases for SDN, such as carrier and enterprise networks.

REFERENCES

[1] Kompella K. New take on SDN: does MPLS make sense in cloud data centers? SDN Central, December 11, 2012. Retrieved from: http://www.sdncentral.com/use-cases/does-mpls-make-sense-in-cloud-data-centers/2012/12/ .

[2] Mahalingam M, Dutt D, Duda K, Agarwal P, Kreeger L, Sridhar T, et al. VXLAN: a framework for overlaying virtualized layer 2 networks over layer 3 networks. Internet Draft, Internet Engineering Task Force, August 26, 2011.

[3] Sridharan M, et al. NVGRE: network virtualization using generic routing encapsulation. Internet Draft, Internet Engineering Task Force, September 2011.

[4] Davie B, Gross J. STT: a stateless transport tunneling protocol for network virtualization (STT). Internet Draft, Internet Engineering Task Force, March 2012.

[5] Farinacci D, Li T, Hanks S, D M, Traina P. Generic Routing Encapsulation (GRE). RFC 2784, Internet Engineering Task Force, March 2000.

[6] Dommety G. Key and sequence number extensions to GRE. RFC 2890, Internet Engineering Task Force, September 2000.

[7] IEEE standards for local and metropolitan area networks—virtual bridged local area networks. IEEE 802.1Q-2005. New York: IEEE; 2005.

[8] Information technology—telecommunications and information exchange between systems—intermediate system to intermediate system intra-domain routing information exchange protocol for use in conjunction with the protocol for providing the connectionless-mode network service (ISO 8473). ISO/IEC 10589:2002(E), Switzerland, November 2002.

[9] Thaler D, Hopps C. Multipath issues in unicast and multicast next-hop selection. RFC 2991, Internet Engineering Task Force, November 2000.

[10] Hopps C. Analysis of an equal-cost multi-path algorithm. RFC 2992, Internet Engineering Task Force, November 2000.

[11] Dijkstra EW. A note on two problems in connection with graphs. Numer Math 1959;1:269–71.

[12] Ammirato J. Ethernet fabric switching for next-generation data centers. Network World, November 03, 2008. Retrieved from: http://www.networkworld.com/news/tech/2008/110308-tech-update.html .

[13] Moriarty E. Juniper Contrail gets deployed by TCP cloud. NetworkStatic, September 22, 2014. Retrieved from: https://www.sdxcentral.com/articles/news/juniper-contrail-sdn-controller-tcp-cloud/2014/09/ .

CHAPTER

SDN IN OTHER ENVIRONMENTS

9

The urgency associated with the growth of data centers has caused SDN to leap to the forefront of solutions considered by IT professionals and CIOs alike. But the data center is not the only environment in which SDN is relevant. This chapter looks at other domains in which SDN will play a major role. The following environments and some accompanying use cases will be examined in this chapter:

- Wide Area Networks (Section 9.1)
- Service Provider and Carrier Networks (Section 9.2)
- Campus Networks (Section 9.3)
- Hospitality Networks (Section 9.4)
- Mobile Networks (Section 9.5)
- Optical Networks (Section 9.6)

As we examine each of these domains, we will explain the most pressing issues and then we consider the ways in which SDN can play a part in helping to resolve them. We will discuss current implementations and *proofs of concept* (PoCs) that have been developed to address needs in these areas. Before we look into each environment in detail, we will review some advantages that accompany an SDN solution that are particular to these environments.

CONSISTENT POLICY CONFIGURATION

One of the major issues currently facing IT departments is the problem of scale, related to having consistent configuration across very large populations of networking devices. In the past this task has fallen to manual configuration or network management, but with SDN there is the promise of great simplification of these tasks, with the associated reduction in operational costs for the enterprise. Some specifics of SDN with respect to configuration are (1) *dealing with large numbers of devices*, (2) *consistency of configuration*, (3) *central policy storage*, (4) *common policy application*, and (5) *granularity of control*.

The sheer number of devices operating in networks today has grown to the point of becoming nearly unmanageable. In large *Wide Area Networks (WANs)* there may be thousands of physical switches deployed. IT personnel today are required to maintain a large number of devices using traditional tools such as CLI, web interfaces, SNMP and other manual, labor-intensive methods. SDN promises to remove many of these tasks. In order for SDN to work and scale, the controllers are being built from the

Software Defined Networks. http://dx.doi.org/10.1016/B978-0-12-804555-8.00009-0

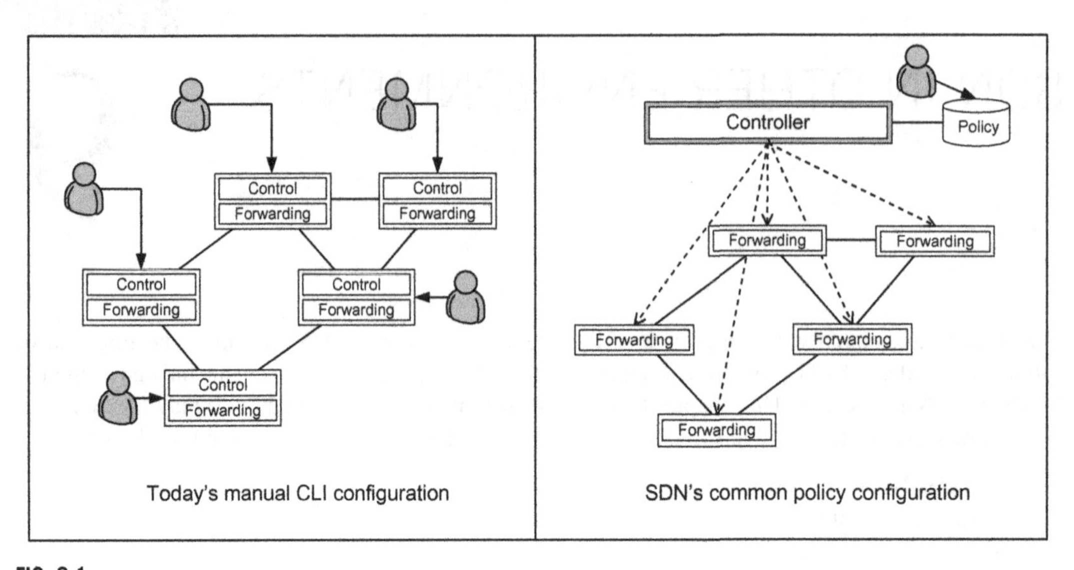

FIG. 9.1

Ensuring consistent policy configuration.

ground up to comprehend and manage a huge number of devices. In addition, this centralization drives more commonality between the devices. The task of managing large numbers of devices is simplified as a direct result of this homogeneity. This simplification of network configuration is one of the major operational cost savings brought about by SDN.

One common issue with large networks today is the difficulty in maintaining consistency across all the devices, which are part of the infrastructure. Fig. 9.1 shows a simple example of the comparison of autonomous devices individually and manually configured versus common policy distributed via flow rules from a central controller. The trivial case depicted in the diagram is compounded by orders of magnitude in real-life networks. Keeping configuration of policy consistent simply cannot be done by purely manual means. For some time now, sophisticated network management tools have existed that purport to provide centralized configuration and policy management. However, as we have described in previous chapters, these tools have only realized marginal success. They are generally based on a model of creating a *driver* that abstracts away vendor and device differences. These tools attempt to standardize at such a high level that the actual low-level vendor and device differences pose a constant challenge. This model has not been able to scale with the amount of change that is endemic in networking. SDN approaches configuration in a radically different way. Configuration is performed in a standard but entirely new and more fine-grained fashion, at the flow level. The configuration is based on the fundamental construct of the flow, which is shared by all networking devices. This permits configuration by a common protocol such as OpenFlow. Rather than trying to band-aid over a configuration morass with a system of the aforementioned drivers, this SDN approach has the ability to scale with the evolution of the network and to implement consistent network-wide configuration and policy. Central policy storage falls out naturally from the centralized controller architecture of SDN.

These attributes stem from SDN's very roots. We recall from Section 3.2.7 that Ethane, the direct precursor to OpenFlow, was specifically targeted at the distribution of complex policy on a network.

GLOBAL NETWORK VIEW

We have mentioned the importance of network-wide visibility in previous chapters, noting this as a real value of controller-based networking. This is true for any type of system that relies on a central management and control system, such as phone systems, power grids, etc. Having complete visibility allows the controlling system to make optimal decisions, as we discuss later. Such a (logically or physically) centralized controller platform can perform global optimizations rather than just per-node optimizations.

The control plane of today's network devices has evolved to the point where each device holds the network topology represented as a graph. This solution is imperfect as the individual graphs held in each device are constructed by propagating changes across the network from neighbor to neighbor. This can result in relatively long periods before the devices converge on a consistent view of the network topology. During these periods of convergence, routing loops can occur. In such networks, it is cumbersome to understand exactly how the network will behave at any point in time. This legacy technology was geared toward maintaining basic connectivity, not the use cases we presented in Chapter 8 and continue to discuss here. In addition, the traditional networks do not have real-time data about traffic loads. Even maximum bandwidth information is not typically shared. A centralized control system is able to gather as much data (traffic loads, device loads, bandwidth limits) as is necessary to make the decisions required at any moment.

Having the aforementioned information makes it possible for the controller to make optimal routing and path decisions. Running on a processor which is many times more powerful than that present on a networking device increases the probability that the controller will be able to perform faster and higher quality analysis in these decisions. Additionally, these decisions can take into account traffic patterns and baselines stored in secondary storage systems, which would be prohibitively expensive for networking devices. The distributed counterpart to this centralized control suffers the additional disadvantage of running a variety of protocol implementations on a selection of weaker processors, which creates a more unpredictable environment. Lastly, if the capabilities of the physical server on which the controller is running are exceeded, the controller can use horizontal scaling approaches, something not possible on a networking device. In this context, horizontal scaling means adding additional controllers and distributing the load across them. Horizontal scaling is not possible on the networking device because such scaling would change the network topology. Changing the network topology does not provide more power to control the same problem, but rather it alters the fundamental problem and creates new ones.

A network-wide view takes the guesswork and randomness out of path decisions and allows the controller to make decisions which are reliable and repeatable. This deterministic behavior is critical in environments which are highly sensitive to traffic patterns and bandwidth efficiency, in which suboptimal behavior based on arbitrary forwarding assignments may cause unacceptable packet delays and congestion.

These benefits of SDN-based networks are generally relevant to the network environments described later. In certain cases they are of particular importance and will be highlighted in those sections.

9.1 WIDE AREA NETWORKS

WANs have historically been used to connect remote islands of networks, such as connecting geographically dispersed LANs at different offices that are part of the same enterprise or organization. A number of non-Ethernet technologies have been used to facilitate this WAN connectivity, including:

- *Leased line*. These point-to-point connections were used in the early days of networking and were very expensive to operate.
- *Circuit switching*. Dial-up connections are an example of this type of connectivity, which had limited bandwidth, but were very inexpensive.
- *Packet switching*. Carrier networks at the center carried traffic from one endpoint to another, using technologies such as X.25 and Frame-Relay.
- *Cell relay*. Technologies such as ATM are still used today, featuring fixed cell sizes and virtual circuits.
- *Cable and digital subscriber loop*. These technologies play a key role in bringing WAN connectivity to the home.

However, recently these technologies have significantly given way to Ethernet-based WAN links, as we describe in greater detail in Section 9.2. Nowadays, companies and institutions are using Ethernet to connect geographically dispersed networks, using either private backbones, or in certain cases, the Internet. When we use the term Ethernet in this context, we are referring to the use of Ethernet framing over the WAN optical links. We do not imply that these links operate as broadcast, CSMA/CD media, like the original Ethernet.

Ethernet has proven to be an excellent solution in areas which were not envisioned in the past. However, the move toward Ethernet brings with it not only the advantages of that technology but also many of those challenges that we have described in detail in this book. The following section addresses some of the major issues that are relevant for WAN environments.

The key issues that face network designers when it comes to WANs are reliability and making the most efficient use of available bandwidth. Due to the cost of long-haul links in the WAN, bandwidth is a scarce commodity as compared to LANs. In environments such as a data center, it is possible to create redundant links with little concern for cost, as cables are relatively inexpensive. Furthermore, bandwidth needs can be met by adding more ports and links. Not so with the WAN, where bandwidth costs are orders of magnitude greater over distance than they are within a data center or campus. Adding redundant ports just exacerbates this problem. As a result, it is important to drive higher utilization and efficiency in the WAN links.

9.1.1 SDN APPLIED TO THE WAN

Overcoming the loss of connectivity on a link using redundancy and failover is common in all forms of networking, including WANs. This technology exists in devices today and has existed for many years. However, routing decisions made during a failover are not predictable during the period of convergence and often are not optimal even after convergence. The suboptimal paths are due to the lack of a central view which sees all paths as well as bandwidth capabilities and other criteria. Such global knowledge permits repeatable and optimal decisions. This aspect of reliability is keenly desired by organizations which rely on wide-area connections to operate their businesses and institutions. As we discussed in

Section 8.4, the SDN controller can have access to all of this information, and as it presumably runs on a high-performance server it can compute optimal paths more quickly and more reliably than the traditional distributed approach.

There are a number of pre-SDN technologies that attempt to enable optimal routing decisions. These include:

- Network Monitoring and Traffic Management Applications. For example, sFlow and NetFlow collect the necessary information so that a network management system can make optimal routing decisions.
- MPLS-Traffic Engineering (MPLS-TE) (see Section 9.2.2) uses traffic engineering mechanisms to choose optimal paths for MPLS *Label Switched Paths* (LSPs). An LSP is a path through an MPLS network. Packets are forwarded along this path by intermediate routers that route the packets by a preassigned mapping of ingress label-port pairs to egress label-port pairs. An LSP is sometimes referred to as an MPLS tunnel.
- The *Path Computation Element (PCE)-based Architecture* [1] determines optimal paths between nodes based on a larger set of network loading conditions than may be considered in simple best-path analysis such as IS-IS.

Even when these pre-SDN technologies attempt to make optimal path decisions based on a global view of the network, they are thwarted by devices that retain their autonomous control planes. We consider an example of this in the following section.

9.1.2 EXAMPLE: MPLS LSPs IN THE GOOGLE WAN

In [2] Google describes how their pre-SDN WAN solution handles the case of a failed link. In this particular scenario the problem revolves around rerouting MPLS LSPs in the event of a link failure.

We depict the pre-SDN situation in Fig. 9.2. In the figure the link on the right labeled *Best route* goes down. When a link goes down and the network needs to reconverge, all the LSPs try to reroute. This

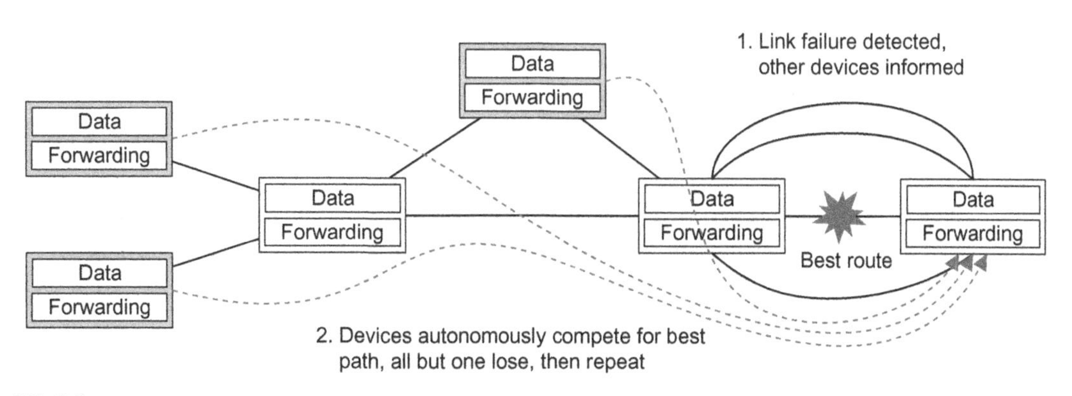

FIG. 9.2

Google without OpenFlow.

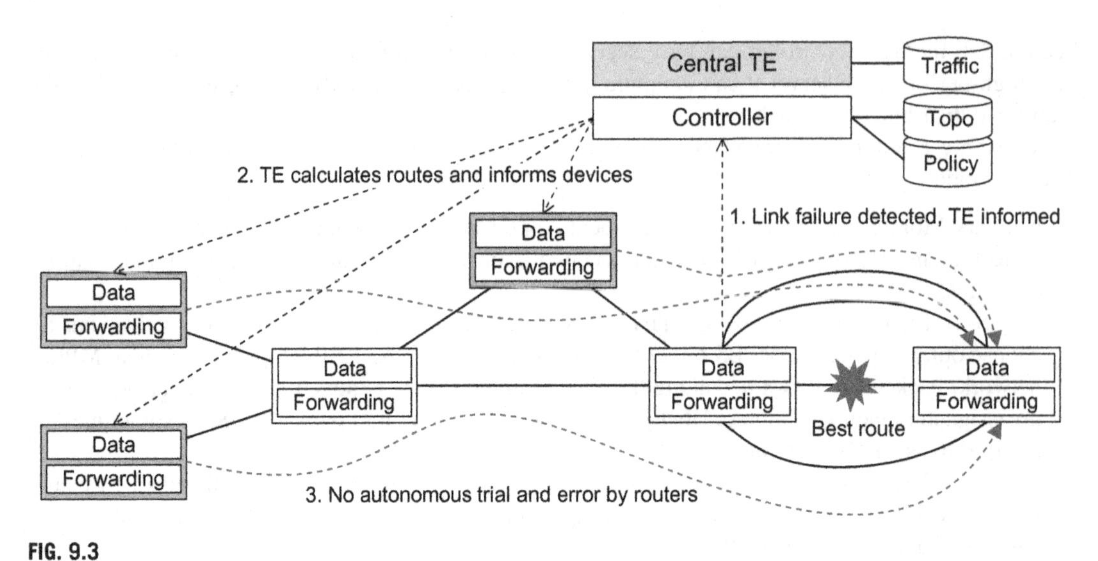

FIG. 9.3

Google with OpenFlow.

process is done autonomously by the routers along the path using the *Resource Reservation Protocol* (RSVP) to establish the path and reserve the bandwidth. When the first LSP is established and reserves the remaining bandwidth on a link, the others have to retry over the remaining paths. Because the process is performed autonomously and *ad hoc*, it can be repeated multiple times with a single winner declared each time. This can iterate for some time until the last LSP is established. It is not deterministic in that it is not known which LSP will be the last in any given scenario. This is a real-life problem and this pre-OpenFlow solution is clearly lacking.

We contrast this with Fig. 9.3, which shows how an SDN system with a centralized controller containing knowledge of all possible routes and current available bandwidth can address this problem much more efficiently. To accomplish this, the controller works in concert with the *Traffic Engineering* (TE) application shown in the figure. The computation of an optimal solution for mapping the LSPs to the remaining network can be made once and then programmed into the devices. This SDN-based solution is deterministic in that the same result will occur each time.

Google has been at the forefront of SDN technology since its outset, and they have made use of the technology to their advantage in managing their WAN networks as well as their data centers. They have connected their data centers using OpenFlow switches connected to a controller which they have designed with the features we just described in mind. The benefits that Google has realized from their conversion to OpenFlow in the WAN are:

- Lower cost of managing devices
- Predictable and deterministic routing decisions in the case of failover
- Optimal routing decisions based on bandwidth and load

DISCUSSION QUESTION

In the example illustrated in Figs. 9.2 and 9.3 is the solution depicted in Fig. 9.3 preferable because the LSPs are ultimately routed over better paths after the failure or merely that they converge on those paths more quickly? What part of this answer cannot be ascertained from the information provided here?

9.2 SERVICE PROVIDER AND CARRIER NETWORKS

Service Provider (SP) and carrier networks are wide-area in nature and are sometimes referred to as *backhaul* networks, since they aggregate edge networks, carrying huge amounts of data and voice traffic across geographies, often on behalf of telecommunications and Internet Service Providers. Examples of Service Providers and carriers are Verizon, AT&T, Sprint, Vodafone, and China Mobile.

A *Network Service Provider* sells networking bandwidth to subscribers, who are often Internet Service Providers. A *carrier* is typically associated with the telecommunications and mobility industries, who in the past have been concerned primarily with voice traffic. The carrier typically owns the physical communications media. The line between carriers and SPs is somewhat blurred. Carriers and SPs now carry a much more diverse traffic mix, including *Voice over IP* (VoIP) and video. In addition to a greater variety of traffic types, the total volume of traffic grows at an ever-increasing pace. Smartphones with data plans have played a large role in increasing both the total amount of traffic as well as its diversity.

Without taking a new approach to managing bandwidth, this growth in diversity and volume of traffic results in costs spiraling out of control. Over-provisioning links to accommodate changing traffic patterns is too costly for the WAN links used by the SPs. Traditionally, bandwidth management is provided via network management platforms which can certainly measure traffic utilization and can perhaps even recommend changes that could be made manually in order to respond to traffic patterns.

Not only does network bandwidth need to be utilized most efficiently, it is also necessary to respond immediately to changes in requirements brought about by service contract upgrades and downgrades. If a customer wishes to have their traffic travel across the network at a higher priority or at greater speeds, SPs would like to be able to implement changes to their service policies immediately, without disruption to existing flows. Being able to provide *bandwidth on demand* is a selling point for SPs. Other requirements include the ability to dynamically change paths to higher bandwidth, lower latency paths, and to re-prioritize packets so that they take precedence in queues as they pass through networking devices.

Another aspect of reducing costs involves the *simplification of devices*. In the large core networks operated by carriers, the OPEX costs of managing a complex device outweigh the increased CAPEX outlay for that device. Thus a simpler device provides cost savings in two ways. This is true as well not only for network devices but also for network appliances which today require specialized hardware, such as load balancers, firewalls, and security systems.

SPs are responsible for taking network traffic from one source, passing it throughout the SP's network, and forwarding it out the remote network edge to the destination. Thus the packets themselves must cross at least two *boundaries*. When more than one SP must be traversed, then the number of boundaries crossed increases. Traffic coming into the SP's network is typically marked with specific tags for VLAN and priority. The SP has needs regarding routing traffic as well, which may require

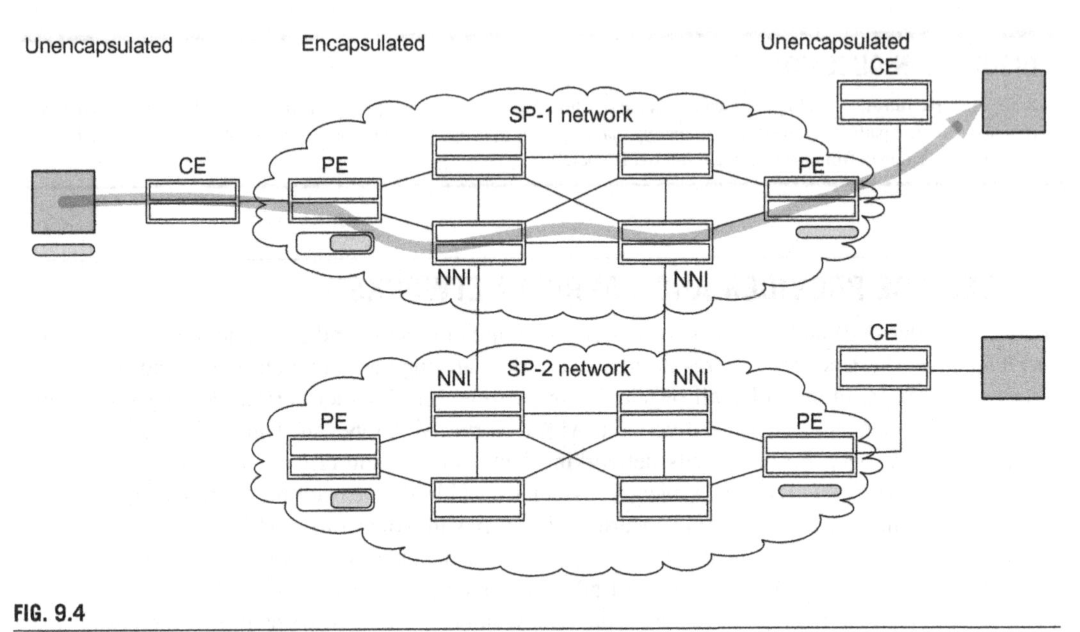

FIG. 9.4

Service provider environment.

the packet to be tagged again. Furthermore, the routing mechanism may be different within the SP network. For example, they may use MPLS or VLAN tagging for internal routing. When these additional tagging mechanisms are used, this entails another layer of encapsulation of the customer data packet. The boundaries that traffic must cross are often referred to as *Customer Edge* (CE) and *Provider Edge* (PE). The *Network to Network Interface* (NNI) is the boundary between two SPs. Since the NNI is an important point for policy enforcement, it is important that policy be easily configurable at these boundaries. Technologies supporting SPs in this way must support these boundary-crossing requirements.

Fig. 9.4 shows a simplified version of a network with customers and a pair of SPs. The two endpoint hosts are attempting to communicate by passing through the CE and then PE. After traversing SP1's network, the packet will egress that network at the PE on the other side. After passing through the CE, the destination receives the packet.

The figure shows a packet traversing the network from the endpoint on the left to the endpoint on the right (see arrow in figure). The original packet as it emanates from the source device is unencapsulated. The packet may acquire a VLAN tag as it passes through the CE (probably provided by the ISP). When the packet passes through the PE, it may be encapsulated using a technology such as PBB that we discussed in Section 5.6.7, or is tagged with another VLAN or MPLS tag. When the packet exits the provider network and passes through the other PE on the right, it is correspondingly either decapsulated or the tag is popped and it is then passed into the destination CE network.

Fig. 9.4 additionally depicts the NNI between SP1 and SP2. If the customer packets were directed to a destination on the SP2 network, they would traverse such an NNI boundary. Policies related to business agreements between the SP1 and SP2 service providers would be enforced at that interface.

9.2.1 SDN APPLIED TO SP AND CARRIER NETWORKS

A key focus of SPs when considering SDN is *monetization*. This refers to the ability to make or save money by using specific techniques and tools. SDN is promoted as a way for providers and carriers to monetize their investments in networking equipment by increasing efficiency, reducing the overhead of management, and rapidly adapting to changes in business policy and relationships.

Some of the ways in which SDN can help improve monetization for providers and carriers are (1) *Bandwidth management*, (2) *CAPEX and OPEX savings*, and (3) *Policy Enforcement at the PE and NNI boundaries*.

SDN exhibits great agility and the ability to maximize the use of existing links using TE and centralized, network-wide awareness of state. The granular and pliable nature of SDN allows changes to be made easily and with improved ability to do so with minimal service interruption. This facilitates the profitable use of bandwidth as well as the ability to adapt the network to changing requirements related to customer needs and *Service-Level Agreements* (SLAs).

SDN can reduce costs in a couple of ways. First, there are CAPEX savings. The cost of white box SDN devices is appreciably lower than the cost for comparable non-SDN equipment. This may be due to *bill of materials* (BOM) reduction as well as the simple fact that the white box vendors are accustomed to a lower-margin business model. The BOM reductions derive from savings such as reduced memory and CPU costs due to the removal from the device of so much CPU- and memory-intensive control software. Second, there are reduced OPEX costs, which come in the form of reduced administrative loads related to the management and configuration of the devices.

As described previously, packets will travel from the customer network across the PE into the provider's network, exiting at the remote PE. OpenFlow supports multiple methods of encapsulating traffic as is required in these circumstances. OpenFlow 1.1 supports the pushing and popping of MPLS and VLAN tags. OpenFlow 1.3 added support for PBB encapsulation. This makes it possible to conform to standard interfaces with Open SDN technology. Additionally, Open SDN provides a versatile mechanism for policy enforcement on the aggregated traffic which traverses NNI boundaries. Since these policies are often tied to fluid business relationships between the SPs, it is essential that the technology supporting policy enforcement be easily manipulated itself. With the additional bandwidth, cost-cutting, visibility, and policy enforcement advantages of SDN, this makes an SDN solution even more compelling.

9.2.2 EXAMPLE: MPLS-TE AND MPLS VPNs

A research project at Stanford [3] demonstrated MPLS-TE using OpenFlow. The authors stated that "while the MPLS data plane is fairly simple, the control planes associated with MPLS-TE and MPLS-VPM are complicated. For instance, in a typical traffic engineered MPLS network, one needs to run OSPF, LDP, RSVP-TE, I-BGP and MP-BGP." They argue that the control plane for a traditional MPLS network that determines routes using traffic information is unnecessarily complex. The amount and complexity of control signaling in the traditional approach is such that when there are frequent changes in the network, there are so many control packets exchanged that some can readily be lost. This results in a cascade of instability in the MPLS-TE control plane. The authors demonstrate that it is simple to build an equivalent network using OpenFlow to push and pop MPLS tags according to the topology and traffic statistics available at the NOX OpenFlow controller. In addition to providing a more stable

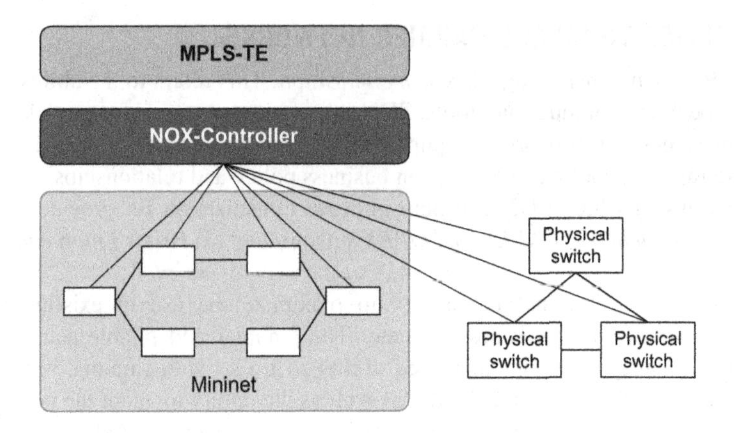

FIG. 9.5

Service provider and SDN: MPLS.

(Reproduced with permission from Sharafat A, Das S, Parulkar G, McKeown N. MPLS-TE and MPLS VPNs with OpenFlow. In: SIGCOMM'11. Toronto; 2011.)

and predictable alternative to the traditional approach, the OpenFlow-based solution was able to be implemented in less than 2000 lines of code, which is far more simple than the current systems.

Fig. 9.5 shows the architecture of the research project, replacing many routing and TE protocols on the devices with an MPLS-TE application running on the OpenFlow controller, setting flows consistent with the connectivity and QoS requirements. The project demonstrated this both on the Mininet network simulator as well as on physical switches. This OpenFlow-based MPLS solution obviates the need for a number of other protocols and functions that would otherwise have had to be implemented by the individual devices.

MPLS-TE remains an active area of research for the application of SDN. For example, Ericsson has published related research in using OpenFlow for MPLS-TE in [4].

DISCUSSION QUESTION

We indicate that the solution illustrated in Fig. 9.5 replaces many routing and TE protocols that would be needed to implement this solution in legacy networks. Name five examples.

9.2.3 EXAMPLE: CLOUD BURSTING WITH SERVICE PROVIDERS

Cloud bursting allows an SP to expand an enterprise's private cloud (data center) capacity on demand by dynamically allocating the compute and storage resources in the SP's data center to that enterprise. This additionally entails allocating the network resources to allow the free flow of data between the original, private cloud, and its dynamic extension in the SP. In [5], the author proposes to accomplish this cloud bursting via an SDN network. Controllers in the enterprise's private cloud make requests to the SP's SDN controller for the needed facilities, and the SP SDN controller allocates the needed networking, compute and storage components to that enterprise. This model facilitates a major business opportunity for SPs. Verizon, in collaboration with HP and Intel, describes an SDN-based proof of concept project [6,7] to implement a solution for cloud bursting. In this PoC, an Intel private cloud in Oregon is dynamically augmented by the capacity in a Verizon public cloud in Massachusetts. Without manual intervention, network bandwidth is increased to handle the surge in data between the two data centers. The security that Intel requires on the Verizon VMs can be implemented without the need for physical appliances, using SDN techniques such as those that we discuss in Chapter 10. This PoC demonstrates a method where an SP can obtain improved CAPEX *Return on Investment* (ROI), lower operating expenses and improved business agility through the use of SDN.

9.3 CAMPUS NETWORKS

Campus networks are a collection of LANs in a concentrated geographical area. Usually the networking equipment and communications links belong to the owner of the campus. This may be a university, a private enterprise or a government office, among other entities. Campus end-users can connect through wireless access points (APs) or through wired links. They can connect using desktop computers, laptop computers, shared computers, or mobile devices, such as tablets and smartphones. The devices with which they connect to the network may be owned by their organization or by individuals. Furthermore, those individually owned devices may be running some form of access software from the IT department or the devices may be completely independent.

There are a number of networking requirements that pertain specifically to campus networks. These include (1) *differentiated levels of access*, (2) *bring your own device* (BYOD), (3) *access control and security*, (4) *service discovery*, and (5) *end-user firewalls*.

Different types of users in the campus will require different levels of access. Day guests should have access to a limited set of services, such as the Internet. More permanent guests may obtain access to more services. Employees should receive access based on the category into which they fall, such as executives or IT staff. These differentiated levels of access can be in the form of access control (i.e., what they can and cannot have access to) as well as their quality of service, such as traffic prioritization and bandwidth limits.

BYOD is a phenomenon that has arisen from the exponential increase in functionality available in smartphones and tablet computers. Campus network users want to access the network with the devices they are familiar with, rather than with a campus-issued device. In addition to mobile devices, some individuals may prefer Apple or Linux laptops over the corporate-issued Windows systems.

In years past, access to networks was based on physical proximity. If you were in the building and could plug your device into the network, then you were granted access. With the popularity of wireless connectivity and heightened security, it is now more common for employees and guests alike to have to overcome some security hurdle in order to be granted access. That security may be in a more secure form, such as IEEE 802.1X [8], or it may be some more limited form of security, such as authentication based on MAC address or a captive portal web login.

Campus end-users want access to services such as printers or file shares, or perhaps even TVs. Service discovery makes this possible through simple interfaces and automatic discovery of devices and systems which provide these services. In order to achieve this level of simplicity, however, there is a cost in terms of network traffic load associated with protocols providing these services.

One of the dangers of campus networks is the possibility of infected devices introducing unwanted threats to the network. This threat is magnified by the presence of BYOD devices on the network. These may take the form of malicious applications like port scanners which probe the network looking for vulnerable devices. End-user firewalls are needed to protect against such threats.

9.3.1 SDN ON CAMPUS: APPLICATION OF POLICY

At the beginning of this chapter we described the value of centralized policy management and deployment. Campus networks are one area where this capability is readily realized. SDN's flexibility to manipulate flow rules based on policy definitions makes it well suited to the application of policy in the campus. The general idea of policy application is as follows:

- The user connects to the network and attempts to send traffic into the network.
- No policy is in place for this user, so the user's initial packets are forwarded to the controller. At this point, the user has either no access or only limited access to the network.
- The controller consults the policy database to determine the appropriate prioritization and access rights for this user.
- The controller downloads the appropriate flow rules for this user.
- The user now has access which is appropriate to the group to which he or she belongs, as well as other inputs such as the user's location, time of day, etc.

This system provides the differentiated levels of access that are needed in campus networks.

Clearly, as we explained in Section 3.2.2, this functionality is similar to what *Network Access Control* (NAC) products provide today. However, this SDN-based application can be simple, open, software-based, and flexible in that policies can be expressed in the fine-grained and standard language of the flow. We discuss programming an SDN NAC application in greater detail in Section 12.11.

Note that this type of access control using SDN should be applied at the edge of the network, where the number of users and thus, the number of flow entries will not exceed the limits of the actual hardware tables of the edge networking device. We introduced the topic of scaling the number of flow entries in Section 4.3.5. After the edge, the next layer of network concentration is called the *distribution* or *aggregation* layer. Application of policy at this layer of the campus network would take another form. At this level of the network, it is not appropriate or even practical to apply end-user rules which should be applied at the edge, as described previously. Rather, SDN policy in these devices should be related to classes of traffic or traffic that has been aggregated in some other way, such as a tunnel or MPLS LSP.

Flows can also be established for various types of traffic that set priorities appropriately for each traffic type. Examples of different traffic types are HTTP versus email. For example, HTTP traffic might be set at a lower priority during office hours, assuming that the organization's work-based traffic was not HTTP-based.

9.3.2 SDN ON CAMPUS: DEVICE AND USER SECURITY

Technological trends such as BYOD, access control and security can also be addressed by SDN technology. For example, one of the requirements for registering users' BYOD systems and guests involves the use of a *captive portal*. This is the mechanism by which a user's browser request gets redirected to another destination website. This other website can be for the purpose of device registration or guest access. Captive portals are traditionally implemented by encoding the redirection logic into the switch firmware or by in-line appliances. To implement this in a pre-SDN environment could entail upgrading to switches that had this capability, implying more complex devices, or by installing specialized in-line appliances, adding to network complexity. Configuring which users need to be authenticated via the captive portal would entail configuring all these devices where a user could enter the network. Below we describe a simpler SDN solution for a captive portal solution:

- The user connects to the network and attempts to establish network connectivity.
- No access rules are in place for this user, so the SDN controller is notified.
- The SDN controller programs flows in the edge device which will cause the user's HTTP traffic to be redirected to a captive portal.
- The user is redirected to the captive portal and engages in the appropriate exchange to gain access to the network.
- Once the captive portal exchange is complete, the SDN controller is notified to set up the user's access rules appropriately.
- The user and/or BYOD device now has the appropriate level of access.

Fig. 9.6 shows an SDN-based captive portal-based application. The network edge device is initially programmed to route ARP, DNS, and DHCP requests to the appropriate server. In the figure, the end-user connects to the network and makes a DHCP request to obtain an IP address. When the

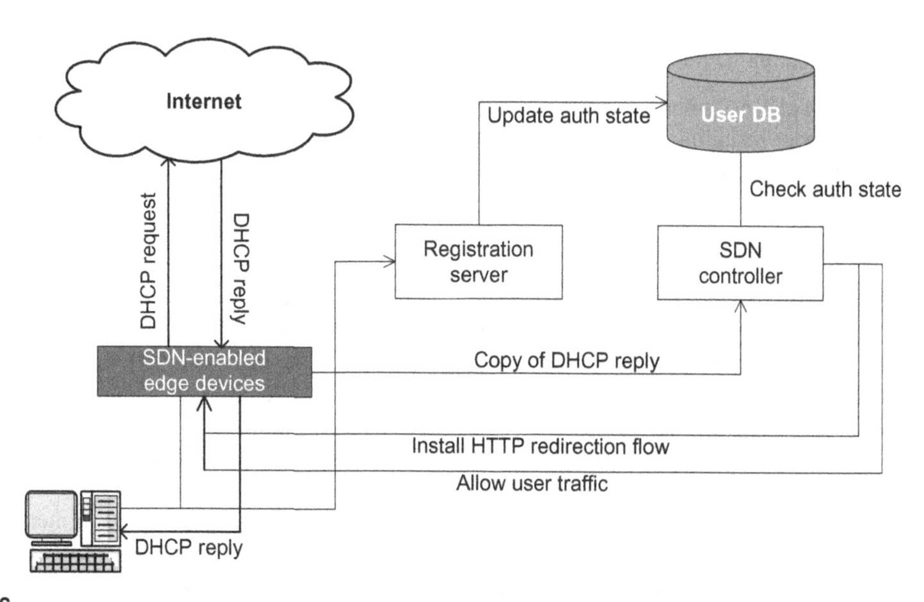

FIG. 9.6

NAC captive portal application.

DHCP reply is returned to the user, a copy is sent to the SDN controller. Using the end-user MAC address as a lookup key, the controller consults the database of users. If the user device is currently registered, it is allowed into the network. If it is not, then OpenFlow rules are programmed to forward that user's HTTP traffic to the controller. When the unauthenticated user's HTTP traffic is received at the controller, that web session is redirected to the captive portal web server. After completing the user authentication or device registration, the controller updates the user's flow(s) so that the packets will be allowed into the network. Note that this is also the appropriate time to configure rules related to the user's policy which can include levels of access and priority, among other items.

This type of functionality, like policy, is appropriately applied at the edge of the network. The further into the network this type of application is attempted, the more problematic it becomes, with flow table overflow becoming a real issue.

HP currently offers a commercial SDN application called Sentinel [9], which provides secure BYOD access. The approach of securing the network from end-user infection by installing anti-malware software on end-user devices is not viable in the BYOD case since they are not under the control of the network administration. The Sentinel SDN application provides a solution by turning the entire network infrastructure into a security-enforcement device using mechanisms that are hardened and more sophisticated versions of the basic methods for campus security we described previously. Sentinel is an SDN application that works with HP's TippingPoint threat database to secure and apply policy to the intranet as follows:

1. Identify *botnets* on the intranet and neutralize them by blocking DNS requests and IP connections to their *botnet masters* in the Internet.
2. Identify and prevent users or machines from accessing infected or prohibited sites on the Internet by monitoring and blocking DNS requests and IP connections (see Section 9.3.3).
3. Identify and quarantine infected machines on the intranet so that they can be remediated.

While traditional NAC solutions provide similar functionality, as an SDN application Sentinel can be deployed without installing additional hardware and it can provide protection at the edge of the network. Sentinel is a sophisticated example of the *blacklist* technology discussed in the next section.

9.3.3 SDN ON CAMPUS: TRAFFIC SUPPRESSION

Having flow table rules at the edge of the network carries the additional benefit of being able to suppress unwanted traffic. That unwanted traffic could be benign, such as service discovery multicasts, or it could be malicious, such as infected devices bringing viruses into the network. With flow tables at the edge of the network, service discovery traffic can be captured by the edge device and forwarded to the controller. The controller can then keep a repository of services and service requestors, which can be used to satisfy requests without having to forward all those multicasts upstream to flood the possibly overloaded network.

The ability to set policy and security for end-users with highly granular flow-based mechanisms facilitates effective per-user firewalls. Only certain types of traffic (e.g., traffic destined for certain UDP or TCP ports) will be allowed to pass the edge device. All other types of traffic will be dropped at the very edge of the network. This is an effective means of blocking certain classes of malicious traffic. Note that the implementation of per-user firewalls requires that the controller know who the user is. If an NAC solution is running in the SDN controller, then this is not a problem. If the NAC solution, such as RADIUS, is implemented outside of SDN, then there is a challenge as the relationship between that external NAC solution and OpenFlow is neither standardized nor well-understood.

One facet of a per-user firewall is an application known as a blacklist. Blacklist technology blocks attempts by end-users to access known malicious or harmful hostnames and IP addresses. Traditional blacklist solutions rely on in-line appliances which *snoop* all traffic and attempt to trap and block attempts to reach these bad destinations. SDN is able to implement a blacklist solution without the insertion of additional appliances into the network. This provides both CAPEX savings due to less network equipment and OPEX savings due to a network that is easier to administer. An SDN application for blocking attempts to reach specific hostnames entails setting up flow table rules in edge devices to capture DNS requests and sending them to the controller. The SDN controller consults a database of undesirable hostnames, and would block the DNS request from being sent out when a blacklisted hostname is encountered.

Fig. 9.7 shows the simple manner in which a DNS blacklist could be implemented. The controller sets flows in the edge devices directing them to forward all DNS requests to the controller. When DNS requests arrive at the controller, it consults a local or remote database of known malicious sites. If the result is that the hostname is clean, then the controller returns the DNS request with instructions to the edge device to forward the packet as it normally would. If the hostname is deemed to be unsafe, the controller instructs the edge device to drop the packet, denying the user access to that host.

A savvy user can circumvent this first blacklist application if he/she knows the IP address of the destination host. By specifying the IP address directly instead of the hostname, no DNS request is sent. Fig. 9.8 shows a simple SDN implementation for this variant of the blacklist problem. In this second

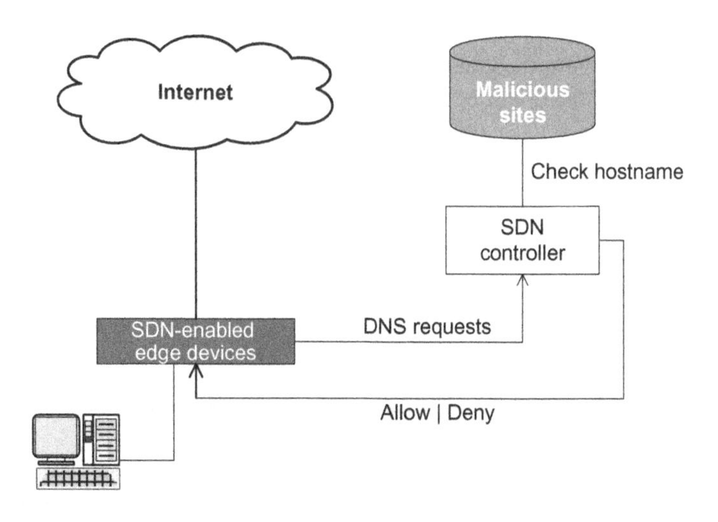

FIG. 9.7

DNS blacklist application.

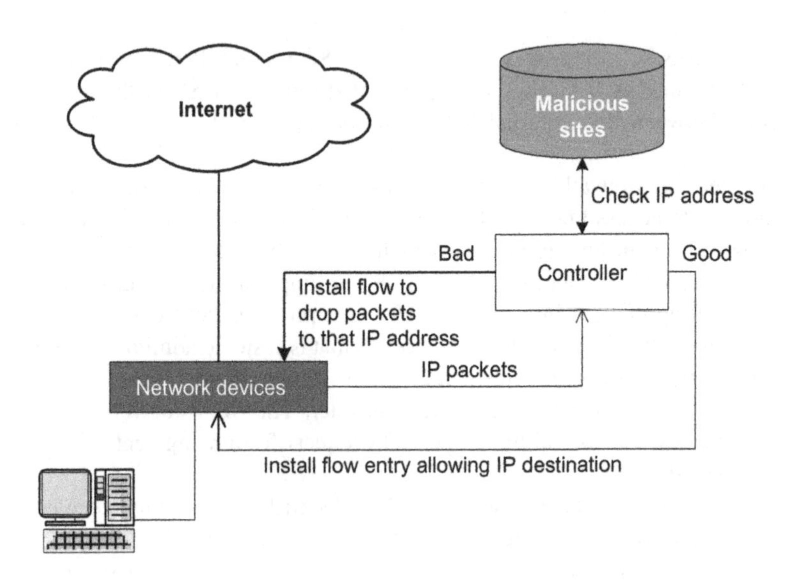

FIG. 9.8

IP address blacklist application.

solution, packets with destination IP addresses of unknown virtue are forwarded to the controller, which inspects the packets and either allows the IP address (so that packets to and from that host will now automatically be allowed) or else the device is programmed to drop this and future packets sent to that bad address. If the packet is allowed, the application will install a transient flow allowing requests to that destination IP address.

Note with this latter solution application design is critical. If destination IP address *allow* rules have aging timers which are too short, then they will age out in between requests and, consequently, every time a destination is visited, there will be delay and overhead associated with verifying and allowing the packets to that destination. If those rules have aging timers which are too long, then the device runs the risk of being overloaded with flow entries, possibly even exceeding the maximum size of the flow table.

We provide a detailed analysis with accompanying source code for an Open SDN blacklist application in Section 12.6.

Blacklist is really a simple host-level firewall. HP's Sentinel application described earlier provides a blacklist feature as a component of its commercial SDN security application.

DISCUSSION QUESTION

Is the blacklist application depicted in Figs. 9.7 and 9.8 a reactive or proactive application?

9.4 HOSPITALITY NETWORKS

Hospitality networks are found in hotels, airports, coffee shops, and fast food restaurants. There is a fair degree of overlap between the user requirements in campus networks and those of hospitality networks. The primary end-user in a hospitality network is a guest who is using the network either through the largesse of the establishment or through the purchase of a certain amount of time. One class of the for-purchase situation is end-users who have purchased connectivity for a certain duration, such as a month or a year, and who are thus able to connect to the network without a direct financial exchange each time they connect.

The application of SDN for captive portals in the campus discussed in Section 9.3.2 applies similarly to hospitality networks. The difference is that in the case of hospitality networks the user is paying directly or indirectly for the access, so a corresponding form of identification will likely be involved in authenticating with the registration server. These could include providing hotel room or credit card information.

Hospitality networks frequently offer WiFi access. We describe a more general application of SDN to WiFi access in the next section on mobile networks.

9.5 MOBILE NETWORKS

Mobile networking vendors, such as AT&T, Verizon, and Sprint, compete for customers to attach to their networks. The customers use smartphones or tablets to connect using the available cellular service, whether that is 3G, 4G, LTE, or another cellular technology.

When mobile customers use traditional WiFi hotspots to connect to the Internet, those mobile vendors effectively lose control of their customers. This is because the users' traffic enters the Internet directly from the hotspot. Since this completely circumvents the mobile vendor's network, the vendor is not even aware of the volume of traffic that the user sends and receives, and certainly cannot enforce any policy on that connection. When their customers' traffic circumvents the mobile provider's network, the provider loses a revenue-generating opportunity. Nevertheless, since their cellular capacities are continually being stretched, from that perspective it is advantageous for the mobile vendors to offload traffic to WiFi networks when possible. Thus the mobile provider is interested in a solution that would allow their customers to access their networks via public WiFi hotspots *without their losing control of and visibility to their customers' traffic*. The owner of such hotspots may wish for multiple vendors to share the WiFi resource offered by the hotspot. The multitenant hotspot we describe here is somewhat analogous to network virtualization in the data center. Just as SDN shines in that data center environment, so can it play a pivotal role in implementing such multitenant hotspots?

9.5.1 SDN APPLIED TO MOBILE NETWORKS

Mobile vendors interested in gaining access to users who are attaching to the Internet via WiFi hotspots require a mechanism to control their users' traffic. Control, in this context, may simply mean being able to measure how much traffic that user generates. It may mean the application of some policy regarding

QoS. It may mean diverting the user traffic before it enters the public Internet and redirecting that traffic through their own network. SDN technology can play a role in such a scheme in the following ways:

- Captive portals
- Tunneling back to the mobile network
- Application of policy

We discussed captive portals and access control in Section 9.3.1. This functionality can be applied to mobile networks as well. This requires allowing users to register for access based on their mobile credentials. Once valid credentials are processed, the user is granted appropriate levels of access.

One of the mechanisms for capturing and managing mobile user traffic is through the establishment of tunnels from the user's location back to the mobile vendor's network. The tunnel would be established using one of several available tunneling mechanisms. By programming SDN flows appropriately, that user's traffic would be forwarded into a tunnel and diverted to the mobile vendor's network. Usage charges could be applied by the mobile provider. In addition to charging for this traffic, other user-specific policies could be enforced. Such policies could be applied at WiFi hotspots where the user attaches to the network. SDN-enabled access points can receive policy, either from the controller of the mobile vendor or from a controller on the premises.

As an example of the utilization of SDN and OpenFlow technology as it relates to the needs of mobile networks and their SPs, consider Fig. 9.9.

The example depicted in Fig. 9.9 shows the basic means by which SDN and OpenFlow can be used to grant SP-specific access and private or public access from mobile devices to the Internet. In Fig. 9.9 the customers on the left want to access the Internet and each set has a different SP though which they

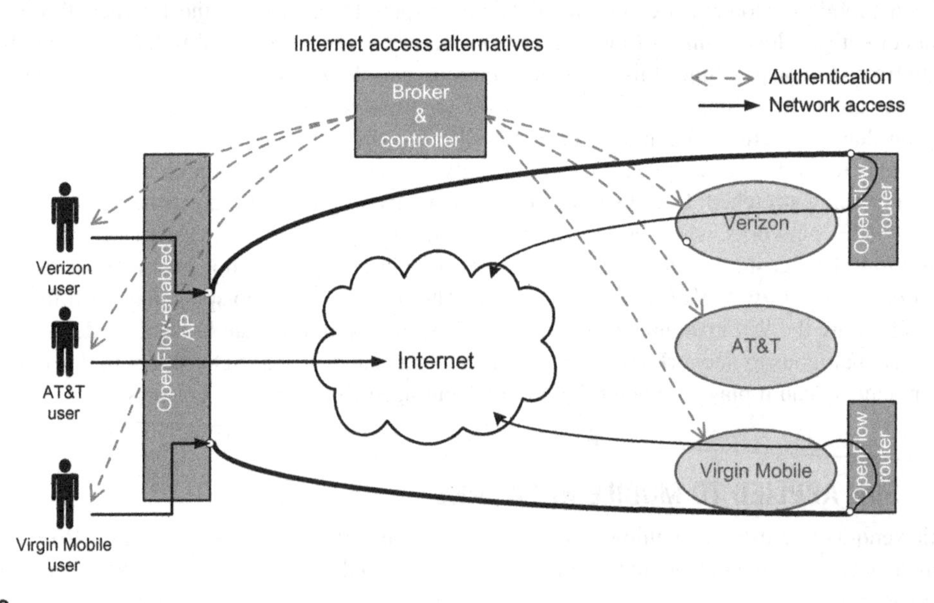

FIG. 9.9

Mobile service providers.

gain network access. Connecting through an OpenFlow-enabled wireless hotspot, they are directed through a broker who acts as an OpenFlow controller. Based on their SP, they are directed to the Internet in various ways, depending on the level of service and mechanism set up by the SP. In the example, AT&T users gain access directly to the Internet, while Verizon and Virgin Mobile users access the Internet by being directed by OpenFlow through a tunnel to the SP's network. Both tunnels start in the OpenFlow-enabled AP. One tunnel ends in an OpenFlow-enabled router belonging to Verizon, the other in an OpenFlow-enabled router belonging to Virgin Mobile. These two routers then redirect their respective customers' traffic back into the public Internet. In so doing, the customers gain the Internet access they desire and two of the mobile providers achieve the WiFi offload they need while maintaining visibility and control over their users' traffic. In order to facilitate such a system there is presumably a business relationship between the providers and the hotspot owner whereby the hotspot owner is compensated for allowing the three providers' users to access the hotspot. There would likely be a higher fee charged Verizon and Virgin Mobile for the service that allows them to retain control of their customers' traffic.

In 2013, the ONF's Wireless and Mobile Working Group published a number of OpenFlow-based use cases. These included:

- Flexible and Scalable Packet Core
- Dynamic Resource Management for Wireless Backhaul
- Mobile Traffic Management
- Management of Secured Flows in LTE
- Media-Independent Handover
- Energy Efficiency in Mobile Backhaul Networks
- Security and Backhaul Optimization
- Unified Equipment Management and Control
- Network-Based Mobility Management
- SDN-Based Mobility Management in LTE
- Unified Access Network for Enterprise and Large Campus

These use cases are very detailed and specific applications relevant to mobile operators. In a number of cases, implementing them would require extensions to OpenFlow. The mobile use cases listed above as well as others are described in [10].

DISCUSSION QUESTION

In the hypothetical example shown in Fig. 9.9, what potential benefits does AT&T forego by not tunneling its users' traffic back through the AT&T network?

9.6 OPTICAL NETWORKS

An *Optical Transport Network* (OTN) is an interconnection of optical switches and optical fiber links. The optical switches are layer one devices. They transmit bits using various encoding and multiplexing techniques. The fact that such optical networks transmit data over a lightwave-based *channel* as opposed to treating each packet as an individually route-able entity lends itself naturally to the SDN concept of a flow. In the past, data traffic was transported over optical fiber using protocols such as *Synchronous Optical Networking* (SONET) and *Synchronous Digital Hierarchy* (SDH). More recently, however, OTN has become a replacement for those technologies. Some companies involved with both OTN and SDN are Ciena, Cyan (now acquired by Ciena), and Infinera. Some vendors are creating optical devices tailored for use in data centers. Calient, for example, uses optical technology for fast links between racks of servers.

9.6.1 SDN APPLIED TO OPTICAL NETWORKS

In multiple-use networks, there often arise certain traffic flows which make intense use of network bandwidth, sometimes to the point of starving other traffic flows. These are often called *elephant flows* due to their sizable nature. The flows are characterized by being of relatively long duration yet having a discrete beginning and end. They may occur due to bulk data transfer, such as backups that happen between the same two endpoints at regular intervals. These characteristics can make it possible to predict or schedule these flows. Once detected, the goal is to re-route that traffic onto some type of equipment, such as an all-optical network which is provisioned specifically for large data offloads such as this. OTNs are tailor-made for these huge volumes of packets traveling from one endpoint to another. Packet switches' ability to route such elephant flows at packet-level granularity is of no benefit, yet the burden an elephant flow places on the packet switching network's links is intense. Combining a packet switching network with an OTN into the kind of *hybrid* network shown in Fig. 9.10 provides an effective mechanism for handling elephant flows.

In Fig. 9.10 we depict normal endpoints (A1, A2) connected through *Top-of-Rack* (ToR) switches ToR-1 and ToR-2, communicating through the normal path, which traverses the packet-based network fabric. The other elephant devices (B1, B2) are transferring huge amounts of data from one to the other; hence, they have been shunted over to the optical circuit switch, thus protecting the bulk of the users from such a large consumer of bandwidth.

The mechanism for this shunting or *offload* entails the following steps:

1. The elephant flow is detected between endpoints in the network. Note that, depending on the flow, detecting the presence of an elephant flow is itself a difficult problem. Simply observing a sudden surge in the data flow between two endpoints in no way serves to predict the longevity of that flow. If the flow is going to end in 500 ms, then this is not an elephant flow and we would not want to incur any additional overhead to set up special processing for it. This is not trivial to know or predict. Normally, some additional contextual information is required to know that an elephant flow has begun. An obvious example is the case of a regularly scheduled backup that occurs across the network. This topic is beyond the scope of this work, and we direct the interested reader to [11,12].
2. The information regarding the endpoints' attaching network devices is noted, including the uplinks (U1, U2) which pass traffic from the endpoints up into the overloaded network core.
3. The SDN controller program flows in ToR-1 and ToR-2 to forward traffic to and from the endpoints (B1, B2) out an appropriate offload port (O1, O2), rather than the normal port (U1, U2). Those offload ports are connected to the optical offload fabric.

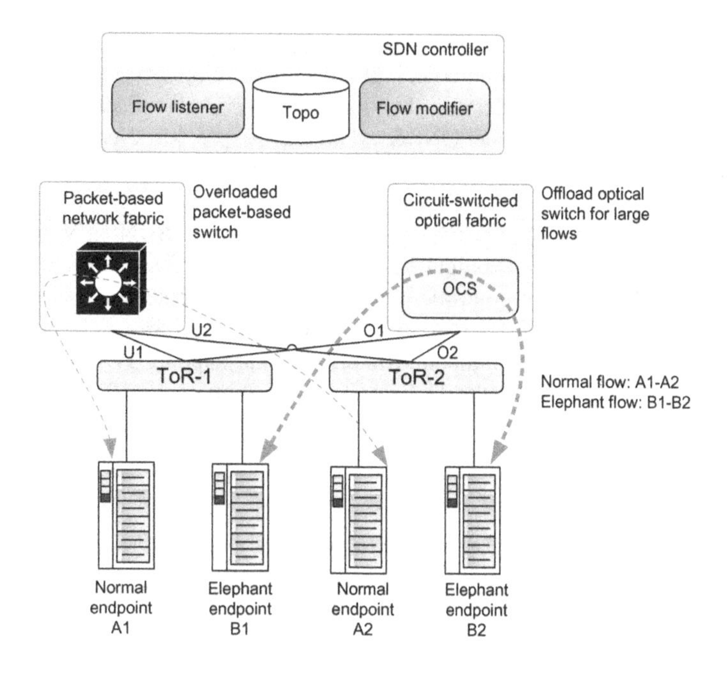

FIG. 9.10

Optical offload application overview.

4. The SDN controller programs flows on the SDN-enabled optical offload fabric to patch traffic between B1 and B2 on the two offload links (O1, O2). At this point, the re-routing is complete and the offload path has been established between the two endpoints through the OTN.
5. The elephant flow eventually returns to normal and the offload path is removed from both connecting network devices and from the optical offload device. Subsequent packets from B1 to B2 traverse the packet-based network fabric.

We discuss details of programming an offload application in Section 12.10.

9.6.2 EXAMPLE: FUJITSU'S USE OF SDN IN OPTICAL NETWORKS

Fujitsu has been an early adopter of SDN and focused on optical networks. One of their first forays into optical SDN was to leverage the technology to accelerate network storage access [13]. The SDN controller observes the storage access (storage flow) in the network on *Fiber Channel over Ethernet* (FCoE) and performs flow manipulation. Fujitsu separated the storage flow detection and storage flow manipulation from the functions needed for FCoE data relays. They then created a converged fabric switch with a software interface to the centralized controller. Fujitsu reported that this SDN implementation resulted in a twofold increase in throughput.

Other Fujitsu optical SDN efforts are targeted toward OTN. Fujitsu, a founding partner of the *Open Network Operating System* (ONOS) community, recently demonstrated a use case of packet-over-optical transport [14]. The ONOS *Cardinal* release was used to demonstrate this packet-over-optical use case, which is central to the application of SDN to OTN. With Cardinal, Fujitsu was able to leverage new southbound plugins to develop *transaction language 1* (TL1) interfaces from the ONOS controller to the *FLASHWAVE 9500* Packet Optical Networking Platform. These interfaces allowed the ONOS controller to provide *Dense Wavelength Division Multiplexing* (DWDM) services such as on-demand bandwidth, bandwidth calendaring, and multilayer optimization.

Through these SDN efforts, Fujitsu has expanded its *Virtuora* SDN/NFV platform [15]. This platform has been built on the OpenDaylight (ODL) controller, but Fujitsu has purposefully ensured that its platform is portable to other controllers. For instance, they note that the Virtuora platform is easily portable to ONOS. The Virtuora NC 3.0 SDN framework was recently launched and it is based on ODL [16]. This framework has southbound interfaces that support TL1 and NETCONF. Based on Fujitsu's optical work with ONOS, the TL1 interface can support the DWDM services previously mentioned.

DISCUSSION QUESTION

If you were asked to map the Fujitsu FLASHWAVE 9500 onto the hypothetical example in Fig. 9.10, which functional box would it be?

9.7 SDN VS P2P/OVERLAY NETWORKS

At a conceptual level *P2P/Overlay networks* resemble the overlay networks presented in detail throughout this book. Just as the data center virtual networks are overlaid on a physical infrastructure the details of which are masked from the virtual network, so also is the P2P/Overlay network overlaid over the public Internet without concern or knowledge of the underlying network topology. Such networks are comprised of a usually ad hoc collection of host computers in diverse locations owned and operated by separate entities, with each host connected to the Internet in either permanent or temporary fashion. The peer-to-peer (and hence the name P2P) connections between these hosts are usually TCP connections. Thus all of the hosts in the network are directly connected. Napster is the earliest well-known example of a PTP/Overlay network.

We introduce P2P/Overlay networks here primarily to distinguish them from the overlay networks we describe in SDN via Hypervisor-based Overlays. Although the nature of the overlay itself is different, it is interesting to consider where there might be some overlap between the two technologies. Just as scaling SDN will ultimately require coordination of controllers across controlled environments, there is a need for coordination between P2P/Overlay devices. SDN helps move up the abstraction of network control, but there will never be a single controller for the entire universe, and thus there will still need to be coordination between controllers and controlled environments. P2P/Overlay peers also must coordinate among each other, but they do so in a topology independent way by creating an overlay network. A big distinction is that Open SDN can also control the underlay network. The only real parallel between these two technologies is that at some scaling point, they must coordinate and control in a distributed fashion. The layers at which this is applied are totally different, however.

9.8 CONCLUSION

Through examples, we have illustrated in this chapter and the preceding chapter that SDN can be applied to a very wide range of networking problems. We remind the reader of the points made in Section 4.1.3 that SDN provides a high-level abstraction for detailed and explicit programming of network behavior. It is this ease of programmability that allows SDN to address these varied use cases. One of the most exciting aspects of SDN is that the very flexibility that has rendered it capable of crisply addressing traditional network problems will also make it adaptable to solve yet-to-be conceived networking challenges. In Chapter 10 we examine a closely related technology, *Network Functions Virtualization* (NFV), and consider some of the use cases where NFV and SDN overlap.

REFERENCES

[1] Farrel A, Vasseur JP, Ash J. A path computation element (PCE)-based architecture. RFC 4655, Internet Engineering Task Force; 2006.
[2] Hoelzle U. OpenFlow@Google. Open networking summit. Santa Clara, CA: Google; 2012.
[3] Sharafat A, Das S, Parulkar G, McKeown N. MPLS-TE and MPLS VPNs with OpenFlow. In: SIGCOMM'11. Toronto; 2011.
[4] Green H, Kempf J, Thorelli S, Takacs A. MPLS OpenFlow and the split router architecture: a research approach. In: MPLS, 2010. Washington, DC: Ericsson Research; 2010.
[5] McDysan D. Cloud bursting use case. Internet draft. Internet Engineering Task Force; 2011.
[6] Verizon to demonstrate software defined networking principles with collaborative lab trials. Verizon Press Release; 2012. Retrieved from: http://newscenter2.verizon.com/press-releases/verizon/2012/verizon-to-demonstrate.html.
[7] Schooler R, Sen P. Transforming networks with NFV & SDN. Open networking summit. Santa Clara, CA: Intel/Verizon; 2013, p. 24–9.
[8] IEEE Standard for local and metropolitan area networks: port-based network access control. New York: IEEE; 2010. IEEE 802.1X-2010.
[9] Case study Ballarat grammar secures BYOD with HP Sentinel SDN. Hewlett-Packard Case Study; Retrieved from: http://h20195.www2.hp.com/v2/GetPDF.aspx/4AA4-7496ENW.pdf.
[10] Open Networking Foundation. Wireless and mobile working group charter application: use cases; 2013.
[11] Platenkamp R. Early identification of elephant flows in Internet traffic. In: 6th Twente Student Conference on IT, University of Twente. 2007. Retrieved from: http://referaat.cs.utwente.nl/conference/6/paper/6797/early-identification-of-elephant-flows-in-internet-traffic.pdf.
[12] Rivillo J, Hernandez J, Phillips I. On the efficient detection of elephant flows in aggregated network traffic. In: Networks and Control Group, Research School of Informatics. Loughborough University; Retrieved from: http://www.ee.ucl.ac.uk/lcs/previous/LCS2005/49.pdf.
[13] Fujitsu develops SDN technology to accelerate network storage access. Fujitsu Press Release; 2013. Retrieved from: http://www.fujitsu.com/global/about/resources/news/press-releases/2013/1209-01.html.
[14] Fujitsu successfully demonstrates ONOS interoperability. Fujitsu Press Release; 2015. Retrieved from: http://www.fujitsu.com/us/about/resources/news/press-releases/2015/fnc-20150615.html.
[15] Fujitsu product overview of SDN/NFV. Fujitsu Product Page; 2016. Retrieved from: http://www.fujitsu.com/us/products/network/technologies/software-defined-networking-and-network-functions-virtualization/index.html.
[16] Burt J. Fujitsu launches multilayer SDN suite for service providers. eWeek 2016; Retrieved from: http://www.eweek.com/networking/fujitsu-launches-multi-layer-sdn-suite-for-service-providers.html.

CHAPTER

NETWORK FUNCTIONS VIRTUALIZATION

10

Many casual followers of SDN sometimes confound *Network Functions Virtualization* (NFV) with SDN. Indeed these two technologies are closely related though they are not the same. While this work is about SDN, the symbiosis between it and NFV drive the need for a more detailed look at this sister technology. NFV was introduced in a presentation titled "Network Functions Virtualisation; an introduction, benefits, enablers, challenges and call for action" in 2012 at the SDN and OpenFlow World Congress [1]. Shortly after it was introduced, the *European Telecommunications Standards Institute* (ETSI) took the lead on NFV. In this chapter, we will define NFV and compare and contrast it with SDN so that our reader will understand its relationship to the focus of our work.

10.1 DEFINITION OF NFV

The following definition is provided in [1]:

> Network Functions Virtualisation aims to transform the way that network operators architect networks by evolving standard IT virtualisation technology to consolidate many network equipment types onto industry standard high volume servers, switches and storage, which could be located in Data Centers, Network Nodes and in the end user premises... It involves the implementation of network functions in software that can run on a range of industry standard server hardware, and that can be moved to, or instantiated in, various locations in the network as required, without the need for installation of new equipment.

This definition made reference to a figure which we reproduce in Fig. 10.1. This vision and definition can be further decomposed. ETSI has created an approach [1] to migrate a *Physical Network Function* (PNF) to a *Virtual Network Function* (VNF). The core concept is that those physical functions run on commodity hardware with virtualized resources driven by software. As we described in Section 3.7, the IT world has been virtualizing compute, storage, and to a lesser extent networks for many years.

Many vendors offer network virtualization (NV) products, and their technologies vary. ETSI has created an NFV framework that will lead to standard solutions from a plethora of network vendors. This framework defines an *NFV Infrastructure* (NFVI), where VNFs are created and managed by an *NFV Orchestrator* (NVFO) and VNF Manager. The concept of the orchestrator is similar to a hypervisor in cloud computing. The orchestrator will be responsible for instantiating resources such as compute, memory, and storage to support the virtual machine hosting the VNF. It is also responsible for grouping NFVs into *service chains*, a suite of connected network services over a single network connection.

Software Defined Networks. http://dx.doi.org/10.1016/B978-0-12-804555-8.00010-7

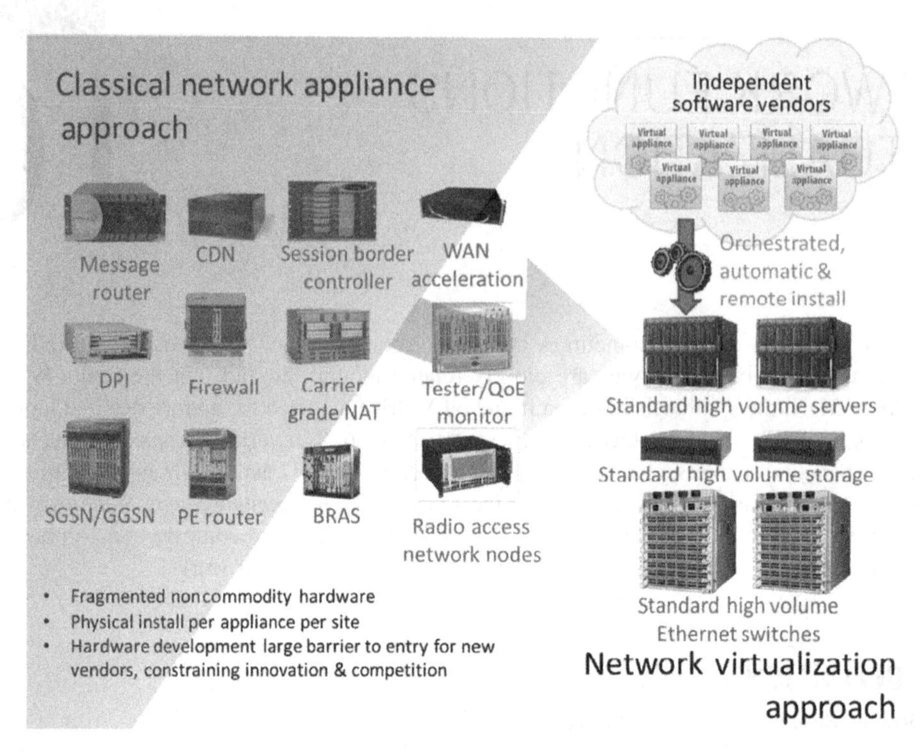

FIG. 10.1

Vision for NFV.

(Courtesy of Network Functions Virtualisation; an introduction, benefits, enablers, challenges & call for action. In: SDN and OpenFlow World Congress. Germany: Darmstadt; 2012. Retrieved from: http://portal.etsi.org/NFV/NFV_White_Paper.pdf.)

The concept of the VNF Manager is to monitor and manage the health of the VNF running on the virtual machine. The ETSI NFV group expects that the VNF will implement a network entity with well-defined, standards-based behavior and interfaces. In brief, NFV formalizes the concept of taking a network function that runs on a physical device and placing the function on a virtual machine running on a virtual server infrastructure [2].

The ETSI model for NFV has defined a management and orchestration architecture and nicknamed it MANO. In reality, MANO is the most challenging part of virtualizing PNFs. Considering the size of networks in carriers like BT, AT&T, Verizon, and others, management and orchestration is a daunting task. Bringing up thousands of virtual machines with network functions running on them could easily result in chaos. As such, the need for management in the NFV solution was a focus of the early efforts. The MANO [3] framework has the following functional blocks:

- *NFVO*: This is used for on-boarding of new *Network Service* (NS), *VNF Forwarding Graph* (VNF-FG), and VNF Packages, NS lifecycle management (including instantiation, scale-out/in, performance measurements, event correlation, termination) global resource management, validation and authorization of NFVI resource requests, and policy management for NS instances.

- *VNF Manager*: This provides lifecycle management of VNF instances, and overall coordination and adaptation for configuration and event reporting between NFVI, the *Element Management System* (EMS), and the *Network Management System* (NMS).
- *Virtualized Infrastructure Manager*: This controls and manages the NFVI compute, storage, and network resources within one operator's infrastructure subdomain. It is responsible for the collection and forwarding of performance measurements and events.

Network virtualization (NV) has existed for some time. It is not unusual to have trouble distinguishing NV from NFV, so we will attempt to clarify this here. NV creates an overlay of the physical network. Instead of connecting two different domains with physical wiring in a network, NV creates tunnels through the existing network. This saves time and effort for network administrators and technicians. NV is well-suited to providing connectivity between virtual machines. On the other hand, NFV virtualizes layer four through seven functions. Examples include firewalls, load balancers, *Intrusion Detection Systems* (IDS), *Intrusion Protection Systems* (IPS), and others. IT cloud administrators spin up VMs by pointing and clicking. NFV and NV allow network administrators to have the same capabilities for the network. For example, Embrane, recently bought by Cisco, is an example of a company that has an IDS/IPS service that can be instantiated as an NFV.

10.2 WHAT CAN WE VIRTUALIZE?

Newcomers to NFV may reasonably ask "what can we virtualize?" While an answer of "anything" might appear glib, it is not far from the truth. Almost any PNF can be virtualized. As an example, the list [4] of ETSI NFV use cases includes a wide range of functions:

- AR: Enterprise Access Router/Enterprise CPE
- PE: Provider Edge Router
- FW: Enterprise Firewall
- NG-FW: Enterprise NG-FW
- WOC: Enterprise WAN Optimization Controller
- DPI: Deep Packet Inspection (Appliance or a function)
- IPS: Intrusion Prevention System and other Security appliances
- Network Performance Monitoring

Another example in [4] is the virtualization of the *IP multimedia subsystem* (IMS). The IMS is a session control architecture to support provisioning of multimedia services over the *Evolved Packet Core* (EPC) and other IP-based networks. Some of the network functions being virtualized include the *Proxy Call Session Control Function* (P-CSCF), *Serving Call Session Control Function* (S-CSCF), *Home Subscriber Server* (HSS), and *Policy and Charging Rules Function* (PCRF). These prototype efforts quickly accelerated the delivery of real products to the market. As early as 2014, Alcatel-Lucent [5] announced a portfolio of VNFs including the EPC, IMS, and the *Radio Access Network* (RAN). A list of some possible NFV *elements* [6] is shown in Fig. 10.2.[1] NFV elements are the

[1]For the curious reader, there is no actual association between the physical elements shown in Fig. 10.2 and the elements of NFV. The physical elements are for visual effect only.

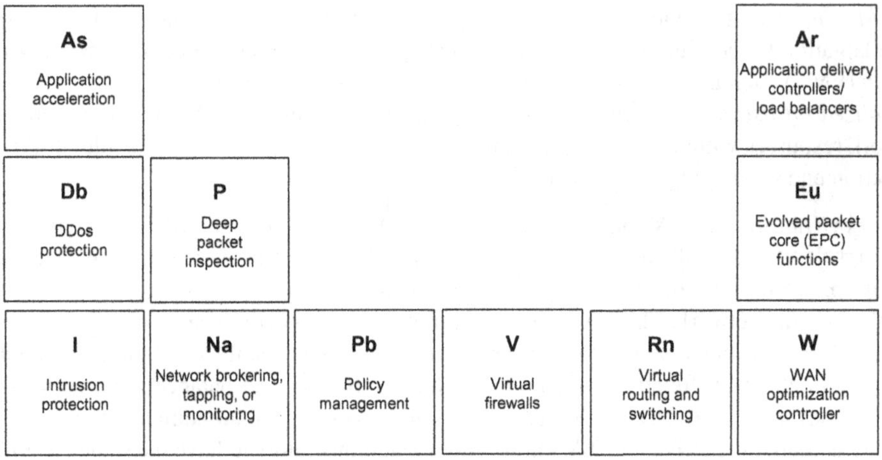

As
Application acceleration

Ar
Application delivery controllers/ load balancers

Db
DDos protection

P
Deep packet inspection

Eu
Evolved packet core (EPC) functions

I
Intrusion protection

Na
Network brokering, tapping, or monitoring

Pb
Policy management

V
Virtual firewalls

Rn
Virtual routing and switching

W
WAN optimization controller

FIG. 10.2

Elements of NFV.

discrete hardware and software requirements that are managed in an NFV installation to provide new communication and application services that have traditionally been performed by application-specific appliances. Table 10.1 lists companies actively selling NFV products.

10.3 STANDARDS

ETSI is the *de facto* standards body for NFV. After being chartered as the standard bearer for NFV in 2012, ETSI published the first five specifications for NFV in 2013 [7]. As of this writing, ETSI has 20 specifications [8] freely available for download.

DISCUSSION QUESTION

When you visit the ETSI website, you will see that they are involved in many different standards and activities outside of NFV. Do you think it would be beneficial to have NFV separated into its own organization singularly focused on NFV?

10.4 OPNFV

Open platform for NFV (OPNFV) is an open source project focused on accelerating NFV's evolution through an integrated, open platform. Using the ETSI architectural framework as a launch pad, OPNFV was created to promote an open source platform to "accelerate the introduction of NFV products and services" [9]. The ETSI NFV efforts sought to standardize solutions from NFV vendors who previously were creating proprietary implementations. The next evolutionary step is to bring open source products

Table 10.1 Leading NFV Vendors

Company	Carrier NFV	Wireless Carriers	Enterprise and Data Centers	Products
Arista	Yes	Yes	Yes	Distribution, core, and access switches
Blue Coat			Yes	Forward and reverse proxy—ProxySG
Brocade	Yes			Distribution, core, access switches, firewall, load balancer, WAN edge, application accelerator
Checkpoint			Yes	Edge/perimeter firewalls
Cisco	Yes	Yes	Yes	UCS fabric interconnect, edge/perimeter firewalls, routers, load balancer, application acceleration, wireless LAN controller, WAN optimization (vWAAS)
Citrix			Yes	Load balancer
Ericsson	Yes	Yes		Carrier aggregation
F5	Yes			Application accelerators
Fortinet			Yes	Edge/perimeter firewall
HP	Yes		Yes	
Hitachi	Yes		Yes	Access switches
Huawei	Yes	Yes		Universal service routers
Lenovo/IBM			Yes	
Juniper	Yes		Yes	Edge/perimeter firewall, compute firewall VGW
Palo Alto Networks	Yes			Device role, edge/perimeter firewall
Radware			Yes	Load balancer/ADC
Riverbed			Yes	Load balancer/ADC, WAN optimization
VMware			Yes	Edge gateway, virtual access switches, vCenter
Nokia[a]	Yes	Yes		
Dell				Access switches
Force10				Access switches

[a]*Alcatel-Lucent is now part of Nokia.*

to the table. Thus OPNFV is very complementary to the ETSI NFV efforts. OPNFV includes an initial build of the NFVI and *Virtual Infrastructure Manager* (VIM) components. The diagram of their second release Brahmaputra is shown in Fig. 10.3.

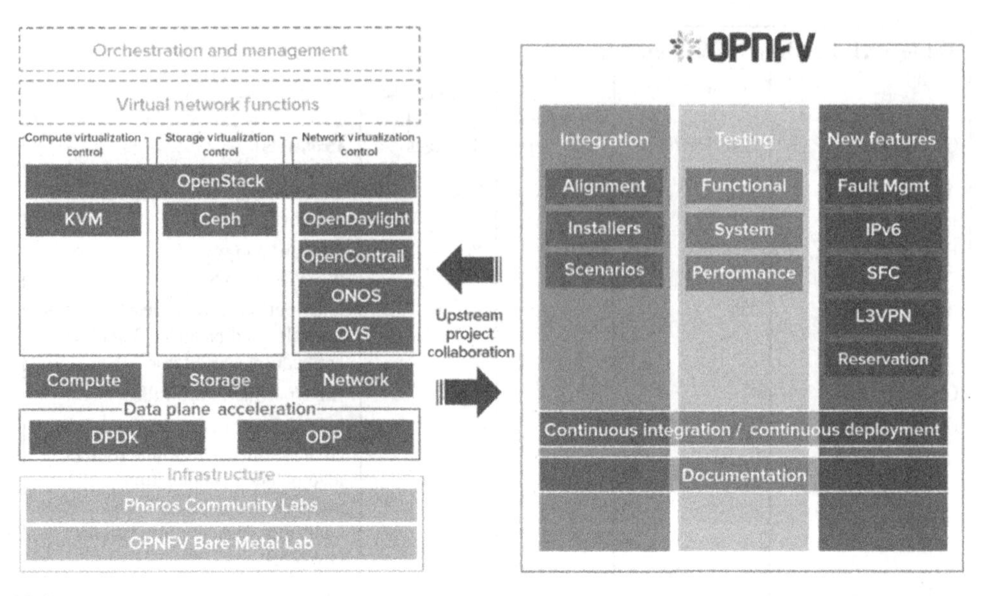

FIG. 10.3

OPNFV second release Brahmaputra.

(Courtesy of OPNFV. Linux foundation: collaborative projects. Retrieved from: https://www.opnfv.org.)

10.5 LEADING NFV VENDORS

Each of the companies we list in Table 10.1 have multiple offerings in the NFV arena. As this is a rapidly evolving field, the data contained in this table really only reflects the marketplace at the time of writing of this work. Looking across these vendors, they each bring some expertise to bear. The table indicates whether or not the vendor is a major participant in the NFV markets of *Carrier NFV*, *Wireless Carriers*, and *Enterprise/Data Centers*. If the company is not listed under one of these categories, this does not imply that they have no relevant offerings, but merely that they are not a major player. A real-time list of various vendors involved in SDN, NFV, and NV is available in [10].

10.6 SDN VS NFV

Open SDN makes the network programmable by separating the control plane (i.e., telling the network what goes where) from the data plane (i.e., the sending of packets to specific destinations). It relies on switches that can be programmed through an SDN controller using an industry standard control protocol, such as OpenFlow. As described previously, NV and NFV add virtual tunnels and functions, respectively, to the physical network. SDN, on the other hand, changes the forwarding behavior of the physical network. The business need for SDN arises from the need to provide faster and more flexible service fulfillment [11]. SDN's scope is to be able to control and manage anything that could contribute to a service. This includes bringing up VNFs under the management of the VNFI. NFV, conversely, is about reducing the cost and time to provide network functions that contribute to the delivery of a service. As previously described, NFV deploys network functions as software on general purpose machines.

DISCUSSION QUESTION

We have discussed NFV and SDN. Since they are complementary, should the organizations be combined? What are the pros and cons of a combined organization?

The SDN controller views a VNF as just another resource. ETSI's NFV architecture framework includes components that *turn up*[2] VNFs and manage their lifecycle. Since SDN and NFV are complementary, network engineers need not choose between them, but rather understand how they might work together to address their needs. For this synergy to be effective, SDN and NFV need to know what the other is doing. The ETSI framework stipulates that there is a VNF manager that can turn up or turn down a VNF. If that VNF is part of an active service being provided to a customer, then the framework should not turn down the VNF. This mandates a common repository that SDN and NFV may access to understand the state of a service. Ideally, some type of repository holding state for an individual or set of VNFs should exist. SDN should be able to communicate to the NFV framework to bring up VNFs to support a provisioned service. The state of this service should be maintained so that the NFV framework and its MANO capabilities understand what is in use and when it can turn down a VNF. Fig. 10.4 depicts a scenario where SDN and NFV would collaborate. The controller would contact the NFV MANO to spin up or tear down VNFs and remain aware of the state of the VNF through a common data store.

10.6.1 WHEN SHOULD NFV BE USED WITH SDN?

NFV should be used with SDN when there are many physical network elements that you wish to virtualize. If one only had a few VNF's, it would not make sense to implement the overhead of the ETSI VNF reference architecture. However, the complexity of management, monitoring, and orchestrating a large number of VNFs and their service chains creates a need for the NFV framework. The ETSI VNF reference architecture is well suited to managing the complexity. As previously mentioned, SDN would view the VNFs as just another resource and readily manage the complex environment supporting vast numbers of VNFs. SDN is not constrained to managing just VNFs, as the scope of SDN is much broader than NFV. The promise of SDN applications relying on external information to help manage the network will enhance new service offerings.

10.7 IN-LINE NETWORK FUNCTIONS

In-Line Network Functions typically include functionality such as load balancers, firewalls, IDS, and IPS. These services must be able to inspect the packets passing through them. These functions are

[2]We will use the terms *turn up* and *turn down* interchangeably with *spin up* and *spin down*, respectively, in our discussion on NFV.

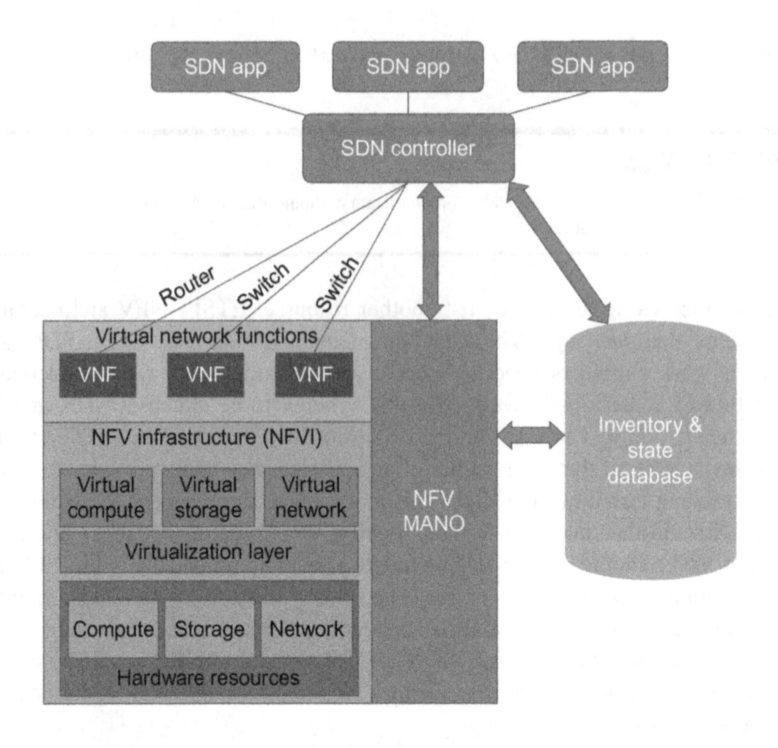

FIG. 10.4

SDN, NFV collaborating with VNFs.

often delivered in specialized appliances which can be very expensive. These appliances are *bumps-in-the-wire*, meaning that they are inserted into the data plane. In some cases, the *bump* shunts traffic to an out-of-band processor, which can result in cost savings. In other cases, the value-added service performed by the appliance is performed at line rate and the increased cost of such devices is necessary if performance is of paramount concern. Data centers and SPs which must employ these types of services welcome any novel means to drive down the costs associated with these devices.

There are many schemes for load balancing traffic, but all involve some level of packet inspection, choosing the destination in order to evenly distribute the load across the entire set of servers.

Firewalls serve the role of admitting or denying incoming packets. The decision criteria involve packet inspection and can be based on factors such as destination or source IP address, destination or source TCP or UDP port, or something deeper into the payload, such as the destination URL.

IDS and IPS solutions analyze packets for malicious content. This entails performing process-intensive analysis of the packet contents to determine if there is a threat present. While an IPS is a bump-in-the-wire, an IDS can run in parallel with the data plane by looking at mirrored traffic. Such systems require some traffic mirroring mechanism. A hybrid system can be implemented in the SDN paradigm by shunting only those packets requiring deeper inspection to an appliance that is off of the data path. This approach has the advantage of removing the bump-in-the-wire, as packets that are not subject to deeper inspection suffer no additional latency at all. Also, since not all packets need to be processed by the appliance, fewer or less powerful appliances are needed in the network. This can be realized in SDN by simply programming flow rules, which is much simpler than the special *Switch Port ANalyzer* (SPAN) ports that are in use on contemporary traditional switches.

In the following three sections, we study examples of SDN being used to virtualize three in-line network functions: (1) Server Load Balancing, (2) Firewalls, and (3) IDS. These simple examples illustrate cases where there is strong overlap between NFV and SDN.

10.7.1 SDN APPLIED TO SERVER LOAD-BALANCING

Load balancers must take incoming packets and forward them to the appropriate servers. Using SDN and OpenFlow technology, it is straightforward to imagine a switch or other networking device having rules in place which would cause it to act as a load balancer appliance at a fraction of the cost of purchasing additional, specialized hardware.

OpenFlow 1.0 supports matching against the basic 12-tuple of input fields as described in Section 5.3.3. A load balancer has a rich set of input options for determining how to forward traffic. For example, as shown in Fig. 10.5, the forwarding decision could be based on the source IP address. This would afford server continuity for the duration of time that that particular endpoint was attempting to communicate with the service behind the load balancer. This would be necessary if the user's transaction with the service required multiple packet exchanges and if the server needed to maintain state across those transactions.

The fact that the controller will have the benefit of understanding the load on each of the links to the servers, as well as information about the load on the various servers behind the firewall, opens up other possibilities. The controller could make load balancing decisions based on a broader set of criteria than would be possible for a single network appliance.

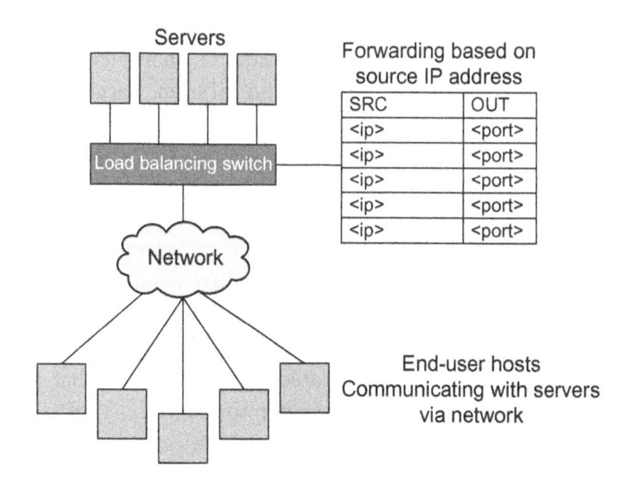

FIG. 10.5

Load balancers using OpenFlow.

Many people confuse network load balancing with server load balancing. Network load balancing is simple and quite fast because it relies on simple parameters like IP address and the TCP port. Application server load balancing is different and more complicated. This is where SDN's centralized knowledge along with SDN applications can play a major role. True application server load balancing must take into consideration many more variables outside of the network. These include the load on the server and processing time, among others. SDN's centralized, global network view can be augmented by an SDN application that can gather many more variables outside of the IP address and TCP port. This includes the server load, its processing time for transactions, etc. The SDN application can then assist the SDN controller by helping it take into consideration these other factors when routing transactions to be processed to particular servers.

10.7.2 SDN APPLIED TO FIREWALLS

Firewalls take incoming packets and forward them, drop them or take some other action such as forwarding the packet to another destination or to an IDS for further analysis. Like load balancers, firewalls using SDN and OpenFlow have the ability to use any of the standard 12 match fields for making these forward or drop decisions. Packets matching certain protocol types (e.g., HTTPS, HTTP, Microsoft Exchange) are forwarded to the destination and those that do not match are dropped. Simple firewalls that are ideally suited to an SDN-based solution include firewalls that are based on blocking or allowing specific IP addresses or specific TCP/UDP ports. As with load balancers, SDN-based firewalls may be limited by a lack of statefulness and inability for deeper packet inspection.

One SDN use case applied to firewalls is to insert an OpenFlow-enabled switch in front of a battery of firewalls. One benefit of this is that the SDN switch can load balance between the firewall devices, permitting greater scale of firewall processing. More importantly, the SDN switch can shunt *known-safe* traffic around the firewalls. This is possible since OpenFlow rules can be programmed such that we offload traffic from the firewalls that is known to be problem-free. The benefit of this is that less firewall equipment and/or power is needed and less traffic is subjected to the latency imposed by firewall processing.

Another example of an SDN firewall-type implementation that slices the network into different departments is FlowVisor [12]. It slices the network into different parts to be managed by different controllers for each network slice. This allows the FlowVisor OpenFlow controller to enforce isolation between different networks on the same campus.

10.7.3 SDN APPLIED TO INTRUSION DETECTION

IDS systems have typically been passive security devices that would listen to all traffic (egress and ingress) to identify suspicious traffic and potential attacks. An IDS looks for application protocol violations and uses rules that analyze statistics and traffic patterns that may indicate an attack. Just as in the case of a firewall, an SDN application can be created to support a VNF that would act as a traditional *network tap* and passively route packets to the application for analysis. There is some concern whether the performance of VMs in this role can compete with customized hardware optimized for this task. It is noteworthy that Cisco has commercialized virtual IPSs and IDSs [13].

An IDS is intended to observe network traffic in order to detect network intrusions. An IPS additionally attempts to prevent such intrusions. The range of intrusions that these systems are intended to recognize is very broad and we do not attempt to enumerate them here. We will, however, consider two particular IDS functions that lend themselves well to SDN-based solutions.

The first type of intrusion detection we consider requires deep packet inspection of payloads, searching for signs of viruses, trojans, worms, and other malware. Since this inspection is processing-intensive, it must take place somewhere outside the data plane of the network. Typical solutions utilize a network tap, which is configured to capture traffic from a particular source, such as a VLAN or physical port, and copy that traffic and forward it to an IDS for analysis. Physical taps such as this are mostly port-based.

One pre-SDN solution is based on a network patch-panel device which can be configured to take traffic from an arbitrary device port and forward it to the analysis system. This requires patch panel technology capable of copying traffic from any relevant network port. This technology is available from companies such as *Gigamon* and *cPacket Networks*. In large networks the cost of such approaches may be prohibitive. Forwarding traffic on uplinks is also sensitive to the volume of traffic, which may exceed the capacity of the uplink data path.

Another pre-SDN solution for IDS is to use configuration tools on the network devices to configure SPAN ports. This allows the switch to operate as a virtual tap, where it copies and forwards traffic to the IDS for analysis. Taps such as this can be based on port or VLAN, or even some other packet-based filter such as device address.

OpenFlow is well-suited to this application since it is founded on the idea of creating flow entries that match certain criteria (e.g., device address, VLAN, ingress port) and performing some action. A tap system using OpenFlow can be programmed to set up actions on the device to forward packets of interest either out through a monitor port or to the IP address of an IDS.

At the end of Section 10.7.2 we described an SDN switch being used to front-end a battery of firewalls. This strategy may also be applied to a battery of IPS appliances. The same rationale of shunting safe traffic around the IPS appliances while allowing the system to scale by using multiple IPS units in parallel applies here as well.

The FlowScale open source project done at InCNTRE is an example of this kind of front-end load balancing with the SPAN concept. FlowScale sends *mirrored* traffic to a battery of IDS hosts, load-balancing the traffic across them.

10.8 CONCLUSION

For the reader interested in additional use cases of NFV, a concrete example of an NFV implementation of DPI is provided in [14]. A deeper technical treatment of an NFV solution called *ClickOS* can be found in [15]. ClickOS is a high-performance, virtualized network appliance platform that can be used to implement a number of the NFV use cases [4] we cited earlier.

NFV complemented by SDN will have a large impact on data centers, networks, and carriers. As the network is virtualized, cost savings will be dramatic due to reduced labor costs and policy-enforced configuration through automated solutions. Under ETSI's leadership, more and more solutions from vendors will be interoperable. Further, the industry solutions will be less costly as open source is increasingly embraced by customers and NFV vendors. The promise of OPNFV is to promote the use of open source products within the ETSI NFV architectural framework. Given the huge cost savings for carriers and other customers, we expect the adoption of NFV to accelerate over the next few years.

Addressing the different use cases which we have described in the last three chapters requires a wide variety of controller applications. In Chapter 12 we will examine some sample SDN applications in detail and see how domain-specific problems are addressed within the general network programming paradigm of SDN. First, though, we should acknowledge that this important movement could only take place because of key players that have pushed it forward. For that reason, in Chapter 11 we will take a step back from technology and review the individuals, institutions and enterprises that have had the greatest influence on SDN since its inception.

REFERENCES

[1] Network Functions Virtualisation; an introduction, benefits, enablers, challenges & call for action. In: SDN and OpenFlow World Congress. Germany: Darmstadt; 2012. Retrieved from: http://portal.etsi.org/NFV/NFV_White_Paper.pdf.

[2] McCouch B. SDN, network virtualization, and NFV in a nutshell. Network Computer; 2014. Retrieved from: http://www.networkcomputing.com/networking/sdn-network-virtualization-and-nfv-in-a-nutshell/a/d-id/1315755.

[3] Network Functions Virtualization: Mano—management and orchestration. Retrieved from: http://www.network-functions-virtualization.com/mano.html.

[4] Network Functions Virtualisation (NFV); use cases. Retrieved from: http://www.etsi.org/deliver/etsi_gs/NFV/001_099/001/01.01.01_60/gs_NFV001v010101p.pdf.

[5] Alcatel-Lucent delivers suite of virtualized network functions, ushering in the next phase of mobile ultra-broadband for service providers. Paris: Alcatel-Lucent Press Release; 2014. Retrieved from: https://www.alcatel-lucent.com/press/2014/alcatel-lucent-delivers-suite-virtualized-network-functions-ushering-next-phase-mobile.

[6] SDX Central. An overview of NFV elements. Retrieved from: https://www.sdxcentral.com/resources/nfv/nfv-elements-overview/.

[7] ETSI. ETSI publishes first specifications for network functions virtualisation. France: ETSI; 2013. Retrieved from: http://www.etsi.org/news-events/news/700-2013-10-etsi-publishes-first-nfv-specifications?highlight=YToxOntpOjA7czozOiJuZnYiO30=.

[8] ETSI. Our standards. Retrieved from: http://www.etsi.org/standards.

[9] OPNFV. Linux foundation: collaborative projects. Retrieved from: https://www.opnfv.org.

[10] SDX Central. SDN & NFV Services Directory. Retrieved from: https://www.sdxcentral.com/nfv-sdn-services-directory/.

[11] Open Networking Foundation. TR-518 relationship of SDN and NFV, issue 1; 2015. Retrieved from: https://www.opennetworking.org/images/stories/downloads/sdn-resources/technical-reports/onf2015.310_Architectural_comparison.08-2.pdf.

[12] Al-Shabibi A. Flowvisor. Open Networking Lab; 2013. Retrieved from: https://github.com/OPENNETWORKINGLAB/flowvisor/wiki.

[13] Cisco. Cisco NGIPSv for VMware data sheet. Cisco FirePOWER 7000 Series Appliances. Retrieved from: http://www.cisco.com/c/en/us/products/collateral/security/firepower-7000-series-appliances/datasheet-c78-733165.html.

[14] Bremler-Barr A, Harchol Y, Hay D, Koral Y. Deep packet inspection as a service. In: Proceedings of the 10th ACM international conference on emerging networking experiments and technologies (CoNEXT '14). New York, NY: ACM; p. 271–82.

[15] Martins J, Ahmed M, Raiciu C, Olteanu V, Honda M, Bilfulco R, et al. ClickOS and the art of network function virtualization. In: Proceedings of the 11th USENIX conference on networked systems design and implementation (NSDI'14). Berkeley, CA: USENIX Association; p. 459–73. Retrieved from: https://www.usenix.org/system/files/conference/nsdi14/nsdi14-paper-martins.pdf.

CHAPTER

PLAYERS IN THE SDN ECOSYSTEM

11

A technological shift as radical and ambitious as SDN does not happen without many players contributing to its evolution. Beyond the luminaries behind the early concepts of SDN there are many organizations that continue to influence the direction in which SDN is headed and its rate of growth. Their impact is sometimes positive and at other times contrary to the growth of SDN, but are in any case influential in the evolution of SDN. We classify the various organizations that have had the most influence on SDN into the following categories:

- Academic Researchers
- Industry Research Labs
- Network Equipment Manufacturers
- Software Vendors
- Merchant Silicon Vendors
- Original Device Manufacturers
- Enterprises and Service Providers
- Standards Bodies
- Industry and Open Source Alliances

Fig. 11.1 depicts these major groupings of SDN players and the synergies and conflicts that exist between them. We also show three super-categories in the figure. Academic and industry-based basic research, standards bodies and industry alliances, and the *white-box* ecosystem make up these super-categories. In the context of this book, a white-box switch is a hardware platform that is purpose-built to easily incorporate a *Network Operating System* (NOS) from another vendor or from open source. In the context of this book, an NOS is normally OpenFlow device software that would come in the form of an OpenFlow implementation such as Switch Light or OVS. In this chapter we will discuss each of these categories, how they have contributed to the evolution of SDN, and the various forces that draw them together or place them into conflict.

11.1 ACADEMIC RESEARCH INSTITUTIONS

Universities worldwide have engaged in research related to SDN. While many academic institutions have contributed to the advancement of SDN, we mention here three institutions of particular interest. They are Stanford University, UC Berkeley, and Indiana University. Stanford and UC Berkeley were at the forefront of SDN since its very inception and have employed an impressive number of the luminaries that created this new model of networking.

Software Defined Networks. http://dx.doi.org/10.1016/B978-0-12-804555-8.00011-9

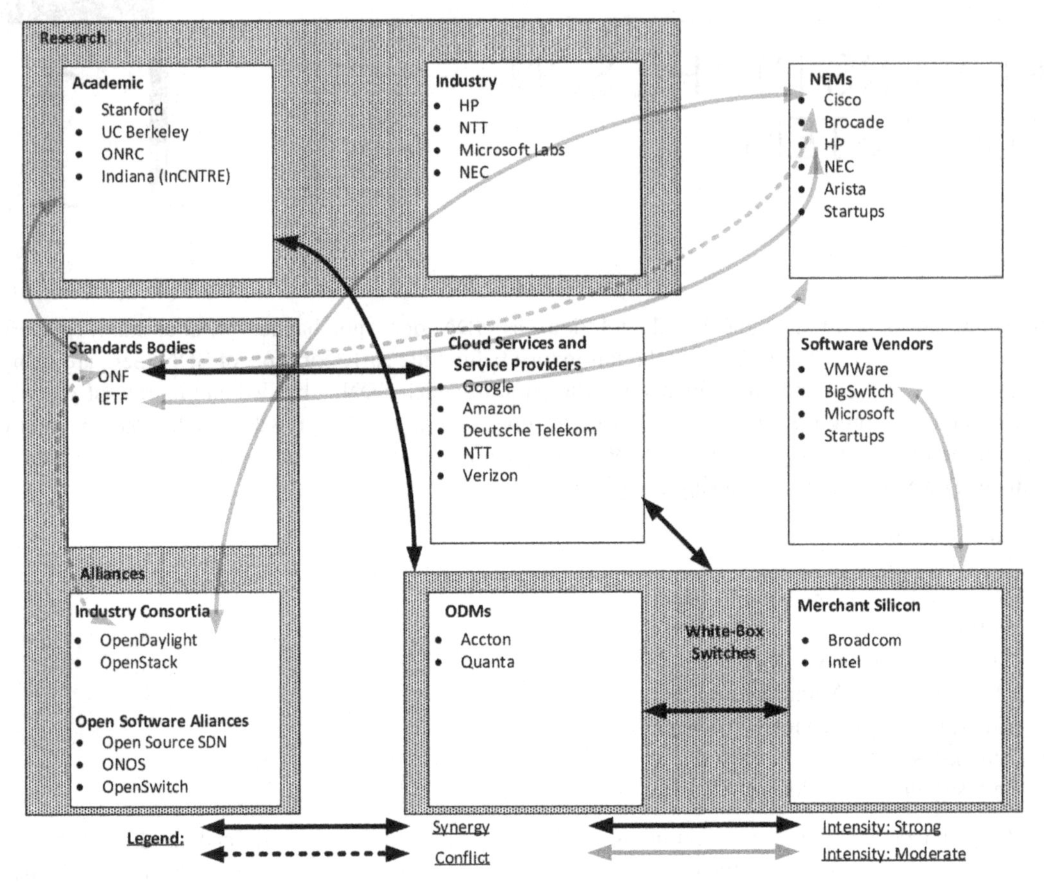

FIG. 11.1

SDN players ecosystem.

As we saw in Section 3.2.6, events at Stanford University led directly to the creation of the *Clean Slate* [1] program, including forwarding and control plane separation and, notably, the OpenFlow protocol. The *Open Networking Research Center* (ONRC) began as a joint project by Stanford and UC Berkeley [2], receiving funding from a number of major vendors such as Cisco, Intel, Google, VMware, and others. Its stated intention is to *open up the Internet infrastructure for innovations*. The ONRC has produced a number of research projects [3]. There is an affiliated organization called ON.Lab which seeks to develop, deploy, and support various SDN and OpenFlow-related tools and platforms. The ONRC is more research oriented while the ON.Lab is more focused on the practical application of SDN.

Indiana University has been instrumental in promoting SDN and OpenFlow, with a number of research projects and implementations, including the *Indiana Center for Network Translational Research and Education* (InCNTRE) [4]. Their goal is to *advance development, increase knowledge, and encourage adoption of OpenFlow and other standards-based SDN technologies*. One of the most notable activities at InCNTRE is a plugfest hosted by the *Open Networking Foundation* (ONF), where vendors bring their networking devices and controllers for interoperability testing. InCNTRE is the first certified lab selected by ONF for their conformance testing program [5].

In addition to the research institutions themselves, numerous academic conferences provide a fertile environment for the discussion of the future of SDN. These include *SIGCOMM*, *HotSDN*, *The Symposium on SDN Research* (SOSR), and *CoNEXT*, among others.

DISCUSSION QUESTION

Fig. 11.1 shows potential conflict between standards bodies and NEMs. Explain what the source of this conflict might be?

11.1.1 KEY CONTRIBUTORS TO SDN FROM ACADEMIA

A significant number of SDN pioneers have worked or studied at Stanford and UC Berkeley, some of whom we list as follows:

- Nick McKeown is currently a professor of Electrical Engineering and Computer Science at Stanford. He has established multiple technology companies which have been acquired by larger companies in the high-tech industry. He is credited with helping to ignite the SDN movement along with Martin Casado and Scott Shenker.
- Martin Casado is a member of the High Performance Networking research group at Stanford University. He co-founded Nicira with Nick McKeown and Scott Shenker and is currently working with the Andreessen Horowitz venture capital firm. He was initially one of the creators of OpenFlow through the Ethane project [6].
- Scott Shenker is a professor in the School of Electrical Engineering and Computer Science at UC Berkeley and was involved with Nick McKeown and Martin Casado in the creation of Nicira.
- Guru Parulkar is a consulting professor at Stanford as well as having been Executive Director of the Clean Slate program at that university. He is also Executive Director of the ONRC and ON.Lab. He is a strong advocate of Open SDN and a vocal opponent of *SDN washing* [7]. SDN washing means the use of the term SDN to refer to related technologies other than Open SDN, such as SDN via APIs and SDN via Hypervisor-based Overlays.
- Guido Appenzeller is co-founder of Big Switch Networks and is another participant in Stanford's Clean Slate program. In his role at Big Switch he is one of the strongest advocates for the Open SDN approach to reinventing networks.
- David Erickson and Rob Sherwood are the authors of many of the modules for the Beacon SDN controller [8]. As most of the current open source SDN controllers were forked from that original Beacon source code, they have had significant influence on the design of contemporary Open SDN controllers. Floodlight, for example, was derived from Beacon.

DISCUSSION QUESTION

Why do you think that Fig. 11.1 shows a high level of synergy between SDN academic researchers and the white-box ecosystem?

11.2 INDUSTRY RESEARCH LABS

Quite often major enterprises and vendors support their own labs which are performing basic research. We distinguish such pure research from that which is directed toward specific product development. These corporations often participate in research organizations and conferences and contribute papers based on their findings. For example, some research labs that have presented papers on SDN at SIGCOMM conferences are listed as follows:

- Telekom Innovation Laboratories (research arm of Deutsche Telekom)
- NTT Innovation Institute
- Microsoft Research
- Hewlett-Packard (HP) Laboratories[1]
- Fujitsu Laboratories Ltd.
- IBM Research
- NEC Labs

These companies contribute to the furthering of Software Defined Networking through research into topics such as scalability, deployment alternatives, ASIC development, and many other topics central to SDN.

11.3 NETWORK EQUIPMENT MANUFACTURERS

A number of networking vendors were quick to join the SDN movement, some of them contributing OpenFlow-enabled devices to researchers for testing even before the first official OpenFlow standard document was published. Others have provided funding for SDN-related standards organizations, industry alliances, or research institutions. Some have made small acquisitions to gain access to SDN technology. While some have begun to support Open SDN, others promote an SDN via APIs or an SDN via Overlays strategy.

Virtually every NEM has SDN as part of its story today. This is a marketing necessity considering all the hype surrounding SDN, yet much of it constitutes the SDN washing described earlier in Section 11.1.1. It is not our intent in this chapter to document all of the vendors' SDN claims. We focus on specific NEMs that, either because of their size or their commitment to SDN, seem most likely to influence the future course of SDN, for better or worse. In Table 11.1 we list some of these SDN networking vendors and the products they offer in the areas of SDN device, SDN controller, and SDN application. The NEMs called out in the table are not the only ones to have brought SDN products to the market as of this writing, but we believe that they reflect the most significant contributions by NEMs thus far. We provide some brief background on the NEMs cited in Table 11.1.

Cisco was not involved with the early work on OpenFlow and began to support OpenFlow in its devices in 2012. Cisco is now heavily involved with SDN. While only a lukewarm supporter of the ONF, Cisco is the primary driving force behind the OpenDaylight (ODL) Project, focused on an open-source SDN controller which we discuss in Section 11.9.2. In addition to their open source efforts

[1]In Nov. 2015 HP split into two separate companies, Hewlett-Packard Enterprise (HPE) and HP Inc. As the networking business will stay with HPE, our remarks about HP regarding events after November 2015 actually pertain to HPE. We will not make that distinction elsewhere in the text, though, due to our readers' long familiarity with the name HP.

Table 11.1 SDN Commercial Products

Vendor	SDN Devices	SDN Controller	SDN Applications
Cisco	SDN via APIs OpenFlow: Nexus	COSC APIC-DC APIC-EM	WAE
Hewlett-Packard	OpenFlow: 3500, 5400, and 8200	VAN	Security (Sentinel)
Brocade	OpenFlow: NetIron CER, CES SDN via APIs	Brocade SDN Controller (Vyatta)	
VMware	OpenFlow: Open vSwitch (OVS)	NVP (NSX)	NVP (NSX)
Big Switch	OpenFlow: Indigo OpenFlow: Switch Light	Big Network Controller	Big Cloud Fabric Big Tap
IBM	OpenFlow: RackSwitch and Flex System	Programmable Network Controller	SDN VE DOVE
NEC	Programmable Flow 5240 and 5820	Programmable Flow Controller	Virtual Tenant Networks Firewall, others
Extreme	OpenFlow	ADARA	
Juniper	OpenFlow: MX, EX series	Contrail	
Alcatel-Lucent[a]	SDN via APIs	Nuage VSC	
Arista	OpenFlow: 7050 series SDN via APIs	VMware NVP Nuage VSC	

[a]*Alcatel-Lucent is now part of Nokia.*

in the ODL project, Cisco strongly promotes SDN via their proprietary APIs. They claim to support APIs and programmability to more of the networking stack than is available with only OpenFlow. We presented this architecture in detail in Section 7.2. With regard to commercial controllers, Cisco emphasizes their *Application Policy Infrastructure Controller* (APIC). They also offer the *Cisco Open SDN Controller* (COSC), which is their commercial version of the ODL controller. As the dominant NEM, Cisco will undoubtedly play a large part in shaping the role that SDN will assume in industry. As Cisco creates APIs, other vendors will pragmatically follow suit and provide the same or very similar APIs in order to sell into Cisco markets, creating de facto standards. This process is unfolding within the ODL project. Examples of this are the ODL BGP-LS/PCEP plugin described in Section 7.2 and the MD-SAL architecture discussed in Section 7.3.2.

Note that Cisco's acquisitions of *spin-in* Insieme, Cariden, Tail-f, and Embrane that we discuss in Chapter 14 is further evidence of their growing interest in SDN.

Brocade has been active for a number of years in providing OpenFlow-supporting networking devices to researchers. However, they also provide SDN via APIs support in their devices and have promoted their RESTful API SDN support in use cases with various customers. This dual-pronged approach allows Brocade to offer short-term transition plans to their customers by offering SDN APIs on their devices as a migration path, as well as supporting Open SDN as a longer-term strategy. Brocade is currently a major supporter of the ODL project. In Section 14.8.3 we will discuss Brocade's acquisition of Vyatta, another SDN-related startup.

NEC and IBM have partnered together from the early days of the OpenFlow project. Their OpenFlow switch and controller implementations have similar capabilities. Both vendors have been stalwart contributors to the OpenFlow research community of OpenFlow-enabled switches. Both vendors have created applications for their controllers, including virtualization applications which implement overlay technology. Both companies are committed to OpenFlow and support SDN via Overlays using OpenFlow as the southbound API from the controller. As founding members of ODL, IBM and NEC have ported their virtualization applications to run on that controller. NEC contributed a version of their own network virtualization application, called *Virtual Tenant Networks*. IBM contributed a version of their network virtualization application, *SDN Virtual Environments*, which is called *Distributed Overlay Virtual Ethernet* (DOVE) in the ODL environment (see Section 6.3.3).

Like NEC and IBM, Hewlett-Packard has contributed OpenFlow-supporting switches to the OpenFlow research community long before the first ratified version of OpenFlow was approved. HP continues to support OpenFlow in many switch product lines and claims to have shipped millions of OpenFlow-supporting switch ports in the past few years. Among NEMs, HP stands out for its unequivocal support of OpenFlow and enthusiastic participation in the ONF.

HP offers the commercial OpenFlow-based VAN SDN controller. At ONS 2013 HP presented a number of SDN applications using OpenFlow to implement security and traffic-prioritization features. HP currently offers the Sentinel SDN security application that we described in Section 9.3.3. Because of its large server business, HP is also heavily involved in OpenStack and, consequently, HP will likely provide some type of network virtualization support through that project.

In Table 11.1 we see several NEMs listed that we have not yet mentioned. Extreme Networks, a long-time NEM, is partnering with ADARA, a provider of infrastructure orchestration products for SDN and cloud computing [9]. Extreme hopes that this partnership will allow their customers to migrate their existing networks to SDN without wholesale hardware upgrades. Alcatel-Lucent provides SDN via APIs in their networking devices. Juniper is headed toward a strategy aligned with SDN via APIs and has also acquired the startup Contrail. The Contrail controller provides network virtualization technology via overlays. Arista is very vocal in SDN forums and ships commercial SDN products. While Arista does offer some OpenFlow support, their emphasis is on SDN via APIs.

11.4 SOFTWARE VENDORS

The move toward network virtualization has opened the door for software vendors to play a large role in the networking component of the data center. Some of the software vendors that have become significant players in the SDN space include VMware, Microsoft, Big Switch, as well as a number of startups. We will try to put their various contributions into context in the following paragraphs.

VMware, long a dominant player in virtualization software for the data center, has contributed significantly to the interest and demand for SDN in the enterprise. VMware boldly altered the SDN landscape when it acquired Nicira. VMware's purchase of Nicira has turned VMware into a networking vendor. Nicira's roots, as we explained in Section 11.1.1, come directly from pioneers in the Open SDN research community.

VMware's current offerings include *Open vSwitch*, (OVS) as well as the *Network Virtualization Platform* (NVP) acquired through Nicira. (NVP is now marketed as VMware NSX.) NVP uses OpenFlow (with some extensions) to program forwarding information into its subordinate OVS switches.

VMware marketing communications claim that SDN is complete with only SDN via Overlays. OpenFlow is their southbound API of choice, but the emphasis is on network virtualization via overlay networks, not on what can be achieved with OpenFlow in generalized networking environments. This is a very logical business position for VMware. The holy grail for VMware is not about academic arguments but about garnering as much of the data center networking market as it can. This is best achieved by promoting VMware's virtualization strengths and avoiding esoteric disputes about one southbound API's virtues versus other approaches.

Other vendors and enterprises have begun to see the need to interoperate with the Nicira solution. Although VMware's solution does not address physical devices, vendors who wish to create overlay solutions that work with VMware environments will likely need to implement these Nicira-specific APIs as well.

While VMware and Cisco were co-definers of the VXLAN [10] standard, they now appear to be diverging. Cisco is focused on their ODL and APIC controller strategy, and VMware on NSX. With respect to tunneling technologies, Cisco promotes VXLAN which is designed for software and hardware networking devices, while VMware promotes both VXLAN and STT. STT [11] has a potential performance advantage in the software switch environment customary for VMware. This advantage derives from STT's ability to use the server NIC's TCP hardware acceleration to improve network speed and reduce CPU load.

As a board-level participant in the ONF, Microsoft is instrumental in driving the evolution of OpenFlow. Microsoft's current initiative and effort regarding SDN has been around the Azure public cloud project. Similar to VMware, Microsoft has its own server virtualization software called *Hyper-V* and is utilizing SDN via Overlays in order to virtualize the network as well. Their solution uses NVGRE [12] as the tunneling protocol, providing multitenancy and the other benefits described in Section 8.3.2.

Big Switch Networks was founded in 2010 by Guido Appenzeller with Rob Sherwood as CTO of controller technology, both members of our SDN hall of fame. Big Switch is one of the primary proponents of OpenFlow and Open SDN. Big Switch has Open SDN technology at the device, controller, and application levels. Big Switch created an open source OpenFlow switch code base called Indigo. Indigo is the basis for Big Switch's commercial OpenFlow switch software, which they market as *Switch Light*. The Switch Light initiative is a collaboration of switch, ASIC and SDN software vendors to create a simple and cost-effective OpenFlow-enabled switch. Big Switch provides a version of Switch Light intended to run as a virtual switch, called *Switch Light for Linux*, as well as one targeted for the white-box hardware market, called *Switch Light for Broadcom*. We discuss the white-box switch concept in Section 11.5.

Big Switch provides both open source and commercial versions of an SDN controller. The commercial version is called *Big Network Controller*, which is based on its popular open source controller called *Floodlight*. Big Switch also offers complete SDN solutions, notably *Big Cloud Fabric*, which provides network virtualization through overlays using OpenFlow virtual and physical devices.

There are numerous software startups that are minor players in the SDN space. Both Nicira and Big Switch were startups. Nicira, as we have mentioned, was the target of a major VMware acquisition. Big Switch has received a large amount of venture capital funding. Both of these companies have become major forces on the SDN playing field. There are a number of other SDN software startups that have received varying amounts of funding and in some cases have been acquired. Except for Nicira and Big Switch, we do not feel that any of these have yet to become major voices in the SDN dialogue. Since they have been influential in terms of being recipients of much venture capital attention and investment, and may in the future become major players in SDN, we will discuss in Chapter 14 startups such as Insieme, PLUMgrid, and Midokura and their business ramifications for SDN.

11.5 WHITE-BOX SWITCHES

We earlier defined a white-box switch as a hardware platform that is purpose-built to be easily loaded with an NOS such as Switch Light or OVS. The goal is to create a simple, inexpensive device which can be controlled by an OpenFlow controller and the SDN applications that run on top of it. The control software that network vendors typically put into their devices is largely absent from a white-box switch.

The natural alliances in the white-box switch ecosystem are depicted in Fig. 11.1. Enterprises that spend fortunes on networking equipment are naturally drawn to a technology that would allow them to populate the racks of their mega-data centers with low-cost switching gear. A company like Big Switch that bets its future on the increasing demand for its Big Cloud Fabric due to an explosion in the number of OpenFlow-enabled switches clearly wishes to foster the white-box model [13]. Merchant silicon vendors and ODMs form the nexus of the physical white-box switch manufacturing ecosystem.

The merchant silicon vendors that make the switching chips are increasingly aligned with capabilities in the OpenFlow standards, though this process is unfolding far more slowly than Open SDN enthusiasts would desire. Traditionally, much of the hardware platforms sold by the major NEMs, particularly lower-end devices, are actually manufactured by *Original Device Manufacturers* (ODMs). The ODMs, in turn, are focused on the most cost-effective manufacturing of hardware platforms whose hardware logic largely consists of switching chips from the merchant silicon vendors plus commodity CPUs and memory. Industry consolidation means that the hardware switching platforms look ever more similar. The major NEMs have traditionally distinguished their products by control and management software and marketed those features plus the support afforded by the large organization. Now that the control and management software can be provided by software vendors such as Big Switch, the possibility exists for the marriage of such an NOS with a white-box device. As we explained in Section 6.4, this situation is reminiscent of the PC, tablet, and smartphone markets where ODMs manufacture the hardware platforms, while operating systems, such as Windows and Linux, are loaded onto them in a generic fashion.

DISCUSSION QUESTION

We have said the loading an NOS (e.g., general purpose OpenFlow device software) onto a white-box switch is reminiscent of the Wintel ecosystem in the PC world. Which SDN players play an analogous role to Microsoft in this ballet? To Intel? Thinking in the other direction, provide some examples of PC ODMs that played similar roles to those played by Accton and Quanta in the SDN ecosystem?

Obviously, while this ecosystem intimately involves the merchant silicon vendors, the ODMs, and the software vendors, the role of the traditional NEMs is diminished. It is important to recognize, though, that the white-box switch concept does not readily apply to all networking markets. When the customer is large and sophisticated, such as a major cloud services vendor like Google, the support provided by the traditional NEM is of less value. Many customers, though, will continue to need the technical support that comes from the traditional NEM even if they choose to migrate to the OpenFlow model.

One example of a white-box ecosystem is the Switch Light Initiative formed by Big Switch. As the originator of Switch Light, Big Switch offers its OpenFlow device software to switch manufacturers, with the goal of developing a broader market for its OpenFlow controllers. While conceptually this switch software can be adapted to run with any merchant silicon switching chips, the initial merchant silicon vendor involved is Broadcom, and there is now a version of the Switch Light software available called *Switch Light for Broadcom*, as we mentioned earlier. ODMs are a natural target for this software. The coupling between Switch Light and the white-box switch can be very loose. Delivered with a boot loader like the *Open Network Install Environment* (ONIE) the white-box switch leaves the manufacturing floor agnostic as to with which NOS it will be used. When ultimately deployed at a customer site as part of a Big Cloud Fabric the union between the white-box hardware and the Switch Light NOS is consummated.

The startup Pica8's entire business model is based on a white-box strategy wherein Pica8 pairs the open source OpenFlow software with white-box hardware and sells the bundled package. Their operating system and software run on white-box switching hardware available from multiple ODM partners and the switches are controlled using the OpenFlow-capable OVS open source switching code. We discuss Pica8 further in Section 14.9.1.

Cumulus Networks is another startup that offers software ready-made for white-box hardware from ODMs like Accton and Quanta. The Cumulus focus is closer to the idea of *opening up the device* that we introduced in Section 6.4 than to Open SDN. While an OpenFlow-capable system can be loaded on top of the Cumulus operating system, their system is not coupled to OpenFlow. Indeed Cumulus touts that it is possible to use their system with white-box switches without using a centralized controller at all.

11.6 MERCHANT SILICON VENDORS

Merchant silicon vendors provide specialty chips that are assembled into finished hardware products by NEMs or by ODMs who then sell the bare-metal boxes to NEMs who market the hardware under their own brand, usually with their own software. Merchant silicon vendors are interested in any new approach that increases the volume of chips that they sell, so a higher-volume, lower-cost model for switching or server hardware is in their interest. To this end, merchant silicon vendors who make switching chips that are compatible with OpenFlow have been naturally interested in SDN and, in particular, in the white-box switch ecosystem. Two major merchant silicon vendors who have been particularly active in the white-box arena are Intel and Broadcom.

Intel is a strong proponent of network device and server NICs that are designed to handle the flow matching and processing needs of OpenFlow 1.1 and beyond. This Intel strategy is focused on their *Data Plane Development Kit* (DPDK) which we introduced in Section 6.4.

For many years Broadcom has been a vendor of switching chips that ODMs have used to build wired switches. Broadcom's dominance in this area is evidenced by the fact that Big Switch's Switch Light currently comes in one version for hardware devices and that version is called *Switch Light for Broadcom*.

A lesser known player in merchant silicon for SDN is Mellanox. Mellanox, long known for its InfiniBand products, has also entered the market for OpenFlow-enabled switching silicon [14,15]. We explained in Section 5.11 that the most recent versions of OpenFlow have surpassed the ability of extant silicon to support the full feature-set of those specifications. It is a widely held belief that the only real solution to this is to design switching chips specifically to support advanced versions of OpenFlow rather than trying to adapt older designs to do something for which they were not designed. As of this writing, Broadcom, Intel, Mellanox [16], and Corsa [17] all offer commercial silicon supporting OpenFlow 1.3.

11.7 ORIGINAL DEVICE MANUFACTURERS

ODMs are focused on the most cost-effective manufacturing of hardware platforms whose hardware logic is mostly embodied in switching chips from the merchant silicon vendors and commodity CPUs and memory. They largely tend to be Taiwanese, Korean, and Chinese low-cost manufacturers. They excel at high-volume, low-cost manufacturing. They have traditionally relied on their NEM customers to provide end-customer support and to develop complex sales channels. Because of this, they have been relegated to a relatively low-margin business of selling bare-metal hardware to NEMs who brand these boxes and sell them through their mature marketing and support organizations. The ODMs desire to move up the value chain and enter a higher margin business model closer to the customer. This process should be aided by customers who are less reliant on NEMs and would rather purchase hardware directly from the ODMs with the expectation of lower CAPEX costs. Two examples of ODMs that are particularly interested in the white-box switch model for SDN are Accton and Quanta.

DISCUSSION QUESTION

Fig. 11.1 depicts synergy between software vendors, merchant silicon vendors, and the ODMs that manufacture white-box switches. Describe the ecosystem that results in this synergy.

11.8 CLOUD SERVICES AND SERVICE PROVIDERS

A significant amount of the momentum behind SDN has come from large customers who purchase networking gear from the likes of Cisco, Juniper, Brocade, and others. These businesses are interested in stimulating the evolution of networking toward something more functional and cost-effective. As a consequence, these companies have made significant contributions to SDN over the last few years.

There are two subcategories of these businesses that have displayed the greatest interest in SDN: cloud services and carriers. Cloud services companies such as Google started by providing public Internet access to specialized data that was a target of wide consumer interest. The carriers traditionally provided network infrastructure. Increasingly, though, both categories of enterprise offer *Everything as a Service* (XaaS), providing large data centers that offer a vast number of tenants dynamic access to shared compute, storage, and networking infrastructure. Such large data centers have provided the most fertile ground for making SDN inroads. Hence, these two classes of companies have taken prominent roles in the ONF. The synergy between the cloud services, carriers, and the ONF, which we discuss further in Section 11.9.1, was reflected in Fig. 11.1.

Google has intense interest in improving the capabilities and lowering the cost of networking, and has thus created their own simplified set of networking devices which use OpenFlow and a controller to manage their inter-datacenter WAN connections. We described such an example of Google using SDN in the WAN in Section 9.1.2.

NTT has long been involved in SDN and OpenFlow research. NTT has created its own SDN controller called *Ryu* which can network devices using all releases of OpenFlow through V.1.5. Ryu uses the *OpenFlow Configuration and Management Protocol* (OF-Config) [18] for device configuration. Note that OF-Config itself utilizes NETCONF [19] as the transport protocol and YANG as the data model. As described in Section 9.2.3, Verizon has been involved in large SDN proofs of concept related to cloud bursting. Verizon has also been a leader in the related *Network Functions Virtualization* (NFV) technology, as they chair the NFV committee for the *European Telecommunications Standards Institute* (ETSI), which is currently taking the lead in NFV research and standardization [20].

There are other large enterprises such as banks that run large data centers that also have significant interest in SDN even though they are not part of either of the two major categories discussed previously. For example, Goldman Sachs is on the board of the ONF and is very public about their support for OpenFlow.

11.9 STANDARDS BODIES AND INDUSTRY ALLIANCES

In order for new technologies to become widely adopted, standards are required. In this section we will discuss two different kinds of organizations that produce standards related to SDN. The most obvious genre is a true standards body, such as the IETF or IEEE. A less obvious sort is what we refer to here as an *Open Industry Alliance*. Such an alliance is a group of companies with a shared interest surrounding a technology. A good historical example of an industry alliance and a standards body working on the same technology is the WiFi Alliance and the IEEE 802. The WiFi Alliance created important standards documents (e.g., WPA, WPA2) but was an industry alliance first, not a standards body. In the instance of SDN, it is the industry alliances that have taken the lead with standards more so than the standards bodies. The four most important alliances for SDN are the ONF, ODL, ONOS, and OpenStack. The general idea of OpenStack is to create an abstraction layer above compute, network, and storage resources. While the network component of this abstraction layer has led to a relationship between OpenStack and SDN, SDN is not its primary focus, so we will direct our attention first to the ONF, ODL, and ONOS, three industry groups that are indeed laser-focused on SDN.

The ONF and ODL look at SDN from two very different perspectives. The driving force behind the ONF and the standards it creates are the board members, who represent enterprises such as Deutsche Telekom, Facebook, Goldman Sachs, Google, Microsoft, NTT Communications, Verizon, and Yahoo!, as well as some university researchers. There is not a single NEM represented on the board. ODL, on the other hand, was founded by NEMs and looks at SDN as a business opportunity and/or challenge. Fig. 11.2 depicts the board composition of the three main SDN industry alliances. It is noteworthy that the ONOS board does overlap somewhat with each of the other two, yet there is no overlap between ODL and ONF.

It is easy to fall victim to the temptation of thinking that an alliance controlled by NEMs like ODL will favor protecting its members' business interests over innovating in a way that most helps their user community. There are downsides to an enterprise-driven alliance like ONF as well, though.

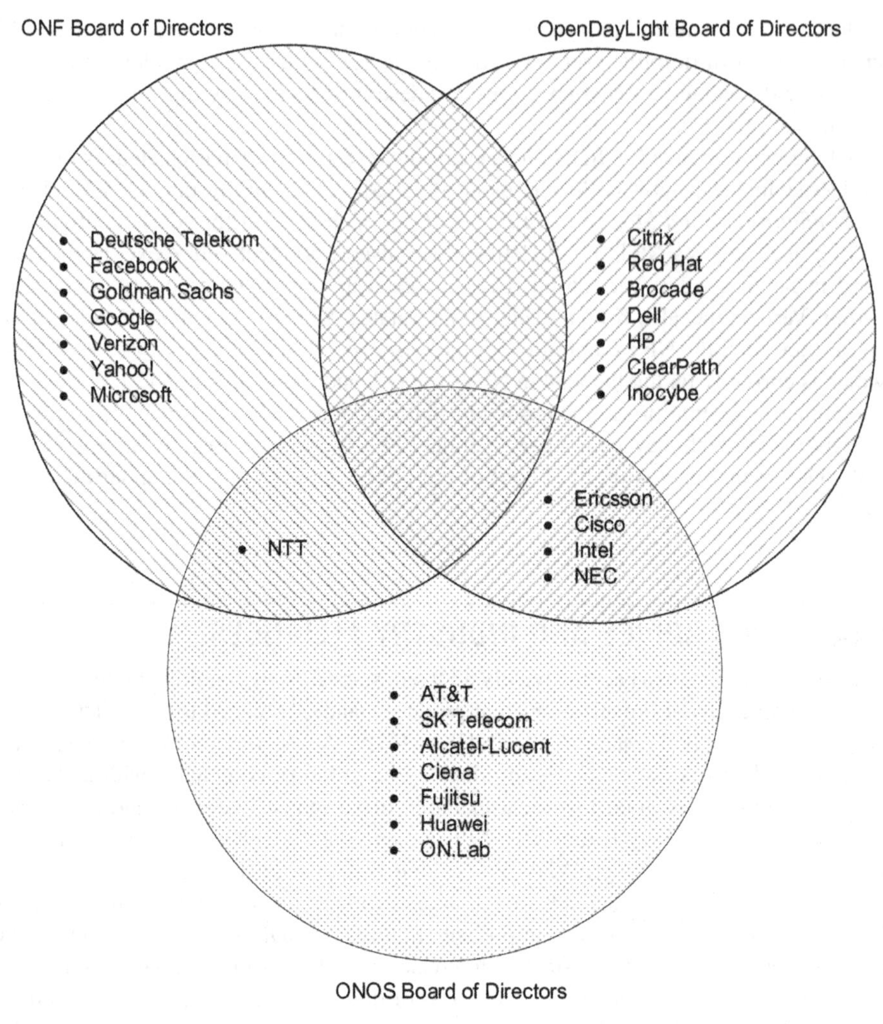

FIG. 11.2

ONF, ODL, and ONOS corporate board membership.

There is the real possibility that the standards that are created will be out of sync with the hardware capabilities of networking devices. To a certain degree this was true with OpenFlow V.1.1 and has become more exaggerated with its later versions. This explains in part the reluctance of the vendors to implement these subsequent revisions of the OpenFlow standard. Conversely, being unencumbered by vendor limitations has resulted in OpenFlow defining what the industry needs, rather than what the vendors can immediately provide. This is the genesis of the conflict between ODL and ONF reflected in Fig. 11.1.

11.9.1 OPEN NETWORKING FOUNDATION

The ONF occupies a preeminent place within the set of standards bodies contributing to SDN. The Open Networking Foundation was created in 2011 and its stated purpose is *the promotion and adoption of Software Defined Networking (SDN) through open standards development* [21]. The ONF is the owner of the OpenFlow standards process. To date, OpenFlow is the most influential work produced by the ONF. The various councils, areas, and working groups active within the ONF as of this writing were shown in Fig. 3.8.

11.9.2 OPENDAYLIGHT

ODL's mission is to *facilitate a community-led, industry-supported open source framework, including code and architecture, to accelerate and advance a common, robust Software Defined Networking platform* [22]. The project is part of the Linux Foundation of Collaborative Projects and boasts many major networking vendors as members (e.g., Brocade, Cisco, IBM, HP, Huawei, Intel, NEC, VMware, and others).

ODL welcomes source code contributions from its members, and has a fairly rigorous process for software review and inclusion. As can be seen by the list of contributing engineers as well as the ODL controller interface, much of the controller software comes from Cisco. In addition to the many Cisco contributors, many other contributors have added software to the ODL open source code base. Plexxi has provided an API allowing the controller and applications to collaborate using an abstraction of the underlying network infrastructure that is independent of the particular switching equipment used. Software to prevent *Distributed Denial of Service* (DDoS) attacks has been contributed by Radware. Ericsson and IBM collaborated with Cisco to provide OpenFlow plugins to the controller. Pantheon has provided a version of OpenFlow 1.3. We provide more details about ODL in Section 13.8.2.

The ONF focuses on defining a specific protocol (OpenFlow) between network devices and a controller to move the control plane from those devices to the controller. ODL focuses not on a single control protocol or on mandating that network devices conform to that protocol. Rather, ODL is focused on providing an open source controller platform, playing a role for SDN controllers much like the role Linux plays in servers. We depict this very different focus between the ONF and ODL consortia in Fig. 11.3. Many different controller-to-device protocols are supported under this umbrella of open-source functionality. Although there was initial skepticism whether ODL would just be a repackaging of Cisco's XNC controller, it has emerged as one of the fastest-growing open source projects in any domain.

11.9.3 ONOS

ONOS is a recent arrival on the SDN scene, but its presence has drastically altered the SDN controller landscape compared to just a few years ago. While many controllers have played their part in the evolution of SDN, the only two significant open source controllers as of this writing are ODL and ONOS. As can be seen in Fig. 11.2, ONOS has attracted more board participation from service providers and NEMs focused on that market than does ODL. ONOS is firmly committed to the principles of Open SDN, which is not strictly true of ODL. Whereas ODL has silver, gold, and bronze levels of participation and corresponding influence, members' influence on ONOS is ostensibly a function of their level of participation rather an amount paid for membership. The ONOS mission is

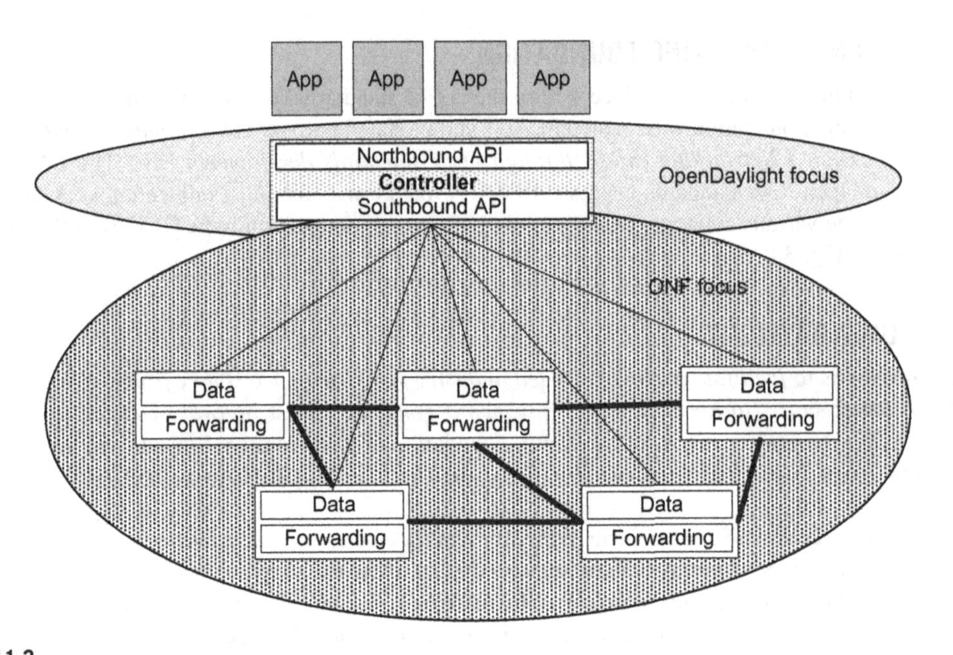

FIG. 11.3

Focus of ONF vs OpenDaylight.

heavily influenced by one of its founding entities, ON.Lab, which, at its core, is a research organization. As we said in Section 11.1.1, Guru Parulkar, the Executive Director of ON.Lab and now ONOS, is strongly opposed to the dilution of SDN from its founding principles. The mission statement *To produce the Open Source Network Operating System that will enable service providers to build real Software Defined Networks* [23], clearly exposes the service provider orientation of this group. ONOS emphasizes the use of OpenFlow and the high reliability features that are required for service provider use. Unlike ODL, it does not prioritize the repurposing of legacy equipment and protocols to achieve SDN goals. Since their controller software was first open-sourced in 2014, the ONOS community has rapidly risen to become an influential player with respect to SDN controllers. We provide more details about ONOS in Section 13.8.3.

DISCUSSION QUESTION

Fig. 11.2 shows that the intersection of the corporate board members of the ONF and ODL is the empty set. Explain how this is a reflection of the different focus of these two organizations.

11.9.4 OpenStack

The OpenStack project was started by Rackspace and NASA in 2010 as an open source IaaS toolset. In a sense, OpenStack can be viewed as the Linux of cloud computing. The interest in this open source project quickly grew, culminating in the establishment in September 2012 of the nonprofit OpenStack

Foundation. The mission of the OpenStack Foundation is to *promote, protect, and empower OpenStack software and its community*. As of this writing, the foundation has more than 6700 member companies spread over 83 countries. The largest current deployment of OpenStack remains in one of its founding companies, Rackspace, which is a major provider of cloud computing.

The general idea of OpenStack is to create an abstraction layer above compute, network, and storage resources. This abstraction layer provides APIs which applications can use to interact with the hardware below, independent of the source of that hardware. The interfaces for compute, storage, and network present pools of resources for use by the applications mentioned previously.

The OpenStack networking definition states that SDN is not required, but that *administrators can take advantage of SDN technology like OpenFlow to allow for high levels of multitenancy and massive scale* [24]. The network virtualization component of OpenStack is called Neutron and, thus, is the component most relevant to our discussion of SDN. Neutron was formerly called Quantum. Neutron provides an application-level networking abstraction. By abstracting the details of the underlying physical network devices, Neutron is intended to provide users of cloud computing with the ability to create overlay networks that are easier to manage and understand by the cloud tenants. It attempts to provide a *Network-as-a-Service* (NaaS), by allowing the cloud tenants to interconnect *virtual network interfaces* (vNICs) [25]. The actual mapping of these virtual network interfaces to the physical network requires the use of Neutron plugins. This is where the relationship between OpenFlow and OpenStack exists. One Neutron plugin could be an interface to an OpenFlow controller that would control physical switches [26]. ODL currently offers a Neutron northbound API. We discuss OpenStack architecture further in Section 13.12.

Although this link between OpenStack and SDN is thin, OpenStack's impact as an industry alliance means that it is an important SDN player despite the peripheral role played by SDN in the overall OpenStack architecture.

11.9.5 OpenSwitch

OpenSwitch [27] is a new open source community started by HP in 2015 with the goal of providing high reliability open source SDN switch code. In this sense it is in direct competition with Switch Light and OVS. We describe the organization's founding members, its goals and licensing in Section 13.7. As it is very new, it is difficult to assess at this point how significant a player this consortium is likely to be, but the strong push it is getting from HP augurs well for its future.

11.9.6 THE OPEN SOURCE SDN COMMUNITY

The *Open Source SDN* (OSSDN) community [28] was introduced in 2015 and differs from the other open source communities we have mentioned in this section in that its participants are not companies but individuals. As of this writing, OSSDN's only funding is from its sponsor, the ONF. Rather than trying to shape a policy that will influence the direction of an industry, its goal is to provide an online gathering place for individuals that wish to contribute to open source projects related to SDN. We describe the OSSDN projects that are currently active in Section 13.9.1. The fact that it is *the* open source community sponsored by the ONF implies that this organization will likely play a significant role on the SDN stage.

11.9.7 IETF

As a late arrival to the SDN playground, the IETF now has a *software-driven networks birds of a feather* (BoF) group and an SDN research group (SDNRG). While there are no current standardization efforts in the IETF directed solely at SDN, the IETF is active in standards that are indirectly related to SDN, such as the existing standards for network virtualization including VXLAN, NVGRE, and STT, which we discussed in Section 8.3 as SDN technologies for overlays and tunneling. Others include the *Path Computation Element* (PCE) [29] and *Network Virtualization Overlays* (nvo3) [30] standards. While the IEEE was not active in the early days of SDN, this important standards body now hosts the IEEE SDN initiative [31], so we expect to see more activity emanating from it in the near future.

11.10 CONCLUSION

In the course of presenting the major players in SDN, this chapter reads somewhat like a *Who's Who* in networking. At times it does appear that the world is jumping on the SDN bandwagon in order to not be left behind. We must temper this by saying that not every NEM or even every large enterprise customer is certain that SDN lies in their future. As an example, for all of the enthusiasm for SDN emanating from Google and other cloud services enterprises, Amazon is notable by its absence from any type of membership whatsoever in the ONF and by their relative silence about building their own SDN solution. Amazon is a huge enterprise with large data centers and the need for the advances in networking technology that have the rest of the industry talking about SDN, and one would expect them to have similar mega-datacenter needs as Google and Yahoo! The industry consensus [32] is that Amazon is indeed using SDN in some form for its enormous *Amazon Web Services* (AWS) business, but is doing so in stealth mode. Only time will tell whether Amazon and other SDN backbenchers turn out to be latecomers to the SDN party or whether they were wise to be patient during the *peak of unreasonable expectations* that may have occurred with respect to SDN.

We have stated more than once in this work that openness is a key aspect of what we define as SDN. In formulating this chapter on the key players in SDN, however, it was not possible to talk about open source directly as it is not an entity, and thus not within the scope of this chapter. Indeed many of the SDN players we have discussed here have been significant largely *because* of their open source contributions to SDN. In any event, openness and open source are key facets of SDN as we have defined it and, as such, Chapter 13 focuses entirely on open source software related to SDN. First, though, in the next chapter we provide a tutorial on writing Open SDN applications.

REFERENCES

[1] Clean Slate: a interdisciplinary research program. Stanford University. Retrieved from: http://cleanslate.stanford.edu.
[2] ONRC research. Retrieved from: http://onrc.stanford.edu/.
[3] ONRC research videos. Retrieved from: http://onrc.stanford.edu/videos.html.
[4] InCNTRE. Indiana University. Retrieved from: http://incntre.iu.edu/.

[5] Open Networking Foundation sees high growth in commercialization of SDN and OpenFlow at PlugFest. Open Networking Foundation; 2012. Retrieved from: https://www.opennetworking.org/news-and-events/press-releases/248-high-growth-in-commercialization-of-sdn.
[6] Ethane: a security management architecture. Stanford University. Retrieved from: http://yuba.stanford.edu/ethane/.
[7] Parulkar G. Keynote session—opening talk (video). Santa Clara, CA: Open Networking Summit; 2013. Retrieved from: http://www.opennetsummit.org/archives-april2013/.
[8] Beacon 1.0.2 API. Stanford University. Retrieved from: https://openflow.stanford.edu/static/beacon/releases/1.0.2/apidocs/.
[9] Ramel D. Cisco, Extreme, Big Switch form SDN partnerships. Virtualization Review; 2015. Retrieved from: https://virtualizationreview.com/articles/2015/09/10/sdn-partnerships.aspx.
[10] Mahalingam M, Dutt D, Duda K, Agarwal P, Kreeger L, Sridhar T, et al. VXLAN: a framework for overlaying virtualized layer 2 networks over layer 3 networks. Internet Draft. Internet Engineering Task Force; 2011.
[11] Davie B, Gross J. STT: a stateless transport tunneling protocol for network virtualization (STT). Internet Draft. Internet Engineering Task Force; 2012.
[12] Sridharan M, et al. NVGRE: network virtualization using generic routing encapsulation. Internet Draft. Internet Engineering Task Force; 2011.
[13] Banks E. Big Switch leaves OpenDaylight, Touts White-Box Future. Network Computing; 2013. Retrieved from: http://www.networkcomputing.com/data-networking-management/big-switch-leaves-opendaylight-touts-whi/240156153.
[14] Marvyn. Mellanox brings software-defined networking to SwitchX-2 chips. Inside HPC; 2012. Retrieved from: http://insidehpc.com/2012/10/mellanox-brings-software-defined-networking-to-switchx-2-chips/.
[15] Mellanox OpenStack and SDN/OpenFlow solution reference architecture. Rev. 1.2. Mellanox Technologies; 2013. Retrieved from: http://www.mellanox.com/sdn/pdf/Mellanox-OpenStack-OpenFlow-Solution.pdf.
[16] Pitt D. OCP summit. ONF members prominent; 2014. Retrieved from: https://www.opennetworking.org.
[17] Matsumoto C. Corsa builds a purely OpenFlow data plane. SDX Central; 2014. Retrieved from: https://www.sdxcentral.com/articles/news/corsa-builds-purely-openflow-data-plane/2014/02/.
[18] OpenFlow management and configuration protocol, Version 1.1.1. Open Networking Foundation; 2013. Retrieved from: https://www.opennetworking.org/sdn-resources/onf-specifications.
[19] Enns R, Bjorklund M, Schoenwaelder J, Bierman A. Network Configuration Protocol (NETCONF). RFC 6241. Internet Engineering Task Force; 2011.
[20] ETSI. Network Functions Virtualization. European Telecommunications Standards Institute. Retrieved from: http://www.etsi.org/technologies-clusters/technologies/nfv.
[21] Open Networking Foundation. ONF overview. Retrieved from: https://www.opennetworking.org/about/onf-overview.
[22] Welcome to OpenDaylight. Retrieved from: http://www.opendaylight.org.
[23] ONOS mission. Retrieved from: http://onosproject.org/mission/.
[24] OpenStack networking. OpenStack Cloud Software. Retrieved from: http://www.openstack.org/software/openstack-networking/.
[25] OpenStack. Neutron. Retrieved from: https://wiki.openstack.org/wiki/Neutron.
[26] Miniman S. SDN, OpenFlow and OpenStack Quantum. Wikibon; 2013. Retrieved from: http://wikibon.org/wiki/v/SDN,_OpenFlow_and_OpenStack_Quantum.
[27] OpenSwitch. Retrieved from: http://openswitch.net.
[28] Open source SDN sponsored software development. Retrieved from: http://opensourcesdn.org/ossdn-projects.
[29] Path computation element (PCE) working group. Internet Engineering Task Force. Retrieved from: https://datatracker.ietf.org/wg/pce/.
[30] Network Virtualization Overlays (NVO3) working group. Internet Engineering Task Force. Retrieved from: https://datatracker.ietf.org/wg/nvo3/.

[31] IEEE software defined networks initiative launches newsletter highlighting global industry developments. Business Wire; 2015. Retrieved from: http://www.businesswire.com/news/home/20151103005536/en/IEEE-Software-Defined-Networks-Initiative-Launches-Newsletter.
[32] Ma C. SDN secrets of Amazon and Google. InfoWorld—New Tech Forum; 2014. Retrieved from: http://www.infoworld.com/article/2608106/sdn/sdn-secrets-of-amazon-and-google.html.

CHAPTER

SDN APPLICATIONS

12

We now descend from the conceptual discussions about SDN use cases and take a deeper dive into how SDN applications are actually implemented. This chapter focuses on SDN application development in both an open SDN and an SDN via APIs environment. Most SDN via Overlays offerings today are bundled solutions which come ready-made with a full set of functionality, generally without the ability to add applications to those solutions. As such, we will not contemplate that type of application development.

In particular, we will examine attributes and considerations related to creating reactive versus proactive applications, as well as creating internal versus external applications. These different application types were introduced in Section 7.4.1. Since the creation of internal, reactive applications is a more ambitious endeavor, we present a sample internal, reactive application, running inside the Floodlight controller framework. Although Floodlight today generates less interest than some newer controllers, the principles for creating this type of application are fairly consistent across controllers in common use, and Floodlight is the most basic of these. That reason, combined with the fact that it evolved out of the original OpenFlowJ [1] libraries make Floodlight a suitable training environment.

It is worth noting that the crisp distinction between reactive and proactive applications that we make in this chapter stems from the original use of these terms within the OpenFlow paradigm. With some of the non-OpenFlow SDN alternatives introduced in Chapter 7, these distinctions may not be absolute. In particular, there may be applications that exhibit both reactive and proactive characteristics.

Many detailed functional block diagrams are included in this chapter, as well as a fair amount of sample source code which serve as helpful examples for the network programmer.

12.1 TERMINOLOGY

Consistent terminology is elusive in a dynamic technology such as SDN. Consequently, certain terms are used in different ways, depending on the context. To avoid confusion, in this chapter we will adhere to the following definitions:

- *Network device*: When referring generically to a device such as a router, switch, or wireless access point, in this chapter we will use the term *network device* or *switch*. These are the common terms used in the industry when referring to these objects. Note that the Floodlight controller uses the term *device* to refer to *end user device*. Hence, in sections of code using Floodlight or supporting text, the word *device* used alone means *end user node*.

Software Defined Networks. http://dx.doi.org/10.1016/B978-0-12-804555-8.00012-0

- *End user node*: There are many examples of end user devices, including desktop computers, laptops, printers, tablets, and servers. The most common terms for these end user devices are *host*, and *end user node*; we will use *end user node* throughout this chapter.
- *Flow*: In the early days of SDN, the only controller-device protocol was OpenFlow, and the only way to affect the forwarding behavior of network devices was through setting flows. With the recent trends toward NETCONF and other protocols to effect forwarding behavior by setting routes and paths, there is now no single term for that which the controller changes in the device. The controller's actions decompose into the fundamentals of create, modify, and delete actions, but the entity acted upon may be a flow, a path, a route, or, as we shall see, an optical channel. For the sake of simplicity, we sometimes refer to all these things generically as *flows*. The reader should understand this to mean either setting flows, or configuring static routes, modifying the RIB, setting MPLS LSPs, among other things, depending on the context.

12.2 BEFORE YOU BEGIN

Before embarking on a project to build an SDN application, one must consider a number of questions. Answering these questions will help the developer to know how to construct the application, including what APIs are required, which controller might be best suited for the application, and what switch considerations should be taken into account. Some of these questions are:

- *What is the basic nature of the application?* Will it spend most of its time reacting to incoming packets which have been forwarded to it from the switches in the network? Or will it typically set flows proactively on the switches and only respond to changes in the network which require some type of reconfiguration?
- *What is the type and nature of network with which the application will be dealing?* For example, an access control application for a campus network might be designed differently than a similar application for a data center network.
- *What is the intended deployment of the controller with respect to the switches it controls?* A controller which is physically removed from the switches it controls (e.g., in the cloud) will clearly not be able to efficiently support an SDN application which is strongly reactive in nature because of the latency involved with forwarding packets to the controller and receiving a response.
- *Are the SDN needs purely related to the data center and virtualization?* If so, perhaps an SDN Overlay solution is sufficient to meet current needs and future needs related to the physical underlay network can be met later, as OpenFlow-related technologies mature.
- *Will the application run in a green-field environment or will it need to deal with legacy switches which do not support SDN?* Only in rare situations is one able to build an Open SDN application from the ground up. Typically, it is required to use existing equipment in the solution. It is important to understand if the existing switches can be upgraded to have OpenFlow capability and whether or not they will have hybrid capability, supporting both OpenFlow and legacy operation. If the switches do not support OpenFlow, then it is important to understand the level of SDN via APIs support that is available.
- *What type of APIs are available on legacy switches and routers? What is their level of capability?* If the application requires NETCONF, what YANG models does each type of switch or router

support, and are the different YANG models consistent? The answer to these questions will dictate the capabilities of your SDN application, and the extent to which you will be able to implement SDN capabilities in your legacy network.

- *What is the level of programming ability present in your development team(s)?* Only software teams with deep experience in the native language of the controller will be able to develop internal SDN applications; this may be especially true if OpenDaylight is your controller, since you must develop applications in a *model-driven service abstraction layer* (MD-SAL) environment. Since the native language of current mainstream controllers is Java, this currently implies that the team have deep experience with Java. If your software team is more comfortable developing applications in a language such as Python, then they will gravitate toward external applications.
- *How many switches will be managed by your application?* Scalability is more of an issue with reactive applications, since packets will periodically be forwarded to your application for processing, and this limits the number of switches or routers that can be controlled by a single application/controller instance.

The answers to these and similar questions will help the developer to understand the nature of the SDN application. Assuming that an Open SDN application will be developed, the developer will need to decide between two general styles of SDN applications, *reactive* and *proactive*.

DISCUSSION QUESTION

We have listed a number of factors to be considered when determining how to build your SDN application. Can you think of any other considerations? Do you believe that certain of these considerations are more important than others, and if so, which ones?

12.3 APPLICATION TYPES

In Section 7.4.1 we introduced the notion of reactive versus proactive applications, and internal vs external applications. In the sections that follow, we take a deeper look at these application types, and do so specifically from the perspective of SDN application development.

12.3.1 REACTIVE VS PROACTIVE APPLICATIONS

The reader should recall from Section 7.4.1 that the main difference between these two application paradigms is that reactive applications will periodically receive packets forwarded from a switch in order for the SDN application to process the packet. The reactive application then sends instructions back to the switch prescribing how to handle this packet, and whether flow changes are required. In the case of reactive applications, the communication between the switch and the controller will typically scale with the number of new flows injected into the network. The switches may often have relatively few flow entries in their flow tables, as the flow idle timers are set to match the expected duration of each flow in order to limit the growth of the flow table. This can result in the switch frequently receiving packets, which match no rules. In the reactive model, those packets are encapsulated in PACKET_IN messages and forwarded to the controller and thence to an application. The application inspects the packet and determines its disposition. The outcome is usually to program a new flow entry in the switch(es) so that the next time this type of packet arrives, it can be handled locally by the switch itself. The application will often program multiple switches at the same time so that each switch along the path of the flow will have a consistent set of flow entries for that flow.[1] The original packet will often be returned to the switch so that the switch can handle it via the newly installed flow entry. The kind of applications that naturally lend themselves to the reactive model are those that need to see and respond to new flows being created. Examples of such applications include per-user firewalling and security applications that need to identify and process new flows.

On the other hand, proactive SDN applications feature less communication emanating from the switch toward the controller. The proactive SDN application sets up the switches in the network with either flow entries or configuration attributes which are appropriate to deal with incoming traffic before it arrives at the switch. Events which trigger changes to the flow table or configuration changes to switches may come from mechanisms which are outside the scope of the switch-to-controller secure channel. For example, some external traffic monitoring module will generate an event which is received by the SDN application, which will then adjust the flows on the switches appropriately. Applications that naturally lend themselves to the proactive model usually need to manage or control the topology of the network. Examples of such applications include new alternatives to spanning tree and multipath forwarding. With the addition of support for legacy protocols via protocol plugins such as NETCONF and BGP-LS/PCE-P, configuration of paths, routes, and traffic shaping are possible via proactive applications.

Reactive programming may be more vulnerable to service disruption if connectivity to the controller is lost. Specifically, when the failure mode is set to *fail secure* the switch is more likely to be able to continue normal operation since many flow table entries will have been prepopulated in the proactive paradigm. Fail secure mode specifies that OpenFlow packet processing should continue despite the loss of connectivity to the controller.[2]

Another advantage of the proactive model is that there is no additional latency for flow setup as they are prepopulated. A drawback of the proactive model is that most flows will be wildcard-style, implying aggregated flows, and thus less-fine granularity of control. We discuss these issues in more detail in the next subsections.

12.3.2 INTERNAL VS EXTERNAL APPLICATIONS

We saw in Section 7.4.3 that internal applications are those that run *inside* the OSGi container of the controller. Conversely, external applications run outside that container, meaning they can run anywhere, and typically use the RESTful APIs provided by the controller. One of the easiest ways to compare internal and external applications is via a table of their limitations and capabilities. This comparison is summarized in Table 12.1.

---

[1]Proper synchronization of flow programming on a series of switches can be a challenge with OpenFlow. This is one of the reasons for the growing interest in the intents-based style of northbound API on the controller, as discussed in Section 7.3.5.
[2]See Section 5.4.5 for an explanation of *failure mode*. If the failure mode is set to *fail stand-alone* mode then loss of controller connectivity will cause the flow tables to be flushed independent of the application style.

Table 12.1 Internal vs External Applications

Feature	Internal	External
Programming language	Java	Any language
APIs used	Java	REST
Deployment	Inside controller	Anywhere
Performance	High	Low
Reactive apps	Yes	No
Proactive apps	Yes	Yes
Failure affects controller	Yes	No

The following points provide background for the entries in Table 12.1:

- Internal applications must be written in Java (assuming the controller's native language is Java, which is true of almost all controllers); external applications can be written in any language.
- Internal applications will use Java APIs on the controller; external applications will use RESTful APIs.
- Internal applications must be deployed inside the environment (typically OSGi) of the controller, running locally with the controller; external applications can run anywhere, even in the cloud if so desired.
- Internal applications are relatively faster; whereas external applications are slower due to: (1) using the REST API, (2) possibly being written in a slower language such as Python, and (3) potentially running remotely from the controller.
- Internal applications are capable of running reactive applications (receiving packets forwarded from an OpenFlow switch); external applications are not.
- Both internal and external applications are capable of running proactive applications.
- Internal applications are running inside the controller, and failures can have negative effects on the operation of the controller; for external applications, this danger is less severe.

12.3.3 DETAILS: REACTIVE SDN APPLICATIONS

We have seen that reactive applications are capable of receiving packets that have been forwarded to the controller from switches in the network. As of this writing, the only SDN protocol that performs that function is OpenFlow, hence, a reactive application is implicitly an OpenFlow application.

Since reactive applications must also run inside the controller in order to receive asynchronous notification of forwarded packets, they must also be internal applications, running inside the OSGi container of the controller. External applications using the RESTful APIs are unable to receive these asynchronous notifications, since REST is a unidirectional protocol. And while certain mechanisms can be put in place for bidirectional communication (e.g. web sockets), the latency involved is impractical for the prompt handling of the possibly huge number of incoming packets. Reactive applications have the ability to register listeners, which are able to receive notifications from the controller when certain events occur. In Section 12.4 we provide a brief historical overview of controllers that have played an important role in the evolution of SDN. Floodlight and Beacon are notable in this list. Some important listeners available in these two historically popular controller packages are:

- *SwitchListener*. Switch listeners receive notifications whenever a switch is added, removed or has a change in port status.
- *DeviceListener*. Device listeners are notified whenever a device (an end-user node) has been added, removed, moved (attached to another switch) or has changed its IP address or VLAN membership.
- *MessageListener*. Message listeners get notifications whenever a packet has been received by the controller and the application then has a chance to examine it and take appropriate action.

These listeners allow the SDN application to react to events which occur in the network and to take action based on those events.

When a reactive SDN application is informed of an event, such as a packet which has been forwarded to the controller, change of port state, or the entrance of a new network device or host into the network, the application has a chance to take some type of *action*. The most frequent event coming into the application would normally be a packet arriving at the controller from a switch, resulting in an action. Such actions include:

- *Packet-specific actions*. The controller can tell the switch to drop the packet, to flood the packet, to send the packet out a specific port, or to forward the packet through the NORMAL non-OpenFlow packet processing pipeline as described in Section 5.3.4.
- *Flow-specific actions*. The controller can program new flow entries on the switch, intended to allow the switch to handle certain future packets locally without requiring intervention by the controller.

Other actions are possible, some of which may take place outside the normal OpenFlow control path, but the packet-specific and flow-specific actions constitute the predominant behavior of a reactive SDN application.

Fig. 12.1 shows the general design of a reactive application. Notice that the controller has a listener interface that allows the application to provide listeners for switch, device (end user node), and message (incoming packet) events. Typically a reactive application will have a module to handle packets incoming to the controller which have been forwarded through the message listener. This packet processing can then act on the packet. Typical actions include returning the request to the switch, telling it what to do with the packet (e.g., forward out a specific port, forward NORMAL, or drop the packet). Other actions taken by the application can involve setting flows on the switch in response to the received packet, which will inform the switch what to do the next time it sees a packet of this nature.

For reactive applications, the last flow entry will normally be programmed to match any packet and to direct the switch to forward that otherwise unmatched packet to the controller. This methodology is precisely what makes the application *reactive*. When a packet not matching any existing rule is encountered, it is forwarded to the controller so that the controller can react to it via some appropriate action. A packet may also be forwarded to the controller in the event that it matches a flow entry and the associated action stipulates that the packet be passed to the controller.

In the reactive model the flow tables tend to continually evolve based on the packets being processed by the switch and by flows aging out. Performance considerations arise if the flows need to be reprogrammed too frequently, so care must be taken to appropriately set the idle timeouts for the flows. The point of an idle timer is to clear out flow entries after they have terminated or gone inactive, so it is important to consider the nature of the flow and what constitutes inactivity when setting the timeout.

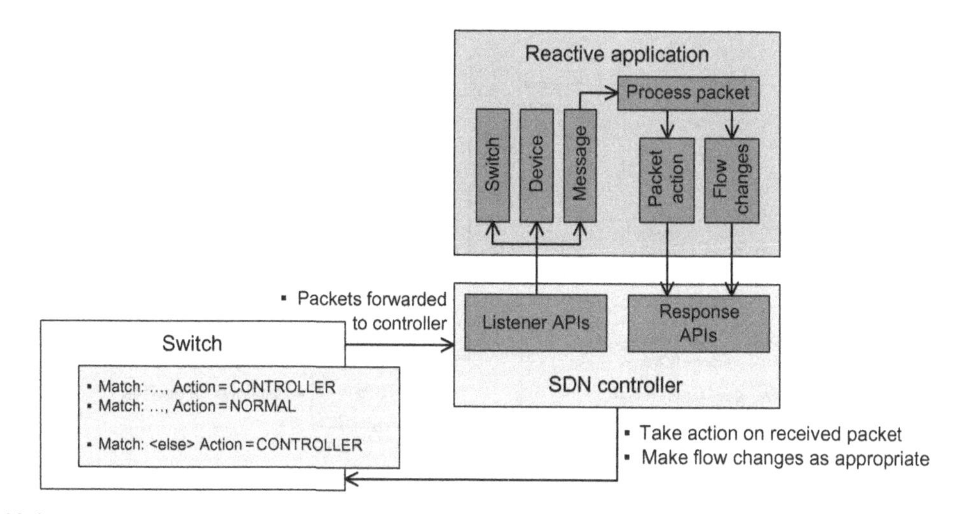

FIG. 12.1

Reactive application design.

Unduly short idle timeouts will result in too many packets sent to the controller. Excessively long timeouts can result in flow table overflow. Knowing an appropriate value to set a timeout requires familiarity with the application driving the flow. As a rough rule of thumb, though, timeouts would be set on the order of tens of seconds rather than milliseconds or minutes.

12.3.4 DETAILS: PROACTIVE SDN APPLICATIONS

Proactive applications can be implemented using either the native API (e.g., Java) or using RESTful APIs, but most often these proactive applications are also external applications using the REST interface.

One important point to remember is that proactive applications are *dynamic*. That is, they can regularly make modifications to networking behavior. Fig. 12.2 shows that proactive applications take notifications from external events, and make changes to the behavior of network devices in the network. In the figure, we show the application setting flows at a relatively low level on the network device (i.e., the *forwarding information base* (FIB) level as described in Section 7.2.1). Proactive applications can also alter behavior of networking devices at the network-wide *routing information base* (RIB) and config layers as well.

In contrast to reactive APIs, these controller RESTful APIs can operate at different levels, depending on the northbound API being used. They can be primitive, basically allowing the application developer to configure OpenFlow flow entries on the switches, or to configure static routes directly on routers using NETCONF. These RESTful APIs can also be high-level, providing a level of abstraction above the network devices, so that the SDN application interacts with network components such as virtual networks, abstract network device models or with the RIB, rather than with the physical switches themselves.

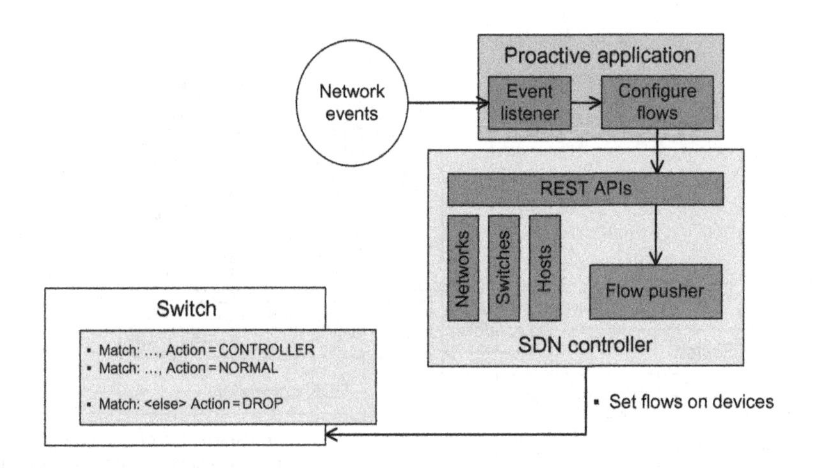

FIG. 12.2

Proactive application design.

Another difference between this type of application and a reactive one is that using RESTful APIs means there is no need for *listening* functionality. In the reactive application the controller asynchronously invokes methods in the SDN application stimulated by events such as incoming packets and the discovery of network devices. This type of asynchronous notification is not possible with a unidirectional RESTful API, which is invoked on the controller by the application rather than the other way around. While it is conceivable to implement an asynchronous notification mechanism from the controller toward a proactive application, generally, proactive applications are stimulated by sources external to the OpenFlow control path. The traffic monitor described in Section 12.12 is an example of such an external source.

Fig. 12.2 shows the general design of a purely proactive application. In such an application there are no *packet* listeners receiving packets that have been forwarded by switches in the network. Such listeners are not a natural fit with a one-way RESTful API, wherein the application makes periodic calls to the API and the API has no means of initiating communication back to the application. Proactive applications instead rely on stimulus from external *network* or *user events*. Such stimuli may originate from traffic monitors including an *intrusion detection system* (IDS) such as SNORT, or external applications like a server virtualization service, which notifies the application of the movement of a virtual machine from one physical server to another. Stimuli may also be in the form of user-initiated input which causes changes to the manner in which packets are handled—for example, a request for certain traffic types to be expedited or rerouted.

The RESTful APIs are still able to *retrieve* data about the network, such as domains, subnets, switches, and hosts, from the SDN controller. This information may be gathered by the controller through various protocols, including OpenFlow, *Link Layer Discovery Protocol* (LLDP), BGP, and BGP-LS. SDN controllers typically will collect and correlate this discovery and topology information, and make it available via RESTful APIs to SDN applications. One of the major advantages of running SDN applications on top of a controller's API is that the difficult but indispensable task of discovering topological information about a network is done by the controller on which your application is running.

As shown in Fig. 12.2, in an OpenFlow context the last flow entry will typically be to DROP unmatched packets. This is because proactive applications attempt to *anticipate* all traffic and program flow entries accordingly. As a consequence, packets which do not match the configured set of rules are discarded. We remind the reader that the match criteria for flow entries can be programmed such that most arriving packet types are expected and match some entry before this final DROP entry. If this were not the case, the purely proactive model could become an expensive exercise in dropping packets!

As mentioned earlier, there will be hybrid reactive-proactive applications which utilize a language such as Java for listeners and communicate via some external means to the proactive part of the application. The proactive component would then program new flow entries on the switch. As applications become more complex, it is likely that this hybrid model will become more common. The comment made above about arriving packets in a proactive application matching some flow entry before the final DROP entry is also germane to the case of hybrid applications. The proactive part of such hybrid applications will have programmed higher-priority flow entries so that only a reasonable number of packets match the final DROP entry. In such a hybrid model, the final entry will send all unmatched packets to the controller and so it is important to limit this to a rate that the controller can handle.

12.3.5 DETAILS: INTERNAL SDN APPLICATIONS

Internal SDN applications on the currently dominant controllers are all Java-based and share the following traits:

- *OSGi*: Internal applications on Java-based controllers will need to run in some type of Java environment supporting multiple modules, and OSGi is the most common framework. Apache Karaf [2] is the most common OSGi implementation for SDN controllers today.
- *Reactive application support*: Controllers providing capabilities today (e.g., Floodlight, HP, ODL, and ONOS) provide the ability to create *packet listeners* for receiving packets forwarded from switches.
- *Use of archetypes*: Considering the complexity of the predominant controllers' internal design, application development would be challenging without a software project management system tailored to such environments. Maven [3] is such a system, and is used for controller application development within ODL, ONOS, and the HP VAN controller. Maven includes the *archetype* [4] functionality to build a template project in order to bootstrap an application development effort.

Most current controllers allowing internal SDN applications share the same conceptual foundation as their common ancestor, Beacon. The OpenFlow APIs provided with these controllers have evolved to support newer Java idioms and current versions of OpenFlow. Examples of controllers falling into this category include Floodlight, the HP VAN SDN Controller, and ONOS. Even the original version of ODL, Hydrogen, followed this pattern. However, with the releases of Helium and Lithium, ODL embarked on the new MD-SAL internal architecture.

MD-SAL presents a new architecture for creating SDN applications and internal modules. From a software development point of view, the main aspects of MD-SAL are as follows:

- *YANG*: ODL has created the requirement that every MD-SAL application must define itself using a YANG model. We described YANG models in Section 7.2.2. Unlike the use of YANG by the

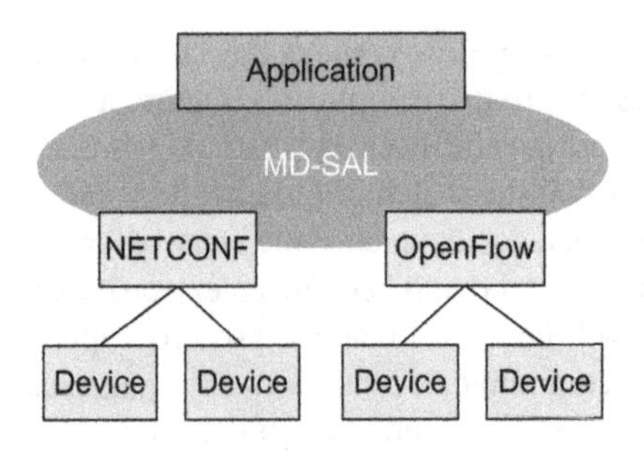

FIG. 12.3

MD-SAL overview.

NETCONF protocol, with MD-SAL YANG is used to define the application's data, for *remote procedure calls* (RPCs), and for notifications. Therefore, the first step in the process of application development is the definition of the relevant YANG model.

- *Model-based network device communication*: ODL incents application developers to interact with *models* of network devices rather than directly with the network devices themselves. This level of abstraction is intended to allow the controller APIs to be independent of the protocols beneath them. Fig. 12.3 shows a single application using APIs allowing communication via either NETCONF or OpenFlow, because it is communicating with the network device model, rather than the network device itself.
- *Standard APIs*: MD-SAL enforces a particular style of communication, using either an auto-generated RESTful API for the application, or through the use of a common YANG data tree. The RESTful APIs are created as part of the standard MD-SAL build process, and hence are common and open to other applications. The YANG data tree holds all of the information for the entire controller, and application data resides in one branch of that tree. Through this consistent use of YANG for all applications running in the controller, applications can access and make modifications to other applications' public data, providing a common and secure means of interapplication communication.

Because of the complexity of the MD-SAL environment, building internal applications on ODL confronts the developer with a relatively steep learning curve. More information on MD-SAL is available in [5,6]. It is important to consider the constraints and benefits presented previously when choosing between the internal and external application paradigms in ODL.

12.3.6 DETAILS: EXTERNAL SDN APPLICATIONS

External SDN applications are the simplest to create, since (a) they can be written in any programming language, (b) they can run in many places in the network, (c) they can use many types of southbound protocol plugins (e.g., OpenFlow, NETCONF, BGP-LS/PCE-P), and (d) they use simple RESTful APIs provided by the controller. External applications can be created more expeditiously than their internal counterparts and are useful for prototyping as well as for production use. There are no strict rules to follow regarding the design of an external SDN application. The application developer is entirely in control of the application's execution environment. The developer may choose to instantiate the application on the controller, on a remote system or even in the cloud. Hence, there are no detailed guidelines for external SDN application development. For these reasons, the external application paradigm has surpassed the internal model as a means of creating SDN solutions.

DISCUSSION QUESTION

In the beginning years of SDN, internal reactive applications attracted the most attention. Why do you think this is? Do you think that these types of applications are still the most important or the most interesting?

12.4 A BRIEF HISTORY OF SDN CONTROLLERS

The Beacon controller is truly a seminal controller in that much of the basic OpenFlow controller code in Floodlight and OpenDaylight was derived directly from Beacon. As we pointed out in Section 11.1.1, Beacon itself was based on early work performed at Stanford by David Erickson and Rob Sherwood. Both Beacon and Floodlight are based on OpenFlowJ for the core Java OpenFlow implementation. Floodlight is maintained by engineers from BigSwitch. While the distributions of Beacon and Floodlight are now distinct and different applications are packaged with each, Beacon remains close enough to Floodlight that we will not elaborate on it further at the detailed implementation level presented here.

We should point out that there are a number of other OpenFlow controllers available. For example, the NOX controller has been used in a considerable number of research projects. We provide basic information about a number of open source controller alternatives in Section 13.8. As of this writing, two controllers are predominant, OpenDaylight, and ONOS. We will explore these controllers in more detail in Section 12.7.

12.5 USING FLOODLIGHT FOR TRAINING PURPOSES

The application examples we provide in this chapter use the Floodlight controller. We have chosen Floodlight because it has historically been one of the most popular open source SDN controllers. Floodlight is not as popular today as it was in the earlier days of SDN, but its APIs are similar enough in nature to other Java OpenFlow APIs as to be useful as a training tool. It is possible to download the source code [7] and examine the details of the controller itself. Floodlight also comes with a number of sample applications, such as a learning switch and a load balancer, which provide excellent additional examples for the interested reader. The core modules of the Floodlight controller come from the seminal Beacon controller [8]. These components are denoted as *org.openflow* packages within the Floodlight source code. Since Beacon is the basis of the OpenFlow functionality present in a number of SDN controllers, the OpenFlow-related controller interaction described in the following sections would apply to those Beacon-derived controllers as well. With respect to ODL, while early versions incorporated much of the Beacon OpenFlow implementation, subsequent versions have evolved in order to implement OpenFlow as an MD-SAL module. For those later versions of ODL, the following sections would not directly apply.

In Section 12.10 we provide an example of proactive applications which use the RESTful *flow pusher* APIs. For this and other examples, remember that our use of *flows* does not restrict these designs to OpenFlow. As we explained in Section 12.1 we use the term flow in a generic sense, referring to the forwarding behavior of network devices, whether controlled by OpenFlow, NETCONF, PCE-P, or any other suitable protocol or protocol plugin.

In our first example, however, we will look in detail at the source code involved in writing a reactive SDN application. We will use this example as an opportunity to walk through SDN application source code in detail and explore how to implement the tight application-controller coupling that is characteristic of reactive applications.

12.6 A SIMPLE REACTIVE JAVA APPLICATION

In the common vernacular, a *blacklist* is *a list of people, products, or locations viewed with suspicion or disapproval.* In computer networking a blacklist is a list of hostnames or IP addresses, which are known or suspected to be malicious or undesirable in some way. Companies may have blacklists, which protect their employees from accessing dangerous sites. A school may have blacklists to protect children from attempting to visit undesirable web pages.

As implemented today, most blacklist products sit in the data path, examining all traffic at choke points in the network, looking for blacklisted hostnames or IP addresses. This clearly has disadvantages in terms of the latency it introduces to traffic as it all passes through that one blacklist appliance. There are also issues of scalability and performance if the number of packets or the size of the network is quite large. Cost would also be an issue since provisioning specialized equipment at choke points adds significant network equipment expense.

It would be preferable to examine traffic at the edge of the network, where the users connect. But putting blacklist-supporting appliances at every edge switch is impractical from a maintenance and cost perspective. A simple and cost-effective alternative is to use OpenFlow-supporting edge switches or wireless access points (APs), with a blacklist application running above the OpenFlow controller. In our simple application, we will examine both hostnames and IP addresses. We describe the application in the following section.

12.6.1 BLACKLISTING HOSTNAMES

We introduced an SDN solution to the blacklist problem in Section 9.3.3. We provide further details of that solution here. The first part of the blacklist application deals with hostnames. Checking for malicious or undesirable hostnames works as follows:

- A single default flow is set on the edge switches to forward all DNS traffic to the OpenFlow controller. (Note that this is an example of a proactive rule. Thus, while blacklist is primarily a reactive application, it does exhibit the hybrid reactive-proactive characteristic discussed in Section 12.3.4.)
- The blacklist application running on the controller listens for incoming DNS requests which have been forwarded to the controller.
- When a DNS request is received by the blacklist application it is parsed to extract the hostnames.

- The hostnames are compared against a database of known malicious or undesirable hosts.
- If any hostname in the DNS request is found to be bad, the switch is instructed to drop the packet.
- If all hostnames in the DNS request are deemed to be safe, then the switch is instructed to forward the DNS request normally.

In this way, with one simple flow entry in the edge switches and one simple SDN application on the controller, end users are protected from intentionally or inadvertently attempting to access bad hosts and malicious websites.

12.6.2 BLACKLISTING IP ADDRESSES

As described earlier in Section 9.3.3, websites sometimes have embedded IP addresses, or end users will attempt to access undesirable locations by providing IP-specific addresses directly, rather than a hostname, thus circumventing a DNS lookup. Our simple blacklist application works as follows:

- A single default flow is set on the edge switches, at a low priority, to forward all IP traffic that does not match other flows to the OpenFlow controller.
- The blacklist application listens for IP packets coming to the controller from the edge switches.
- When an IP packet is received by the blacklist application, the destination IP address is compared against a database of known malicious or undesirable IP addresses.
- If the destination IP address is found to be bad, the switch is instructed to drop the packet.
- If the destination IP address in the IP packet is deemed to be safe, then the switch or AP is instructed to forward the IP packet normally, and a higher-priority flow entry is placed in the switch which will explicitly allow IP packets destined for that IP address.
- The IP destination address flow entry was programmed with an idle-timeout which will cause it to be removed from the flow table after some amount of inactivity.

Note that a brute force approach to the blacklist problem would be to explicitly program rules for all known bad IP addresses such that any packet matching those addresses is directly dropped by the switch. While this works in principle, it is not realistic as the number of blacklisted IP addresses is potentially very large, and as we have pointed out, the number of available flow entries, while varying from switch to switch, is limited. This is exacerbated by the fact the flow entries for the blacklisted IP addresses generally should not expire, so they more or less permanently occupy valuable flow table space.

There are three major components to this application:

- *Listeners*. The module that listens for events from the controller.
- *Packet handler*. The module that handles incoming packet events and decides what to do with the received packet.
- *Flow manager*. The module that sets flows on the switches in response to the incoming packet.

Fig. 12.4 provides a more detailed view of the blacklist application introduced in Section 9.3.3. As a reactive application, Blacklist first programs the switch to forward unmatched packets to the controller. The figure depicts three listeners for network events: the switch listener, the device (end user node) listener, and the message listener. We show the device listener as it is one of the three basic

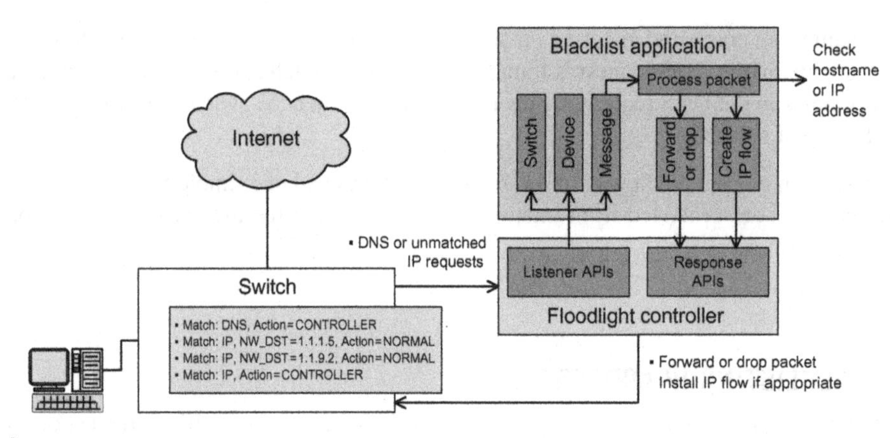

FIG. 12.4

Blacklist application design.

classes of listeners for reactive applications but as it has no special role in Blacklist we do not mention it further here. The *Message Listener* will forward incoming packets to the *Process Packet* module, which processes the packet and takes the appropriate action. Such actions include forwarding or dropping the packet and programming flow entries in the switch. We discuss the Message Listener and the Switch Listener in greater detail in Section 12.6.3.

The flow entries depicted in the switch in Fig. 12.4 show that DNS requests are always forwarded to the controller. If the destination IP address returned by DNS has already been checked, the application will program the appropriate high-priority flow entries. In our example, such entries are shown as the two flow entries matching 1.1.1.5 and 1.1.9.2. The reader should understand that the flows for destination 1.1.1.5 and 1.1.9.2 were programmed earlier by the controller when packets for these two destinations were first processed. If the IP address on the incoming packet has not been seen earlier or has aged out, then there will be no flow entry for that address and the IP request will be forwarded to the controller for blacklist processing.

Source code for the Listener, Packet Handler, and Flow Manager components is provided in Appendix B. Code snippets will also be provided throughout the following analysis. The reader should refer to the appendix to place these snippets in context. We include in the appendix the SDN-relevant source code for Blacklist in part to illustrate that a nontrivial network security application can be written in just 15 pages of Java, evidence of the relative ease and power of developing applications in the Open SDN model.

We now delve more deeply into the components of Blacklist.

12.6.3 BLACKLIST: LISTENERS

Our blacklist application uses two important listeners. The first is the *SwitchListener*. The *Switch-Listener* receives notifications from the controller whenever a switch is discovered or when it has disconnected. The *SwitchListener* also receives notifications when the switch's port configuration has changed (e.g., when ports have been added or removed). The exact frequency of such notifications depends on the particular switch's environment. Once steady-state operation has been achieved, however, it is unlikely that these notifications will occur more often than once every few seconds. The two most common SDN switch-type entities in this example are wired switches and APs.

When the application first learns that a switch has been discovered, the application programs it with two default rules, one to forward DNS requests to the controller and the other a low-priority flow entry to forward IP packets to the controller. For nonblacklisted IP addresses, this low-priority IP flow entry will be matched the first time that IP destination is seen. Subsequently, a higher priority flow is added that allows packets matching that description to be forwarded normally without involving the controller. If that flow entry idle timer expires, the flow entry is removed and the next packet sent to that IP address will once again be forwarded to the controller, starting this process anew.

Examination of the actual source code in Appendix B reveals that we add one other default flow for ARP requests, which are forwarded normally. ARP requests pose no threat from the perspective of the blacklist application. So, by default, all new switches will have three flows: DNS (send to controller), IP (send to controller), and ARP (forward NORMAL).

The other important listener is the *MessageListener*. The most important of the various methods in this listener is the *receive* method. The receive method is called whenever a packet arrives at the controller. The method then passes the packet to the blacklist SDN application. When the *PACKET_IN* message is received, our application invokes the *PacketHandler* to process the packet. In Section 5.3.5 we described how the switch uses the *PACKET_IN* message to communicate to the controller.

The code for the *MessageListener* module is listed in Appendix B.1. We examine part of that module below. First, we see that our class is derived from the Floodlight interface called *IOFMessageListener*:

```
//- - - - - - - - - - - - - - - - - - - - - - - - - - - - -
public class MessageListener implements IOFMessageListener
//- - - - - - - - - - - - - - - - - - - - - - - - - - - - -
```

Next is the registration of our *MessageListener* method:

```
//- - - - - - - - - - - - - - - - - - - - - - - - - - - - - - -
public void startUp()
{
    // Register class as MessageListener for PACKET_IN messages.
    mProvider.addOFMessageListener( OFType.PACKET_IN, this );
}
//- - - - - - - - - - - - - - - - - - - - - - - - - - - - - - -
```

This code makes a call into Floodlight (*mProvider*), invoking the method *addOFMessageListener* and telling it we wish to listen for events of type *PACKET_IN*.

The next piece of code to consider is our *receive* method, which we provide to Floodlight to call when a packet is received by the listener. We are passed a reference to the switch, the message itself, and the Floodlight context reference.

```
//- - - - - - - - - - - - - - - - - - - - - - - - - - - - - -
@Override
public Command receive( final IOFSwitch         ofSwitch,
                        final OFMessage         msg,
                        final FloodlightContext context )
{
    switch( msg.getType() )
    {
    case PACKET_IN:  // Handle incoming packets here

        // Create packethandler object for receiving packet in
        PacketHandler ph = new PacketHandler( ofSwitch,
                                              msg, context);

        // Invoke processPacket() method of our packet handler
        // and return the value returned to us by processPacket
        return ph.processPacket();

    default: break;  // If not a PACKET_IN, just return

    }

    return Command.CONTINUE;
}
//- - - - - - - - - - - - - - - - - - - - - - - - - - - - - -
```

In the code mentioned previously, we see a Java *switch* statement based on the message type. We are only interested in messages of type *PACKET_IN*. Upon receipt of that message, we create a *PacketHandler* object to process this incoming packet. We then invoke a method within that new object to process the packet.

12.6.4 BLACKLIST: PACKET HANDLERS

In our application there is one *PacketHandler* class which handles all types of incoming packets. Recall that our flow entries on the switches have been set up to forward DNS requests and unmatched IP packets to the controller. The *processPacket* method of the PacketHandler shown in Fig. 12.4 is called to handle these incoming packets, where it performs the following steps:

1. *IP packet*. Examine the destination IP address and, if it is on the blacklist, drop the packet. *Note that we do not attempt to explicitly program a DROP flow for this bad address. If another attempt is made to access this blacklisted address, the packet will be diverted to the controller again. This is a conscious design decision on our part, though there may be arguments that support the approach of programming blacklisted flows explicitly.*
2. *DNS request*. If the IP packet is a DNS request, examine the hostnames, and if any are on the blacklist, drop the packet.
3. *Forward packet*. If the packet has not been dropped, forward it.

4. *Create IP flow*. If it is not a DNS request, create a flow for this destination IP address. Note that there is no need to set a flow for the destination IP address of the DNS server because all DNS requests are caught by the higher-priority DNS flow and sent to the controller.

In our blacklist application, the *Create IP flow* step mentioned previously will actually invoke a method in the *FlowMgr* class which has a method for creating the Floodlight OpenFlow objects that are required to program flow entries on the switches.

The source code for the *PacketHandler* class can be found in Appendix B.2. We describe some of the more important portions of that code in the following paragraphs.

PacketHandler is a *Plain Old Java Object* (POJO), which means that it does not inherit from any Floodlight superclass. The object's constructor stores the objects passed to it, including: (1) references to the switch that sent the packet, (2) the message itself (the packet data), and (3) the floodlight context, which is necessary for telling Floodlight what to do with the packet once we have finished processing it.

The first piece of code is the beginning of the *processPacket* method, which was called by the *MessageListener*.

```
//- - - - - - - - - - - - - - - - - - - - - - - - - - - - -
public Command processPacket()
{
    // First, get the OFMatch object from the incoming packet
    final OFMatch ofMatch = new OFMatch();
    ofMatch.loadFromPacket( mPacketIn.getPacketData(),
                            mPacketIn.getInPort()     );

//- - - - - - - - - - - - - - - - - - - - - - - - - - - - -
```

This is our first look at the *OFMatch* object provided by Floodlight, which derives from the same OpenFlow Java source code that can be found in Beacon and, thus, in OpenDaylight. In the code mentioned previously, we create an *OFMatch* object and then we load the object from the packet that we have received. This allows us to use our own *ofMatch* variable to examine the header portion of the packet we have just received.

In different parts of *processPacket* we look at various portions of the incoming packet header. The first example uses *ofMatch.getNetworkDestination()* to get the destination IP address, which is used to compare against our IP blacklist.

```
//- - - - - - - - - - - - - - - - - - - - - - - - - - - - -
IPv4.toIPv4AddressBytes( ofMatch.getNetworkDestination() ) );
//- - - - - - - - - - - - - - - - - - - - - - - - - - - - -
```

The next example determines whether this is a DNS request:

```
//- - - - - - - - - - - - - - - - - - - - - - - - - - - - - -
if( ofMatch.getNetworkProtocol()      == IPv4.PROTOCOL_UDP &&
    ofMatch.getTransportDestination() == DNS_QUERY_DEST_PORT )
//- - - - - - - - - - - - - - - - - - - - - - - - - - - - - -
```

Much of the other code in the module in Appendix B.2 deals with doing the work of the blacklist application. This includes checking the destination IP address, parsing the DNS request payload and checking those names as well. The most relevant SDN-related code is the creation of an action list to hand to the controller, which it will use to instruct the switch as to what to do with the packet that was received. In *forwardPacket()* we create the action list with the following code:

```
//- - - - - - - - - - - - - - - - - - - - - - - - - - - - - - -
final List<OFAction> actions = new ArrayList<OFAction>();
actions.add( new OFActionOutput( outputPort ) );
//- - - - - - - - - - - - - - - - - - - - - - - - - - - - - - -
```

This code creates an empty list of actions, then adds one action to the list. That action is a Floodlight object called *OFActionOutput*, which instructs the switch to send the packet out the given port.

The *FlowManager* object is responsible for the actual interaction with Floodlight to perform both the sending of the packet out of the port as well as the possible creation of the new flow entry. We describe this object in the following section.

12.6.5 BLACKLIST: FLOW MANAGEMENT

The *FlowMgr* class deals with the details of the OpenFlow classes that are part of the Floodlight controller. These classes simplify the process of passing OpenFlow details to the controller when our application sends responses to the OpenFlow switches. These classes make it possible to use simpler Java class structures for communicating information to and from the controller. However, there is considerable detail that must be accounted for in these APIs, especially compared to their RESTful counterparts.

We now examine some of the more important Java objects that are used to convey information about flows to and from the controller:

- *IOFSwitch*. This interface represents an OpenFlow switch and, as such, it contains information about the switch, such as ports, port names, IP addresses, etc. The *IOFSwitch* object is passed to the *MessageListener* when a packet is received and it is used to invoke the *write* method when an OpenFlow packet is sent back to the switch in response to the request that was received.
- *OFMatch*. This object is part of the *PACKET_IN* message that is received from the controller that was originated by the switch. It is also used in the response that the application will send to the switch via the controller to establish a new flow entry. When used with the *PACKET_IN* message, the match fields represent the header fields of the incoming packet. When used with a reply to program a flow entry, they represent the fields to be matched and may include wildcards.
- *OFPacketIn and OFPacketOut*. These objects hold the actual packets received and sent, including payloads.
- *OFAction*. This object is used to send actions to the switch, such as NORMAL forwarding or FLOOD. The switch is passed a list of these actions so that multiple items such as *modify destination network address* and then NORMAL forwarding may be concatenated.

Code for the *FlowMgr* class can be found in Appendix B.3. We will walk the reader through some important logic from *FlowMgr* using the following code snippets. One of the methods in *FlowMgr* is *setDefaultFlows* which is called from the *SwitchListener* class whenever a new switch is discovered by Floodlight.

```
//- - - - - - - - - - - - - - - - - - - - - - - - - -
public void setDefaultFlows(final IOFSwitch ofSwitch)
{
    // Set the intitial 'static' or 'proactive' flows
    setDnsQueryFlow( ofSwitch );
    setIpFlow( ofSwitch );
    setArpFlow( ofSwitch );
}
//- - - - - - - - - - - - - - - - - - - - - - - - - -
```

As an example of these internal methods, consider the code to set the DNS default flow. In the following code, we see *OFMatch* and *OFActionOutput* that were introduced in Section 12.6.4. The *OFMatch* match object is set to match packets which are Ethernet, UDP, and destined for the DNS UDP port.

```
//- - - - - - - - - - - - - - - - - - - - - - - - - - - - - - - -
private void setDnsQueryFlow( final IOFSwitch ofSwitch )
{
    // Create match object to only match DNS requests
    OFMatch ofMatch = new OFMatch();
    ofMatch.setWildcards( allExclude( OFPFW_TP_DST,
                                      OFPFW_NW_PROTO,
                                      OFPFW_DL_TYPE ) )
           .setDataLayerType( Ethernet.TYPE_IPv4 )
           .setNetworkProtocol( IPv4.PROTOCOL_UDP )
           .setTransportDestination( DNS_QUERY_DEST_PORT );

    // Create output action to forward to controller.
    OFActionOutput ofAction  = new OFActionOutput(
                             OFPort.OFPP_CONTROLLER.getValue(),
                             (short) 65535                   );

    // Create our action list and add this action to it
    List<OFAction> ofActions = new ArrayList<OFAction>();
    ofActions.add(ofAction);

    sendFlowModMessage( ofSwitch,
                        OFFlowMod.OFPFC_ADD,
                        ofMatch,
                        ofActions,
                        PRIORITY_DNS_PACKETS,
                        NO_IDLE_TIMEOUT,
                        BUFFER_ID_NONE );
}
//- - - - - - - - - - - - - - - - - - - - - - - - - - - - - - - -
```

The flow we are setting instructs the switch to forward the packet to the controller by setting the output port to *OFPort.OFPP_CONTROLLER.getValue()*. Then we call the internal method *sendFlowModMessage*. The code for that method is shown as follows:

```
//- - - - - - - - - - - - - - - - - - - - - - - - - - - - - - - - - -
private void sendFlowModMessage( final IOFSwitch      ofSwitch,
                                 final short          command,
                                 final OFMatch        ofMatch,
                                 final List<OFAction> actions,
                                 final short          priority,
                                 final short          idleTimeout,
                                 final int            bufferId )
{
    // Get a flow modification message from factory.
    final OFFlowMod ofm = (OFFlowMod) mProvider
                                 .getOFMessageFactory()
                                 .getMessage(OFType.FLOW_MOD);

    // Set our new flow mod object with the values that have
    // been passed to us.
    ofm.setCommand( command ).setIdleTimeout( idleTimeout )
                             .setPriority( priority )
                             .setMatch( ofMatch.clone() )
                             .setBufferId( bufferId )
                             .setOutPort( OFPort.OFPP_NONE )
                             .setActions( actions )
                             .setXid( ofSwitch
                                .getNextTransactionId() );

    // Calculate the length of the request, and set it.
    int actionsLength = 0;
    for( final OFAction action : actions )
       { actionsLength += action.getLengthU(); }
    ofm.setLengthU(OFFlowMod.MINIMUM_LENGTH + actionsLength);

    // Now send out the flow mod message we have created.
    try
    {
        ofSwitch.write( ofm, null );
        ofSwitch.flush();
    }
    catch (final IOException e)
    {
        // Handle errors with the request
    }
}
//- - - - - - - - - - - - - - - - - - - - - - - - - - - - - - - - - -
```

Items to note in the code mentioned previously are as follows:

- The *OFFlowMod* object is created and used to hold the information regarding how to construct the flow, including items such as idle timeout, priority, match field, and actions.

- The length must still be calculated by this method. The length could be provided by the OpenFlow package, but as of this writing it is not.
- The switch is sent the flow modification message using the *write* method.

There is additional code available for the reader to examine in Appendix B.3. For example, consider the FlowMgr methods *sendPacketOut* and *createDataStreamFlow* for more examples of interacting with the Floodlight controller.

The next section looks at the two dominant controllers available today and examines the APIs that they make available to the SDN application developer.

DISCUSSION QUESTION

We have described an internal, reactive application, in some detail. Does the level of software development expertise required seem overwhelming or reasonable? What part of the blacklist application is the most difficult to understand, and why?

12.7 CONTROLLER CONSIDERATIONS

In the earlier days of SDN, controllers were created for open application development by researchers (e.g., Beacon at Stanford University), by enterprises (e.g., Ryu by NTT Communications), and by network vendors (e.g., XNC by Cisco, VAN SDN Controller by HP, Trema by NEC). As the SDN technology has evolved and matured, these controllers still exist but have less momentum. As evidenced by various open conferences on SDN such as the Open Networking Summit, the emphases for SDN application development have begun to focus on two controllers: ODL and ONOS. In this section we consider each of these controllers and their suitability for SDN application development.

12.7.1 OpenDaylight

With respect to the number of applications available on an SDN controller, ODL currently holds the lead over ONOS, as it has a longer history. As such, it has a head start on attracting implementation efforts by vendors (e.g., Cisco [9], Brocade [10]) and customers (e.g., AT&T [11], Fujitsu [12]) alike. From a development point of view, the main arguments for choosing ODL are as follows.

- *Multiprotocol*: You should consider ODL if your interests lie in creating SDN solutions based on protocols such as NETCONF, BGP, MPLS, or other existing protocols. ODL has the lead in supporting these protocols, and the key backers of ODL such as Cisco have a vested interest in creating SDN solutions that work with them. ODL also provides a platform for integrating your own protocol into the ODL environment, opening the door for applications written by customers to utilize ODL to control your network devices using your device-specific protocol. ONOS has projects working on existing protocol solutions, but is behind ODL in this area.
- *Minimal risk*: You should consider ODL if you are planning on creating SDN solutions which cause a minimum of disruption to the status quo of existing networks, as ODL helps pave the way for slower, less risky evolution toward an eventual SDN network.
- *Controller confidence*: ODL is a Linux Foundation project, is backed by a host of networking vendors, and the number of customers creating SDN solutions on ODL is growing, making it a

confident choice for the future. There is little risk that ODL is going to cease to exist, which is not true for many of the controllers cited in Section 12.4 that are no longer garnering investment of time and money. In addition, there is a very large community of developers of code, both for ODL itself and for applications running on the controller. This promotes added stability and helps guarantee the longevity of the controller.

ODL offers both RESTful APIs for creating external applications as well as the MD-SAL environment for creating internal applications. External, RESTful applications require limited software development expertise and are easy to develop. Internal, MD-SAL applications require significant Java software development expertise and present a fairly steep learning curve. Thus caution is called for before opting for the internal application model. The benefits of creating MD-SAL applications may, however, outweigh these considerations.

12.7.2 ONOS

ONOS is a relative newcomer, having first been released in late 2014, but it has gained traction with service providers such as AT&T [13] and NEMs like Ciena [14] and Fujitsu [15]. As such, ONOS warrants consideration by prospective SDN application developers.

- *OpenFlow*: An SDN network engineer should consider ONOS if his/her interests lie primarily in creating OpenFlow-based applications, as this is the focus of ONOS. ONOS is a project emanating from the research communities at Stanford and other universities, with more of a focus on OpenFlow-based experimentation and development.
- *Service provider*: ONOS is explicitly targeted at service providers, and has gone to lengths to recruit them to join the ONOS project.
- *Intents*: For the developer interested in creating applications which reside above a layer of intents-based abstraction, ONOS has been built with this type of API in mind and provides a rich environment for applications that operate at this level.

ONOS offers both RESTful APIs for creating external applications, as well as Java APIs for internal application development. Both the RESTful APIs and Java APIs follow the general models that have evolved alongside with the Beacon, Floodlight, and HP VAN SDN controllers, and their derivatives. SDN developers familiar with those APIs will find similarities with the ONOS APIs. However, ONOS additionally provides intents-based APIs, which are new to SDN in general, and will require some amount of adjustment to this new, albeit improved, model.

DISCUSSION QUESTION

We have discussed different types of SDN applications, and two significant controllers, ODL and ONOS. Which controller would be best suited for which type of application? Why might you be inclined to use one versus the other?

12.8 NETWORK DEVICE CONSIDERATIONS

One of the most important considerations when building an SDN application is the nature of the network devices that are to be controlled by the application. These considerations vary depending on your intended type of SDN application, as well as whether yours is an OpenFlow versus a non-OpenFlow-based strategy. We examine these in turn.

12.8.1 OpenFlow DEVICE CONSIDERATIONS

Creating applications for network devices using OpenFlow requires an understanding of the capabilities of the devices you wish to control. Some of the considerations in this area are:

- Do all the switches support OpenFlow? What version of OpenFlow do they claim to support? How much of that version do they in fact support?
- Is there a mix of OpenFlow and non-OpenFlow switches? If so, will some non-OpenFlow mechanism be used to provide the best simulation of an SDN environment? For example, do the non-OpenFlow switches support some type of SDN API which allows a controller to configure flows in the switch?
- For OpenFlow-supporting switches, are they hardware or software switches or both? If both, what differences in behavior and support will need to be considered?
- For hardware OpenFlow switches, what are their hardware flow-table sizes? If the switches are capable of handling flow table overflow in software, at what point are they forced to begin processing flows in software, and what is the performance impact of doing so? Is it even possible to process flows in software? What feature limitations are introduced because of the hardware?
- What is the mechanism by which the switch-controller communications channel is secured? Is this the responsibility of the network programmer?

Hardware switches vary in their flow capacity, feature limitations and performance, and these should be considerations when creating an SDN application. The major issue regarding flow capacity is how many flow entries the hardware ASIC is capable of holding. Some switches support less than 1000 flow entries, while some, such as the NEC PF5240 [16], support greater than 150,000. Note that for some switches the maximum number of flows depends on the nature of the flows supported. For example, the NEC PF5820 switch [17] can support 80,000 layer 2-only flows but only 750 full 12-tuple flows. (We described matching against the basic 12-tuple of input fields in Section 5.3.3.) A layer 2 flow entry only requires matching on the MAC address, so the amount of logic and memory occupied per flow entry is much less. Some applications, such as a traffic prioritization application or a TCP-port-specific application, may use relatively few flow table entries. Others, such as an access control application or a per-user application, will be much more flow-entry hungry. The topological placement of the network device may have a bearing on the maximum number of flow entries as well. For example, an access control application being applied in a switch or AP at the user edge of the network will require far fewer flow entries per network device than the same application controlling an upstream or core switch.

Hardware implementations often have feature limitations because, when the OpenFlow specification is mapped to the capabilities of the hardware, generally some features cannot be supported. For example, some ASICs are not able to perform NORMAL forwarding on the same packet that has had its header modified in some way. While it is true that the list of features supported by the network device can be learned via OpenFlow *Features Request* and *Features Reply* messages (see Section 5.3.5), this does not tell you the specific nuances and limitations that may be present in any one switch's implementation of each feature. Unfortunately, this must often be discovered via trial and error. Information about these issues may be available on the Internet sites of interested SDN developers. The application developer must be aware of these types of limitations before designing the solution.

Some hardware switches will implement some functions in hardware and others in software. It is wise to understand this distinction so that an application does not cause unexpected performance degradation by unknowingly manipulating flows such that the switch processes packets in software.

In general software implementations of OpenFlow are able to implement the full set of functionality defined by the OpenFlow specification for which they claim support. They implement flow tables in standard computer memory. Thus the feature limitations and table size issues described previously are not a major concern for application developers. However, performance will still be a consideration, since implementations in hardware will generally be faster.

12.8.2 NON-OpenFlow DEVICE CONSIDERATIONS

Creating applications for network devices using NETCONF or BGP-LS/PCE-P requires knowledge of the devices you wish to control. We discuss considerations for each of these protocols as follows.

NETCONF

The NETCONF protocol is standard and devices supporting it may be compatible. YANG also is a standard, and should be consistent among vendors and across product families. There are, however, potential pitfalls:

- *NETCONF without YANG*: There are devices which support an earlier version of NETCONF that does not support YANG models. These will likely be proprietary implementations and the data models for each must be understood in order to communicate successfully with each device. This is a particular challenge for service providers, as they tended to be early adopters of NETCONF and hence are forced to contend with pre-YANG versions of the protocol.
- *YANG models*: The critical issue regarding NETCONF/YANG control of network devices is the definition of the data models for individual functional areas such as routing, policy, and security. Today these models vary between vendors, and even vary among different products from a single vendor. It should be noted, however, that standardization efforts are underway in bodies such as the IETF for creating common YANG models for interfaces, routing, and other functional aspects of networking devices.

BGP-LS/PCE-P

The BGP-LS/PCE-P project in ODL includes plugins for BGP (which includes BGP-LS and BGP-FS) and PCE-P. OSPF or IS-IS topology information is received by the BGP plugin via the BGP-LS protocol. PCE-P is then employed to set the desired MPLS LSPs for routing traffic through the network. ODL supports other protocols as well. Some examples include the *Locator/ID Separation Protocol* (LISP), SNMP, and the *Interface to the Routing System* (I2RS).

We now will provide a number of high-level designs for some additional simple applications relevant to several different environments. These include the data center, the campus, and service provider environments. We restrict ourselves to high-level design issues in the following sections. It is our hope that the reader will be able to extend the detailed source code analysis performed for the blacklist application in Section 12.6 and apply those learnings to the cases we cover next.

12.9 CREATING NETWORK VIRTUALIZATION TUNNELS

As explained in Section 8.3, data center multitenancy, MAC-table overflow, and VLAN exhaustion are being addressed with new technology making use of tunnels, the most common being the MAC-in-IP tunnels provided by protocols such as VXLAN, NVGRE, and STT. Fig. 12.5 shows a possible software application design for creating and managing overlay tunnels in such an environment. We see in the figure the switch, device, and message listeners commonly associated with reactive applications. This reactive application runs on the controller and listens for IP packets destined for new IP destination addresses for which no flow entries exist. When the message listener receives such an unmatched packet, the application's tunnel manager is consulted to determine which endpoint switch is responsible for this destination host's IP address.

The information about the hosts and their tunnel endpoints is maintained by the application in the associated host and tunnel databases. The information itself can be gathered either outside of the domain of OpenFlow or it could be generated by observing source IP addresses at every switch. Whichever way this is done, the information needs to be available to the application.

When a request is processed, the tunnel information is retrieved from the database and, if the tunnel does not already exist, it will be created. Note that there may be other hosts communicating between these same two switches acting as tunnel endpoints, and so there may be a tunnel already in existence. The tunnel creation can be done in a number of ways, all of which are outside the scope of the OpenFlow specification. The tunnel creation is performed on the OpenFlow device and a mapping is established between that tunnel and a virtual port, which is how tunnels are represented in flow entries. These methods are often proprietary and specific to a particular vendor. For the reader interested in configuring VXLAN or NVGRE tunnels on the OVS switch, a useful guide can be found in [18].

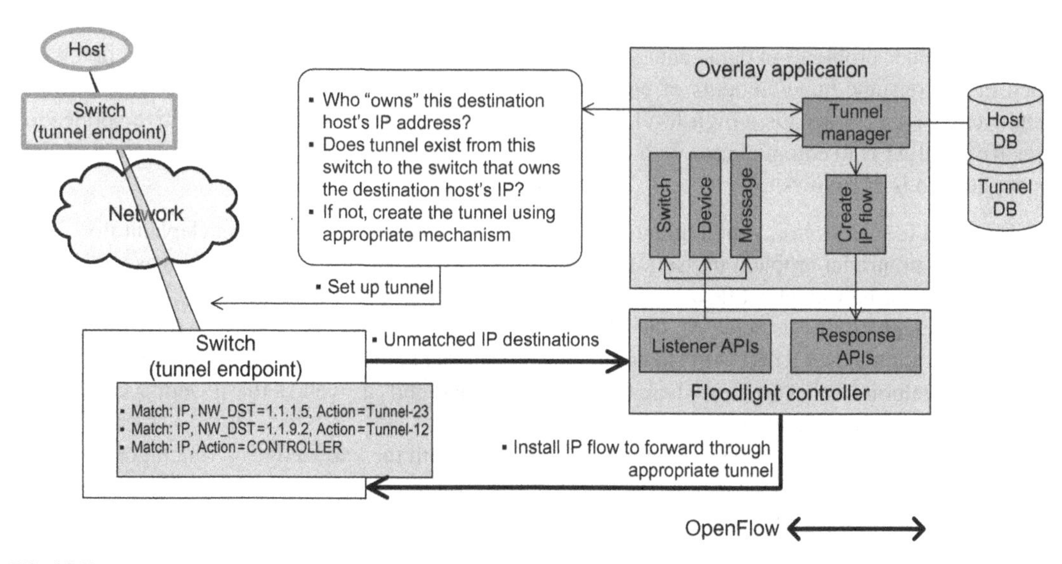

FIG. 12.5

Tunnels application design.

Once the tunnel is either created or identified, the tunnel application can set up flows on the local switch to direct traffic heading toward that destination IP address into the appropriate tunnel. Once the virtual port exists, it is straightforward to create flow entries such as those shown in the figure where we see the tunneled packets directed to virtual ports TUNNEL-23 and TUNNEL-12. The complementary flow will need to be set on the switch hosting the remote tunnel endpoint as well.

This design attempts to minimize tunnel creation by only creating tunnels as needed, and removing them after they become idle. A different, proactive approach would be to create the tunnels a priori, obviating the need for *PACKET_IN* events.

Note that the preceding high-level design is unicast-centric. It does not explain the various mechanisms for dealing with broadcast packets such as ARPs. These could be dealt with by a global ARP service provided by the controller. The idea behind such a controller-based service is that the controller learns IP-MAC relationships by listening to ARP replies and, when it has already cached the answer to an ARP request, it does not forward the ARP request but instead replies automatically, thus suppressing ARPs from using up broadcast bandwidth. This process is fairly easy to embody in the SDN controller as it learns the IP and MAC addresses of unmatched packets which by default are forwarded to the controller. Thus the controller is often able to trap ARP requests and provide the reply without propagating the broadcast.

There are also ways of handling these requests via the VXLAN, NVGRE, and STT protocols. The methods of dealing with broadcasts and multicasts in the tunneling protocols are specific to each protocol and beyond the scope of this book. For details we refer the reader to the references we have previously provided for these three protocols.

12.10 OFFLOADING FLOWS IN THE DATA CENTER

One of the major problems in data centers is the elephant flows, which we introduced in Section 9.6.1. Such flows consume huge amounts of bandwidth, sometimes to the point of starving other flows of bandwidth. Fig. 12.6 depicts a high-level design of an offload application. This proactive application uses RESTful APIs to communicate with the controller. The general overview of the operation of such an application is as follows:

- There is a Flow Monitor, which monitors flows and detects the beginning of an elephant flow. (This is a nontrivial problem in itself. In Section 9.6.1 we provided simple examples as well as references for how elephant flows can be scheduled, predicted, or detected.)
- Once an elephant flow is detected, the Flow Monitor sends a notification message (possibly using the application's own REST API) to communicate details about the flow. This includes identifying the two endpoint ToR switches where the flow begins and ends as well as the IP addresses of the starting (1.1.1.5) and ending (2.2.2.10) hosts.
- The offload application has been previously configured with the knowledge of which ports on the ToR switches connect to the offload Switch.
- Armed with this topology information, the offload application is able to set flows appropriately on the ToR switches and the offload switch.
- As shown in Fig. 12.6 a new flow rule is added in this ToR switch. This is the flow shown in italics in the figure. It is important that this be a high-priority flow and match exactly on the source and

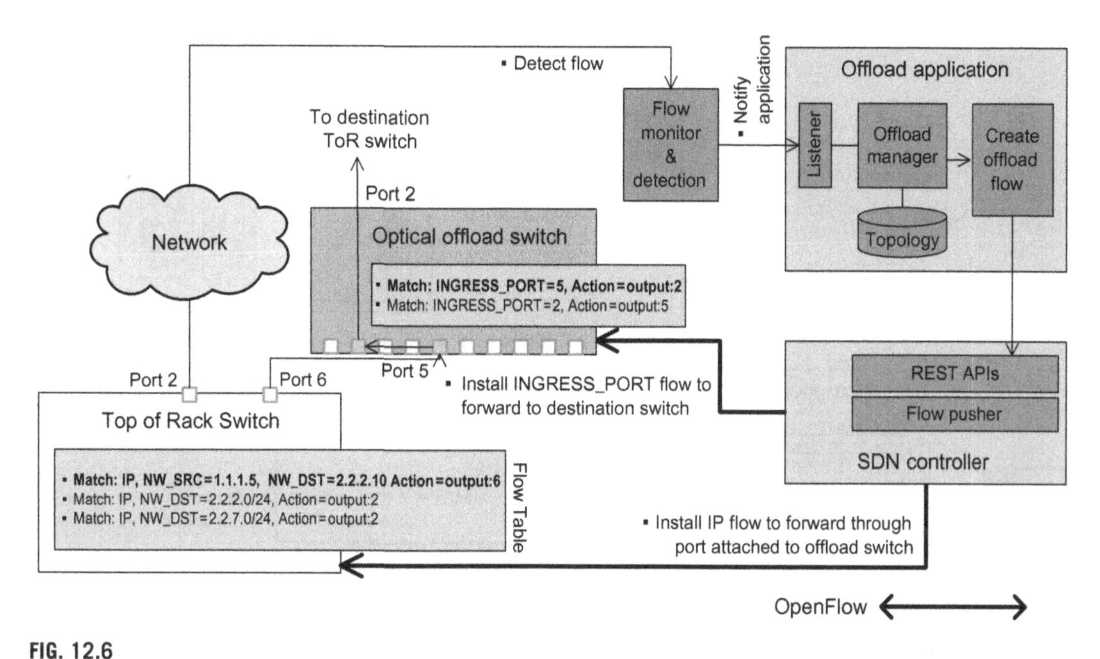

FIG. 12.6

Offload application design.

destination host addresses. The next lower-priority flow entry matches on the *network* address 2.2.2.0/24, but, as that is a lower-priority entry, packets belonging to the elephant flow will match the higher priority entry first and those packets will be directed out port 6 to the optical offload switch. Note that traffic other than the elephant flow will match the lower-priority entries and will be transmitted via the normal port.

- On the optical offload switch, flow rules are programmed which connect the two ports which will be used to forward this elephant flow between the switches involved. For the sake of simplicity, we will assume that an elephant flow is unidirectional. If there is bulk traffic in both directions, the steps we describe here can be duplicated to instantiate a separate OTN flow in each direction.
- Once the elephant flow ceases, the offload application will be informed by the flow monitor and the special flows will be removed.

12.11 ACCESS CONTROL FOR THE CAMPUS

In Section 9.3.1 we introduced the idea of an SDN *network access control* (NAC) application for the campus. Fig. 12.7 shows a high-level design of a campus NAC application. It should be apparent from the diagram that this application is a reactive application because it needs to be notified of unauthenticated users attaching to the network. The initial state of the switch is to have flows which allow ARP and DNS, but which forward DHCP requests and responses to the controller. The DHCP responses will be used to build a database of MAC and IP addresses for the end user nodes attaching to the network.

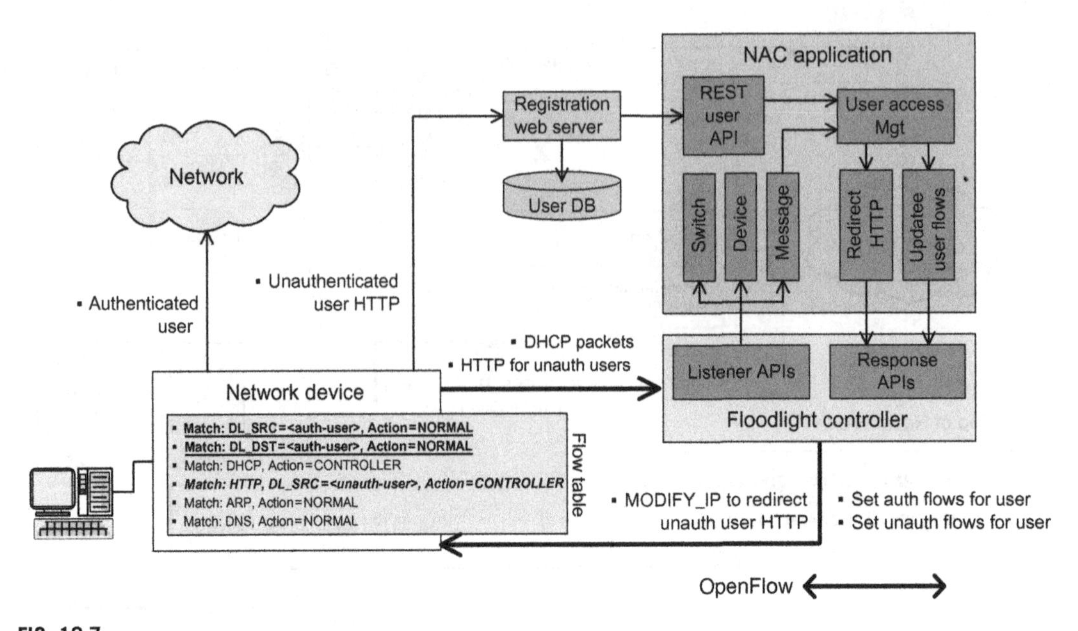

FIG. 12.7

NAC application design.

When the DHCP response is observed, if this is a new user, the NAC application adds the user to the database and the user's state is *unauthenticated*. The application then programs a flow entry in the switch based on the user's MAC address. This flow in the nonunderlined italics text in Fig. 12.7 is programmed to forward the user's HTTP requests to the controller. Thus, when a user opens a browser, in typical captive portal behavior, the application will cause the HTTP request to be forwarded to the captive portal web server where user authentication takes place. This is easily achieved by modifying the destination IP address (and possibly the destination MAC address as well) of the request so that when the switch forwards the packet it will go to the captive portal *registration web server* shown in Fig. 12.7 rather than to the intended destination.

Normal captive portal behavior is to force an HTTP redirect which sends the user's browser to the desired captive portal registration, authentication or guest login page. The controller will have programmed the switch with flows which will shunt HTTP traffic to the captive portal. No other traffic type needs to be forwarded as the authentication takes place via the user's web browser. Thus the user engages in an authentication exchange with the captive portal. This could entail payment, registration or some related mechanisms. When complete, the captive portal informs the NAC application that the user is authenticated. Note that the registration server uses the application's RESTful APIs to notify the NAC application of changes in the user's status. At this point the application removes the nonunderlined italics flow entry shown in Fig. 12.7 and replaces it with the two bold-underlined flow entries, thus allowing the user unfettered access to the network. When the user later connects to the network, if he is still registered, then he will be allowed into the network without needing to go through the registration/authentication step, unless these flow entries have timed out.

The reader should note that in this example and in the accompanying Fig. 12.7 we stop part-way through the complicated process of HTTP redirection. The full process entails sending an HTTP 302 message back to the user in order to redirect the user's browser to the registration server. These details are not germane to our example so we exclude them here.

12.12 TRAFFIC ENGINEERING FOR SERVICE PROVIDERS

In Section 9.2.1 we explained how traffic engineering is a key tool for service providers to optimize their return on investment in the large networks they operate for their customers. Fig. 12.8 shows a simplified design for a proactive traffic engineering application that takes traffic monitoring information into account. As can be seen in the figure, the flow rules in the router(s) are very simple. The design of the application likewise is simple. The complexity resides inside the path computation manager component doing the path calculations and optimizing traffic over the available links.

This problem can also be addressed using reactive application design. Fig. 12.9 reveals a high-level design of such an approach. The key difference between this design and the proactive one lies in the flow tables on the routers which feature a final flow entry that forwards unmatched packets to the controller. In this solution, initially there are no flows on the switch except for the CONTROLLER flow. The routers are not preconfigured with any specific flows other than to forward unmatched packets to the controller, as is typical for a reactive application. When these packets are received at the controller, the application's listener passes them to the path control manager. The path control manager then calculates the best path to the destination given the current traffic loads. Thus the flows are established in an on-demand manner. When traffic to that destination eventually stops, the flow is

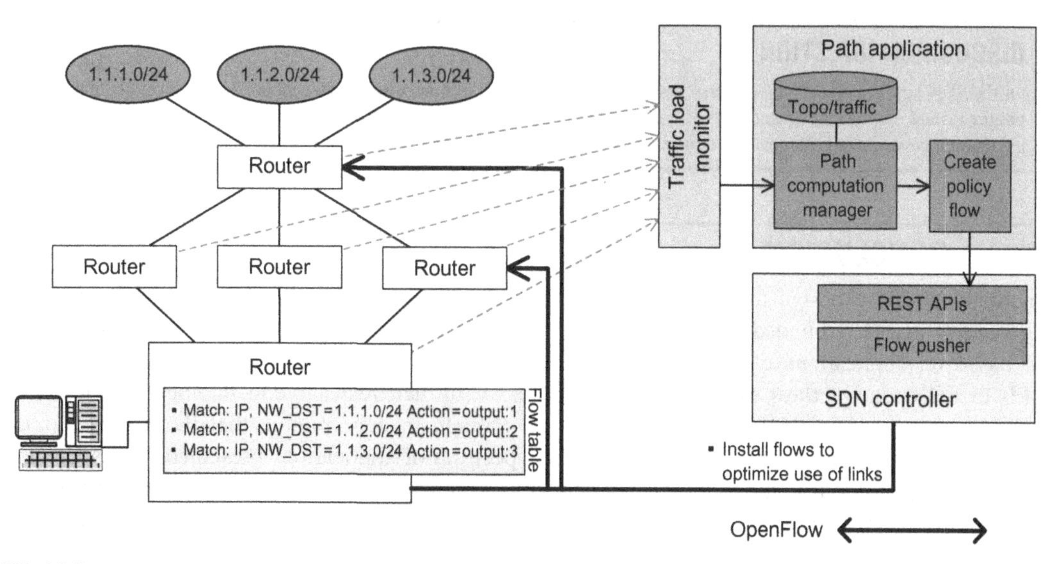

FIG. 12.8

Traffic engineering proactive application design.

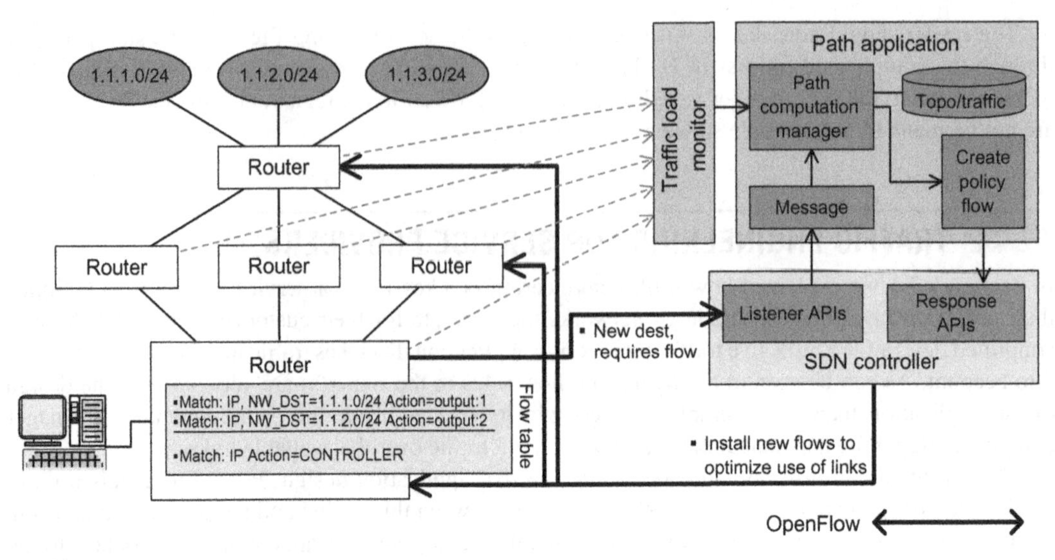

FIG. 12.9

Traffic engineering reactive application design.

removed. The flow could be removed implicitly by aging out or an external traffic monitoring function could explicitly inform the application that the flow has stopped and then the flow could be deleted by the application.

DISCUSSION QUESTION

Of the SDN application designs we have presented, which do you feel is the most innovative? Which would be the most pragmatic and easy to implement? Can you think of other designs that would be interesting to discuss? Can you think of any current networking problems that demand an SDN solution such as these?

12.13 CONCLUSION

In this chapter we bridge the OpenFlow specifics of Chapter 5 with the conceptual use cases of Chapters 8 and 9, grounded with concrete design and programming examples. Our examples are all based on applications of Open SDN. Certainly, network programmers familiar with SDN via APIs or SDN via Overlays would be able to address with their respective technologies some of the scenarios presented in this chapter. It is unlikely, though, in our opinion, that those technologies would be so easily and simply adapted to the full spectrum of applications presented in this chapter. Having said that, though, the viability of a technology depends not only on its extensibility from a programmer's perspective, but on other factors as well. Such issues include: (1) the cost, sophistication, and availability of ready-made solutions, (2) the market forces that push and pull on various alternatives, and (3) how rapidly needed future innovations are likely to emerge from different alternatives. We will spend the final three chapters of this book addressing these three aspects of SDN.

REFERENCES

[1] OpenFlowJ. Retrieved from: http://bitbucket.org/openflowj/openflowj.
[2] Apache Karaf. Retrieved from: http://karaf.apache.org.
[3] Apache Maven. Retrieved from: http://maven.apache.org.
[4] Maven archetypes. Retrieved from: http://maven.apache.org/archetype/.
[5] OpenDaylight Controller: MD-SAL. Retrieved from: https://wiki.opendaylight.org/view/OpenDaylight_Controller:MD-SAL.
[6] Controller core functionality tutorials. Retrieved from: https://wiki.opendaylight.org/view/Controller_Core_Functionality_Tutorials:Tutorials:Starting_A_Project.
[7] Project Floodlight Download. Project Floodlight. Retrieved from: http://www.projectfloodlight.org/download/.
[8] Erikson D. Beacon. OpenFlow @ Stanford; February 2013. Retrieved from: https://openflow.stanford.edu/display/Beacon/Home.
[9] Cisco Open SDN controller. Retrieved from: http://www.cisco.com/c/en/us/products/cloud-systems-management/open-sdn-controller/index.html.
[10] Brocade SDN controller. Retrieved from: http://www.brocade.com/en/products-services/software-networking/sdn-controllers-applications/sdn-controller.html.
[11] How AT&T is using OpenDaylight. Retrieved from: https://www.opendaylight.org/news/user-story/2015/05/how-att-using-opendaylight.
[12] Fujitsu Virtuora NC. Retrieved from: http://www.fujitsu.com/us/products/network/technologies/software-defined-networking-and-network-functions-virtualization/.
[13] ONOS and AT&T team up to deliver CORD. SDX Central. Retrieved from: https://www.sdxcentral.com/articles/news/cord-onos-att/2015/06/.
[14] ONOS framework builds out SDN ecosystem. Retrieved from: http://www.ciena.com/connect/blog/ONOS-framework-builds-out-SDN-ecosystem.html.
[15] Fujitsu successfully demonstrates ONOS interoperability. Retrieved from: http://www.fujitsu.com/us/about/resources/news/press-releases/2015/fnc-20150615.html.
[16] ProgrammableFlow PF5240 Switch. NEC datasheet; Retrieved from: http://www.necam.com/SDN/.
[17] ProgrammableFlow PF5820 Switch. NEC datasheet; Retrieved from: http://www.necam.com/SDN/.
[18] Salisbury B. Configuring VXLAN and GRE tunnels on OpenvSwitch. NetworkStatic; 2012. Retrieved from: http://networkstatic.net/configuring-vxlan-and-gre-tunnels-on-openvswitch/.

CHAPTER

SDN OPEN SOURCE

13

13.1 SDN OPEN SOURCE LANDSCAPE

In the first 5 years of SDN, the predominant investment of time and effort came from the research community, focused on OpenFlow. As we have discussed in Chapter 7 and elsewhere in this book, the years between 2012 and 2015 have seen not only continuing investment in OpenFlow, but also investment in the use of other protocols like BGP and NETCONF to accomplish SDN goals.

With respect to OpenFlow-based devices, controllers, and applications, there is a rich collection of open source software available. For legacy protocols, however, the related solutions run primarily on devices using proprietary software, and hence the open source development is centered on controllers, The preeminent example of this pattern is OpenDaylight (ODL). There is, however, open source available for some of the protocols the controller uses to communicate with the proprietary devices. Examples of such protocols include NETCONF and BGP.

In this chapter we will consider both the OpenFlow-based open source efforts around devices, controllers, and applications, as well as looking at emerging open source controllers that support multiple protocols, such as ODL and ONOS.

13.2 THE OpenFlow OPEN SOURCE ENVIRONMENT

In Chapter 5 we presented an overview of OpenFlow, arguably the single most important and fully open component of the SDN framework. The OpenFlow standard is developed by the ONF; however, the ONF does not provide a working implementation of either the switch or the controller. In this chapter we will provide an inventory of a number of open source components that are available for use in research, including source code for OpenFlow itself. Some of these are available under licenses that make them suitable for commercial exploitation. Many of these initiatives will speed the standardization and rate of adoption of SDN. In this chapter, we provide a survey of these initiatives and explain their relationship and roles within the broader SDN theme.

We have noted in earlier chapters that accepted use of the term SDN has grown well beyond its original scope. Indeed the definition of SDN has blurred into most of the current work on network virtualization. Much of this work, though, particularly in Open SDN, has occurred in the academic world. As such, there is a significant library of open source available for those wishing to experiment or build commercial products in the areas of SDN and network virtualization. Beyond academic contributions, many companies are contributing open source to SDN projects. Some examples that have

Software Defined Networks. http://dx.doi.org/10.1016/B978-0-12-804555-8.00013-2

been mentioned in earlier chapters are Big Switch's Indigo, VMware's OVS, and the many companies cited in Section 11.9.2 contributing to ODL. Our survey of SDN-related open source projects was greatly aided by the compilations already performed by SDN Central [1] and others [2,3].

In Fig. 13.1 we map this wide selection of open source to the various network components. The major components, from the bottom up, are:

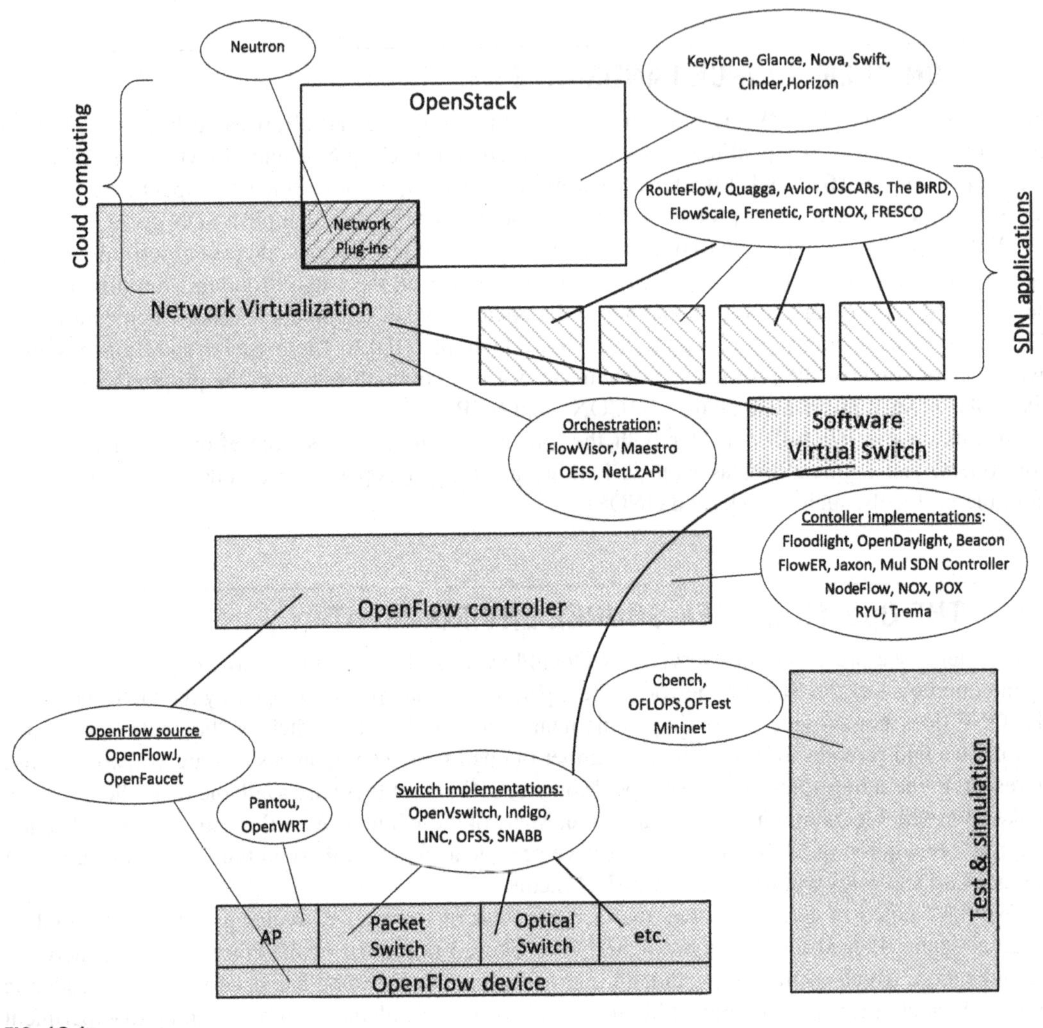

FIG. 13.1

SDN open source landscape.

- OpenFlow Device
- OpenFlow Controller
- Applications
- Network Virtualization
- Test and Simulation

The last category, Test and Simulation, is shown to the side of the diagram as it does not fit neatly into a bottom-to-top hierarchy. For each of these categories we show in the call-out ovals the notable available open source that contributes to that category. In the sections that follow we provide an inventory of this software. Some of these have surfaced as the basis for commercial deployments of network virtualization and SDN and in those cases we will explore the related open source product in greater depth. As we move through the different network components in Sections 13.6–13.12, we encourage the reader to look back at Fig. 13.1 to place those open source projects in the context of the SDN ecosystem.

The manner in which open source software may be commercially exploited is dependent on the nature of the open source license under which it is provided and these licenses are by no means created equal. So, before discussing the software modules themselves, we will examine the prevalent types of open source licenses in use today.

13.3 CHAPTER-SPECIFIC TERMINOLOGY

In this chapter we introduce a term commonly used by the open source development community, *forking*. If we view a source code base as the trunk of a tree, there are development branches in different locations around the world where innovation and experimentation occur. In many cases, successful and important innovations are contributed back to the evolving master copy of the code and appear in a future release higher up on the trunk of the source code tree, to continue with the analogy. When a new code base is forked from an existing release, it cuts the connection back to the trunk. This means that the continuing innovation on that fork will not be contributed back to the original code base.

13.4 OPEN SOURCE LICENSING ISSUES

Unless a person has been intimately involved in building a commercial product that incorporates open source software, it is unlikely that he or she understands the complexity of this field of open source licensing. There is a large number of different open source licenses extant today. At the time of this writing, *gnu.org* [4] lists a total of 92 free software licenses.

There are many nuances between these free software licenses that may seem arcane to a nonlegal professional. Slight differences about permissions or requirements about using the name of the originator of the license or requirements about complying with US export laws may be some of the only differences between them. Enterprises that offered free source code in the past did so with their organization's name in the license, where the name itself may be the only substantive difference between their license and another already-existing license. We will provide a brief description of the different licenses under which the source mentioned in this chapter has been released, paying particular attention to the business ramifications of choosing to incorporate the software under each of these licenses in a commercial product.

An accepted common denominator of open source is that it is source code provided without charge which you may use and modify for a defined set of purposes. There is open source available that only allows noncommercial use. We believe that this greatly limits the impact that such software will have, so we will not discuss this further here. Even when commercial use is allowed, however, there are significant differences in the obligations implicit in the use of such software.

Some major categories of open source licensing in widespread commercial use today are the GNU *General Public License* (GPL), *BSD-style*, and *Apache*. There are different organizations that opine on whether a given license is truly open source. These include the *Free Software Foundation* (FSF) and the *Open Source Initiative*. They do not always agree.

GPL is an extremely popular licensing form. The Linux operating system is distributed under the GPL license. It allows users to copy and modify the software for virtually any purpose. For many companies, though, it is too restrictive in one major facet. It incorporates the notion of *copyleft* whereby if any derivative works are distributed they must be distributed under the same license. From a practical standpoint, this means that if a commercial company extends a GPL open source code base to provide some new function that it wishes to offer to the market, the modifications that are made must be made freely available to the open source community under GPL terms. If the feature in question here embodies core intellectual property that is central to that company's value proposition, this is rarely acceptable. If the function provided by the GPL open source can be used with little added intellectual property, then GPL code may be suitable in commercial products. Ensuring that this distinction is kept clear can have massive consequences on the valuation that a startup receives from investors. For this reason, GPL open source is used with caution in industry. Nonetheless, GPL remains the most widely used free software license. There is a GPL V.2 and GPL V.3 that are somewhat more restrictive than the initial version.

The BSD-family licensing model is more permissive. This style of license was first used for the *Berkeley Software Distribution* (BSD), a Unix-like operating system. There have been a number of variants of this original license since its first use, hence we use here the term BSD-family of licenses. Unlike GPL code, the user is under no obligation to contribute derivative works back to the open source community. This makes it much less risky to incorporate BSD-licensed open source into commercial products. In some versions of this style of license there is an *advertising clause*. This was originally required to acknowledge the use of UC Berkeley code in any advertisement of the product. While this requirement has been eliminated in the latest versions of the BSD license, there are still many free software products licensed with a similar restriction. An example of this is the widely used OpenSSL encryption library, which requires acknowledgment of the author in product advertising.

The Apache license is another permissive license form for open source. The license was developed by the *Apache Software Foundation* (ASF) and is used for software produced by the ASF as well as some non-ASF entities. Software provided under the Apache license is unrestricted in use except for the requirement that the copyright notice and disclaimer be maintained in the source code itself. As this has minimal impact on potential commercialization of such software, many commercial entities build products today freely incorporating Apache-licensed source code.

The *Eclipse Public License* (EPL) [5] can be viewed as a compromise between the strong *copyleft* aspects of the GPL and the commercially friendly Apache license. Under certain circumstances, using EPL-licensed source code can result in an obligation to contribute one's own modifications back to the open source community. This is not a universal requirement, though, so this software can often be incorporated by developers into commercial products without risk of being forced to expose their own intellectual property to the entire open source community. The requirement comes into effect when the modifications are not a separate software module but contain a significant part of the EPL-licensed original and are thus a derivative work. Even when the requirement to release extensions to the original is not in effect, the user is required to grant a license to anyone receiving their modification to any patent the user may hold over those modifications.

The Stanford University *Free and Open Source Software* (FOSS) license [6] is noteworthy in the manner in which it attempts to address the common problem with commercial use of GPL-licensed source code due to its copyleft provision. Stanford's motivation in developing this concept was to avoid the situation where developers would hesitate to develop applications for the Beacon controller (see Section 13.8). Because of the way in which applications interface with the Beacon controller, if it were licensed under GPL without the FOSS exception, such developers would be obliged to contribute their work as GPL open source. This license is specific to the Beacon controller, though the concept can be applied more generally. In essence, Stanford wants to release the Beacon controller code under the terms of GPL. FOSS specifies that a user of the code must continue to treat the Beacon code itself under the terms of GPL, but the code they write as a Beacon application can be released under any of a number of approved FOSS licenses.

These approved licenses are the most common of the open source licenses discussed in this chapter. Notably an open source application can be released under a license that is more commercially friendly, such as Apache, which does not require that the derivative works on the application code be released to the open source community, allowing the developer to guard his newly generated intellectual property.

DISCUSSION QUESTION

We have discussed many different types of open source license in the preceding section. Of all of these, one particular license is held up as an example of being too restrictive to be of general use when building a software-based business. Which license type is that and what is the single problematic characteristic that causes concern for many businesses?

13.5 PROFILES OF SDN OPEN SOURCE USERS

When judging the relevance of open source to the SDN needs of one's own organization, the answer will be highly dependent on the role to which the organization aspires on the SDN playing field. To be sure, the early SDN adopters were almost exclusively researchers, many of whom worked in an academic environment. For such researchers, keeping flexibility and the ability to continually tinker with the code is always paramount compared to concerns about support or protection of intellectual property. For this reason, open source OpenFlow controllers and switch code were and continue to be very important in research environments. Similarly, a startup hoping to build a new OpenFlow switch or controller will find it very tempting to leverage the work present in an existing open source implementation, provided that does not inhibit the development and commercial exploitation of the company's *own* intellectual property. If one is a cloud computing operator, tinkering with the OpenFlow controller is far less important than having a reliable support organization that one can call in the event of network failure.

Thus we propose three broad categories of likely users of SDN-related open source software:

- Research
- Commercial developers
- Network operators

Assuming that a given piece of open source software functionally meets a user's requirements, each of these user classes will have a different overriding concern. A developer in a commercial enterprise will focus on ensuring that he or she does not incorporate any open source that will limit his/her company's ability to exploit their products commercially and that it will protect their intellectual property. The GPL is often problematic when the open source will be modified and the modifications must thus be made publicly available under GPL. Generally, researchers have the greatest flexibility and simply desire that the source code be available free of charge and can be modified and distributed for any purpose. Often, though, researchers are also strong proponents of the growth of open source and, thus, may prefer a copyleft license that obliges the users to contribute their modifications back to the open source community. Finally, the network operator will be most concerned about maintaining the network in an operational state, and therefore will focus on how the open source product will be supported and how robust it is.

Because of the fundamental noncommercial nature of open source, it will always be a challenge to identify whom to hold responsible when a bug surfaces. This will give rise to companies that will effectively commercialize the open source, where the value they add is not development of new features but to provide well tested releases and configurations as well as a call desk for support. Some operators will simply fall back on the old habit of purchasing technology from the mainstream NEMs. Whether or not their NEM provider uses open source is almost irrelevant to the operator as they really are purchasing equipment plus an insurance policy and it matters little where the NEM obtained this technology. On the other hand, while a cloud operator may not want to assume responsibility for support of their switching infrastructure, the open source applications, test or monitoring software might serve as three useful starting points for locally supported projects.

In terms of the software projects discussed in this chapter, some examples of how different user types might use some of the software include:

- A hardware switch company wishing to convert a legacy switch into an OpenFlow-enabled switch may use the OpenFlow switch code presented in Section 13.7 as a starting point.
- A company that wishes to offer a commercial OpenFlow controller might start their project with one of the OpenFlow controllers we list in Section 13.8, add features to it and harden the code.
- A cloud operator might use an open source SDN application such as those we describe in Section 13.9 to begin to experiment with OpenFlow technology prior to deciding to invest in this technology.

As we review several different categories of software projects in the following sections, for each we will list which of these three classes of users are likely to adopt a given software offering. We will now provide an inventory of some of the SDN-specific open source that is used by researchers, developers, and operators. Since the number of these open source projects is quite large, we group them into the basic categories of switches, controllers, orchestration, applications, and simulation and test that were depicted in Fig. 13.1. We summarize the different open source contributions in each of these categories in a number of tables in the following sections. Only in those instances where the particular project has played an unusually important role in the evolution of SDN do we describe it in further detail.

13.6 OpenFlow SOURCE CODE

As Fig. 13.2 shows, both the device and controller can leverage an implementation of the OpenFlow protocol. There are multiple open source implementations of OpenFlow. We use a format of two paired tables to provide a succinct description of these. In Table 13.1 we provide a basic description of available implementations and in Table 13.2 we list detailed information regarding the provider of the code, the type of license, the source code in which it is written (where relevant) and the likely class(es) of users. We have assigned these target classes of users using a number of criteria, including: (1) the nature of the license, (2) the maturity of the code, and (3) actual users, when known. OpenFlowJ is

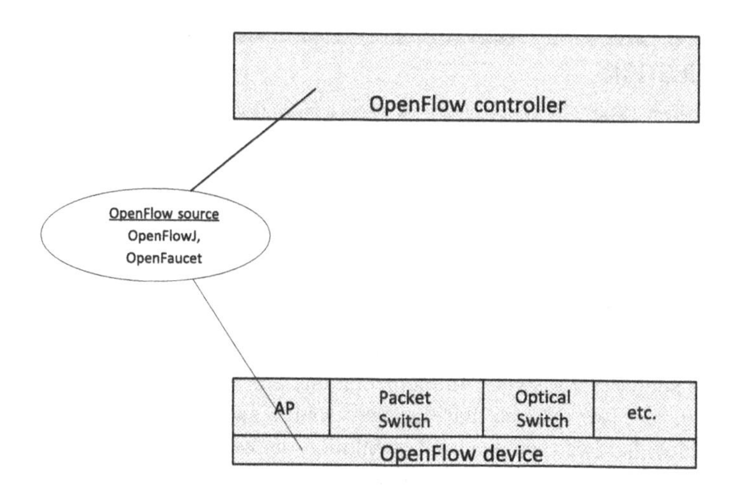

FIG. 13.2

Applications of OpenFlow source.

Table 13.1 Open Source Implementations of OpenFlow: Description

Product Name	Description
OpenFlow Reference	Reference code base that tracks the specification
OpenFaucet	Source code implementation of OpenFlow protocol 1.0, used in both controllers and switches
OpenFlowJ	Source code implementation of OpenFlow protocol. Both Beacon and FlowVisor incorporate this code

Table 13.2 Open Source Implementations of OpenFlow: Details

Product Name	Source	License	Language	Target User
OpenFlow Reference	Stanford	BSD-style	C	Research
OpenFaucet	Midokura	Apache	Python	Research
OpenFlowJ	Stanford	Apache	Java	Research

notable in that the Beacon controller, itself the fount of a number of other controller implementations, uses the OpenFlowJ code.

We will continue to use this format of pairing two tables to provide the description of notable open source projects in each of the major categories we discuss in this chapter. In some cases the license or language may not be known. We indicate this by placing a hyphen in the corresponding table cell. We will also mark the table cell with a hyphen for languages in the event that the project uses multiple languages.

DISCUSSION QUESTION

Name at least four different classes of networking devices where it might make sense to use OpenFlow source code when building the software for that device.

13.7 SWITCH IMPLEMENTATIONS

In this section we discuss some of the more important open source SDN switch implementations available. The open source SDN switch implementations are summarized in Tables 13.3 and 13.4. Looking back at Fig. 13.1 we can see that the open source switch implementations can potentially apply both to OpenFlow hardware device implementations as well as software-based virtual switches. In the following paragraphs we examine the three most widely discussed switch implementations, Open vSwitch (OVS), Indigo, and OpenSwitch, as well as the only OpenFlow wireless access point code base listed, Pantou.

As of this writing, OVS is the most mature and feature-complete open source implementation of the switch-side of an OpenFlow implementation. OVS has been used as the control logic inside genuine hardware switches, as well as a virtual switch running within a hypervisor to provide network virtualization. It is in the latter context that the developer of OVS, Nicira, has been shipping for some time OVS with its NVP product. Since the acquisition of Nicira by VMware, the NVP product is now marketed under the name NSX. OVS uses OpenFlow for control of flows and OVSDB for configuration. OVS uses the sFlow protocol for packet sampling and monitoring. The code has been ported to work with multiple switching chipsets and has thus found a home in several hardware switches as well. Earlier, in Section 6.3.3, we presented an example of OVS being used for SDN via Hypervisor-based Overlays.

Table 13.3 Open Source OpenFlow Switches: Description

Project	Description
Open vSwitch	Production-quality multilayer virtual switch designed for programmatic network automation using standard management interfaces as well as OpenFlow. Also called OVS
Indigo	OpenFlow switch implementation designed to run on physical switches and use their ASICs to run OpenFlow at line rates. Note that Switch Light is Big Switch's commercial switch software product and is based on the second generation of the Indigo platform
OpenSwitch	Full-featured network operating system, built on Linux, designed to provide a rich set of switch features
LINC	OpenFlow 1.2 Softswitch
OFSS	OpenFlow 1.1 Softswitch
ofl3softswitch	User-space OpenFlow 1.3 Softswitch. Based on Ericsson TrafficLabs 1.0 Softswitch code
Pantou	Wireless router implementation based on OpenWRT with OpenFlow capabilities (see Section 6.4 for more information about OpenWRT)

Table 13.4 Open Source OpenFlow Switches: Details

Product Name	Source	License	Language	Target User
Open vSwitch	Nicira	Apache 2.0	C	Researchers, developers
Indigo	Big Switch	EPL	–	Developers
OpenSwitch	OpenSwitch	Apache 2.0	C++, Go, Java	Researchers, developers
LINC	Infoblox	Apache 2	Erlang	Developers
OFSS	Ericsson	BSD-like	C	Research
of13softswitch	CPqD	BSD-like	C	Research
Pantou	Stanford	–	–	Research

Big Switch Networks offers the *Indigo* switch code base under the Eclipse public license. This was a natural move for Big Switch in its desire to create an ecosystem of hardware switches that work well with its commercial OpenFlow controller product, Big Network Controller. Like OVS, Indigo is targeted for use both on physical switches as well as an OpenFlow hypervisor switch for network virtualization environments. In particular, this code can be used to convert a legacy layer 2 or 3 switch into an OpenFlow switch. By making this code base available to switch manufacturers, especially smaller manufacturers that may lack the means to develop their own OpenFlow switch-side control code, Big Switch hopes to expand the market for its controller.[1] Because Indigo is integrated with Ethernet switch ASICs, it is possible to switch flows under the OpenFlow paradigm at line rates.

[1] Big Switch is now marketing a commercial version of Indigo called Switch Light for this express purpose.

Another distinction between OVS and Indigo is that Indigo is implemented specifically to support OpenFlow, whereas OVS can support other control mechanisms such as OVSDB. In addition, OVS has broader support for affiliated network virtualization features and includes richer contributions from the open source community than does Indigo. From a feature-set point of view, OVS seems to be a superset of Indigo's features.

OpenSwitch [7] is a new community started by HP, which includes contributions from Accton, Broadcom, Intel, VMware, Qosmos, and Arista. OpenSwitch is released under the Apache 2.0 license [8]. The differentiating features of OpenSwitch are (1) it is a fully featured layer 2 and layer 3 switching platform, (2) it is intended to be modular, highly available, and built using modern tools providing greater reliability, and (3) it natively supports OVSDB, OpenFlow, and sFlow. As of this writing, OpenSwitch is relatively new and it will be interesting to see if this open networking device initiative gains traction in the next few years.

Pantou is actually designed to turn commodity APs into an OpenFlow-enabled AP. This is accomplished by integrating the NOX OpenFlow implementation with OpenWRT. We explained in Section 6.4 that OpenWRT is an open source project that applies the SDN principle of *opening up the device* to a wide selection of low-cost consumer AP hardware. This union of OpenFlow with basic wireless AP functionality available on low-cost hardware is a boon to researchers wishing to experiment with the largely virgin field of SDN applied to 802.11 networks.

DISCUSSION QUESTION

We have discussed the two primary OpenFlow-based implementations, Open vSwitch and Indigo, and have introduced a new open source project, OpenSwitch. Compare and contrast these three implementations of switching technology. Which do you think will prevail?

13.8 CONTROLLER IMPLEMENTATIONS

13.8.1 HISTORICAL BACKGROUND

In this section we discuss some of the more important open source SDN controller implementations that have been and are currently available for research, development, and operations. The open source SDN controller implementations are summarized in Tables 13.5 and 13.6.

The Beacon [6] controller was based on OpenFlowJ, an early open source implementation of OpenFlow written in Java. The OpenFlowJ project is hosted at Stanford University. Beacon is a highly influential controller, both for the large amount of early OpenFlow research and development that was done on that controller as well as being the code base from which the Floodlight controller source code was forked.

Beacon is a cross-platform, modular OpenFlow controller. By 2013, Beacon had been successfully deployed in an experimental data center consisting of 20 physical switches and 100 virtual switches. It runs on many different platforms, including high-end Linux servers. As mentioned earlier, while the core controller code is protected with GPL-like terms, the FOSS license exception allows developers to extend the controller with applications that may be licensed under more commercially favorable terms. Beacon's stability [9] distinguishes it from other controllers used primarily for research purposes.

Table 13.5 Open Source Controllers: Description

Project	Description
NOX	The first OpenFlow controller, used as the basis for many subsequent implementations
POX	Python version of NOX
Beacon	Java-based OpenFlow controller which is multithreaded, fast, and modular; uses OpenFlowJ
Floodlight	Java-based OpenFlow controller, derived from Beacon, supported by large developer community
Ryu	Network controller with APIs for creating applications; integrates with OpenStack. Manages network devices not only through OpenFlow but alternatives such as NETCONF and OF-Config. As this is more than just a controller, it is sometimes referred to as a Network Operating System
OpenDaylight	This controller supports multiple southbound protocols (including OpenFlow), is currently the predominant open source SDN controller, and provides the MD-SAL abstraction for internal applications
ONOS	This controller focuses on OpenFlow, is targeted at service providers, and is a new and major competitor to OpenDaylight. Supports an "intents"-based northbound API
FlowER	OpenFlow controller (still in development)
Jaxon	Thin Java interface to NOX controller
Mul SDN Controller	OpenFlow controller
NodeFlow	OpenFlow controller
Trema	OpenFlow controller, integrated test and debug. Extensible to distributed controllers

Modules of Beacon can be started, stopped, and even installed while the other components of Beacon continue operating. Evidence of the impact that this code base has had on the OpenFlow controller industry is found in the names of two of the more influential open source controllers being used by industry today, *Floodlight* and *OpenDaylight*, both echoing the illumination connotation of the name Beacon. In Section 12.5 we discussed the use of Floodlight as a training tool when developing SDN applications.

Floodlight was forked from Beacon prior to Beacon being offered under the current GPL/FOSS licensing model [10]. As Floodlight itself is licensed under the Apache 2.0 license, it is not subject to the copyleft provisions of the Beacon license and the source is thus more likely to be used by recipients of the code who need to protect the intellectual property of their derivative works on the code base. Floodlight is also integrated with OpenStack (see Section 13.12). A number of commercial OpenFlow controllers are being developed using Floodlight as the starting point. This includes Big Switch's own commercial controller, Big Network controller. Big Switch's business strategy [11] surrounding the open source controller is complex. They offer a family of applications that work with the Floodlight controller. While their commercial controller will not be open source, they promise to maintain compatibility at the interface level between the commercial and open source versions. This reflects a fairly classical approach to building a software business, which is to donate an open source version of functionality in order to bootstrap a community of users of the technology, hoping that it

Table 13.6 Historical and Current Open Source Controllers

Product Name	Source	License	Language	Target User
OpenDaylight	OpenDaylight	EPL	Java	Research, developers, operators
ONOS	ON.Lab	Apache 2.0	Java	Research, developers, operators
NOX	ICSI	GPL	C++	Research, operators
POX	ICSI	GPL	Python	Research
Beacon	Stanford University	GPLv2, Stanford FOSS, Exception v1.0	Java	Research
Floodlight	Big Switch Networks	Apache 2.0	Java	Research, developers, operators
Ryu	NTT Communications	Apache 2.0	Python	Research, developers, operators
FlowER	Travelping GmbH	MIT License	Erlang	Operators
Jaxon	University of Tsukuba	GPLv3	Java	Research
Mul SDN	Kulcloud	GPLv2	C	Operators
NodeFlow	Gary Berger	–	Javascript	Research
Trema	NEC	GPLv2	Ruby, C	Research

gains traction, and then offer a commercial version that can be *up-sold* to well-heeled users that require commercial support and extended features. Big Switch joined the OpenDaylight project, donating their Floodlight controller to the effort, in hopes that this would further propagate Floodlight. These hopes were ultimately not realized, as we explain later. Indeed, as of this writing, Big Switch has tweaked this original business strategy. We discuss their *big pivot* in Section 14.9.1.

13.8.2 OpenDaylight

The OpenDaylight project [12] was formed in early 2013 to provide an open source SDN framework. The platinum-level sponsors currently include Cisco, Brocade, Ericsson, Citrix, Intel, HP, Dell, and Red Hat. Unlike the Open Networking Foundation (ONF), ODL only considers OpenFlow to be one of many alternatives to provide software control of networking gear. Part of the ODL offering includes an open source controller, which is why we include ODL in this section. Big Switch initially joined and donated their Floodlight controller to ODL, anticipating Floodlight would become the key OpenFlow component of the project. This expectation was quickly shattered, as ODL decided to make the Cisco *eXtensible Network Controller* (XNC) the centerpiece and to merely add Floodlight technology to that core. Subsequent to this, Big Switch reduced its involvement in the ODL project and ultimately withdrew completely [13].

It is thus noteworthy that both SDN via APIs and Open SDN are present in the same ODL controller. Originally this fact was a source of friction between ODL and the ONF, whose focus is primarily OpenFlow. However, as of 2015 ONF board members had begun to attend and participate in ODL

Summit panels and discussions [14,15]. With significant momentum behind ODL from the vendor and customer community, most SDN players are participating and contributing to the effort.

Although many of the core contributors to ODL's code base are associated with major networking vendors [16], ODL does now boast a large number of projects and contributors from many different vendors. As such it has clearly been a huge success as an open source project, and has served to redirect the focus of SDN by introducing solutions around legacy protocols and existing vendor equipment.

One of the major statistics for evaluation of an open source project is the number of contributors. The ODL *Spectrometer* project [16] provides an organized set of statistics on the subject. Another source of information are the *GitHub* software version control pages for open source projects. As of early 2016, ODL's GitHub page [17] shows over 60 projects, with many contributors for each. While not all of these are mainstream ODL controller projects, this nonetheless provides an indication of the community support for this controller.

13.8.3 ONOS

As of this writing, the only true challenger to ODL's dominance is ON.Lab's ONOS project. Many of the same major service providers and transport vendors investing in ODL development are also investing in ONOS. Indeed many of these companies were instrumental in the very formation of ONOS as a competitor to ODL. It is thus unsurprising that ONOS has quickly risen to become a major factor in the debate over open source SDN controllers.

Like ODL, ONOS is supported by a number of major vendors and service providers, including Alcatel-Lucent, AT&T, China Unicom, Ciena, Cisco, Ericsson, Fujitsu, Huawei, Intel, NEC, NTT Communications, among others. The Linux Foundation has also entered into a strategic partnership with the ONOS project. Hence the foundation supports *both* ODL and ONOS. The two controllers have fundamentally different focus areas. ODL emphasizes legacy protocols while ONOS emphasizes OpenFlow.

One way to compare activity levels of ODL versus ONOS is to look at the amount of code and the number of contributors to their respective open source projects. BlackDuck *Open HUB* [18] tracks open source projects, calculating statistics such as those cited previously. As of the time of this writing, the numbers for both projects *for the preceding 12 months* are shown in Table 13.7. As is apparent from the table, ODL has a significant lead in number of contributors and lines of code, although it is also true that ONOS has been around for less time than ODL.

Table 13.7 2015 Contribution Statistics for OpenDaylight vs ONOS

Controller	Commits	Lines Added	Lines Removed	Contributors
OpenDaylight	13,305	3,050,094	2,496,450	392
ONOS	3534	884,274	440,253	116

DISCUSSION QUESTION

We have listed many SDN controller implementations, and then highlighted the two currently most prominent, OpenDaylight and ONOS. Describe what you feel to be the strengths and weaknesses of each, specifically related to open source. Do you think that one will prevail? Do you think it is better to have just one preeminent controller implementation or more than one, and why?

13.9 SDN APPLICATIONS

Chapter 12 was dedicated to an analysis of how SDN applications are designed, how they interact with controllers, and provided a number of sample designs of different SDN applications. If the reader wishes to actually embark on the journey of writing his or her own SDN application, it is clearly advantageous to have a starting point that can be used as a model for new work. To that end, in this section we present some of the more important open source SDN applications available. The open source SDN applications are summarized in Tables 13.8 and 13.9. The term *applications* by its very nature

Table 13.8 Open Source SDN Applications: Description

Product Name	Description
Routeflow	Integrates IP routing with OpenFlow controller based on the earlier Quagflow project (see Section 13.13)
Quagga	Provides IP routing protocols (e.g., BGP, OSPF)
Avior	Management application for Floodlight
OSCARS	On-demand secure circuits and advance reservation system used for optical channel assignment by SDN; predates SDN
The BIRD	Provides IP routing protocols (e.g., BGP, OSPF)
FlowScale	Traffic load balancer as a service using OpenFlow
Frenetic	Provides language to program OpenFlow controller abstracting low-level details related to monitoring, specifying, and updating packet forwarding policies
FortNOX	Security framework, originally coupled with NOX controller, now integrated in SE-Floodlight; extends OpenFlow controller into a security mediation service; can reconcile new rules against established policies
FRESCO	Security application integrated with FortNOX provides security-specific scripting language to rapidly prototype security detection and mitigation modules
Atrium	Integrates standalone open source components
PIF	Protocol independent forwarding intermediate representation for datapaths
Boulder	Intents-based northbound interface (NBI)
Aspen	Real-time media interface specification

Table 13.9 Open Source SDN Applications: Details

Product Name	Source	License	Language	Target Users
Routeflow	CPqD (Brazil)	–	–	Research, developers
Quagga	Quagga Routing Project	GPL	C	Research, developers
Avior	Marist College	MIT License	Java	Research
OSCARS	Energy Services Network (US Department of Energy)	New BSD	Java	Research
The BIRD	CERN	GPL	–	Research
FlowScale	InCNTRE	Apache 2.0	Java	Research
Frenetic	Princeton	GPL	Python	Research
FortNOX	SRI International	–	–	Research
FRESCO	SRI International	–	–	Research
Atrium	Open Source SDN	Apache 2.0 & EPL	Java	Research, developers
PIF	Open Source SDN	Apache 2.0 & EPL	Java	Research, developers
Boulder	Open Source SDN	Apache 2.0 & EPL	Java	Research, developers
Aspen	Open Source SDN	Apache 2.0 & EPL	Java	Research, developers

is general, so it is not possible to list all the possible categories of SDN applications. Nevertheless, four major themes surface that have attracted the most early attention to SDN applications. These are security, routing, network management, and *Network Functions Virtualization* (NFV).

In Table 13.8 there are three examples related to routing, *The BIRD*, *Quagga*, which is a general purpose open source routing implementation that is suitable for use in SDN environments, and *Routeflow*, which is SDN-specific. In Section 13.13 we will describe a specific example of open source being used to implement a complete SDN network where Routeflow and Quagga are used together. *Avior* is a network management application for the Floodlight controller. OpenFlow has great potential to provide the next generation of network security [19]. The two examples of this in Tables 13.8 and 13.9 are *FortNOX* and *Fresco*. As we described in Chapter 10, NFV applications is the class of applications that virtualize a network services function that is performed by a stand-alone appliance in legacy networks. Examples of such devices are traffic load balancers and intrusion detection systems. FlowScale is an NFV application that implements a traffic load balancer as an OpenFlow controller application.

Some of the open source applications covered in this section are OpenFlow applications, where domain-specific problems are addressed through applications communicating with the OpenFlow controller via its northbound interface. Not all SDN applications are OpenFlow applications, however.

13.9.1 OPEN SOURCE SDN COMMUNITY PROJECTS

Of recent interest is the set of projects from the *Open Source SDN* (OSSDN) community, whose intent is to *sponsor and develop open SDN solutions to provide greater adoption of SDN* [20]. These projects are listed in the last four entries of Tables 13.8 and 13.9. Projects from this community currently include: (1) *Protocol Independent Forwarding* (PIF), (2) the Boulder project (intents-based *Northbound Interface*

(NBI)), (3) the Atrium project (open source SDN distribution), and (4) the Aspen project (real-time media interface specification). These projects are intended to help bootstrap the development of SDN solutions.

Open source SDN includes leadership from the ONF, Tail-f/Cisco, Big Switch Networks, Infoblox, and other organizations [21]. This is a new initiative and may contribute to the adoption of SDN on a broader scale in the years ahead.

DISCUSSION QUESTION

There are many private as well as open source SDN applications being written for SDN controllers today. Will the new "Open Source SDN" projects be helpful to the promotion of SDN? To the promotion of Open SDN? Or do you believe that only applications designed for OpenDaylight will have relevance in the next few years?

13.10 ORCHESTRATION AND NETWORK VIRTUALIZATION

We introduced the concept of orchestration in Section 3.2.3. There are many compute, storage, and networking resources available in the data center. When virtualization is applied to these resources, the number of assignable resources explodes to enormous scale. Orchestration is the technology that provides the ability to program automated behaviors in a network to coordinate these assignable resources to support applications and services.

In this section we list some of the more important open source implementations available that provide orchestration functionality. Network orchestration is directly related to network virtualization. The term orchestration came into use as it became clear that virtualizing the networking portion of a complex data center, already running its compute and storage components virtually, involved the precise coordination of many independent parts. Unlike legacy networking, where the independent parts function in a truly distributed and independent fashion, network virtualization required that a centralized entity coordinate their activities at a very fine granularity—much like a conductor

Table 13.10 Open Source Orchestration Solutions: Description

Product Name	Description
FlowVisor	Creates slices of network resources; delegates control of each slice, that is, allows multiple OpenFlow controllers to share a set of physical switches
Maestro	Provides interfaces for network control applications to access and modify a network
OESS	Provides user-controlled VLAN provisioning using OpenFlow switches
NetL2API	Provides generic API to control layer 2 switches via vendors' CLIs, not OpenFlow; usable for non-OpenFlow network virtualization
Neutron	OpenStack's operating system's networking component which supports multiple network plugins, including OpenFlow

Table 13.11 Open Source Orchestration Solutions: Details

Product Name	Source	License	Language	Likely Users
FlowVisor	ON.Lab	–	Java	Research
Maestro	Rice University	LGPL	Java	Research
OESS	Internet2 Indiana University	Apache 2.0	–	Research
NetL2API	Locaweb	Apache	Python	Research, developers, operators
Neutron	OpenStack Foundation	Apache	–	Operators

coordinates the exact timing of the thundering percussion interlude with the bold entry of a brass section in a symphony orchestra. The open source SDN orchestration and network virtualization solutions are summarized in Tables 13.10 and 13.11.

13.11 SIMULATION, TESTING, AND TOOLS

In this section we discuss some of the more important open source implementations available related to network simulation, test and tools pertinent to SDN. The open source SDN solutions available in this area are summarized in Tables 13.12 and 13.13. Some of these projects offer software that may have relevance in nonresearch environments, particularly *Cbench*, *OFLOPS*, and *OFTEST*. Mininet has seen wide use by SDN researchers to emulate the large networks of switches and hosts and the traffic that can cause a controller, such as an OpenFlow controller, to create, modify, and tear down large numbers of flows.

Table 13.12 Open Source Test and Simulation: Description

Product Name	Description
Cbench	OpenFlow controller benchmarker. Emulates a variable number of switches, sends PACKET_IN messages to controller under test from the switches and observes responses from controller
OFLOPS	OpenFlow switch benchmarker. A stand-alone controller that sends and receives messages to/from an OpenFlow switch to characterize its performance and observes responses from the controller
Mininet	Simulates large networks of switches and hosts. Not SDN-specific, but widely used by SDN researchers who simulate OpenFlow switches and produce traffic for OpenFlow controllers
OFTest	Tests switch compliance with OpenFlow protocol versions up to 1.2

Table 13.13 Open Source Test and Simulation: Details

Product Name	Source	License	Language	Likely Users
Cbench	Stanford University	GPL V.2	C	Research, developers, operators
OFLOPS	Stanford University	Stanford License	C	Research, developers, operators
Mininet	Stanford University	Stanford License	Python	Research
OFTest	Big Switch Networks	Stanford License	Python	Research, developers, operators

13.12 OPEN SOURCE CLOUD SOFTWARE

13.12.1 OpenStack

We introduced the OpenStack Foundation in Section 11.9.4. The OpenStack open source is a broad open source platform for cloud computing, released under the Apache license. We show the role of OpenStack as well as its components in Fig. 13.3. OpenStack provides virtualization of the three main components of the data center, compute, storage, and networking. The compute function is called

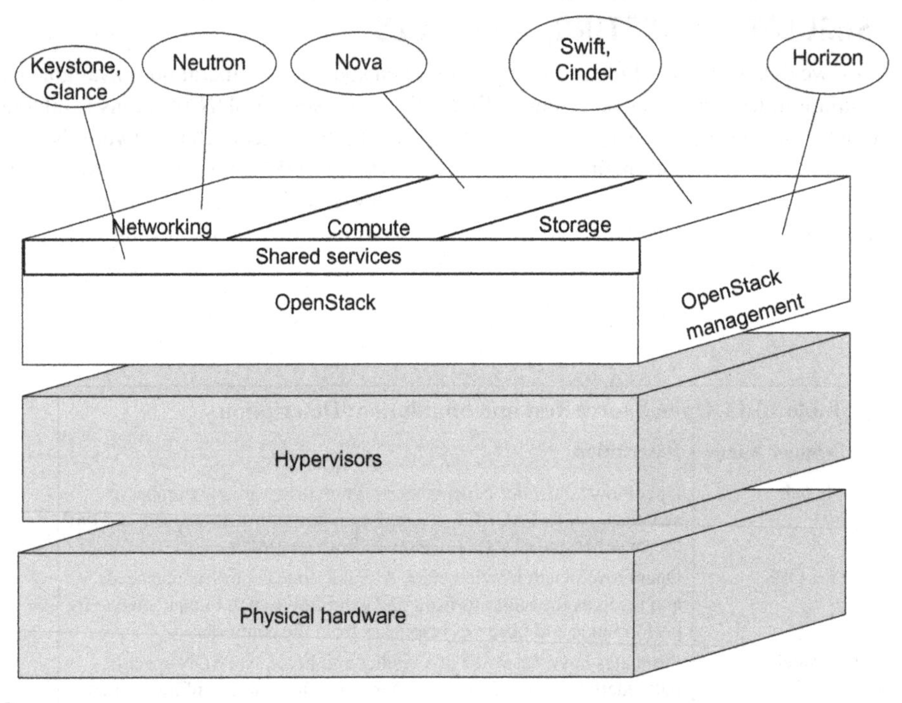

FIG. 13.3

OpenStack components and roles.

Nova. Nova works with the major available hypervisors to manage pools of virtual machines. Examples of the hypervisors that can be used with OpenStack include KVM, XenServer and VMware among others. The storage functions are *Swift* and *Cinder*. Swift provides redundant storage such that storage servers can be replicated or restored at will with minimum dependency on the commodity storage drives that provide the actual physical storage. Cinder provides OpenStack compute instances with access to file and block storage devices. This access can be used with most of the prevalent storage platforms found in cloud computing today. *Horizon* provides a dashboard to access, provision, and manage the cloud-based resources in an OpenStack environment. There are two shared services, *Keystone* and *Glance*. Keystone provides a user authentication service and can integrate with existing identity services such as LDAP. Glance provides the ability to copy and save a server image such that it can be used to replicate compute or storage servers as services are expanded. It also provides a fundamental backup capability for these images. The network virtualization component of OpenStack is provided by *Neutron* and, thus, is the component most relevant to our discussion of SDN. Note that Neutron was formerly known as Quantum.

Architecturally, Neutron's role in OpenStack is embodied as a number of plugins that provide the interface between the network and the balance of the OpenStack cloud computing components. While OpenStack is not limited to using Open SDN as its network interface, Open SDN is included as one of the networking options. Looking back at Fig. 13.1 we see that the Neutron plugin can interface to

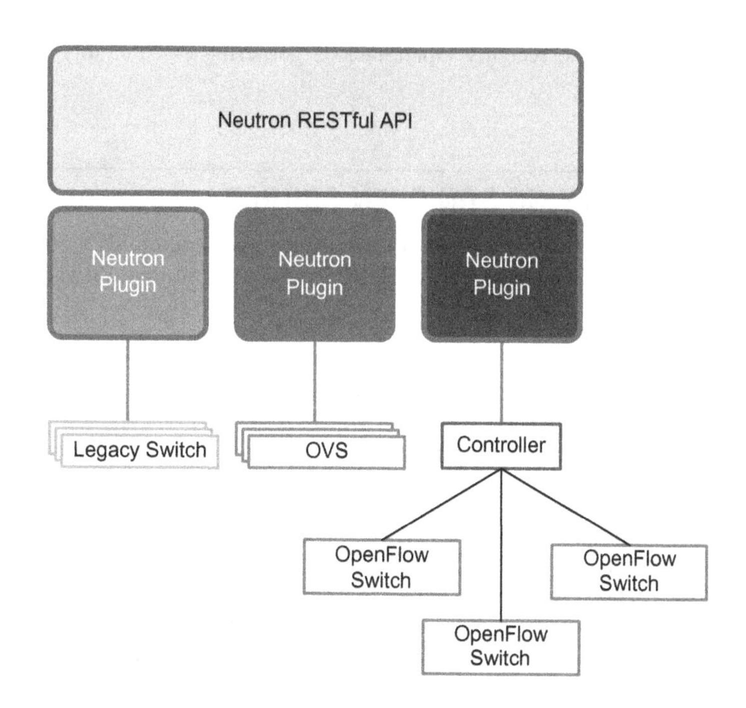

FIG. 13.4

OpenStack plugins.

the northbound API of an OpenFlow controller. Neutron can thus provide the network abstraction layer for an OpenFlow-enabled network. Just as OpenFlow can work with many different network-control applications via a northbound API, so can OpenStack's Neutron have many different kinds of network plugins. Thus OpenStack and OpenFlow may be married to provide a wholistic network solution for cloud computing, but neither is exclusively tied to the other. As shown in Fig. 13.4, OpenStack can use Neutron plugins to control (1) legacy network devices, (2) an OpenFlow controller that controls OpenFlow-enabled physical switches, or (3) virtual switches such as OVS [22]. The implementation of the OVS interface, for example, consists of the plugin itself, which supports the standard Neutron northbound APIs, and an agent that resides on the Nova compute nodes in the OpenStack architecture. An OVS instance runs locally on that compute node and is controlled via that agent.

OpenStack exposes an abstraction of a *virtual pool of networks*. This is closely related to the virtual networks abstraction we discussed in relation to the SDN via Overlays solution. So, for example, with OpenStack one could create a network and use that for one specific tenant, which maps quite well to the SDN via Overlays solutions concept. OpenStack has plugins for many of the existing overlay solutions.

13.12.2 CloudStack

CloudStack is Apache Foundation's alternative to OpenStack. Similar to OpenStack's Neutron plugin described previously, CloudStack supports a native plugin for the OVS switch [23]. This provides direct support of Open SDN within CloudStack. While the two competing open source cloud implementations have coexisted for several years, recently OpenStack is garnering considerably more support than CloudStack [24].

13.13 EXAMPLE: APPLYING SDN OPEN SOURCE

The *Routeflow* project [25] provides an excellent use case for open source SDN code. The Routeflow project has developed a complete SDN test network using only open source software. The goals of the Routeflow project are to demonstrate:

- a potential migration path from traditional layer 3 networks to OpenFlow-based layer 3 networks,
- a pure open source framework supporting different aspects of network virtualization,
- IP routing-as-a-service, and
- legacy routing equipment interoperating with simplified intra and interdomain routing implementations.

Fig. 13.5 depicts the network topology the project uses to demonstrate their proposed migration path from networks of legacy routers to OpenFlow. This is particularly important because it demonstrates a practical method of integrating the complex world of routing protocols with the OpenFlow paradigm. Today's Internet is utterly dependent on this family of routing protocols so, without a clean mechanism to integrate the information they carry into an OpenFlow network, the use of OpenFlow will remain constrained to isolated data center deployments. In Fig. 13.5 we see the test network connected to the Internet cloud with BGP routing information injected into the test network via that connection. The test

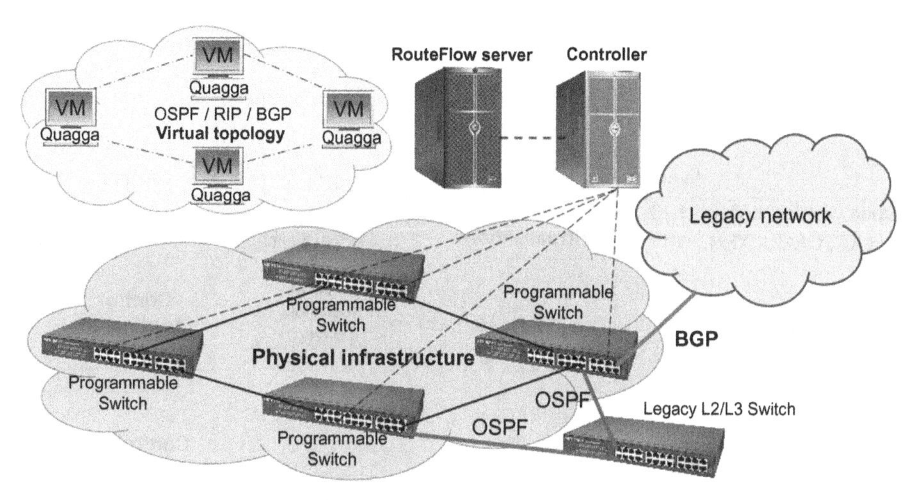

FIG. 13.5

Routeflow network topology.

(Source: CPqD. Retrieved from https://sites.google.com/site/routeflow/home.)

network itself includes one legacy layer 2/layer 3 switch which communicates routing information to the OpenFlow part of the test network via OSPF. In the test network we see a cloud of four OpenFlow-enabled switches under the control of an OpenFlow controller. This OpenFlow controller is connected to the RouteFlow server. The RouteFlow server, in turn, collects routing table information from the virtual switches in the adjacent cloud. Fig. 13.6 shows these same components in a system architecture view. It is significant that all of the components presented in Fig. 13.6 are derived from open source projects, though not all are SDN-specific.

The major components described in the figure are listed as follows, along with the source of the software used to build the component:

- Network (OpenFlow) controller: NOX, POX, Floodlight, Ryu
- Routeflow server: developed in Routeflow project
- Virtual topology Routeflow client developed in Routeflow project
- Virtual topology routing engine: Quagga routing engine, XORP
- Virtual topology switches: OVS
- OpenFlow-enabled hardware switches: for example, NetFPGA
- VMs in virtual topology: Linux virtual machines

The virtual routers shown in Fig. 13.6 correspond to the VMs in Fig. 13.5. Each virtual router is comprised of an OVS instance that forwards routing updates to the routing engine in the VM. This routing engine processes the OSPF or BGP updates in the legacy fashion and propagates the route and ARP tables in that VM accordingly. The key to the Routeflow solution is the Routeflow client that runs on the same VM instance that is in each virtual router. As shown in Fig. 13.6, the Routeflow client gathers the routing information in its local VM routing tables and communicates this to the

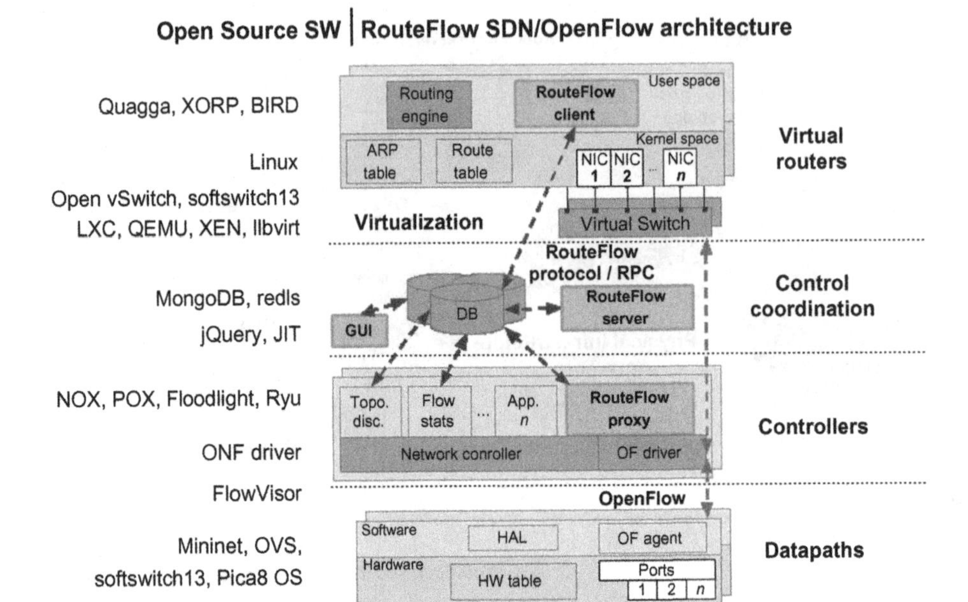

FIG. 13.6

Routeflow architecture.

(Source: CPqD. Retrieved from https://sites.google.com/site/routeflow/home.)

Routeflow server via the Routeflow protocol. The Routeflow server thus maintains global knowledge of the individual distributed routing tables in each of the virtual routers and can use this information to map flows to the real topology of OpenFlow-enabled switches. This information is communicated from the Routeflow server to the Routeflow proxy, which is an OpenFlow controller application running on the same machine as the OpenFlow controller. At this point, the routing information has been translated to OpenFlow tuples which can be programmed directly into the OpenFlow switches using the OpenFlow protocol. The process results in the OpenFlow-enabled hardware switches forwarding packets just as they would if they followed local routing tables populated by local instances of OSPF and BGP running on each switch, as is the case with legacy layer 3 switches. Routeflow is an example of the sort of proactive application that we described in Section 12.3. The flows are programmed proactively as the result of external route information. We contrast this to a hypothetical reactive version of Routeflow, wherein an unmatched packet would be forwarded to the controller, triggering the determination of the correct path for that packet, and the subsequent programming of a flow in the switch.

We have presented this test network showing OpenFlow-enabled hardware switches. Routeflow additionally supports software switches such as OVS for use with network virtualization applications. That is, while we described switches depicted in Fig. 13.5 as OpenFlow-enabled hardware switches, the Routeflow project has also demonstrated that these switches may be a softswitch that runs in the ToR alongside a hypervisor.

13.14 CONCLUSION

SDN's roots are squarely aligned with the open source movement. Indeed what we call Open SDN in this book remains closely linked with the open source community. In addition to reviewing the better-known SDN-related open software projects, we have provided insight into the many variants of open source licenses in common use. Different open source licenses may be more or less suitable for a given organization. We have defined three broad classes of users that have very different requirements and goals when they use open source software and we have explained why certain open source licenses might be inappropriate for certain classes of users. Depending on the class of user with which our reader associates himself or herself, we hope that it is now clear where the reader might find open source applicable for his/her SDN project. Whatever the commercial future of open source software is within the expanding SDN ecosystem, it is clear that the cauldron of creativity that spawned SDN itself could never have existed without open source software and the sharing of new ideas that accompanies it.

The open source movement may be a powerful one, but as the potential financial ramifications of SDN have become more apparent, other very strong forces have arisen as well. Some of these may be in concert with the open source origins of SDN while others may be in discord. In the next chapter we look at how those other forces are shaping the future of SDN.

REFERENCES

[1] Comprehensive list of open source SDN projects. SDNCentral. Retrieved from: http://www.sdncentral.com/comprehensive-list-of-open-source-sdn-projects/.

[2] Sorensen S. Top open source SDN projects to keep your eyes on. O'Reilly Community; 2012. Retrieved from: http://broadcast.oreilly.com/2012/08/top-open-source-sdn-projects-t.html.

[3] Casado M. List of OpenFlow software projects (that I know of). Retrieved from: http://yuba.stanford.edu/casado/of-sw.html.

[4] Various licenses and comments about them. GNU Operating System. Retrieved from: http://www.gnu.org/licenses/license-list.html.

[5] Wilson R. The Eclipse Public License. OSS Watch; 2012. Retrieved from: http://www.oss-watch.ac.uk/resources/epl.

[6] Erikson D. Beacon. OpenFlow @ Stanford; 2013. Retrieved from: https://openflow.stanford.edu/display/Beacon/Home.

[7] OpenSwitch. Retrieved from: http://openswitch.net.

[8] Willis N. Permissive licenses, community, and copyleft. LWN.net; 2015. Retrieved from: https://lwn.net/Articles/660428/.

[9] Muntaner G. Evaluation of OpenFlow controllers; 2012. Retrieved from: http://www.valleytalk.org/wp-content/uploads/2013/02/Evaluation_Of_OF_Controllers.pdf.

[10] Floodlight documentation. Project Floodlight. Retrieved from: http://docs.projectfloodlight.org/display/floodlightcontroller/.

[11] Jacobs D. Floodlight primer: an OpenFlow controller. Retrieved from: http://searchsdn.techtarget.com/tip/Floodlight-primer-An-OpenFlow-controller.

[12] Lawson S. Network heavy hitters to pool SDN efforts in OpenDaylight project. Network World; 2013. Retrieved from: http://www.networkworld.com/news/2013/040813-network-heavy-hitters-to-pool-268479.html.

[13] Duffy J. Big Switch's Big Pivot. Network World; September 11, 2013. Retrieved from: http://www.networkworld.com/community/blog/big-switchs-big-pivot.
[14] 2015 OpenDaylight Summit. Retrieved from: http://events.linuxfoundation.org/events/archive/2015/opendaylight-summit.
[15] Pitt D. ODL Summit. Open Collaboration. ONF Blog. Retrieved from: https://www.opennetworking.org/?p=1820&option=com_wordpress&Itemid=316.
[16] OpenDaylight Commit statistics. Retrieved from: http://spectrometer.opendaylight.org/.
[17] OpenDaylight GitHub Main Page. Retrieved from: https://github.com/opendaylight.
[18] Black Duck—OpenHUB. Retrieved from: https://www.openhub.net.
[19] Now available: our SDN security suite. OpenFlowSec.org. Retrieved from: http://www.openflowsec.org.
[20] Open source SDN sponsored software development. Retrieved from: http://opensourcesdn.org/ossdn-projects.
[21] Open source SDN leadership council. Retrieved from: http://opensourcesdn.org.
[22] Miniman S. SDN, OpenFlow and OpenStack Quantum. Wikibon; 2013. Retrieved from: http://wikibon.org/wiki/v/SDN,_OpenFlow_and_OpenStack_Quantum.
[23] The OVS Plugin. CloudStack Documentation. Retrieved from: http://docs.cloudstack.apache.org/en/latest/networking/ovs-plugin.html.
[24] Linthicum D. CloudStack, losing to OpenStack, takes its ball and goes home. InfoWorld–Cloud Computing; 2014. Retrieved from: http://www.infoworld.com/article/2608995/openstack/cloudstack-losing-to-openstack-takes-its-ball-and-goes-home.html.
[25] Welcome to the RouteFlow Project. Routeflow. Retrieved from: https://sites.google.com/site/routeflow/.

CHAPTER

BUSINESS RAMIFICATIONS

14

The fact that the very definition of SDN is vague confounds any attempt to assess its full business impact. If we narrow our focus to Open SDN, the business impact so far is minimal. The largest financial transactions to date, however, have revolved around technologies that only fall under broader definitions of SDN. This larger SDN umbrella encompasses any attempt to perform networking in a novel way that involves separation of control and data planes, unlike classical layers 2 and 3 switching. It also includes the creative use of existing protocols and platforms to provide solutions equally relevant and nearly as flexible as their Open SDN counterparts. By casting this wider net, we include companies and technologies that are often totally proprietary and whose solution is tightly focused on a specific network problem, such as network virtualization in the data center. But, as the saying goes, this has been where the money has gone, so this is where the greatest business impact has been.

There seems to be a growing consensus in the networking business community that SDN is here to stay and that its impact will be significant. This is supported by the sheer number of venture capital companies that have invested in SDN startups over the past 9 years, the magnitude of those investments, as well as the size of the acquisitions that have occurred since 2012 in this space. Unusually in the networking field, this particular paradigm-shift does not hinge on any breakthroughs in hardware technology. In general SDN is a software-based solution. This means that technical hurdles are more easily overcome than having to pass some gate-density or bits-per-second threshold. Customers generally view the advent of SDN with excitement and the hope of more capable networking equipment for lower costs. At first, entrenched NEMs may view it with trepidation as it is a disruptive force, and then usually find a way to jump on the SDN bandwagon by incorporating some form of the technology into their product offering. We will look at the dynamics of these various trends in the following sections.

In Sections 14.1–14.6 we review forecasts and trends related to the transition to SDN. When the data is historical, the reader is assured that it is not conjecture. With respect to the future, however, these remain just that—*forecasts*—and we encourage the reader to consider them accordingly. The SDN marketplace is in near constant flux. A sensible forecast one year can easily be proven folly just a couple of years hence.

14.1 EVERYTHING AS A SERVICE

In Fig. 14.1 we show that the ratio of *capital expenditures* (CAPEX) to *operational expenditures* (OPEX) on network equipment will *decrease* as we move through the first few years of SDN deployments. This is because the growth of SDN coincides with a fundamental shift in the way

Software Defined Networks. http://dx.doi.org/10.1016/B978-0-12-804555-8.00014-4

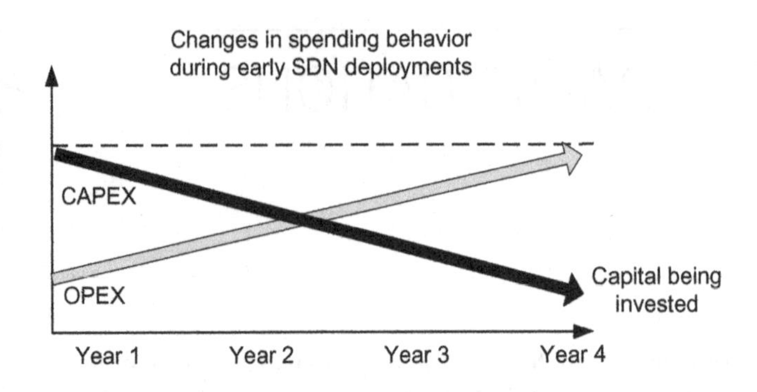

FIG. 14.1

CAPEX on network equipment migrates to OPEX.

that data centers and other large IT consumers will pay for their network equipment and services. New licensing models, subscription-based models and even usage-based models are all becoming increasingly common. The rapid expansion of new service delivery models such as *Software as a Service* (SaaS) and *Infrastructure as a Service* (IaaS) is testimony to this shift. The old model of enterprises purchasing network boxes from NEMs, amortizing them as a capital expense and then repeating the cycle with fork-lift upgrades every few years was a great business model for the NEMs but is gradually disappearing. These new models, where virtually every aspect of the networking business is available as a service and treated as an operational expense, will likely continue to displace the traditional network spending patterns. How quickly this process will unfold is a topic of hot debate.

14.2 MARKET SIZING

The net impact of a move to SDN technology will be influenced more than anything by the *total available market* (TAM) for SDN. At this early stage, forecasts are notoriously unreliable, but some effort has been expended to try to quantify what the total opportunity may be. Based on the studies reflected in [1], while SDN revenues in 2012 were less than 200 million dollars, they are likely to grow to more than 35 billion dollars by 2018. This is forecast to derive 60% from layers 2 and 3 SDN equipment sales and 40% from services equipment (layers 4–7). This growth will be primarily driven by a growth in network virtualization. It does not represent new growth in networking expenditures due to SDN, but rather the displacement of other network spending on SDN technology. Another study [2] predicts the SDN market to grow sixfold between 2014 and 2018, with SDN-enabled switches and appliances comprising the bulk of that revenue. Considering the SDN and NFV markets as a whole, more recent forecasts show the combined SDN and NFV market value ranging from $11 billion by 2018 [3] to $105 billion by 2020 [4].

14.3 CLASSIFYING SDN VENDORS

Network virtualization is happening and the term SDN is being blurred with this, for better or worse. What kind of company dominates network virtualization remains an open question, though. Certainly, there are at least three major classes of companies that are well positioned to make this transition. First, the NEMs themselves as by definition, already dominate the network equipment business. Secondly, since the server virtualization companies like VMware and Citrix already have a firm grasp on compute and storage virtualization, it is possible for them to make a horizontal move into network virtualization. Finally, since success in this area will likely depend on being successful at an SaaS business model, it would be wrong to discount the possibility of a software giant such as Microsoft dominating here. At this time, though, it appears that the battle for network virtualization will be fought between the titans of the network equipment industry and those of server virtualization. We take a deeper look at the impact of SDN on incumbent NEMs in Section 14.4. We first look at the server virtualization companies most likely to be impacted by SDN.

14.3.1 SERVER VIRTUALIZATION INCUMBENTS AND SDN

Since so much of the focus on SDN has been related to network virtualization in the data center, it is probable that incumbents that have traditionally been involved in compute and server virtualization will be drawn to extend their offerings to include network virtualization. The most salient example, VMware, has already surfaced many times earlier in this book. Already the world leader in server virtualization, VMware extended its reach into network virtualization with the acquisition of Nicira. Other companies that could follow suit include HP's server division, NEC and Citrix, among others. Not unusually, HP is a wild card here. HP is already both a major server manufacturer as well as a leading network equipment manufacturer. Of the larger incumbent NEMs, HP seems to be embracing Open SDN more enthusiastically than some of its competitors. Since HP's server division is already a powerful force in compute and storage servers, it would seem that HP was particularly well positioned to bring these two divisions together in a comprehensive networking-compute-storage virtualization story. HP will have to execute on this better than it has on most of its recent major initiatives, so it is impossible to predict what the outcome will be here.

14.3.2 VALUE-ADDED RESELLERS

One class of vendor that must not be neglected in this discussion is the *Value-Added Reseller* (VAR). Such companies, including well-known examples such as Presidio and Compucom, as well as numerous smaller companies, have long played a critical role in providing the customer face for the NEM. From a revenue perspective, this was possible because the network equipment sold had sufficient margins that the VAR could run a viable business based on their share of those margins. As illustrated in Fig. 14.1, part of the premise of SDN is that the hardware boxes become a smaller percentage of the overall network expenditure. Also, the *box sale* will not involve as much customer touch as it has traditionally and, therefore, the traditional value added by the VAR is not available to be added. It thus seems likely that in order to survive in an increasingly SDN world, these VARs will have to evolve into a service model of business. Indeed, their value was always in the service that they provided to their customers, but now this will not be at the box level but at the network or service level. This problem will be exacerbated as the NEMs like Cisco, due to the thinner margins, increasingly sell their boxes directly to the customers. SingleDigits and SpotOnNetworks are classic examples of companies that will need to shift with this change in business model. With SDN, such VARs will have the opportunity to innovate in specific markets of interest to their customer base, not merely sell the features that are provided to them by the NEMs.

We believe that these arguments will likely dictate the long-term direction of the VARs' business, but this process may unfold slowly. The NEMs will justifiably attempt to keep such companies selling their products via the old model. Because of the fact that they are so entrenched, incumbent NEMs have many tools at their disposal. These include aggressive, targeted pricing designed to keep big deals from swinging toward the new paradigm. Also, in order to switch to the service model described previously, the VAR needs to sell future flexibility to the customer *today*. This is never an easy sale. Whatever the timing of this process turns out to be, the companies most likely to be heavily impacted by a major transition to SDN are the incumbent NEMs.

DISCUSSION QUESTION

There is much talk about *Everything as a Service*. IaaS and SaaS are examples of this. Why does this present a challenge to the business models of the established NEMs?

14.4 IMPACT ON INCUMBENT NEMs

Incumbent NEMs are extremely astute marketing organizations. When they intuit that a tectonic shift in network purchasing patterns is about to occur, they generally will find a way to redefine themselves or the problem set so that their current solutions are in lockstep with the paradigm shift. In some cases, this will be just an instance of slide-ware and, in many other cases, there will be genuine shifts within the NEMs to align themselves in *some* way with the new paradigm.

This transition will not come without pain. There are already signs within some NEMs of belt-tightening attributed to market shifts related to SDN. For example, in early 2013, Cisco laid off 500 employees in part in an effort to realign with the advent of SDN [5]. The growth of SDN, along with network virtualization, will reduce the demand for legacy *purpose-built* routers and switches that have been the mainstay of the large NEMs' revenues for many years.

14.4.1 PROTECT MARKET SHARE

Large, entrenched NEMs, such as Cisco and Juniper, have to walk a fine line between protecting their market share and missing out on a new wave of innovation. For example, the high valuation given Nicira by VMware shown in Table 14.2 later in this chapter can clearly put Cisco, Juniper, and their ilk on the defensive. The NEMs are extremely competent in both technical and marketing terms and they have the financial resources to make sure they have a competitive answer to the network virtualization offering from VMware.

When an incumbent has a large war chest of cash, one defensive answer to a new, competing technology is to simply purchase one of the startups offering it and incorporate that product into its solutions. However, if this is not done carefully, this can actually *erode* market share. While the startup has the luxury of being able to tout disruptive technology as a boon to consumers, the incumbent must somehow portray that technology as complementary to its existing offerings. Failing to respond in this way, the company runs the risk of reducing sales of the traditional product line while the nascent technology is still gaining a foothold.

14.4.2 INNOVATE A LITTLE

While an established NEM can readily incorporate *adjacent* technologies into its product line and even gain market share as a result, doing so with a directly confrontational, paradigm-shifting technology like SDN is a much more delicate proposition. In order to avoid getting left behind, the incumbent will need to ostensibly adopt the new technology, especially in name. An incumbent can provide a solution that is competitive to the paradigm shift by adhering to the following model:

- purport to solve the same problem (e.g., network virtualization);
- remain based on the incumbent's then-current technology;
- innovate by layering some aspects of the new technology on top of the existing technology;
- employ widespread use of the buzz-words of the new technology in all marketing materials (e.g., SDN washing).

The SDN via APIs approach, which we discussed in detail in Section 6.2, epitomizes the layering of the new approach on top of legacy technology. We recall that the underlying switching hardware of the leading NEMs has been optimized by custom hardware designs over many years. These hardware platforms may have inherent performance advantages over low-cost, *white-box* switches. If the incumbent provides a veneer of SDN over their already-superior switching hardware, this may present formidable competition to a new SDN paradigm of a separate and sophisticated control plane controlling low-cost commodity switches. Indeed, it is possible that from a pure performance metric such as packets switched per second, the pure SDN model may suffer when compared to high-end legacy switches. This veneer of SDN need not be just a stop-gap measure. As this evolves over time it can have the very positive effect of providing technologies that help their customers evolve to SDN. This can provide many of the benefits of SDN, albeit on proprietary technology. Most customers simply will not tolerate a fork-lift upgrade being imposed on them.

Following the model mentioned previously is a virtual necessity for large, established NEMs. This model provides them with margin preservation during a time of possibly great transition. If the new paradigm truly does become dominant, they are able to ride the coattails of its success, while slowing down the transition such that they are able to maintain robust margins without interruption. As we mentioned previously, they have the technical and marketing know-how to manage favorably this kind of transition. They also have something else, which may be even more important: *the bulk of the enterprise customers*. Such customers do not shift their mission-critical compute, storage or networking vendors willy-nilly, unless there are drastic technical shortcomings in their product offerings. By managing their customers' expectations carefully and innovating just enough, the advantage of owning the customer may be unbeatable.

A recent example of an NEM doing just this very successfully is Cisco's support of the BGP-LS/PCE-P plugin for OpenDaylight that we described in Section 7.2. Through this creative reuse of existing protocols and platforms, associated with an open source SDN controller project, Cisco has managed to create an offering that addresses many of the same needs purported to be solvable by Open SDN. While not as open or flexible as a comparable Open SDN solution, this plugin avoids the disadvantage of too much change too quickly, one of the potential pitfalls of Open SDN discussed in Section 6.1.1.

14.5 IMPACT ON ENTERPRISE CONSUMERS

Inevitably some of this book may read like a sort of SDN evangelist. The vast potential benefits of SDN have been inventoried and explained. It is easy to underestimate the risks that enterprises will run by severing their historic umbilical cords with the NEMs. These umbilical cords have been unwieldy and expensive, to be sure, but the security that comes with this safety net has undoubtedly provided countless IT managers with many a good night's sleep. We, the authors, are in fact SDN evangelists and believe that, at least in certain parts of networking infrastructure, SDN's power to innovate and to mitigate costs will indeed be a historically positive force. Nonetheless, it would be naive to trivialize the birthing pains of a major paradigm shift in such a highly visible area. In this section we will present an overview of both the positive business ramifications of this new technology paradigm (e.g., reduced costs, faster innovation) and also the downside of diluted responsibility when the network goes down.

The most prevalent solution described in the SDN context is network virtualization, which we have discussed at length throughout this book. While many of the solutions for network virtualization are not based on Open SDN, network virtualization via the varied flavors of SDN is a hot area of growth. According to [4], the portion of network purchases influenced by network virtualization is anticipated to increase from 16% in 2015 to almost 80% by 2020.

One challenge is that while these same customers understand compute and storage virtualization, for many, network virtualization remains a snappy catch phrase. Customers need considerable sophistication to understand how network virtualization will work for them.

14.5.1 REDUCED EQUIPMENT COSTS

In the situation of an existing data center, the hoped-for reduction in equipment costs by migrating to SDN may prove elusive or at least delayed. A solution like VMware's NSX touts the fact that its separate control plane can achieve network virtualization without switching out the legacy networking equipment. While this is desirable in the sense that it produces less equipment churn in the data center racks, it does nothing to lower the money being spent on the switching equipment itself. Under this scenario, the customer continues to pay for highly capable and costly layers 2 and 3 switches but only uses them as dumb switches. It is hard to see a fast path to reduced equipment costs in this scenario. Over the longer term and as the legacy switching equipment becomes fully amortized, the opportunity will arise to replace that legacy underlay network with lower cost OpenFlow-enabled switches.

In a greenfield environment, the Open SDN model prescribes intelligent software control of low-cost, OpenFlow-enabled switches. While this scenario is conceivable, that division of the control and data plane brings with it a concomitant dilution of responsibilities. In the past, when your data center network went down, the IT department could usually point their finger at their primary NEM, whether it be Cisco, Juniper, HP, or another. In a perfect, pure SDN world, the white-box switches are just hardware devices that perform flawlessly according to the programming received from their SDN controller. No one with operational experience with a big network would believe that things will actually play out according to that script. The customer will still want *one throat to choke*, as the colloquial saying goes. Getting one company to offer up that single throat to choke may be difficult when the network is instantiated on virtual and physical switches as well as a controller, all running on hardware purchased from three different vendors.

14.5.2 AVOIDING CHAOS

Who ya gonna call…(when the network goes down)? When it boils down to a CIO or IT manager making the decision to migrate any major part of the network from the legacy gear to SDN, the hype about greater ease and granularity of control, higher equipment utilization rate and the other pro-SDN arguments will run into a brick wall if that decision maker fears that network reliability may be jeopardized by making the transition. Because of this, it is a virtual certainty that any migration to SDN in typical enterprise customers will be a slow and cautious one. NEMs may incorporate fully integrated SDN-based solutions into their product portfolio, much like what has occurred with Linux in the world of servers. This will still provide the proverbial throat to choke while using SDN to meet specific customer needs.

We note that certain immense enterprises like Google and Yahoo! are outliers here in that they may have as much networking expertise as the large NEMs themselves. This explains why these organizations have been at the forefront of OpenFlow innovation along with selected research universities. As we explained in Section 11.9, this is evidenced by the composition of the board of the ONF. There are no NEMs on that board, but Yahoo! and Google are represented. For the more typical enterprise, however, this transition to SDN will be viewed with much more caution. Experiments will be done carefully with nonmission-critical parts of the network.

While these pro-SDN rationales are all cogent, a web search on *SDN Purchases* in 2016 yielded a list of acquisitions of SDN startups rather than major investments in the technology by users. In other words, there is more hype about companies buying SDN companies than about customers buying SDN products. This indicates that the process described previously is still in its very early stages. In general we should assume that large-scale transition to SDN will take much longer than most pundits have claimed. This will provide more time for incumbent NEMs to adapt in order to avoid a real shake-up in the networking industry.

DISCUSSION QUESTION

There are some signs that large NEMs like Cisco are transforming themselves and the very definition of SDN to protect their market position. There are examples of this strategy succeeding as well as failing. Give one example of each.

14.6 TURMOIL IN THE NETWORKING INDUSTRY

14.6.1 FEWER AND FEWER LARGE NEMs

One of the advantages of an ecosystem where a small number of NEMs make large profits off of a market they dominate was that the super-sized profits did in fact drive continued innovation. Each NEM tried to use the next generation of features to garner market share from its competitors. Generally, this next generation of features addressed real problems that their customers faced and innovation proceeded apace. A precept of Open SDN is that since users, via the open source community, will be able to *directly* drive this innovation, the innovation will be both more pertinent and less expensive. This thesis has yet to be proven in practice, though. That is, will the creativity of the open source community be able to match the self-serving creativity of the NEMs themselves? The NEMs' ability to maintain margins through innovation has resulted in much of the progress in data networking that our society has enjoyed over the past two decades, so the answer to this question is not an obvious one. We forecast that the NEMs will strive to perform a balancing act between innovating sufficiently to retain their market control and their desire to maintain the status quo.

14.6.2 MIGRATION TO CLOUD COMPUTING

If we take as a given that the need for network virtualization in large data centers is the prime driver behind SDN, then we must acknowledge that the growth of cloud computing is one of the major factors that has forced those data centers to grow ever larger and more complex. From the end user's perspective, cloud computing has grown more relevant due to transitions in a number of related elements:

- Usage-based payment models
- Digital media and service providers
- Digital media access and sharing providers
- Online marketplaces
- Mobile phone network providers
- Web hosting

More and more enterprises avail themselves of the cloud computing capabilities of Google and Amazon for a variety of applications. Apple's *iCloud* takes local management of your smartphone, tablets, and laptops and handles it in the cloud. Sharing services like *DropBox* and *GoogleDocs* offer media sharing via the cloud. Google offers a plethora of business applications via the cloud. Amazon Web Services offers web hosting on an unprecedented scale. As more users depend on the cloud for an ever-increasing percentage of their applications' needs, both the quantity and reliability of data centers are under constant pressure to increase. This force creates the pressure on the data center providers and the vendors that supply equipment to them to innovate to keep up with this relentless growth. The nature of the networks in these data centers is virgin soil for a new networking paradigm in network virtualization if it can help manage this growth in a controlled fashion [4].

IaaS provides networking capacity via the OPEX payment model. The growing popularity of this model in itself contributes to the turmoil in the networking industry where networking capacity has traditionally been obtained by purchasing equipment. Many of the SDN startups discussed in this chapter hope to displace incumbent technology by offering their solutions via some kind of subscription model.

14.6.3 CHANNEL DYNAMICS

Among the consequences of the migration to cloud computing are radical changes to the way that customers purchase networking capabilities and to the companies that traditionally had the direct customer touch in those sales. Networking operating systems and applications will be bundled with a mix of private and public cloud offerings. A new set of vendors skilled at cloud-based orchestration solutions will customize, deploy, and manage compute, storage, and networking virtualization functions. Based on the enterprise's migration to cloud services, network solution providers will migrate from selling networking boxes to offering networking services [6]. The VARs that now focus on reselling and supporting network hardware design and implementation will likely consolidate into larger and more comprehensive IT services companies. IT services companies will hire, train, and acquire skilled resources that understand how to design, implement, and manage *Web Operations Functions* (WebOps) and *Development and Operations* (DevOps) specialists on behalf of enterprise customers.

14.7 VENTURE CAPITAL

It is no coincidence that a significant amount of *venture capital* (VC) investments have been placed in SDN startups in the past 7 years. It is the nature of venture capital firms to seek out groups of bright and ambitious entrepreneurs who have aspirations in a space that is likely to see rapid growth within a well-bounded time horizon. Depending on the field, the distance to that time horizon can vary greatly. In the field of medical startups, this horizon may be more than 10 years after the initial investments. In data networking, however, the time horizons tend to be shorter, between 3 and 5 years. SDN technology is the perfect contemporary fit for the VC firm with a history of investing in networking startups. The market, at least in the data center, is clamoring for solutions that help them break what is perceived as a stranglehold by the large incumbent NEMs. The amount of money available to be spent on upgrades of data center networks in the next few years is enormous. SDN's time horizon and the corresponding TAM is like a perfect storm for VCs. SDN may be the largest transformation in the networking world in the past three decades [7]. This is not merely a US-based phenomenon, either. Some other countries, notably Japan and China, see SDN as a means to thwart the multidecade long hegemony that US NEMs have had on switching innovation. In these countries, both adoption by customers as well as offerings by local NEMs are likely to heavily influence the uptake of SDN in the coming years. In Table 14.1 we list VC firms that have been the earliest and most prominent investors in SDN startups.

These investors have furnished the nest from which many companies have been hatched. Some of these companies are still in startup mode and others have been acquired. As of the time of writing of this work, none of the investments listed in Table 14.1 have resulted in *initial public offerings* (IPOs). In the next sections we examine what has happened both with the major acquisitions that have resulted from these investments as well as those companies still striving to determine their future.

14.8 MAJOR SDN ACQUISITIONS

One of the most visible signs of major transformations in a given market is a sudden spate of large acquisitions of startups all purporting to offer the same new technology. Certainly 2012 was such a year for SDN startups. The size of the acquisitions and the concentration of so many in such a relatively short calendar period was an extraordinary indicator of the momentum that SDN had gathered by 2012. In Table 14.2 we provide a listing of the largest SDN-related acquisitions that took place in 2012–15.

Table 14.1 Major VCs Investing in SDN Startups as of 2015

Venture Capital Company	Invested in
Khosla Ventures	Big Switch
Redpoint Ventures	Big Switch
Goldman Sachs	Big Switch
Intel Capital	Big Switch
Benhamou Global Ventures	ConteXtream
Gemini Israel Funds	ConteXtream
Norwest Venture Partners	ConteXtream, Pertino
Sofinnova Ventures	ConteXtream
Comcast Interactive Capital	ConteXtream
Verizon Investments	ConteXtream
Lightspeed Venture Partners	Embrane, Pertino, Plexxi
NEA	Embrane
North Bridge Venture Partners	Embrane, Plexxi
Innovation Network Corporation of Japan	Midokura
NTT Group's DOCOMO Innovations	Midokura
Innovative Ventures Fund Investment (NEC)	Midokura
Jafco Ventures	Pertino
Hummer Winblad	Plumgrid
US Venture Partners	Plumgrid
Vantage Point Capital	Pica8
Battery Ventures	Cumulus
Matrix Partners	Plexxi
SEB Capital	Tail-f
Presidio Ventures	Embrane
New Enterprise Associates	Embrane
Sequoia Capital	Viptela

It is rare to see so many acquisitions in one field of technology of such large dollar value concentrated in so short a period of time. If SDN did not already serve as the honey to attract investors, it surely did by 2012–13. Most of the companies listed in Table 14.2 were already either major players in the SDN landscape at the time of their acquisition or are now part of large companies that plan to become leaders in SDN as a result of their acquisition. In this section, we will attempt to explain the perceived synergies and market dynamics that led to the acquisition of each.

14.8.1 VMware

Nicira's technology and the success they have enjoyed with it has been a recurring theme in this book. Nicira's acquisition by VMware stands out as the largest dollar-value acquisition purporting to be primarily about SDN. VMware was already a dominant force in the business of compute and storage software in large data centers. Prior to the acquisition, VMware was largely leaving the networking part

Table 14.2 Major SDN Acquisitions in 2012–15

Company	Founded	Acquired by	Year	Price	Investors
Nicira	2007	VMware	2012	$1.26B	Andreessen Horowitz Lightspeed Ventures
Meraki	2006	Cisco	2012	$1.2B	Google Felicis Ventures Sequoia Capital DAG Ventures Northgate Capital
Contrail	2012	Juniper	2012	$176M	Khosla Ventures Juniper
Cariden	2001	Cisco	2012	$141M	Not available
Vyatta	2006	Brocade	2012	Undisclosed	JPMorgan Arrowpath Venture Partners Citrix Systems HighBAR Partners
Insieme	2012	Cisco	2013	$863M	Cisco
Tail-f	2005	Cisco	2014	$175M	SEB Capital
Embrane	2009	Cisco	2015	Undisclosed	Cisco Presidio Ventures New Enterprise Associates Lightspeed Venture Partners North Bridge Venture Partners
ConteXtream	2007	HP	2015	Undisclosed	Benhamou Global Ventures Gemini Israel Funds Norwest Venture Partners Sofinnova Ventures Comcast Interactive Capital Verizon Investments
Pertino	2011	Cradlepoint	2015	Undisclosed	Norwest Venture Partners Lightspeed Venture Partners Jafco Ventures
Cyan	2006	Ciena	2015	$400M	Norwest Venture Partners Azure Capital Partners

of the data center to the NEM incumbents, most notably Cisco [8], with whom they have partnered on certain data center technologies such as VXLAN. Since SDN and the growing acceptance of the need for network virtualization were demonstrating that a shift was forthcoming in data center networking, it made perfect sense for VMware to use this moment to make a major foray into the networking part of the data center. Nicira provided the perfect opportunity for that. The explosive growth of cloud computing has created challenges and opportunities for VMware as a leader in the compute and storage virtualization space [9]. VMware needed a strategy to deal with cloud networking alternatives, such as

OpenStack and Amazon Web Services, and Nicira provides that. This competitive landscape is fraught with fuzzy boundaries, though. While VMware and Cisco have worked closely together in the past, and Nicira and Cisco collaborated on OpenStack, it is hard not to imagine the battle for control of the network virtualization in the data center pitting VMware and Cisco against one another in the future.

14.8.2 JUNIPER

Juniper was an early investor in Contrail [10]. This falls into the *spin-in* category of acquisition. Juniper recognized that it needed to offer solutions to address the unique east-west traffic patterns common in the data center, but atypical for its traditional WAN networking customers. By acquiring Contrail, Juniper obtained network virtualization software as well as network-aware applications that help address these data center-specific problems. Juniper additionally gained the Contrail management team, which consisted of senior managers from Google and Aruba as well as others with outstanding pedigrees that are targeted to help Juniper bring SDN solutions to enterprise customers. We described the Contrail data center solution in Section 6.2.5.

Contrail also uses the OpenStack standard for network virtualization. Thus support for OpenStack is a common thread between Cisco, VMware, and now Juniper. It is difficult to predict at this time whether these companies will use this standard as a platform for collaboration and interoperability or look elsewhere to distinguish their network virtualization offerings from one another.

DISCUSSION QUESTION

Considering what we have said about Contrail here and in previous chapters, explain how their technology leverages Juniper's strengths with service provider WAN solutions to create a beachhead for Juniper in the network virtualization marketplace.

14.8.3 BROCADE

Brocade needed a virtual layer 3 switch in order to connect virtual network domains for cloud service providers [11]. Brocade believes it has found the solution to this missing link in its acquisition of Vyatta. Vyatta delivers a virtualized network infrastructure via its on-demand network operating system. The Vyatta operating system is able to connect separate groupings of storage and compute resources. Vyatta's solution works both with traditional network virtualization techniques like VLANs as well as newer SDN-based methods [11]. Like HP, Brocade remains more committed to Open SDN than most other incumbent NEMs.

14.8.4 CISCO

Not surprisingly, in total dollar terms, the biggest SDN-acquirer in 2012 was Cisco. As seen in Table 14.2, Cisco acquired both Cariden and Meraki in 2012. Both of these acquisitions are considered in the networking literature [10,12] to be SDN-related acquisitions, though the Meraki case may be stretching the definition of SDN somewhat. Cariden was founded in 2001, well before the term SDN came into use. Cariden was respected as the developer of IP/MPLS planning and traffic engineering software. Cisco has used Cariden's software for managing networks to help achieve the industry-wide goal of being able to erect virtual networks with the same ease as instantiating a new virtual machine. Within Cisco, the Cariden platform has evolved into the *WAN Automation Engine* (WAE). As for Meraki, the value of the acquisition indicates the importance that Cisco places on this new member of the Cisco family. What distinguishes Meraki from Cisco's preacquisition technologies is that Meraki offers cloud-based control of the wireless APs and wired switches that it controls. This web-centric approach to device management fills a gap in Cisco's portfolio [12], providing it with a mid-market product in a space vulnerable to a competitive SDN-like offering. While we believe that Meraki is really an outlier in terms of the classical definition of an SDN company, the fact that their device control is separated from the data plane on the devices and implemented in the cloud supports their inclusion here as a major SDN acquisition.

Insieme was founded by Cisco in 2012 with plans to build *application-centric infrastructure* [13]. Insieme is a Cisco *spin-in* that received over $100 million in seed funding from Cisco. SDN is a game-changing technology that can be disconcerting to the largest incumbent in the field, Cisco. While Cisco brought to market a family of SDN-related products revolving around its commercial XNC controller, those actions had not convinced the market that Cisco had truly embraced SDN. The Insieme acquisition changed this. The high valuation given to the Insieme acquisition is affirmation of the importance Cisco is now giving to SDN. Understandably, the Insieme product offering attempts to pull potential SDN customers in a purely Cisco direction. While the new products may be configured to interoperate with other vendors' SDN products, they may also be configured in a Cisco-proprietary mode that purportedly works better. Significantly, the Insieme product line does represent a break with Cisco's legacy products. The new router and switch line from Insieme will not smoothly interoperate with legacy Cisco routers [14]. Cisco now emphasizes the Insieme controller, known as the *Application Policy Infrastructure Controller* (APIC), over the older XNC product. The company intends for the APIC to work with its installed base of legacy routers as well as the new equipment from Insieme [15].

Cisco acquired the Swedish-based Tail-f in 2014 to augment its capabilities for network configuration and orchestration. While Tail-f was founded in 2005, well before the hype around SDN and NFV began, it received a significant new round of funding in 2011 and repositioned itself as a provider of network orchestration tools tailored at bridging the gap between legacy *Operational Support Systems* (OSS) and the new SDN and NFV systems. Service providers see these new SDN and NFV technologies as a necessary step toward streamlining their operations and increasing their profitability. Being able to orchestrate the interoperation of these new tools without a radical departure from legacy systems is a significant challenge for service providers. Tail-f's solution is focused on this space. As Tail-f's customer base included more service providers than enterprise customers [16], this acquisition gives Cisco a wedge to drive itself further into the service provider market. The fact that AT&T is one of Tail-f's main customers is particularly significant, as this acquisition gives Cisco a more prominent place in AT&T's Domain 2.0 vendor program, where formerly it was relegated to a secondary role. The Tail-f solution allows a Cisco product to manage a variety of non-Cisco equipment. While it has not traditionally been Cisco's focus to work in a multivendor environment, this may be a requirement to advance its market share with service providers.

The managing of non-Cisco equipment with Cisco management tools is also a theme of another important Cisco acquisition, Embrane [17]. We describe Embranes's technology and business strategy as a stand-alone startup in Section 14.9.6. Embrane makes management software that facilitates the spin-up, shut down, and reconfiguration of NFVs. It has turned out that while NFVs are a formidable concept in and of themselves, the agility that they afford can only be readily harnessed via management systems tailored to this highly dynamic environment. In executing this acquisition in 2015, Cisco asserts that the Embrane technology dovetails well with the recently acquired Insieme product line.

14.8.5 HEWLETT-PACKARD

HP acquired ConteXtream in 2015 to strengthen its position in the NFV market [18]. We discuss ConteXstream's technology and business strategy as a stand-alone startup in Section 14.9.4. As part of its product lineup, ConteXstream offers an OpenDaylight-based controller. HP intends to increase its contributions to the OpenDaylight open source effort via this acquisition.

14.8.6 CIENA

Ciena added significantly to its SDN and NFV portfolio via its 2015 acquisition of Cyan. Through this acquisition, Ciena gains access to Cyan's SDN and NFV products as well as its optical networking hardware [19]. Cyan's product suite includes an NFV orchestration platform that Ciena plans to combine with its *Agility* software portfolio. It is noteworthy that Cyan does not appear in Table 14.3. The reason for this is that Cyan was founded as an optical chip designer long before the SDN movement began and thus was never a genuine SDN startup. More recently, Cyan made a major commitment to the SDN and NFV space which is why it is relevant in this discussion of major SDN acquisitions. Interestingly, Cyan actually had a moderately successful IPO in 2013, and thus Ciena's acquisition was of a publicly traded Cyan. Since we do not consider Cyan a genuine SDN startup, our earlier statement that none of our SDN startups in Table 14.1 have had a successful IPO remains true.

14.8.7 CRADLEPOINT

Pertino was acquired by Cradlepoint in late 2015. Cradlepoint will incorporate Pertino's SDN and cloud networking capabilities into Cradlepoint's 4G LTE offerings [20]. Pertino's SDN technology is expected to not only help Cradlepoint bolster their service provider business, but also to gain a stronger foothold with enterprise networks. In Section 14.9.4 we describe Pertino's strategy prior to this acquisition.

DISCUSSION QUESTION

NEMs with deep pockets often buy NFV and SDN startups to incorporate their innovative technology. Some believe that this slows down innovation from startups. Is there a way to prevent this from happening? If so, how?

14.9 SDN STARTUPS

The frenzy of customer interest in SDN combined with the willingness of investors to provide capital for new business ventures result in the unsurprisingly long yet inevitably incomplete list of SDN startups as of 2015 which we present in Table 14.3. SDN entrepreneurs are arriving so fast upon the scene that we cannot claim that this is an exhaustive list. Rather than attempting to cite every SDN startup, we hope to describe a succinct set of different markets that might be created or disrupted by SDN startups. For each of these, we will provide one or more examples of a new business venture approaching that market. The general categories we define are:

- OpenFlow stalwarts
- Network virtualization for the data center
- Network virtualization for the WAN
- Network functions virtualization
- Optical switching
- Mobility and SDN at the network edge

In deference to SDN purists, we begin with companies squarely focused on products compliant with the OpenFlow standard. Although this may represent to SDN purists the only true SDN, it has not had the financial impact that non-OpenFlow-focused network virtualization solutions have had. Indeed, network virtualization represents the bulk of investments and acquisition dollars expended thus far in SDN. For this reason, we will treat network virtualization in the two separate above-listed subcategories. We will conclude with some markets peripheral to the current foci in SDN that may have considerable future potential.

14.9.1 OpenFlow STALWARTS

One of the tenets of SDN is that by moving the intelligence out of the switching hardware into a general-purpose computer the switching hardware will become commoditized and thus low-cost. Part of the tension created by the emergence of SDN is that such a situation is not favorable to the incumbent NEMs, accustomed to high margins on their switches and routers. OpenFlow zealots generally respond that lower cost *original device manufacturers* (ODMs) will fill this space with inexpensive, OpenFlow-compatible, white-box switches. This, of course, opens the door for startups that view providing the white-box switches as a virgin soil opportunity.

As we described in Section 11.5, this is the basis of the strategy of Chinese-based startup Pica8 [21]. Pica8 provides a network operating system and reference architectures to help their customers build their SDN networks from low-cost switches, the Pica8 operating system and other vendors' OpenFlow-compatible products. Through its white-box ODM partners Accton and Quanta, Pica8 offers several Gigabit Ethernet switch platforms based on OVS that currently support OpenFlow 1.4 [22]. One wonders, though, if there is a way to turn this into a high margin business.

Big Switch also now has a business strategy centered on working with white-box switch ODMs. While Big Switch continues to put OpenFlow support at the cornerstone of their strategy, the company performed what they termed *a big pivot* in 2013 [23]. As one of the major players in the ecosystem, we described some of Big Switch's history in Section 11.4. They continue to offer the same set of commercial application, controller, and switch software products that they have all along. The pivot relates to selling them as bundles that white-box switches can download and self-configure. Their original product strategy included getting a number of switch vendors to become OpenFlow-compatible by facilitating this migration with the free switch code, thus creating a southbound ecosystem of switch vendors compatible with their controller. They also encouraged application partners to develop

Table 14.3 Startup Landscape in 2015

Company	Founded	Investors	Product Focus	Differentiator
Big Switch	2010	Khosla Ventures Redpoint Ventures Intel Capital Goldman Sachs	OpenFlow-based software	
ConteXtream[a]	2007	Benhamou Gbl Ventures Gemini Israel Funds Norwest Venture Ptr. Sofinnova Ventures Verizon Investments Comcast Interactive Cap.	Network virtualization in the data center	Grid computing
Cumulus	2010	Battery Ventures	White-box switch software	
Embrane[a]	2009	Lightspeed Venture Ptr. NEA North Bridge Venture Ptr.	Network virtualization in the data center	Network appliances via SDN
Midokura	2009	Innovation Network Corp. of Japan NTT Investment Ptr.	Network virtualization for Cloud Computing	IaaS
Nuage	2013	Alcatel-Lucent	Network virtualization	
Pertino[a]	2011	Norwest Venture Ptr. Lightspeed Venture Ptr. Jafco Ventures	Network virtualization for the WAN	
Pica8	2009	Vantage Point Cap.	White-box switch software	OpenFlow 1.2 OVS
Plexxi	2010	North Bridge Venture Ptr. Matrix Ptr. Lightspeed Venture Ptr.	Ethernet-Optical Switch Orchestration controller	SDN control of Optical Switches
Plumgrid	2011	Hummer Winblad US Venture Ptr.	Network virtualization	No Openflow
Pluribus	2010	Menlo Ventures New Enterprise Assoc. Mohr Davidow Ventures China Broadband Cap.	Physical switch	Integrated controller
Tallac	2012	–	Cloud-based SDN WiFi providing multitenant services	Open API to enable NFV services for vertical market applications

[a]*Has been acquired (see Table 14.2.)*

Table 14.3 Startup Landscape in 2015—Cont'd

Company	Founded	Investors	Product Focus	Differentiator
Vello	2009	–	Network Operating System (controller) Optical switches White-box Ethernet switches	OpenFlow support
Tail-f	2005	SEB Capital	Configuration and orchestration for SDN and NFV	
Viptela	2012	Sequoia Capital	SD-WAN	

applications to the northbound side of their controller. Such applications could be in the areas of firewalls, access control, monitoring, network virtualization, or any of a long list of other possibilities. The amount of investments they received [8,24] indicates that they were very successful at convincing others that their business model is viable.

After a rocky experience in trying to get production networks configured with different vendors' switches to interoperate with their controller and applications, they concluded that they needed to make the roll-out of the solution much easier than what they were experiencing with all but their most sophisticated customers. Big Switch concluded that by partnering instead with white-box ODMs, rather than established switch vendors like Juniper, Arista, and Brocade, they could better control the total user experience of the solution. This new approach, or pivot, has their customers purchasing white-box switches from Big Switch's ODM partners. These white-box switches are delivered with the *Open Network Install Environment* (ONIE) boot loader that can discover the controller and download Big Switch's *Switch Light* OpenFlow switch code [25]. Switch Light is based on the Indigo open source OpenFlow switch software. The notion is to sell the customer an entire bundle of switch, controller, and application code that is largely auto-configured, providing a particular solution to the customer.

There were two initial solution bundles offered at the time of the pivot, a network monitoring solution *Big Tap Monitoring Fabric*, and a cloud fabric for network virtualization. Significantly, the cloud fabric solution [26], called *Big Cloud Fabric*, is not based on an overlay strategy, but on replacing all of the physical switches with white-box switches running Big Switch's software. Big Switch is betting that by replacing the physical network with the low-cost white-box switches, and by having Big Switch stand behind the entire physical and virtual network as the sole provider, they will have created a compelling alternative to the many overlay alternatives being offered for network virtualization.

DISCUSSION QUESTION

The so-called *white-box strategy* is generally seen as a challenge to incumbent hardware-centric NEMs. While this strategy is symbiotic with Open SDN, can you identify possible weaknesses in the white-box strategy from a technical or business standpoint?

14.9.2 NON-OpenFlow WHITE-BOX VENTURES

Cumulus Networks takes the white-box boot loader concept and generalizes it one step further than Big Switch. As of this writing, Cumulus offers the product closest to the concept of *opening up the device* described in Section 6.4. As we pointed out in our discussion of white-box switches in Section 11.5, any switching code, OpenFlow or non-OpenFlow, controller-based or not, can be loaded into the Cumulus switches if written compatible with the Cumulus bootloader. Cumulus is a switch-centric company, and does offer the broad portfolio of controller and application software as does Big Switch.

14.9.3 AN OpenFlow ASIC?

Another OpenFlow hardware opportunity is to create a switching ASIC that is specifically designed to support OpenFlow 1.3 and beyond. Implementers have found it challenging to fully implement some parts of the advanced OpenFlow specifications in existing ASICs. One such challenge is supporting multiple, large flow tables. Since there is growing industry demand for the feature sets offered by OpenFlow 1.3, this creates an opportunity for semiconductor manufacturers to design an OpenFlow chip from scratch. Some of the incumbent switching chip manufacturers may be slow to displace their own advanced chips, but this is an inviting opportunity for a startup chip manufacturer. In Section 11.6 we described work being done at Mellanox that may be leading to such an advanced OpenFlow-capable ASIC. Intel has also announced switching silicon [27] that is explicitly designed to handle the multiple OpenFlow flow tables. This remains a technologically challenging area, however, and an active area of research [28].

14.9.4 DATA CENTER NETWORK VIRTUALIZATION

The team at ConteXtream has used their grid-computing heritage as the basis for their distributed network virtualization solution [29]. Theirs is an overlay network consisting of L4–L7 virtual switches [30]. It is touted as an SDN product, since the routing of sessions in the data center is controlled by a distributed solution with global knowledge of the network. This solution is very different from the classical OpenFlow approach of centralizing the network routing decisions to a single controller with a small number of backup controllers. In the ConteXtream approach, the control is pushed down to the rack level. If a control element fails, only a single rack fails, not the entire network. This per-rack control element is added to the TOR server. Admittedly, this architecture deviates significantly from the more classical SDN approaches of Big Switch or Nicira. In fact, a distributed algorithm with global knowledge of the network sounds similar to the definition of OSPF or IS-IS. The ConteXtream solution does differ significantly from classical routing in that (1) it can perform switching at the session level, (2) network devices like firewalls and load balancers can be virtualized into their software, and (3) the network administrators have explicit control over the path of each session in the network. ConteXtream's offering also includes provisioning of NFVs, which was the primary motivation cited in Section 14.8.5 behind HP's 2015 acquisition of ConteXtream.

As shown in Table 14.3, PLUMgrid has also received a sizable investment to develop network virtualization solutions. PLUMgrid offers an SDN via Overlays solution that is designed for data centers and integrates with VMware ESX as well as with *Kernel-based Virtual Machine* (KVM). PLUMgrid does not use OpenFlow for directing traffic through its VXLAN tunnels, but instead uses a proprietary mechanism. It also uses a proprietary virtual switch implementation rather than using OVS or Indigo.

PLUMgrid indicates that OpenFlow might be supported in the future [31], whereas it does currently support OpenStack [32].

Midokura [33] is another well-funded startup attacking the network virtualization market. Midokura currently integrates with OpenStack [34] and the CloudStack cloud-software platform [35]. Like other network virtualization companies, their product, MidoNet, creates virtual switches, virtual load balancers, and virtual firewalls. MidoNet is an SDN via Overlays product that runs on existing hardware in the data center. MidoNet claims to allow data center operators to construct public or private cloud environments through the creation of thousands of virtual networks from a single physical network. Since managing this plethora of virtual networks using traditional configuration tools is a huge burden, MidoNet provides a unified network management capability that permits simple network and service configurations of this complex environment. As shown in Table 14.3, Midokura's lead investors are large Japanese institutions, and some of the senior management are battle-hardened data communications veterans. Midokura is further evidence of the enthusiasm in Japan for SDN as is reflected by the incumbents NTT and NEC's interest in this technology [36,37].

14.9.5 WAN NETWORK VIRTUALIZATION: SD-WAN

SDN for the WAN, or SD-WAN as it is commonly known, has attracted a number of startups in recent years. The general idea espoused by these startups is to use an overlay strategy of virtual connections, often via some tunneling mechanism across a set of WAN connections, usually including the Internet itself. The mapping of traffic from branch offices of enterprises over these virtual connections is orchestrated by a centralized controller. This market is forecast to reach $7.5 billion by 2020 [38]. In this section we will review the technology and business of some of the more prominent startups in this field. We will explore the details of SD-WAN technology in greater depth in Section 15.2.

Pertino's offering is very different from the aforementioned data center-focused companies. They believe that there is a large market for *SDN-via-cloud*. Pertino offers a cloud service that provides WAN or LAN connectivity to organizations that wish to dynamically create and tear down secure private networks for their organization [39]. The term SDN-via-cloud is another stretch from the original meaning of the term SDN but, indeed, this concept truly is a *software defined network* in the sense that software configuration changes in Pertino's cloud spin up and spin down their customers' private networks through simple configuration changes. The only technical requirement for the individual users of these virtual networks is that they have Internet connectivity. A handful of branch offices from a small enterprise can easily obtain a secure, private network via the Pertino service as long as each of those branch offices has Internet connectivity. Such a private network can be brought on-line or off-line very easily under the Pertino paradigm. Pertino also attempts to make this service easy to use from a business perspective. This is another IaaS proposition. Their business model is a usage-based service charge, where the service is free for up to three users and after that there is a fixed monthly fee per user. As mentioned in Section 14.8.7, Cradlepoint acquired Pertino in 2015.

Viptela was founded in 2012 as an SD-WAN company. The Viptela *Secure Extensible Network* solution [40] for architecture transformation includes five steps:

1. Enable transport independence
2. Enable security at routing scale
3. Enable network-wide segmentation

4. Centrally enforce policy and business logic
5. Insert layers 4–7 services on demand

Viptela's offering separates the service provided from the underlying physical network by building an overlay network on top of the physical connections that exist. This allows the network transport level to be provisioned and operate independently of the underlying physical WAN connections. The Viptela solution is comprised of three major components: (1) the centralized *vManage* Network Configuration and Monitoring System, (2) the *vSmart* controller, and (3) the *vEdge* router one of which is deployed at every site belonging to the enterprise customer.

Other SD-WAN startups [41] include Aryaka, CloudGenix, Talari, and VeloCloud. Some established vendors in the SD-WAN space include Citrix, FatPipe, Ipanema, Silver Peak, and Riverbed. Unsurprisingly, Cisco is also a major player in SD-WAN via its *Intelligent WAN* (IWAN) offering.

14.9.6 NETWORK FUNCTIONS VIRTUALIZATION

Embrane has a distributed software platform for NFV that runs on commodity server hardware. The company's focus is to virtualize load balancers, firewalls, VPNs, and WAN optimization through the use of *distributed virtual appliances* (DVAs) [42]. Considering that a typical physical server in a data center hosts multiple customers with a pair of firewalls for each customer, the number of physical devices that can be removed by deploying their product, called Heleos, is significant [43]. They emphasize that Heleos functions at network layers 4–7, unlike the OpenFlow focus on layers 2 and 3. While Heleos does work with OpenFlow, it is not required. Indeed, Heleos integrates virtualization technology from multiple hypervisor vendors. One of Embrane's goals is to allow Heleos to coexist with existing network appliances to permit a gradual migration to the virtualized appliance model. In keeping with the growth of IaaS, in addition to the standard annual subscription fee, they support a usage-based model that charges an hourly rate for a certain amount of available bandwidth. As we described in Section 14.8.4, Embrane was acquired by Cisco in 2015.

Pluribus' offering is based on their server-switch hardware platform along with their NetVisor network operating system. Their target market is network virtualization of public and private clouds. The Pluribus product, like that of others mentioned previously, provides services like load balancing and firewalls that formerly required independent devices in the data center [31]. The Pluribus SDN controller is integrated with its server-switches such that their solution is completely distributed across the Pluribus hardware. The Pluribus solution also provides interoperability with OpenFlow controllers and switches by interfacing with the integrated Pluribus SDN controller. It also is compatible with the VMware NSX controller. The combination of the NetVisor network operating system and their proprietary server-switches attempts to provide a complete fabric for handling the complex cloud environment that requires coordination between applications, hypervisors, and the compute virtualization layer [39].

Nuage entered the SDN playing field with its *Virtualized Services Platform* (VSP) [44]. Alcatel-Lucent founded Nuage in 2013. Network operators can use VSP to set policies across the network, at tenant-level granularity. VSP is designed to control both virtualized and nonvirtualized infrastructures and, as such, can help service providers and enterprises implement cloud services. Nuage has also entered the NFV space, integrating with ETSI-standard MANO stacks. Nuage's future will undoubtedly be heavily influenced by the 2015 acquisition of Alcatel-Lucent by Nokia. The NFV and network virtualization talent in Nuage should nicely mesh with Nokia's efforts in those areas. As of this writing, Nuage remains an independent subsidiary under the Nokia-Alcatel-Lucent umbrella.

14.9.7 OPTICAL SWITCHING

Optical switches have been around for more than a decade now. They offer extremely high bandwidth circuit switching services, but have been comparatively unwieldy to reconfigure for dynamic networking environments. The SDN concept of moving the layers 2 and 3 control plane to a separate controller translates readily to the layer 1 control plane appropriate for optical switches. A couple of SDN startups, Plexxi and Vello, are attempting to leverage this natural SDN-optical synergy into successful business ventures. Plexxi has a dual mode switch that is both an Ethernet and an optical switch. The optical ports interconnect the Plexxi switches. The fact that each Plexxi switch is actually an optical switch allows extremely high bandwidth, low latency *direct* connections between many Plexxi switches in a data center. The Plexxi controller platform provides SDN-based network orchestration. This controller has been used to configure and control pure optical switches from other manufacturers such as Calient [45]. In this combined offering, the Plexxi switch's Ethernet ports are used for short, bursty flows and the Plexxi optical multiplexing layer is used for intermediate-sized flows. High volume, persistent flows (elephant flows) are shunted to the dedicated Calient optical switch.

Vello also offers Ethernet and optical switches. The Vello products are OpenFlow-enabled themselves and can interface to other vendors' OpenFlow-compatible switches [31]. While Vello, like most of the other startups discussed previously, addresses network virtualization, they focus on use cases and applications related to storage devices.

14.9.8 MOBILITY AND SDN AT THE NETWORK EDGE

Current SDN efforts have focused primarily on the data center and carrier WAN, but Tallac Networks is expanding that focus with an emphasis on managing services for the Campus LAN [6,46]. Tallac technology offers a new approach called *Software Defined Mobility* (SDM), which enables MSPs to connect users to the network services they demand. One powerful aspect of OpenFlow is the ability to virtualize a network so it can be easily shared by a diverse set of network services. In the context of *Network-as-a-Service* (NaaS), this means creating a multitenant WiFi environment. The fine-grained control of the OpenFlow protocol allows network connections and secure tunnels to be provisioned dynamically. Through a set of technologies combining OpenFlow, access to Network Functions, and standards-based Hotspot 2.0 (Passpoint), Tallac delivers an SDN-enabled, multitenant, service-oriented WiFi network for wireless LAN operators to centrally administer networks and services. Combined with NaaS business frameworks, Tallac's cloud service enables network operators to bring a multitude of applications, services, and network functions right to the users that need them.

14.10 CAREER DISRUPTIONS

Most of this chapter has focused on the potential disruption SDN is likely to cause to the incumbent NEMs and the opportunities it affords startup enterprises. We need to remember that the driving force inspiring this new technology was to simplify the tasks of the network administrators in the data center.

Most of the marketing materials for the startups discussed in this chapter declare that the current network administration environment in the ever-more-complex data center is spiraling out of control. One consequence of this complexity has been a seemingly endless demand for highly skilled network administrators capable of the black art of data center network management. If *any* of the SDN hype presented in this chapter proves to be real, one certain outcome is that there will be less of a need for highly skilled technology-focused network engineers to manually reconfigure network devices than there would be without the advent of SDN [43]. For example, consider that one of the highly sought-after professional certifications today is that of the *Cisco Certified Internetwork Expert* (CCIE). This sort of highly targeted, deep networking knowledge will become less and less necessary as SDN gains a larger foothold in the data center [4,42]. In [47], the author cites the VLAN provisioning technician as an example of a particularly vulnerable career as SDN proliferates.

There will, however, be a growing need in the data center for the kind of professional capable of the sort of agile network service planning depicted in Fig. 14.2. Much like the *Agile Development Methodology* that has turned the software development industry on its head in recent years, the network professionals in an SDN world will be required to rapidly move through the never-ending cycles of *plan*, *build*, *release*, and *run* shown in Fig. 14.2 in order to keep pace with customers' demands for IaaS. *DevOps* is a term now commonly used for this more programming-aware IT workforce that will displace the traditional IT network engineer. This new professional will possess a stronger development operations background than the traditional CLI-oriented network administrator [48]. Since SDN increases how much innovation IT professionals will be able to apply to the network, professionals working in this area today need to embrace that change and ensure that their skills evolve accordingly.

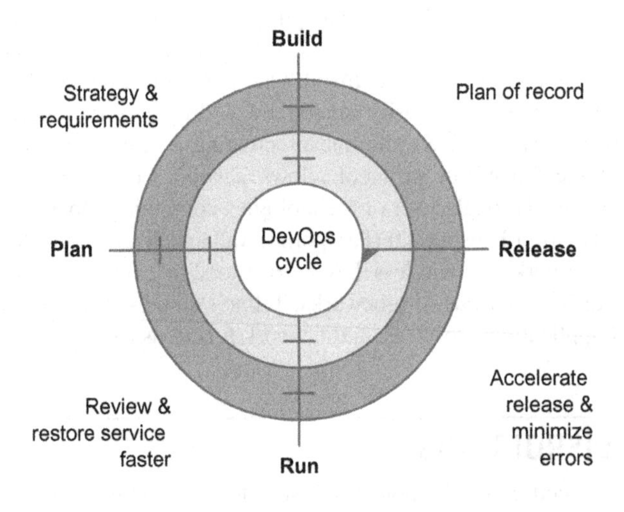

FIG. 14.2

Agile environment for network administrators.

DISCUSSION QUESTION

Describe how the job of a network administrator for a data center is likely to have changed between 2010 and 2020?

14.11 CONCLUSION

There is a historical pattern with new Internet technologies where the hype surrounding the possibilities and financial opportunities reaches a frenzy that is out of proportion to its actual impact. We have seen this happen numerous times over the past two decades. FDDI and WIMAX are but two examples of this trend. While it may be likely that the current hype surrounding SDN is probably near its peak [11], it does seem inevitable that some form of SDN will take root. In particular, de-embedding the control plane from the physical switch for data center products seems to be an unstoppable trend. The technical underpinnings of this idea are very sound and the amount of investment that has poured into this area in the past few years is ample evidence of the confidence in this technology. The big question from a business ramifications standpoint is whether or not there will be a sea-change in the relationship between vendors and customers, or will the degree of vendor-lock that the large NEMs enjoy today persist in a slightly modified form. Idealistic SDN proponents proselytize that the adoption of SDN will take the power from the NEMs and put it in the hands of customers. Whether or not this happens largely hinges on whether or not openness is truly a mandatory part of the SDN paradigm. In the business world today there is no consensus on that and we acknowledge that separation of control plane from data plane need not be performed in an open manner. It is entirely possible that proprietary incarnations of this concept end up dominating the market. This is still an open question and the answer to it will undoubtedly be the biggest determinant of the business ramifications of SDN.

In the next and final chapter we attempt to synthesize the current state of SDN research with current market forces to forecast where SDN is headed in the future.

REFERENCES

[1] Palmer M. SDNCentral exclusive: SDN market size expected to reach $35B by 2018. SDN Central; 2013. Retrieved from: http://www.sdncentral.com/market/sdn-market-sizing/2013/04/.

[2] Burt J. SDN market to grow sixfold over 5 years: Dell'Oro. eWeek; 2013. Retrieved from: http://mobile.eweek.com/blogs/first-read/sdn-market-to-grow-six-fold-over-5-years-delloro.html.

[3] Howard M. Infonetics forecasts carrier SDN and NFV market to reach $11 billion by 2018. Infonetics Research; 2014. Retrieved from: http://www.infonetics.com/pr/2014/Carrier-SDN-NFV-Market-Highlights.asp.

[4] SDxCentral SDN and NFV market size report. SDX Central; 2015. Retrieved from: https://www.sdxcentral.com/wp-content/uploads/2015/05/SDxCentral-SDN-NFV-Market-Size-Report-2015-A.pdf.

[5] Duffy J. Cisco lays off 500: company realigning operations, investments on growth areas. Network World; 2013. Retrieved from: http://www.networkworld.com/news/2013/032713-cisco-layoffs-268165.html.

[6] Geer D. When SDN meets cloud-managed Wi-Fi control, HotSpot 2.0 gets easier. Light Reading. Retrieved from: http://searchsdn.techtarget.com/feature/When-SDN-meets-cloud-managed-Wi-Fi-control-HotSpot-20-gets-easier.

[7] Heavy reading rates the 10 most promising SDN startups. InvestorPoint; 2012. Retrieved from: http://www.investorpoint.com/news/MERGERAC/55405881/.
[8] Williams A. Big Switch raises $6.5M from Intel Capital, gains attention for next generation networking, challenges Cisco. Techcrunch; 2013. Retrieved from: http://techcrunch.com/2013/02/08/big-switch-raises-6-5m-from-intel-capital-gains-attention-for-next-generation-networking-challenges-cisco/.
[9] Williams A. VMware Buys Nicira for $1.26 billion and gives more clues about cloud strategy. Techcrunch; 2012. Retrieved from: http://techcrunch.com/2012/07/23/vmware-buys-nicira-for-1-26-billion-and-gives-more-clues-about-cloud-strategy/.
[10] Higginbotham S. Juniper to buy SDN startup Contrail in deal worth $176M. Gigaom; 2012. Retrieved from: http://gigaom.com/2012/12/12/juniper-to-buy-sdn-startup-contrail-in-deal-worth-176m/.
[11] Hachman M. Brocade buys Vyatta for SDN Tech. Slashdot; 2012. Retrieved from: http://slashdot.org/topic/datacenter/brocade-buys-vyatta-for-sdn-tech/.
[12] Malik O. Here is why Cisco bought Meraki for $1.2 billion in cash. Gigaom; 2012. Retrieved from: http://gigaom.com/2012/11/18/cisco-buys-meraki-for-1-2-billion-in-cash-here-is-why/.
[13] Burt J. Cisco may unveil Insieme SDN Technology, Nov 6. eWeek; 2013. Retrieved from: http://www.eweek.com/blogs/first-read/cisco-may-unveil-insieme-sdn-technology-nov.-6.html.
[14] Bort J. Cisco launches its secret startup Insieme, then buys it for $863 million. Business Insider; 2013. Retrieved from: http://www.businessinsider.com/cisco-buys-insieme-for-863-million-2013-11.
[15] Matsumoto S. Cisco slims its SDN story down to ONE controller. SDX Central; 2014. Retrieved from: https://www.sdxcentral.com/articles/news/cisco-slims-sdn-story-one-controller/2014/03/.
[16] Kerravala Z. Why the Tail-f acquisition is a big win for Cisco. Network World; 2014. Retrieved from: http://www.networkworld.com/article/2364311/cisco-subnet/why-the-tail-f-acquisition-is-a-big-win-for-cisco.html.
[17] Wagner M. Cisco to acquire Embrane for network services. Light Reading; 2015. Retrieved from: http://www.lightreading.com/carrier-sdn/sdn-technology/cisco-to-acquire-embrane-for-network-services-/d/d-id/714828.
[18] Duffy J. HP buying SDN company for NFV. Network World; 2015. Retrieved from: http://www.networkworld.com/article/2926795/cloud-computing/hp-buying-sdn-company-for-nfv.html.
[19] Meyer D. Ciena boosts SDN, NFV platform with $400M Cyan acquisition. RCR Wireless News; 2015. Retrieved from: http://www.rcrwireless.com/20150504/telecom-software/ciena-boosts-sdn-nfv-platform-with-400m-cyan-acquisition-tag2.
[20] Talbot C. Cradlepoint boosts SDN capabilities with Pertino acquisition. Fierce Enterprise Communications; 2015. Retrieved from: http://www.fierceenterprisecommunications.com/story/cradlepoint-boosts-sdn-capabilities-pertino-acquisition/2015-12-09.
[21] Rath J. Pica8 launches Open Data Center Framework. Data Center Knowledge; 2013. Retrieved from: http://www.datacenterknowledge.com/archives/2013/04/15/pica8-launches-open-data-center-framework/.
[22] McGillicuddy S. Pica8 claims to be first vendor to support OpenFlow 1.4. SearchSDN; 2014. Retrieved from: http://searchsdn.techtarget.com/news/2240219935/Pica8-claims-to-be-first-vendor-to-support-OpenFlow-14.
[23] Duffy J. Big Switch's Big Pivot. Network World; 2013. Retrieved from: http://www.networkworld.com/community/blog/big-switchs-big-pivot.
[24] Wauters R. OpenFlow startup Big Switch raises $13.75M from Index, Khosla ventures. TechCrunch; 2011. Retrieved from: http://techcrunch.com/2011/04/22/openflow-startup-big-switch-raises-13-75m-from-index-khosla-ventures/.
[25] Kerner S. Big Switch switches SDN direction, staying independent. Enterprise Networking Planet; 2013. Retrieved from: http://www.enterprisenetworkingplanet.com/datacenter/big-switch-switches-sdn-direction-staying-independent.html.

[26] Jones P. Big Switch goes hyperscale with cloud fabric. Datacenter Dynamics; 2014. Retrieved from: http://www.datacenterdynamics.com/it-networks/big-switch-goes-hyperscale-with-cloud-fabric/88088.fullarticle.
[27] Ozdag R. Intel Ethernet Switch FM6000 Series—software defined networking. Intel White Paper. Retrieved from: http://www.intel.com/content/dam/www/public/us/en/documents/white-papers/ethernet-switch-fm6000-sdn-paper.pdf.
[28] OpenFlow optimized Switch ASIC design. ONRC Research. Retrieved from: http://onrc.berkeley.edu/research_openflow_optimized.html.
[29] Higginbotham S. ConteXtream joins the software defined networking rush. Gigaom; 2011. Retrieved from: http://gigaom.com/2011/12/13/contrextream-joins-the-software-defined-networking-rush/.
[30] Duffy J. SDN company helps build clouds. Network World; 2011. Retrieved from: http://www.networkworld.com/article/2184029/data-center/sdn-company-helps-build-clouds.html?nsdr=true.
[31] Guis I. SDN start-ups you will hear about in 2013—snapshots of Plexxi, Plumgrid, Pluribus and Vello. SDN Central; 2013. Retrieved from: http://www.sdncentral.com/companies/sdn-start-ups-you-will-hear-about-in-2013-pt-2/2013/01/.
[32] Kerner S. PLUMgrid secures OpenStack with networking suite. eWEEK; 2014. Retrieved from: http://www.eweek.com/cloud/plumgrid-secures-openstack-with-networking-suite.html.
[33] Etherington D. Midokura scores $17.3M series A to ramp up its network virtualization offering on a global scale. Techrunch; 2013. Retrieved from: http://techcrunch.com/2013/04/01/midokura-scores-17-3m-series-a-to-ramp-up-its-network-virtualization-offering-on-a-global-scale/.
[34] Midokura announces integration with Mirantis OpenStack for easier, affordable and more scalable networking. OpenStack News; 2015. Retrieved from: https://www.openstack.org/news/view/112/midokura-announces-integration-with-mirantis-openstack-for-easier,-affordable-and-more-scalable-networking.
[35] Dayaratna A. Midokura announces integration with CloudStack. Cloud Computing Today; 2013. Retrieved from: http://cloud-computing-today.com/2013/07/03/midokura-announces-integration-with-cloudstack/.
[36] Pica8 partners with NTT data for end-to-end SDN solutions. BusinessWire; 2013. Retrieved from: http://www.businesswire.com/news/home/20130402005546/en/Pica8-Partners-NTT-Data-End-to-End-SDN-Solutions.
[37] NEC displays SDN leadership at PlugFest. In: InCNTRE. Indiana University; 2012. Retrieved from: http://incntre.iu.edu/featured/NEC-displays-SDN-leadership-at-PlugFest.
[38] Savage M. SD-WAN gaining traction. Network Computing; 2015. Retrieved from: http://www.networkcomputing.com/networking/sd-wan-gaining-traction/1910292861.
[39] Clancy H. Pertino serves up small-business networks, in the cloud. ZDNet; 2013. Retrieved from: http://www.zdnet.com/pertino-serves-up-small-business-networks-in-the-cloud-7000013277/.
[40] Musthaler L. Viptela brings software-defined WANs to the enterprise. Network World; 2015. Retrieved from: http://www.networkworld.com/article/2873608/sdn/viptela-brings-software-defined-wans-to-the-enterprise.html.
[41] Conde D. SD-WAN: the Killer App for enterprise SDN? Network Computing; 2015. Retrieved from: http://www.networkcomputing.com/networking/sd-wan-killer-app-enterprise-sdn/1747434541.
[42] Berndtson C. Startup Embrane looks beyond the SDN trend. CRN; 2012. Retrieved from: http://www.crn.com/news/networking/240004082/startup-embrane-looks-beyond-the-sdn-trend.htm.
[43] Higginbotham S. Embrane's virtual network appliances for an SDN world. Gigaom; 2011. Retrieved from: http://gigaom.com/2011/12/11/embranes-virtual-network-appliances-for-an-sdn-world/.
[44] Wagner M. Will Nokia appreciate AlcaLu's Nuage? Light Reading; 2015. Retrieved from: http://www.lightreading.com/carrier-sdn/sdn-technology/will-nokia-appreciate-alcalus-nuage/d/d-id/715115.
[45] Lightwave Staff, Calient, Plexxi bring SDN to large-scale data center networks. Lightwave; 2013. Retrieved from: http://www.lightwaveonline.com/articles/2013/03/calient-plexxi-bring-sdn-to-large-scale-data-center-networks.html.

[46] Bent K. The 10 coolest networking startups of 2014. CRN; 2014. Retrieved from: http://www.crn.com/slide-shows/networking/300075121/the-10-coolest-networking-startups-of-2014.htm/pgno/0/5.
[47] Pepelnjak I. SDN's casualties. ipSpace; 2013. Retrieved from: http://blog.ioshints.info/2013/06/response-sdns-casualties.html.
[48] Knudson J. The ripple effect of SDN. Enterprise Networking Planet; 2013. Retrieved from: http://www.enterprisenetworkingplanet.com/netsysm/the-ripple-effect-of-sdn.html.

CHAPTER

SDN FUTURES

15

An accurate forecast of the future of any nascent technology calls for a better crystal ball than what is usually available. To attempt such a look into the future requires that we first lift ourselves up out of the details of SDN on which we have focused in the latter half of this book and view the current state of SDN affairs with a broader perspective. We need to consider what lies in store for SDN as a whole as well as individually for each of the SDN alternatives described in Chapter 6. We are optimistic about the future of SDN. Our optimism for the future of SDN is partly founded on the belief that there are many potential areas of application for the technology that lie outside those we explored in Chapters 8–10. For that reason, we dedicate a significant part of this chapter to revealing just a few of the novel use cases and research areas for Open SDN that are beginning to attract attention. Before that, however, we should acknowledge that not all contemporary press releases about SDN are filled with unbridled optimism. In the next section, we look at the turbulence and some of the pessimism surrounding SDN.

15.1 CURRENT STATE OF AFFAIRS

The turbulence that has surrounded the SDN movement during these early years has closely tracked the well-known *Gartner Hype Cycle* [1]. We depict a generic version of this hype cycle in Fig. 15.1. For SDN, the *technology trigger* was the birth of OpenFlow and the coining of the term Software Defined Networks in 2009. [1] The MAC table and VLAN ID exhaustion we have discussed in earlier chapters were manifesting themselves in the data center in the years leading up to 2012. This provided hard evidence that networking was at the breaking point in the data center. The fact that SDN provides a solution to these problems added fuel to the SDN fire. The *peak of inflated expectations* was reflected in the flurry of acquisitions in 2012 and the sizable VC investments preceding and coinciding with that. The *trough of disillusionment*, in our opinion, occurred in early 2013.

Big Switch's fate leading up to 2013 and then in the first half of that year mirrored these trends. Their withdrawal first from board status of the OpenDaylight consortium and subsequent total exit reflected their feeling that forces other than a commitment to Open SDN were influencing that organization's choice of controller technology. That was followed by their *big pivot* which was an admission that their early strategy of open platforms needed to be retooled to be plug-and-play solutions for a broader

[1]When we refer to the first use of the term Software Defined Networks we mean in the specific context in which it is used in this book and in conjunction with the work surrounding OpenFlow at Stanford University. The term itself was used previously in patents in the 1980s.

Software Defined Networks. http://dx.doi.org/10.1016/B978-0-12-804555-8.00015-6

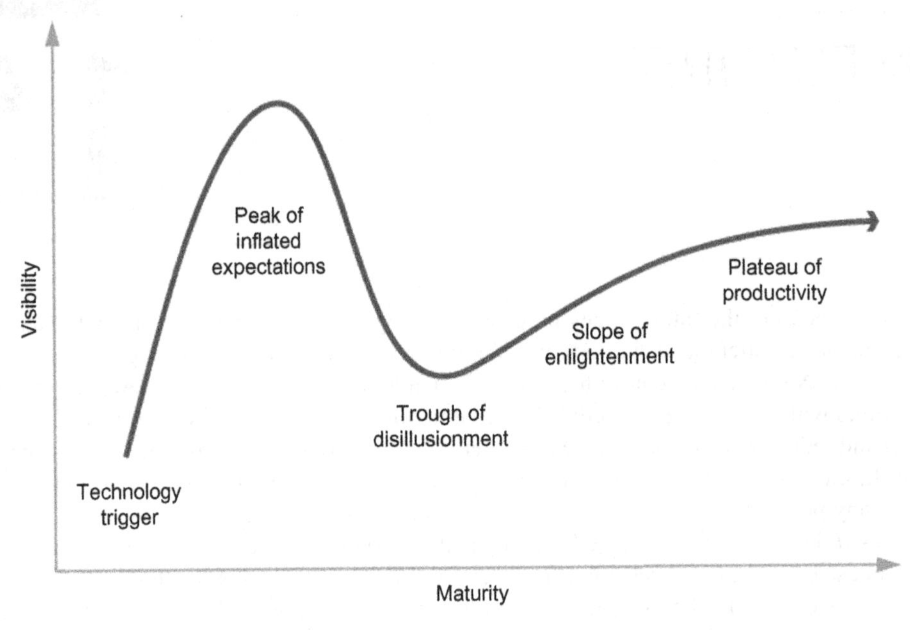

FIG. 15.1

Gartner Hype Cycle.

(Source: Gartner, Inc. Research methodology. Retrieved from: http://www.gartner.com/it/products/research/methodologies/research_hype.jsp.)

customer base (see Section 14.9.1). Big Switch's moves during 2013 were reflective of a more general fragmentation of the SDN movement.

Many feel that the degree of SDN washing that now occurs means that SDN can mean anything to anyone. Indeed, one of the founding fathers of SDN listed in Section 11.1.1, Martin Casado, was quoted in April 2013 as saying "I actually don't know what SDN means anymore, to be honest" [2]. If this is the feeling of one of the core group that coined the very term SDN then one might reasonably fear that SDN has lost much of its original focus.

The Gartner Hype Cycle does not end with the trough of disillusionment, however, nor does Open SDN's story begin to fall apart in 2013. Just as the Hype Cycle predicts that disillusionment is followed by the *slope of enlightenment* and then by the *plateau of productivity*, so do we now begin to see realistic assessments for where SDN will be applied. We believe that this slope of enlightenment in SDN's future reflects a middle ground between the extremes of belief that it will totally unseat the incumbent NEMs and traditional switching technology, and believing that it is largely an academic exercise with applicability restricted to the 1% of customers sophisticated enough to grapple on their own with the technological change. As an example, Big Switch's pivot has moved the company toward a strategy based on OpenFlow-enabled white-box switches. Thus, while their strategy is more tactical than it had been, they are still betting on the long-term future of OpenFlow. Furthermore, SDN has already established a beachhead in the data center that demonstrates that it is there to stay. SDN experts from HP, Dell, and NEC debating the future of SDN in [3] caution against impatience with the rate of penetration of SDN. They point out that server virtualization in the data center has existed for over a decade and only now is reaching 50% market penetration.

DISCUSSION QUESTION

In which of the five phases depicted in Fig. 15.1 do you believe SDN found itself in 2015?

We believe that for SDN to transition from the slope of enlightenment to the plateau of productivity we must first see evidence of an entirely new generation of switching ASICs that have been purposely built with SDN and OpenFlow in mind. Switch ASICs may take 2 years or more to mature and reach production. Perhaps the amount of flow table space required will mandate new silicon geometries and the time to availability may be even longer.

The integration of SDN into the work being performed by the IEEE and other important forums is further evidence of its growing acceptance. Nonetheless, the exact forms of SDN that are here to stay are still debated. In this chapter we will discuss some new forms of SDN that have emerged between 2012 and 2015 as well as areas of research and experimental use cases that provide evidence of the totally unique ways that SDN can address networking problems. The slope of enlightenment embodies a recognition that SDN solves certain networking problems better than any other known technology and that it will indeed be disruptive there. This is a strong vindication of SDN but falls short of being the death knell for traditional networking, as some have argued. It is likely that the reduction in hype will reduce the appeal of applying the SDN moniker to every new networking variant, and the different versions of SDN that gain market traction, while not all OpenFlow-based by any means, will tend increasingly to those characteristics that we used to define Open SDN in Section 4.1.

Among those basic characteristics, one wildcard is that of *openness*. Openness means different things to different members of the networking ecosystem. For an academic researcher, it means open source available to all. On the other extreme, openness to an NEM may mean exposing certain APIs to your product in order to allow user-written applications to fine-tune external control knobs that the NEM exposes. (These two extremes reflect our earlier definitions of Open SDN vs SDN via APIs.) A special report [4] on the history and demise of the *Open Systems Interconnection* (OSI) standards concluded with the following remark: "Perhaps the most important lesson is that 'openness' is full of contradictions. OSI brought to light the deep incompatibility between idealistic visions of openness and the political and economic realities of the international networking industry. And OSI eventually collapsed because it could not reconcile the divergent desires of all the interested parties." This lesson learned the hard way for OSI should not be lost on Open SDN evangelists. Too strict of an interpretation of openness in the definition of SDN could result in a similar fragmentation of interests between the open source community and industry. If major parts of industry cannot align behind the Open SDN movement, then the movement will morph into something that is synergistic with industry's goals or it will be relegated to the halls of academia. This is the *realpolitik* of the Internet.

15.2 SD-WAN

SDN for the WAN, or SD-WAN as it is commonly known, is one of the most promising future areas for the application of the SDN principles presented throughout this book. In Section 14.9.5 we discussed several of the startups that target this potent strategy of using an overlay of virtual connections, often via some tunneling mechanism across a set of WAN connections, including MPLS, Internet, and other WAN transport links [5]. The overlay is achieved by placing specialized edge devices at branch locations. These *Customer Premises Equipment* (CPE) devices are managed by a centralized controller. The key function of these edge devices is the intelligent mapping of different user traffic flows over the most appropriate available tunnel. To a large degree, this is focused on cost savings achieved by placing only the traffic that is highly sensitive to loss, delay, or jitter on the more expensive links offering those guarantees. Normal data traffic such as email and web browsing may be mapped to best-effort, less expensive broadband Internet connections. Some classes of traffic may be routed over less expensive wireless backup links if that traffic can sustain periodic downtime or lower bandwidth.

It is easy to envision such an SD-WAN as a natural extension of the SDN-via-Overlays introduced in Section 4.6.3 and discussed at length in Chapters 6 and 8. SD-WAN is indeed related to our earlier discussions on SDN-via-Overlays in that in the data center the virtual switches in the overlay solution decide over which tunnel to inject traffic over the data center network. The CPE gateways in SD-WAN solutions perform a similar function of intelligently mapping application traffic to a particular tunnel. The criteria by which these mapping decisions are made are much more complex in SD-WAN than in the data center, however. For instance, in the data center overlay solution, there is typically just one tunnel between two hosts, whereas a significant feature of the SD-WAN solution is intelligent selection between tunnel alternatives to optimize cost and performance.

We saw earlier that this overlay approach can be successfully used to address data center problems of large layer 2 domains where the size of the MAC address tables can exceed the capacity of the switches. This technology was also used to limit the propagation of broadcast traffic that is an inherent part of the layer 2 networking so prevalent in the data center. While such issues might not be relevant in binding together a geographically dispersed enterprise over the WAN, the application of overlays in SD-WAN is more motivated by cost savings. While the variations in cost and QoS characteristics of different transport links within the data center are negligible, this is not the case in the WAN. The cost of an MPLS link offering QoS guarantees is significantly higher than a best-effort Internet connection. The provisioning time for such an MPLS connection is far greater than for the Internet connection. The difference is even greater with other wireless WAN link alternatives such as LTE or satellite [6]. Another major difference between the data center overlay paradigm and SD-WAN lies in the fact that SD-WAN CPE devices exercise little control over the actual paths used by the virtual link. Whereas in theory an OpenFlow controller could determine the exact hop-by-hop routing of a virtual link in a data center, within SD-WAN the edge devices view the tunnels they have at their disposal as coarse entities with certain cost, loss, speed, and QoS characteristics. They may choose to map a particular application flow over a given link according to those characteristics, but they are unlikely to be able to dynamically change the characteristics of any one of those links on the fly, including the actual path over which that tunnel, for example, is routed.

The interest in applying SDN to the WAN in some ways is an extension of earlier efforts at WAN optimization, such as data compression, web caching, and *Dynamic Multipoint Virtual Private Networks* (DMVPN) [7]. Indeed, some of the companies offering SD-WAN products today have their roots in WAN optimization. In addition to generating local acknowledgments and caching data, WAN optimization is also related to shaping and steering traffic to where it is best suited. For example, QoS-sensitive traffic like VoIP could be directed to MPLS paths that could offer latency and jitter guarantees while other traffic could be shunted to less expensive Internet links. There were certainly pre-SDN attempts of moving control away from the data plane such as the use of *Route Reflectors* for computing the best paths for MPLS links [8]. This is fully generalized within SD-WAN via the use of a centralized controller in the SDN paradigm. Of the five fundamental traits of SDN that have been a recurring theme throughout this work, SD-WAN solutions exhibit *centralized control*, *plane separation*, and *network automation and virtualization*. To date, there is little evidence in SD-WAN solutions of *a simplified device* or *openness*.

Some facets of the SD-WAN solutions converge on standards-based solutions [9]. In particular, the encryption tunnels and key distribution converge on SSL VPNs and IPSec. Conversely, compression and optimization methods, along with the path selection algorithms, tend to be proprietary and where the unique value-add of the vendors is concentrated.

In general, SD-WAN refers to any WAN managed by software control. The *Open Networking User Group* (ONUG) defines two broad types of SD-WANs [10]:

- Intradomain SD-WANs, where a single administrative domain uses SDN-controlled switches to accomplish various network management tasks, such as provisioning of secure tunnels between multiple geographically distributed portions of a network that are under the control of a single administrative domain.
- Interdomain SD-WANs, where multiple independently operated domains connect to one another via a shared layer-2 switch to accomplish various network management tasks, including inbound traffic engineering and denial-of-service (DoS) attack prevention.

As with any purported technological panacea, there are many challenges confronted in the actual implementation of the SD-WAN paradigm. We discuss these in the next section.

DISCUSSION QUESTION

Which of the five basic SDN traits defined in Section 4.1 do SD-WAN solutions generally exhibit? Give an example of each.

The devil is in the details

Due to the vibrant demand for commercial SD-WAN solutions, the marketplace abounds with contending claims and technologies. A very useful guide for evaluating and comparing SD-WAN solutions is found in [11]. Another is [12], where the author poses 13 questions to ask about how a given SD-WAN solution will address a specific challenge. These questions arose during the *Networking Field Day Event 9* [13]. Two vendors published their responses in [14,15], respectively. We leave it to interested reader to see all 13 questions and the various responses via the provided references. Here we look at three of the questions, one related to the concept of *host routing*, another dealing with *asymmetric routing*, and a third about the issue of *double encryption*.

Host routing question:

> How does the SD-WAN solution get traffic into the system? As in, routers attract traffic by being default gateways or being in the best path for a remote destination. SD-WAN tunnel endpoints need to attract traffic somehow, just like a WAN optimizer would. How is it done? WCCP? PBR? Static routing? (All 3 of those are mostly awful if you think about them for about 2.5 seconds.) Or do the SD-WAN endpoints interact with the underlay routing system with BGP or OSPF and advertise low cost routes across tunnels? Or are they placed inline? Or is some other method used? [12]

The answer depends on whether the CPE equipment is attempting to be standards-based or proprietary. If proprietary, some of the proposed solutions described *application traffic sniffing* whereby the CPE box analyzes the application traffic to determine the nature of the application and thus the appropriate transport link over which it should be mapped. Such a proprietary approach can be stymied if the application traffic is already encrypted by the application. Since most SD-WAN solutions suggest that the tunnels forming the virtual links should themselves be encrypted, this highlights the issue cited in question 8 in [12], which is how to deal with unnecessary double encryption. An alternative to application traffic sniffing in the CPE gateway is to use *host routing*. The key concept here is that the host itself understands that one gateway exiting the local site is preferred over another for that traffic type. A layer 2 alternative to this is for hosts to belong to different VLANs depending on the traffic type. A simple example of this is a data VLAN and a voice VLAN. The voice VLAN's traffic is mapped over the WAN virtual link that offers QoS guarantees and the data VLAN is mapped over the lower cost link that does not offer such guarantees.

Asymmetric routing question:

> Is path symmetry important when traversing an SD-WAN infrastructure? Why or why not? Depending on how the controller handles flow state and reflects it to various endpoints in the tunnel overlay fabric, this could be an interesting answer. [12]

When routing decisions between two endpoints were based on simple criteria such as the least cost path, traffic flowing between those two endpoints would generally follow the same path regardless of which direction it flowed. With more complex routing criteria, the possibility of asymmetric routing becomes reality. Asymmetric routing means that the packets flowing from application A to application B may take a different path than packets flowing from B to A. This may or may not be a problem. It certainly is a documented problem for certain older classes of firewalls. Thus knowing whether or not a given SD-WAN solution may engender asymmetric routing may be an important consideration.

Double encryption question:

> Double-encryption is often a bad thing for application performance. Can certain traffic flows be exempted from encryption? As in, encrypted application traffic is tunneled across the overlay fabric, but not encrypted a second time by the tunnel? [12]

The possibility of double encryption is both a performance consideration as well as a design consideration. If the SD-WAN device purports to determine application flow routing based on the aforementioned sniffing, this will not work in the event that the application has already encrypted the traffic. Possible workarounds here include using unencrypted DNS lookups to *guess* the nature of the application and thus the appropriate mapping of its traffic to a given WAN virtual link [16]. The second consideration of performance is simply the inefficiency of encrypting and decrypting the traffic a second time. Some SD-WAN solutions could offer unencrypted tunnels for WAN virtual links specifically for already-encrypted traffic. Another possible alternative would be to terminate the application encryption at the originating CPE gateway, perform the sniffing, and then reencrypt.

Cisco's *Intelligent WAN* (IWAN) is a hybrid WAN offering and another example of SD-WAN. Cisco's venerable *Integrated Services Router* (ISR) plays the role of the CPE gateway in this solution.

You can automate IWAN feature configuration using the IWAN application that runs in Cisco's *Application Policy Infrastructure Controller—Enterprise Module* (APIC-EM). IWAN is built upon the DMVPN technology mentioned previously. DMVPN provides dynamic secure overlay networks using the established *Multipoint GRE* (mGRE) and *Next-Hop Resolution Protocol* (NHRP) technologies, all of which predate SD-WAN. DMVPN builds a network of tunnels across the Internet to form the SD-WAN virtual links.

Cisco IWAN builds an SD-WAN solution using largely existing Cisco technology as well as technology obtained via its acquisition of Meraki. Many of these are Cisco-proprietary protocols, but they are generally less opaque than the closed systems of the SD-WAN startups. The Cisco IWAN design guide [17] gives detailed instructions about configuration of the HSRP and *Enhanced Interior Gateway Routing Protocol* (EIGRP) routing protocols in order to get the right traffic routed over the right WAN link. The ability to form and send traffic over VPN tunnels over two interfaces provides administrators with a way to load-balance traffic across multiple links. *Policy-based Routing* (PBR) allows IT staff to configure paths for different application flows based on their source and destination IP addresses and ports [18]. IWAN also allows configuration of performance criteria for different classes of traffic. Path decisions are then made on a per-flow basis as a function of which of the available VPN tunnels meet these criteria, which is in turn determined by automatically gathered QoS metrics.

Cisco's IWAN differs from some of the startup alternatives in a few fundamental ways. The mapping of traffic is not based on an application sniffing approach. Instead, APIC-EM tells the border router which egress path to take based on the conditions of the path as well as routing policies, and then intelligently load-balances the application traffic across the available WAN virtual links. According to [17], the variety of the WAN links is limited to MPLS and the Internet, whereas some of the alternatives offer wireless WAN connections as well. The number of WAN links for a given site is limited to a primary and secondary link, whereas the choice between more than two WAN links is feasible with some of the other solutions. It is important to understand that the number of tunnel endpoints may be far greater than the number of offered WAN links, depending on the particular SD-WAN solution. Each tunnel endpoint may map to a different customer premise if the design is based on a mesh topology, or each tunnel could terminate in a centralized hub if the design is based on a *hub and spoke* topology. In addition, tunnels may offer different levels of encryption and potentially different QoS guarantees.

Many of the SD-WAN solutions purport to offer dynamic QoS assessment of links and, based on those assessments, to dynamically shift QoS traffic from one path to another. This assessment can be accomplished by *performance probes* periodically injected into the tunnels and then using them to empirically measure QoS levels. The assessments can then be used to adaptively direct traffic via the most appropriate tunnel. This is certainly feasible but the point is obscured when related marketing statements imply that this allows provision of QoS guarantees over the Internet. While it may be possible to detect current QoS characteristics over a given Internet connection and potentially react to it, this falls far short of the marketing hype that implies that the solution can actually enforce QoS levels.

While this general concept, called *dynamic path selection*, is touted by SD-WAN vendors as novel in SD-WAN, we remind the reader of the example in Section 9.6.1 of dynamically shunting traffic over *Optical Transport Networks* (OTN), which is based on similar principles. Hence, SDN has contemplated this before the advent of SD-WAN.

An interesting adjunct to SD-WAN is the possible application of the *Locator/ID Separation Protocol* (LISP) [19]. LISP uses *Endpoint Identifiers* (EIDs), which correspond to hosts, and *Routing Locators*

(RLOCs), which correspond to routers. An EID indicates who the host is and denotes a static mapping between a device and its owner. The RLOC, on the other hand, indicates *where* that device is at a given time. If one considers the mobility of users coming and going from branch offices of an enterprise, it is clear that integration of this kind of technology is very relevant to the problems purported to be addressed by SD-WAN solutions. Indeed, Cisco's APIC-EM controller integrates LISP with IWAN.

DISCUSSION QUESTION

Cisco's IWAN SD-WAN solution offers the alternatives of MPLS VPN and Internet-based WAN links. One possible configuration shown in [17] depicts a configuration with *no* MPLS links but only two Internet links, each from a different provider. Beyond additional bandwidth, what is one advantage proferred by such a configuration?

15.3 POTENTIAL NOVEL APPLICATIONS OF OPEN SDN

There is considerable active research related to SDN as well as many novel use cases being proposed that illustrate potential new areas where SDN can be applied or improved. In this section we will present a sampling of some of the areas that promise to increase the scope of applications of Open SDN and thus broaden the foundation of its future.

A full survey of SDN-related research is beyond the scope of this work. There are a number of conferences that have focused on SDN research including [20,21]. The proceedings of these conferences provide a good review of current research trends in this area. The survey of the past, present, and future of SDN found in [22] may also provide the reader with broader coverage of current research directions in this field.

We have selected below a number of areas of potential application of Open SDN with the intent of giving the reader a flavor of the areas of application that radically depart from those that we have described during the bulk of this work. It is our hope that this will underscore that this new network programming paradigm opens the doors to many possibilities, only a few of which have already been explored.

15.3.1 MANAGING NONTRADITIONAL PHYSICAL LAYER LINKS

There is growing interest in using OpenFlow to control devices that have flows but are not traditional packet switches. The two most promising areas involve flows over optical and wireless links. In Section 9.6.1 we presented a use case where SDN was used for offloading elephant flows onto optical devices. This represents one example of the general area of using OpenFlow to manage flows over physical layers that are outside the domain of classical LANs and WANs. In addition to the example in Section 9.6.1, we mentioned two startups in Section 14.9.7, Plexxi and Vello, that have brought products to market that marry optical switching with OpenFlow.

In [23], the researchers formalize the study of how to map big data applications to an OpenFlow network. An elephant flow is an example of a big data application. The authors propose mechanisms to build on the existing Hadoop[2] job scheduling method to optimize how and when to schedule jobs over the optical links. They point out that unlike the more classical SDN use cases of cloud network provisioning and WAN traffic engineering, such as those discussed in Chapters 8 and 9, such big data jobs require more rapid and frequent updates of the flow tables on the switches. They conclude that current OpenFlow switches are capable of handling their predicted number and frequency of flow table changes. They believe that there may be challenges in keeping the various flow tables across the network switches synchronized with this rate of flow table changes. This is important as one of the problems in traditional networks that we described in Section 1.5.3 was the relatively slow convergence time in the face of routing changes. At the fine granularity of flow table changes being made to route these big-data flows, slow convergence times may be an issue for optical offload.

Because of the ease of installing wireless backhaul links to bring mobile traffic back to the core from *Radio Access Network* (RAN) cells, wireless backhaul is becoming increasingly popular. One of the challenges in doing this is that the effective bandwidth of a wireless link is not constant as in its wired counterpart. In [24, Section 8.4], OpenFlow is proposed as a mechanism to segregate traffic from different providers and different types into separate flows which are then transmitted over a single shared wireless backhaul medium. In the example, the wireless backhaul technology is IEEE 802.16. While the segregation of different traffic types for differential QoS treatment is well established, in the example the wireless backhaul link is potentially shared by a number of different co-located RAN technologies (e.g., LTE, 3G, WiFi) that may be offered by several different operators. The wireless backhaul provider may have different SLA agreements with each of those operators, and for each operator different levels of service depending on the terminating RAN technology, traffic type, or business agreement. Since the wireless backhaul bandwidth capability itself may vary due to environmental conditions, the process of satisfying all of the various SLA commitments becomes far more complex and can benefit from OpenFlow's ability to segregate the traffic into different flows and route or police those flows dynamically with the changing wireless conditions.

15.3.2 APPLYING PROGRAMMING TECHNIQUES TO NETWORKS

One of the themes of this work has been that the greater network programmability inherent in OpenFlow provides benefits in the richness of policies and in the fine-grained control it offers the network programmer. To be fair, though, when a programming environment gives the programmer a lot of power, the probability of mis-programming the network is likely greater than the legacy situation where the network engineer was constrained by control knobs of limited functionality. Thus part of the ultimate success of OpenFlow is the development of tools that will aid in detecting such programming flaws.

More formally, in [25] the author suggests that Open SDN provides a proper abstraction of the network that allows us to address networking challenges more efficiently. The author draws an analogy to the difference between programming a CPU in today's high-level languages and application development environments versus programming in machine language. A corollary of this is that the network abstraction allows other advanced programming methodologies to be applied, including debuggers, analysis tools, network simulation, verification tools, and others. These tools have enabled

[2]Hadoop is open source software that enables the distributed processing of large data sets across clusters of commodity computers.

tremendous progress in the development of software because of formal analysis and the application of theory. No equivalents have been practical in the networking space because of the lack of the abstraction as described in [25]. We consider later some of the tools that are enabled by this new way of looking at networking.

Network debugger

In Section 4.1.3 we drew an analogy between the control plane of networks and computer languages and CPUs. We explained that we no longer work at the level of assembly language because high-level languages provide us with efficient abstractions to program the CPUs effectively. We argued that the network needs a similar evolution toward a high-level abstraction of network programming, and suggested that OpenFlow is a good example of such an abstraction. Following this same analogy, if OpenFlow provides us with a way of programming the network as a single entity, can we now consider other advanced programming techniques and tools for application in the network? For example, if we view the network as a vast distributed computer, can we envision a debugger for that computer? This is precisely what the work performed in [26] attempts. The authors propose a *network debugger* (ndb) that will implement basic debugger functionality at the packet level. They aspire to implement a full set of debugger actions, including *breakpoint*, *watch*, *backtrace*, *single-step*, and *continue*. In [26] the authors describe an example where they have implemented both breakpoints and packet-backtraces. For instance, a breakpoint could be set in a network switch such that when a packet matches no flow entry in that switch, the breakpoint is hit. At that point the network programmer could request a packet backtrace showing which flow entries were matched by which switch that led to the packet arriving to this switch where no flow entry matched. The fact that such technology can even be contemplated underscores the more deterministic network behavior possible under the centralized programming model of Open SDN.

No bugs in controller execution

It is also naive to believe that a controller programming a large set of switches that are in a constant state of forwarding packets is an *entirely* deterministic system. The architecture of OpenFlow does not tightly synchronize the programming of multiple switches when the controller must simultaneously program flow entries in several switches. Due to the entropy involved in this process, it may not be possible to ensure that all flow entries are in place before the packets belonging to that flow begin to arrive. Thus race conditions and even routing loops are possible for short periods of time. While it is likely impossible to completely eliminate this hysteresis during rapid network reprogramming, a tool that could model the effects of a particular controller program (i.e., application) could help minimize such occurrences and ensure their effects were short lived. In [27] the authors propose *No Bugs in Controller Execution* (NICE), a tool for modeling the behavior of OpenFlow applications to ensure their correctness, including detecting forwarding loops and black holes. Advanced modeling techniques are used to create input scenarios that test all possible execution paths within the application while taking account of variability of the state of the network switches and links. In [27] the authors modeled and debugged three sample applications: a MAC-learning switch, an in-network server load balancer, and energy-efficient traffic engineering. Additional recent work related to troubleshooting bugs in SDN networks can be found in [28].

Veriflow

While there is great appeal to checking for OpenFlow application correctness off-line and before deployment, it may be necessary or desirable to perform real-time formal checking for correct network behavior. In [29] the authors describe Veriflow, a system that resides between the controller and the switches and verifies the correctness of each flow entry update before it is applied. One advantage of the approach described in [29] is that it does not require knowledge of all the network programs themselves as it verifies correctness based on observations of flow rules as they are sent from the controller to the switches. One of the biggest challenges faced in this study is to keep the latency of the correctness checks low enough to avoid becoming a bottleneck in the path between the controller and the switches. Such speed is not possible if one were to take a brute-force approach to reflecting the global network state which is very complex and changes rapidly. The authors claim to have developed algorithms that allow such checks to be performed at 100-μs granularity, which is sufficiently fast to avoid adversely affecting network performance. Their novel work is related to how to represent the portion of network state that could possibly be affected by a given rule change and to develop algorithms that reduce the search space such that the effect of a rule change on network state can be ascertained in real time.

Proof-based verification of SDNs

Like NICE and Veriflow described previously, the authors of [30] propose the use of formal techniques to verify the correctness of SDN programming. The approach in [30] is to encode the controller program and switch functionality in a declarative programming language. The specific formal language used in this work is *NDlog*. They then apply formal proof techniques to confirm the correctness of the programming. Correctness of static configuration includes confirming needed reachability and the absence of loops. Correctness during updates proves that reachability goals and a loop-free topology are maintained across updates. Finally, correct implementation of the controller mandates that internal ports be programmed before ingress ports and that each flow is installed only once and in the correct order.

Verification of correctness of SDN is an important and active research area. Other relevant works in this area include [31,32].

Yet another network controller

Throughout this book we have presented many different SDN controllers. Each new wave of controller innovation seems to carry with it a new API that is fine-tuned to expose the features of that controller. A downside of this paradigm is that application innovation is slowed by the constant retooling of applications to fit with the evolving controllers. Another drawback is that the rapid evolution of the controller platforms has stymied the development of a mature tool and language set that only evolves over a significant period of time. In [33] the authors formalize their premise of *applying operating system principles to SDN controller design* by providing the controller API and tools via a variant of the well-known Linux operating system. This variant is called *Yet Another Network Controller* (YANC). In this model, SDN applications are Linux processes. These processes communicate with the controller via the Linux file system. The internals of the YANC operating system are such that the file input and output correspond to the dialog typical between an SDN controller and its applications. The applications may be written in any language that supports that file system, providing a degree of freedom not typical of SDN controllers. Furthermore, YANC inherits the rich Linux tool set, making for a complete application development environment.

Software defined networks as databases

Just as YANC attempts to leverage learnings from the mature field of operating system design, the authors in [34] attempt to apply database techniques to SDN programming. The authors contend that while writing an SDN application may seem easy, applications exert very weak control over event ordering and the timing when changes are actually implemented. Inadequate control of the correctness of SDN programming is likely to be a significant growth inhibitor for SDN. The authors draw a parallel to databases where the programmer is able to ignore practical considerations related to concurrency and synchronization of distributed components. The work in [34] attempts to abstract the details related to the synchronization of multiple switches' flow tables using established models from the database world.

DISCUSSION QUESTION

In Section 7.6 we introduced a programming language named P4. How does P4 relate to the research discussed here in Section 15.3.2?

In this section we have presented a number of examples applying advanced programming methodologies to SDN networks. The fact that SDN inherently facilitates network abstraction is a key enabler to each example. Beyond the samples presented here, there are higher-level programming abstractions possible with SDN. We refer the interested reader to the work done in [35] using the *Procera* language, that using the *Pyretic* language in [36], and the *intents-based* approach discussed in Section 7.3.5.

15.3.3 SECURITY APPLICATIONS

SDN provides a fertile environment for novel work in the area of network resilience and security. A comprehensive survey of research done in this area can be found in [37]. In this section we delve into a selection of examples of new research in this area.

Hiding IP addresses

Many network attacks are based on identifying active IP addresses in a targeted domain. Protecting hosts by making it impossible for an attacker to identify a host's IP address would be an effective countermeasure. In [38] the authors propose that each protected host be assigned a virtual IP address which is the one exposed to the outside world by DNS lookups. In this method, the OpenFlow controller randomly and at high frequency assigns the virtual IP address to the protected hosts, maintaining the temporary mapping of virtual to physical IP addresses. Earlier systems based on DHCP and NAT exist to change the physical IP address of hosts but these mechanisms change the IP address too infrequently to be a strong defense. In [38] only authorized hosts are allowed to penetrate through to the physical IP address. The translation from the virtual IP address to the physical IP address happens at an OpenFlow switch immediately adjacent to the protected hosts. The unpredictability and speed of the virtual IP address mutation is key to thwarting the attacker's knowledge of the network and the planning of the attacks. While this approach could conceivably be implemented using specialized appliances in a conventional network, OpenFlow provides a flexible infrastructure that makes this approach tractable.

Flowguard

In [39] authors propose *Flowguard*, an SDN firewall capable of filtering not only at the packet level but also at the flow level. An entire flow may violate policy and should thus be rejected. A key theme of the authors is that flow policy changes may be in conflict with firewall policy and resolving these conflicts is the responsibility of an SDN firewall. Thus the authors assert that an SDN firewall is both a packet filter and also a policy checker. In Flowguard, *flow packet violations* are distinguished from a broader class of *flow policy violations*. Flow policy violations may occur when a new flow is installed or when policy is changed. The simpler of the two cases is when a new flow is installed. This may violate existing policy and may thus be rejected. It is more complicated when flow policy is changed as this change can cause a violation of firewall policy. The authors illustrate this via a simple example of an OpenFlow-based load balancer which may alter packet header fields in order to change flow paths. When flows are modified by this load balancer, it is imperative that the packet header fields modified by the OpenFlow rules do not result in a firewall rule being circumvented. In order to prevent this, an SDN firewall must check that flow policy does not conflict with firewall policy. Flowguard decomposes this policy conflict problem into *entire* violations and *partial* violations. Outcomes may include rejection of the policy update, removal of affected flows or, in the case of a partial violation, specific packet blocking. The authors contrast Flowguard with Veriflow which, as described in Section 15.3.2 does not do its flow analysis in real time. Flowguard, however, does work in real time. The authors provide data that shows that Flowguard performs favorably as compared with Floodlight's built-in firewall.

Segregating IPSec traffic in mobile networks

Wireless providers using LTE secure the user and control data between the base station (eNB) and their network core using IPSec tunnels. These tunnels terminate in the core at a *Serving Gateway* (S-GW). As currently specified, there is a single IPSec tunnel encrypting all services. IPSec tunneling carries significant overhead, and significant efficiency gains are possible if traffic that does not require the security afforded by IPSec can be sent in the clear. In two separate proposals [24, Sections 8.5 and 8.12] the authors suggest using OpenFlow to map between different traffic types and the appropriate level of IPSec security. In [24, Section 8.5] the focus is on making the mapping of individual flows to different IPSec tunnels. In [24, Section 8.12] the emphasis is on distinguishing those traffic types requiring security and to only tunnel those. The authors suggest that YouTube video, social media, and software updates are examples of traffic that does not require IPSec encryption and can thus be sent in the clear. While the authors do not describe how such traffic types would be detected, one possible means of doing so that avoids the Deep Packet Inspection problem is to base the decision on server addresses that belong to YouTube, Facebook, Microsoft software updates, and so on. An additional possible benefit of providing flow-specific treatment to these other traffic types is that it is not always necessary to send separate copies of this kind of traffic from the network core. By inserting a *deduplication* processor closer to the mobile users, this redundant traffic can be cached at that location and a twofold bandwidth savings is realized by eliminating the round trip to the S-GW. By providing the operators control over which flows are tunneled and which are not, this allows operators to monetize their investment by offering different plans with different levels of security.

15.3.4 ROAMING IN MOBILE NETWORKS

Mobile traffic offload

In Chapter 9 we presented a number of SDN applications involving traffic steering, where flows are directed toward security systems or load balanced among a set of similar-function servers. In the multiradio environment common for today's mobile operators, a new SDN application is possible in the area of mobile traffic offload. Mobile offload means moving a client *mobile node* (MN) from one *Radio Access Network* (RAN) to another. This may make sense for a number of reasons, but the one most often offered is to shunt the traffic to a RAN where the spectrum is more available or less expensive than the one currently used by the MN. Such offloading has been contemplated for some time by mobile operators but existing approaches have not provided the flexible, fine-grained control offered by Open SDN. In [24, Section 8.3] a use case is described where OpenFlow switches are used in key gateways in the 3GPP architecture. Based on observing flow-related criteria and the location of the MN, an OpenFlow application can redirect the MN's access connection from 3G to a WiFi hotspot, for example. If the MN needs to roam from WiFi to a cellular radio technology the same mechanism may be used. This approach allows operators to flexibly and dynamically apply offloading policies rather than static policies that are not able to adapt to changing network conditions. For example, if a user is currently connected to a heavily loaded WiFi hotspot and located within a lightly loaded LTE cell, it may make sense to do reverse offload and move the user from WiFi back to LTE. Such a decision hinges on observing rapidly changing conditions and could not be put into effect based on static policies. Note that while the basic traffic steering inside the 3GPP packet network is native to OpenFlow here, the control signaling related to the RF aspects of the roam would have to be coordinated with the appropriate 3GPP signaling functions, as those are outside the current scope of OpenFlow.

Media-independent handovers

IEEE 802.21 is an established protocol for media-independent handovers between 802-family *Points of Access* (PoAs). Examples of PoAs are 802.11 access points and 802.16 base stations. The 802.21 *Point of Service* (PoS) is responsible for the messaging to the PoAs to accomplish either make-before-break or break-before-make handovers (roams). A significant part of accomplishing such handovers is to redirect the traffic flow from an MN such that it enters and exits the network from the new PoA. This combined with the fact that OpenFlow is natively media-independent leads to a natural overlap between the role defined for the 802.21 PoS and an OpenFlow controller. This synergy is discussed in [24, Section 8.6] and a proof of concept proposed to design a system where an OpenFlow controller uses 802.21 messages to control roaming between 802.11 access points. The authors suggest that extensions to OpenFlow may be necessary, however.

Infrastructure-controlled roaming in 802.11 networks

In [40] the authors present Odin, an SDN-based framework to facilitate AAA, policy, mobility, and interference management in 802.11 networks. The fact that in 802.11 the client makes the decision regarding with which AP to associate at any given time has presented a major challenge to giving the network explicit control over with which AP a client associates. Controlling the choice and timing of the client-AP association in 802.11 is fundamental to roaming, RF interference and load balancing solutions. The Odin master function, which is an SDN application, controls Odin agents in each of the APs. These agents implement light virtual access points (LVAPs) that appear as physical APs to the client. Each client is assigned a unique LVAP, giving each client a unique BSSID. Since LVAPs may be instantiated on any of the APs under the control of the controller, the client may physically roam to another physical AP by the controller moving its LVAP instantiation to that new physical AP. OpenFlow is used to move the user data flows in accordance with such roaming.

BeHop: SDN for dense WiFi networks

In BeHop, presented in [41], the authors describe a virtualized WiFi architecture intended to provide improved wireless performance in dense WiFi deployments. The authors contend that much of the chaos that results in dense WiFi environments stems from the fact that in 802.11 it is the client who selects an AP for association. These client decisions are often not globally optimal. BeHop incorporates the notion of a single, personal AP that follows the user wherever he goes. This personal AP is a virtual AP, of course, and in this sense reminiscent of the ideas presented previously in Odin. Since the user is always connected to his own personal SSID, he never formally *roams* in the 802.11 sense. BeHop is implemented on standard Netgear APs running OpenWRT and *Open vSwitch* (OVS). The SDN-based coordination this affords allows for intelligent channel assignments, load-balancing users across APs, and energy savings via the powering-off of redundant APs.

15.3.5 TRAFFIC ENGINEERING IN MOBILE NETWORKS

Dynamic assignment of flows to fluctuating backhaul links

In Section 9.2.2 we presented an example of Open SDN being used for traffic engineering in MPLS networks. In mobile networks, there is a novel opportunity to apply traffic engineering to wireless backhaul infrastructure links. Unlike the backhaul example in Section 15.3.1, the wireless links we refer to here connect switching elements that may carry the traffic of multiple base stations (eNBs). The appeal of wireless backhaul links is growing as the number of base stations grows, the locations of the switching elements become more dynamic, and new backhaul capacity is brought online to a more freely chosen set of locations. A downside of wireless backhaul is that the bandwidth of the wireless backhaul is both more limited and, more importantly, less stable than in its wired counterparts. Current resource management practices are static and do not redirect load dynamically based on short-term fluctuations in wireless capacity.

In [24, Section 8.2] the authors propose that an OpenFlow controller be enabled to be aware of the current available bandwidth on the set of wireless links it is managing. It may be managing a hybrid set of wired and wireless backhaul links. If OpenFlow is responsible for assigning user flows to that set of backhaul links, that assignment can be made as a function of the SLAs specific to each flow. High SLA (higher guarantees) traffic can be steered over wired links and low SLA traffic can be steered over a wireless link. If one wireless link is experiencing temporary spectrum perturbation, then OpenFlow can shunt the traffic over a currently stable wireless link. Note that this proposal requires a mechanism by which the SDN application on the OpenFlow controller be made aware of changes in the wireless bandwidth on each of the wireless links it is managing.

Sharing wireless backhaul across multiple operators

The IEEE has chartered a group to study how to enable an Open Mobile Network Interface for omni-Range Access Networks (OmniRAN). The omni-Range aspect implies a unified interface to the multiple radio access network types in the 802 family. These include 802.11 and 802.16, among others. The business model behind this is that multiple operators offering a range of RAN technologies in a well-defined geographic area could benefit from a common backhaul infrastructure shared by all the operators for all the different radio technologies. The cost savings of this approach compared to separate backhaul networks for each (operator, RAN) pair is significant. In [24, Section 8.16] the authors argue that as both OpenFlow and OmniRAN are media-independent that there is a natural synergy in applying OpenFlow as the protocol for controlling and configuring the various IEEE 802 nodes in this architecture, as well as using OpenFlow for its native benefits of fine-grained control of network flows and implementation of network policy. This effort would require extension to OpenFlow for controlling and configuring the IEEE 802 nodes.

SoftMoW

SoftMoW [42] proposes a hierarchy of parent and child SDN controllers to more efficiently route cellular traffic. In particular this approach allows more direct paths between two local mobile stations rather than routing their traffic deep into a rigidly organized hierarchy as is done today. Currently the lack of a nearby *Packet Data Network Gateway* (PGW) often results in unnecessarily long and convoluted paths between the endpoints of mobile user traffic streams. The rise of machine-to-machine traffic in such fields as mobile health will only exacerbate this problem. SoftMoW replaces inflexible and expensive hardware devices, such as the PGWs, with SDN switches. The large cellular network is partitioned into independent and dynamic logical regions, where a child SDN controller manages the data plane of each region. Sophisticated, cross-region traffic engineering is accomplished via a label-swapping concept reminiscent of MPLS. Parent SDN controllers program rules that push global labels onto packets corresponding to traffic groups. When packets reach their local region, these global labels are popped and local labels pushed. Traffic remaining local to one region would never obtain global labels.

An OpenFlow switch on every smartphone!

Today's smartphones generally have multiple radios. For example, it is common to see LTE, WiFi, and 3G radios on the same mobile phone. In the existing model the MN chooses which radio to use based on a static algorithm. This algorithm normally routes voice traffic over the best available cellular connection and data over a WiFi connection if it is available. If no WiFi connection is available, then data is sent via the cellular connection. This is a fairly inefficient and brute force use of the available radio spectrum. For example, if a functioning WiFi connection is poor, it may be much more efficient to route data over a functioning LTE connection. If multiple radios are available, it may be wise to use more than one simultaneously for different data flows. Expanding the horizon for OpenFlow applications to a bold new level, in [24, Section 8.8] there is a proposal to install an instance of OVS on every mobile device. This virtual switch would direct flows over the radio most appropriate for the type and volume of traffic as well as the current spectrum availability for the different radio types. The controller for this OVS instance resides at a gateway in the 3GPP architecture and uses its global knowledge of network state to steer flows in the mobile phone over the most appropriate radio according to the bandwidth and current loading of that particular radio spectrum in that time/location.

We have just described a number of examples of SDN-related research in the areas of roaming and traffic engineering in mobile networks. In general the field of mobile networking offers a ripe environment for SDN research and we do not attempt to provide a complete survey of the relevant work here. The interested reader will find other recent publications in this area in [3,24,43–46].

15.3.6 ENERGY SAVINGS

When asked about their vision for the future role of SDN, an expert panel [3] from HP, Dell, and NEC indicated that energy savings was an important area where SDN can play an increasing role. Data centers have enormous OPEX costs keeping their massive data warehouses cooled and fully and redundantly powered. Companies that produce compute and storage servers now tout their energy savings relative to their equivalents of only a few years ago. While there have been significant improvements in the energy efficiency of servers, the corresponding gains in networking equipment have been smaller.

Between the cooling and the energy consumed by the compute and storage servers, it is estimated [47] that about 70% of the power is allocated to those resources. Beyond that, however, it is estimated that approximately 10–20% of the total power is spent to power networking gear. Considering the vast sums spent on electric bills in the data center, making a dent in this 10–20% would represent significant savings.

ElasticTree

One approach to applying OpenFlow to energy savings in the data center, called *ElasticTree*, is described in [47]. The premise of that work is that data centers' networking infrastructure of links and switches is designed to handle an anticipated peak load and is consuming more power than necessary during normal operation. If a means could be found to only power the minimum necessary subset of switches at any moment, there is an opportunity for significant energy savings. The assertion in [47] is that during periods of less than peak load, OpenFlow can be used to shunt traffic around switches that are candidates for being powered off. Such a system assumes an out-of-band mechanism whereby switches may be powered on and off via the OpenFlow application. Such systems exist and are readily integrated with an OpenFlow application. The authors suggest that by varying the number of powered-on switches their system can provide the ability to fine-tune between energy efficiency, performance, and fault tolerance. The novel work in [47] is related to determining what subset of network switches and links need to be powered on to meet an established set of performance and fault tolerance criteria.

Dynamic adjustment of wireless transmit power levels

A related use case is proposed for wireless backhaul for mobile networks in [24, Section 8.8]. In this case, the mobile operator has the ability to vary the amount of power consumed by the wireless links themselves by varying the transmission power levels. For example, relatively lower traffic loads over a microwave link may require less bandwidth, which may be achieved with a lower transmission power level. If no traffic is flowing over a wireless link, the transmission power may be turned off entirely. As with ElasticTree, the proposal is to use OpenFlow to selectively direct traffic flows over the wireless links such that transmission power levels may be set to globally optimal settings.

More energy efficient switching hardware

A more direct approach to energy savings in an SDN environment is to directly design more energy efficient switches. Well-known methods of reducing power consumption used on servers are already being applied to switch design. At some point, use of electronic circuitry consumes energy, though. One of the most power-hungry components of a modern switch capable of flow-based policy enforcement is the TCAM. In [48] the authors propose prediction circuitry that allows flow identification of a high percentage of packets in order to avoid the power-hungry TCAM lookup. The prediction logic uses a memory cache that exploits *temporal locality* of packets. This temporal locality refers to the fact that numerous packets related to the same application flow often arrive at a switch ingress port close to one another. Each TCAM lookup that can be circumvented lowers the switch's cumulative power consumption. A different approach to such TCAM-based energy savings is proposed in [49] where the authors suggest organizing the TCAM into blocks such that the search space is segmented in accordance with those blocks. When a particular lookup is initiated by the switch, it is possible to identify that subset of internal blocks that may be relevant to the current search and only power on that subset of the TCAM during that search.

15.3.7 SDN-ENABLED SWITCHING CHIPS

In Section 14.9.3 we described two ongoing commercial efforts to build chips that are designed from the ground up to support advanced OpenFlow capability. We asserted in Section 15.1 that the *plateau of productivity* would likely remain elusive until such chips become commercially available. A survey of which vendor is offering what sort of commercial SDN-enhanced switching chip is available in [50]. In addition to these commercial efforts, this is also an active area of research. For example, in [51] the authors show a model of a 256-core programmable network processing unit. This chip is highly programmable and would support the nature and size of flow tables that will be required to exploit the features of OpenFlow 1.3. The main argument of [51] is that such a programmable network processor can be produced and still have the high throughput characteristics more commonly associated with purpose-built, nonprogrammable Ethernet chips. It is reasonable to believe that such a 256 core chip will be commercially available in the not-too-distant future. In [52] the authors describe an existing programmable networking processor with 256 core. While this is a general-purpose processing unit, this provides evidence of the scale that multicore SDN-specific ASICs are likely to achieve in the near future.

In Section 15.3.6 we described a modified ASIC design that would be more energy efficient than existing switching ASICs when performing the flow table lookups required in SDN flow processing. Another advancement is proposed in [53] where the combination of an SDN-enabled ASIC and local general-purpose CPU and memory could be programmed to handle current and yet-unanticipated uses of per-flow counters. This is an important issue, since maintaining and using counters on a per-flow basis is a key part of SDN programming, yet the inability to anticipate all future applications of these counters makes it difficult to build them all into ASIC designs. In [53] the authors describe a hybrid approach where the hardware counters on the ASIC are programmable and may be assigned to different purposes and to different flows. The counter information combines the matching rule number with the byte count of the matched packet. This information is periodically uploaded to the local CPU where the counters are mapped to general-purpose OpenFlow counters. Such a system can enable some of the high-level network programming we describe in Section 15.3.2 whereby complex software triggers based on a counter value being reached can be programmed directly into the general-purpose CPU.

In Chapter 10 we discussed the overlap between SDN and NFV and showed how many network functions can be virtualized in software. We explained how rudimentary *Intrusion Detection Systems* (IDS) can be built from SDN technology. At some point, though, the *Deep Packet Inspection* (DPI) required for some NFV tasks mandates specialized hardware. When this hardware becomes too specialized, the cost savings and simplicity expected of SDN and NFV are sacrificed. Thus it makes sense to enhance a commodity X86 platform such that it has DPI capabilities. In [54] the authors present a novel design for an energy efficient SDN/NFV accelerator chip compatible with low-cost X86 architectures. This is significant in that such an *Application Specific Instruction Set Processor* permits standard X86 software to control high-speed DPI functions.

15.4 CONCLUSION

Ironically, one advantage of SDN washing is that it leads to an easy conclusion that SDN is on a path to a permanent and important part of networking technology for decades to come. This is true simply because the term SDN can be stretched to apply to ANY networking technology that involves a modicum of software control. Virtually all networking technologies involve software to some degree; hence, they are all SDN. This flippant analysis is actually not far removed from what appears in some NEM marketing materials. Thus, interpreted that way, SDN's solid hold on the future of networking is ascertained.

Taking a step in a more serious direction, SDN via Overlays has had such success in the data center that we conclude that this technology, in one form or another, is here for the long run. SDN via Opening up the Device is so nascent with so little track record that any prediction about its future is pure speculation. We feel that SDN via APIs is really a stopgap approach designed to prolong the life of existing equipment and, thus, will not be long-lived. To the extent that the APIs evolve and permit a remote and centralized controller to directly program the data plane of the switch, then SDN via APIs begins to approach Open SDN. It is significant that such API evolution is not merely a matter of developing new software interfaces. Direct programming of the data plane depends on an evolution of the data plane itself. The API evolution, then, is irrevocably caught up in the much longer delays customary with ASIC design cycles.

The founding precepts of the SDN movement are embodied in Open SDN. As we have admitted previously, we, the authors, are indeed Open SDN evangelists and we wish to conclude this work with a serious look at its future. We begin our answer by retreating to the analogy of Open SDN compared to the relationship of Linux with the entrenched world of the PC industry, dominated by the incumbent Microsoft and, to a much smaller degree, Apple. While Linux has actually enjoyed tremendous success, it has not come close to displacing the incumbents from their predominant market. Linux's most striking success has come instead in unexpected areas. The incumbents continue to dominate the PC operating system market, notwithstanding the few zealots who prefer the freedom and openness of Linux. (*How many people do you know who run Linux on their laptops?*) Linux has enjoyed great success, however, in the server and embedded OS markets. The amazing success of the Android operating system, is after all, based on Linux. Thus Linux's greatest success has come in areas related to but orthogonal to its original area of application.

In our opinion, the most likely scenario for Open SDN is to follow that Linux trajectory. For a variety of reasons, it will not soon displace the incumbent NEMS in traditional markets as was posited in 2009 through 2012. It will maintain and expand its beachhead in spaces where its technical superiority simply trumps all alternatives. Whether that beachhead remains restricted to those very large and very sophisticated customers for whom technical superiority of the Open SDN paradigm trumps the safety of the warm embrace of the established NEMs, or expands to a larger market because some NEMs manage to adopt Open SDN without sacrificing profits, is too early to tell. There will be applications and use cases such as those cited in Section 15.3 where Open SDN solves problems for which no good solution exists today. In those areas Open SDN will surely gain traction just as Linux has transformed the world of operating systems.

A Chinese proverb states "May you live in interesting times." For better or worse, due to data centers, cloud, mobility, and the desire to simplify and reduce costs, networking today is in the midst of such "interesting times." How it all plays out in the next few years is up for debate—but, whatever the outcome, *Software Defined Networking* is certain to be in the middle of it.

REFERENCES

[1] Gartner. Research methodology. Retrieved from: http://www.gartner.com/it/products/research/methodologies/research_hype.jsp.

[2] Kerner S. OpenFlow inventor Martin Casado on SDN, VMware, and Software Defined Networking Hype [VIDEO]. Enterprise Networking Planet; 2013. Retrieved from: http://www.enterprisenetworkingplanet.com/netsp/openflow-inventor-martin-casado-sdn-vmware-software-defined-networking-video.html.

[3] Karimzadeh M, Sperotto A, Pras A. Software defined networking to improve mobility management performance. In: Monitoring and Securing Virtualized Networks and Services, Lecture Notes in Computer Science; vol. 8508. Berlin/Heidelberg: Springer; 2014. p. 118–22.

[4] Russell A. The Internet that wasn't. IEEE Spectr 2013;50(8):38–43.

[5] Khan F. The Top 4 SD-WAN Myths. Network Computing; 2016. Retrieved from: http://www.networkcomputing.com/networking/top-4-sd-wan-myths/1072744350?piddl_msgid=313766.

[6] Viprinet SD-WAN. Viprinet. Retrieved from: https://www.viprinet.com/en/why-viprinet/sd-wan .

[7] Conde D. SD-WAN: the Killer App for enterprise SDN? Network Computing; 2015. Retrieved from: http://www.networkcomputing.com/networking/sd-wan-killer-app-enterprise-sdn/1747434541.

[8] Conran M. Viptela—Software Defined WAN (SD-WAN). Viptela. Retrieved from: http://viptela.com/2015/05/viptela-software-defined-wan-sd-wan/.

[9] Ferro G. Concerns about SD-WAN Standards and Interoperability. Ethereal Mind—Infrastructure; 2015. Retrieved from: http://etherealmind.com/concerns-about-sd-wan-standards-and-interoperability/.

[10] Feamster N. Software defined wide-area networks. Open Networking User Group; 2015. Retrieved from: https://opennetworkingusergroup.com/software-defined-wide-area-networks/.

[11] Open Networking User Group. ONUG Software-Defined WAN use case. SD-WAN Working Group; 2014. https://opennetworkingusergroup.com/wp-content/uploads/2015/05/ONUG-SD-WAN-WG-Whitepaper_Final1.pdf .

[12] Banks E. Questions I'm asking myself about SD-WAN solutions. Ethan Banks on Technology; 2015. Retrieved from: http://ethancbanks.com/2015/02/13/questions-im-asking-myself-about-sd-wan-solutions/.

[13] Networking Field Day 9. Silicon Valley: 2015. Retrieved from: http://techfieldday.com/event/nfd9/.

[14] Cloudgenix. 13 Interesting questions about SD-WAN; 2015. Retrieved from: http://www.cloudgenix.com/13-interesting-questions-about-sd-wan/.

[15] Prabagaran R. Answering the profound SD-WAN questions that bother Ethan Banks. Viptela; 2015. Retrieved from: http://viptela.com/2015/03/answering-the-profound-sd-wan-questions-posed-by-ethan-banks/.

[16] Herbert J. Riverbed's take on SD-WAN. Network Computing; 2015. Retrieved from: http://www.networkcomputing.com/networking/riverbeds-take-sd-wan/1591748717.

[17] Cisco Systems. Intelligent WAN technology design guide; 2015. Retrieved from: http://www.cisco.com/c/dam/en/us/td/docs/solutions/CVD/Jan2015/CVD-IWANDesignGuide-JAN15.pdf.

[18] Cisco Meraki. IWAN Deployment Guide. Retrieved from: https://documentation.meraki.com/MX-Z/Deployment_Guides/IWAN_Deployment_Guide.

[19] Farinacci D, Fuller V, Meyer D, Lewis D. The Locator/ID Separation Protocol (LISP). RFC 6830. Internet Engineering Task Force; 2013.

[20] Hot topics in Software Defined Networking (HotSDN). In: SIGCOMM 2012. Helsinki; 2012. Retrieved from: http://conferences.sigcomm.org/sigcomm/2012/hotsdn.php.

[21] ACM SIGCOMM workshop on hot topics in Software Defined Networking (HotSDN). In: SIGCOMM 2013. Hong Kong; 2013. Retrieved from: http://conferences.sigcomm.org/sigcomm/2013/hotsdn.php.

[22] Nunes B, Mendonca M, Nguyen X, Obraczka K, Turletti T. A survey of software-defined networking: past, present and future of programmable networks. IEEE Commun Surv Tutorials 2014;16(3):1617–34.

[23] Wang G, Ng T, Shaikh A. Programming your network at run-time for Big Data applications. In: HotSDN. Helsinki, Finland; 2012. Retrieved from: http://conferences.sigcomm.org/sigcomm/2012/paper/hotsdn/p103.pdf.

[24] Schulz-Zander J, Sarrar N, Schmid S. Towards a scalable and near-sighted control plane architecture for WiFi SDNs. In: Proceedings of the Third Workshop on Hot Topics in Software Defined Networking (HotSDN '14). New York, NY: ACM; 2014. p. 217–8. doi:10.1145/2620728.2620772.

[25] Shenker S. The future of networking, and the past of protocols. Open Networking Summit. Stanford University; 2011.

[26] Handigol N, Heller B, Jeyakumar V, Mazieres D, McKeown N. Where is the Debugger for my Software-Defined Network? In: HotSDN. Helsinki, Finland; 2012. Retrieved from: http://conferences.sigcomm.org/sigcomm/2012/paper/hotsdn/p55.pdf.

[27] Canini M, Venzano D, Peresini P, Kostic D, Rexford J. A nice way to test openflow applications. In: 9th USENIX Symposium on Networked System Design and Implementation. San Jose, CA; 2012. Retrieved from: https://www.usenix.org/system/files/conference/nsdi12/nsdi12-final105.pdf.

[28] Scott C, Wundsam A, Raghavan B, Panda A, Or A, Lai J, et al. Troubleshooting blackbox SDN control software with minimal causal sequences. In: Proceedings of the 2014 ACM conference on SIGCOMM (SIGCOMM'14). New York, NY: ACM; 2014. p. 395–406. Retrieved from: http://people.eecs.berkeley.edu/~rcs/research/sts.pdf.

[29] Khurshid A, Zhou W, Caesar M, Godfrey PB. VeriFlow: verifying network-wide invariants in real time. In: HotSDN. Helsinki, Finland; 2012. Retrieved from: http://conferences.sigcomm.org/sigcomm/2012/paper/hotsdn/p49.pdf.

[30] Chen C, Limin J, Wenchao Z, Thau LB. Proof-based verification of Software Defined Networks. Santa Clara, CA: Open Networking Summit; 2014. Retrieved from: http://www.archives.opennetsummit.org/pdf/2014/Research-Track/ONS2014_Chen_Chen_RESEARCH_TUE1.pdf.

[31] Kazemian P, Varghese G, McKeown N. Header space analysis: static checking for networks. In: Proceedings of the 9th USENIX conference on Networked Systems Design and Implementation (NSDI'12). Berkeley, CA: USENIX Association; 2012. Retrieved from: https://www.usenix.org/system/files/conference/nsdi12/nsdi12-final8.pdf.

[32] Al-Shaer E, Al-Haj S. FlowChecker: configuration analysis and verification of federated openflow infrastructures. In: Proceedings of the 3rd ACM workshop on assurable and usable security configuration (SafeConfig '10). New York, NY: ACM; 2010. p. 37–44.

[33] Monaco M, Michel O, Keller E. Applying operating system principles to SDN controller design. In: Twelfth ACM Workshop on Hot Topics in Networks. SIGCOMM 2013. College Park, MD; 2013. Retrieved from: http://conferences.sigcomm.org/hotnets/2013/papers/hotnets-final97.pdf.

[34] Wang A, Zhou Y, Godfrey B, Caesar M. Software-defined networks as databases. Santa Clara, CA: Open Networking Summit; 2014. Retrieved from: http://www.archives.opennetsummit.org/pdf/2014/Research-Track/ONS2014_Anduo_Wang_RESEARCH_TUE1.pdf.

[35] Voellmy A, Kim H, Feamster N. Procera: a language for high-level reactive network control. In: Proceedings of the first workshop on hot topics in software defined networks (HotSDN '12). New York, NY: ACM; 2012.

[36] Monsanto C, Reich J, Foster N, Rexford J, Walker D. Composing software-defined networks. In: Proceedings of the 10th USENIX conference on Networked Systems Design and Implementation (NSDI'13). Berkeley, CA: USENIX Association; 2013.

[37] Silva AS, Smith P, Mauthe A, Schaeffer-Filho A. Resilience support in software-defined networking: a survey. Comput Netw 2015;92(1):189–207.

[38] Jafarian JH, Al-Shaer E, Duan Q. OpenFlow random host mutation: transparent moving target defense using Software Defined Networking. In: HotSDN. Helsinki, Finland; 2012. Retrieved from: http://conferences.sigcomm.org/sigcomm/2012/paper/hotsdn/p127.pdf.

[39] Hu H, Ahn G, Han W, Zhao Z. Towards a reliable SDN firewall. Santa Clara, CA: Open Networking Summit; 2014. Retrieved from: http://www.archives.opennetsummit.org/pdf/2014/Research-Track/ONS%202014_Hongxin_Hu_RESEARCH_MON.pdf.
[40] Suresh L, Schulz-Zander J, Merz R, Feldmann A, Vazao T. Towards programmable enterprise WLANs with Odin. In: HotSDN. Helsinki, Finland; 2012. Retrieved from: http://conferences.sigcomm.org/sigcomm/2012/paper/hotsdn/p115.pdf.
[41] Yiakoumis Y, Bansal M, Katti S, Van Reijendam J, McKeown N. BeHop: SDN for dense WiFi networks. Santa Clara, CA: Open Networking Summit; 2014. Retrieved from: http://www.archives.opennetsummit.org/pdf/2014/Research-Track/ONS2014_Yiannis_Yiakoumis_RESEARCH_TUE1.pdf.
[42] Moradi M, Li L, Mao Z. SoftMoW: a dynamic and scalable software defined architecture for cellular WANs. Santa Clara, CA: Open Networking Summit; 2014. Retrieved from: http://www.archives.opennetsummit.org/pdf/2014/Research-Track/ONS2014_Mehrdad_Moradi_RESEARCH_TUE2.pdf.
[43] Pupatwibul P, Banjar A, Sabbagh A, Braun R. Developing an application based on OpenFlow to enhance mobile IP networks. In: IEEE 38th Conference on Local Computer Networks Workshops (LCN Workshops); 2013. p. 936–40.
[44] Namal S, Ahmad I, Gurtov A, Ylianttila M. Enabling secure mobility with OpenFlow. In: IEEE SDN for Future Networks and Services (SDN4FNS); 2013. p. 1–5.
[45] Al-Sabbagh A, Pupatwibul P, Banjar A, Braun R. Optimization of the OpenFlow controller in wireless environments for enhancing mobility. In: IEEE 38th conference on Local Computer Networks workshops (LCN workshops); 2013. p. 930–5. doi:10.1109/LCNW.2013.6758534.
[46] Li Y, Wang H, Liu M, Zhang B, Mao H. Software defined networking for distributed mobility management. In: IEEE Globecom workshops (GC Wkshps); 2013. p. 885–9. doi:10.1109/GLOCOMW.2013.
[47] Heller B, Seetharaman S, Mahadevan P, Yiakoumis Y, Sharma P, Banerjee S, et al. ElasticTree: saving energy in data center networks. In: 7th USENIX Symposium on Networked System Design and Implementation. San Jose, CA; 2010. Retrieved from: https://www.usenix.org/legacy/event/nsdi10/tech/full_papers/heller.pdf.
[48] Congdon PT, Mohapatra P, Farrens M, Akella V. Simultaneously reducing latency and power consumption in openflow switches. IEEE/ACM Trans Netw 2014;22:1007–20.
[49] Ma Y, Banerjee S. A smart pre-classifier to reduce power consumption of TCAMs for multi-dimensional packet classification. In: SIGCOMM 2012. Helsinki; 2012. Retrieved from: http://conferences.sigcomm.org/sigcomm/2012/paper/sigcomm/p335.pdf.
[50] Johnson S. Primer: a new generation of programmable ASICs. Tech Target. Retrieved from: http://searchnetworking.techtarget.com/feature/Primer-A-new-generation-of-programmable-ASICs.
[51] Pongracz G, Molnar L, Kis ZL, Turanyi Z. Cheap silicon: a myth or reality? Picking the right data plane hardware for software defined networking. In: SIGCOMM 2013. Hong Kong: HotSDN; 2013. Retrieved from: http://conferences.sigcomm.org/sigcomm/2013/papers/hotsdn/p103.pdf.
[52] Dupont de Dinechin, B. KALRAY: high performance embedded computing on the MPPA single chip many-core processor. In: CERN Seminar. 2013. Retrieved from: http://indico.cern.ch/getFile.py/access?resId=0&materialId=slides&confId=272037.
[53] Mogul J, Congdon P. Hey, you darned counters! Get off my ASIC! In: SIGCOMM 2012. Helsinki: HotSDN; 2012. Retrieved from: http://conferences.sigcomm.org/sigcomm/2012/paper/hotsdn/p25.pdf.
[54] Yamazaki K, Osaka T, Yasuda S, Ohteru S, Miyazaki A. Accelerating SDN/NFV with transparent offloading architecture. Santa Clara, CA: Open Networking Summit; 2014. Retrieved from: http://www.archives.opennetsummit.org/pdf/2014/Research-Track/ONS2014_Koji_Yamazaki_RESEARCH_TUE2.pdf.

APPENDIX

Acronyms and Abbreviations

The acronyms and abbreviations are sorted based on themselves, and not their expansions. In the event that the acronym describes a particular company's product, or is only meaningful in the context of another entity, we cite that company or entity in parentheses.

AAA	Authentication, Authorization and Accounting
AC	Access Controller
ACL	Access Control List
ACS	Auto-Configuration Server
AP	(Wireless) Access Point
API	Application Programming Interface
APIC	Application Policy Infrastructure Controller
APIC-DC	Application Policy Infrastructure Controller—Data Center
APIC-EM	Application Policy Infrastructure Controller—Enterprise Module
AR	Access Router
ARP	Address Resolution Protocol
ARPANET	Advanced Research Projects Agency Network
AS	Autonomous System
ASF	Apache Software Foundation
ASIC	Application-Specific Integrated Circuit
AWS	Amazon Web Services
BGP	Border Gateway Protocol
BGP-LS	Border Gateway Protocol Link State
BMP	BGP Monitoring Protocol
BOM	Bill of Materials
BSD	Berkeley Software Distribution
B-VID	Backbone VLAN ID
BYOD	Bring Your Own Device
CAM	Content-Addressable Memory
CAPEX	Capital Expense
CE	Control Element
CE	Customer Edge
CLA	Contributor License Agreement
CLI	Command Line Interface
COPS	Common Open Policy Service
CoS	Class of Service

COSC	Cisco Open SDN Controller
COTS	Common Off-the-Shelf
CPE	Customer Premises Equipment
CPU	Central Processing Unit
CSMA/CD	Carrier-Sense Multiple-Access/Collision Detect
CSPF	Constrained Shortest Path First
C-VID	Customer VLAN ID
C-VLAN	Customer VLAN ID
DCAN	Devolved Control of ATM Networks
DCO	Developer Certificate of Origin
DDoS	Distributed Denial of Service
DHCP	Dynamic Host Configuration Protocol
DMVPN	Dynamic Multipoint Virtual Private Networks
DNS	Domain Name System
DoS	Denial of Service
DOVE	Distributed Overlay Virtual Ethernet (IBM)
DPDK	Data Plane Development Kit (Intel)
DPI	Deep Packet Inspection
DSCP	Differentiated Services Code Point
DSL	Digital Subscriber Loop (aka Digital Subscriber Line)
DVR	Digital Video Recorder
DWDM	Dense Wavelength Division Multiplexing
EBGP	External Border Gateway Protocol
ECMP	Equal-Cost Multi-Path
EID	Endpoint Identifiers
EIGRP	Enhanced Interior Gateway Routing Protocol
EMS	Element Management System
EPC	Evolved Packet Core
EPL	Eclipse Public License
ETSI	European Telecommunications Standards Institute
EVB	Edge Virtual Bridging
FCoE	Fiber Channel over Ethernet
FDDI	Fiber Distributed Data Interface
FE	Forwarding Element
FHRP	First Hop Redundancy Protocol
ForCES	Forwarding and Control Element Separation
FOSS	Free and Open Source Software
FSF	Free Software Foundation
Gbps	Gigabits per second
GPL	General Public License
GSMP	General Switch Management Protocol
HA	High Availability
HSRP	Hot Standby Router Protocol
HSS	Home Subscriber Server

HTTP	HyperText Transfer Protocol
HTTPS	HyperText Transfer Protocol Secure
I2RS	Interface to the Routing System
IaaS	Infrastructure as a Service
IDS	Intrusion Detection System
IEEE	Institute of Electrical and Electronics Engineers
IETF	Internet Engineering Task Force
IGP	Interior Gateway Protocol
IMS	IP Multimedia Subsystem
InCNTRE	Indiana Center for Network Translational Research and Education
IP	Internet Protocol
IPS	Intrusion Prevention System
IPv4	Internet Protocol Version 4
IPv6	Internet Protocol Version 6
IoT	Internet of Things
ISG	Industry Specification Group (ETSI)
I-SID	Service Instance VLAN ID
IS-IS	Intermediate System to Intermediate System
ISP	Internet Service Provider
IT	Information Technology
ITU	International Telecommunications Union
JSON	JavaScript Object Notation
KVM	Kernel-based Virtual Machine
LAG	Link Aggregation
LAN	Local Area Network
LIFO	Last-In-First-Out
LISP	Locator/ID Separation Protocol
LSP	Label Switched Path
MAC	Media Access Control
Mbps	Megabits per second
MD-SAL	Model-Driven Service Abstraction Layer
mGRE	Multipoint GRE
MIB	Management Information Base
MPLS	Multiprotocol Label Switching
MPLS-TE	MPLS Traffic Engineering
MSDC	Massively Scaled Data Center
MSTP	Multiple Spanning Tree Protocol
MTU	Maximum Transmission Unit
NAC	Network Access Control
NAT	Network Access Translation
NBI	Northbound interface
NE	Network Element
NEM	Network Equipment Manufacturer
NETCONF	Network Configuration Protocol

NFV	Network Functions Virtualization
NFVI	NFV Infrastructure
NFVO	NFV Orchestrator
NHRP	Next-Hop Resolution Protocol
NIC	Network Interface Card
NMS	Network Management System
NOS	Network Operating System
NS	Network Service
NTT	Nippon Telegraph and Telephone
NV	Network Virtualization
NVGRE	Network Virtualization using Generic Routing Encapsulation
nvo3	Network Virtualization Overlays
NVP	Network Virtualization Platform (Nicira)
OCh	Optical Channel
ODM	Original Device Manufacturer
ODU	Optical Data Unit
ODUCLT	Optical Data Unit Client
OEM	Original Equipment Manufacturer
OF-Config	OpenFlow Configuration and Management Protocol
OFLS	OpenFlow Logical Switch
ONE	Open Networking Environment
ONIE	Open Network Install Environment
onePK	Open Networking Environment Platform Kit
ONF	Open Networking Foundation
ONOS	Open Network Operating System
ONRC	Open Networking Research Center
ONUG	Open Networking User Group
OOOD	Out-of-Order Delivery
OPEX	Operational Expense
OSGi	Open Services Gateway initiative
OSI	Open Systems Interconnection
OSS	Operational Support System
OTN	Optical Transport Network
OTWG	Optical Transport Working Group
OVS	Open vSwitch
OVSDB	Open vSwitch Database Management Protocol
PaaS	Platform as a Service
PAN	Personal Area Network
PBR	Policy-Based Routing
PCC	Path Computation Client
PCE	Path Computation Element
PCE-P	Path Computation Element Protocol
PCRF	Policy and Charging Rules Function
P-CSCF	Proxy Call Session Control Function

PE	Provider Edge
PE-VLAN	Provider Edge VLAN
PGW	Packet Data Network Gateway
PHY	Physical Layer and Physical Layer Technology
PKI	Public Key Infrastructure
PNF	Physical Network Function
PoA	Point of Access
PoC	Proof of Concept
POJO	Plain Old Java Object
PoS	Point of Service
QoS	Quality of Service
RADIUS	Remote Authentication Dial in User Service
RAN	Radio Access Network
REST	Representational State Transfer
RESTCONF	REST-based NETCONF
RFC	Request for Comments
RISC	Reduced Instruction Set Computing
RLOC	Routing Locators
ROI	Return on Investment
RPC	Remote Procedure Call
RSTP	Rapid Spanning Tree Protocol
RSVP	Resource Reservation Protocol
SaaS	Software as a Service
SAL	Service Abstraction Layer
S-CSCF	Serving Call Session Control Function
SDH	Synchronous Digital Hierarchy
SDM	Software Defined Mobility
SDN	Software Defined Networking
SDN VE	Software Defined Network for Virtual Environments
S-GW	Serving Gateway
SIP	Session Initiation Protocol
SLA	Service Level Agreement
SMI	Structure of Management Information
SNMP	Simple Network Management Protocol
SONET	Synchronous Optical Networking
SP	Service Provider
SPAN	Switch Port ANalyzer
SPB	Shortest Path Bridging
STP	Spanning Tree Protocol
STT	Stateless Transport Tunneling
S-VID	Service Provider VLAN ID
S-VLAN	Service Provider VLAN ID
TAM	Total Available Market
TCAM	Ternary Content-Addressable Memory

TE	Traffic Engineering
TLV	Type-Length-Value
TL1	Transaction Language 1
ToR	Top-of-Rack
ToS	Type of Service
TRILL	Transparent Interconnection of Lots of Links
TSO	TCP Segmentation Offload
TTI	Trail Trace Identifier
TTL	Time-To-Live
TTP	Table Type Pattern
VAR	Value-Added Reseller
VEPA	Virtual Edge Port Aggregator
VIM	Virtual Infrastructure Manager
VLAN	Virtual Local Area Network
VNF	Virtual Network Function
VNF-FG	VNF Forwarding Graph
VoIP	Voice over IP
VTEP	Virtual Tunnel Endpoint or VXLAN Tunnel Endpoint
VXLAN	Virtual eXtensible Local Area Network
WaaS	WiFi-as-a-Service
WAE	WAN Automation Engine
WAN	Wide Area Network
Wintel	Refers to the collaboration of Microsoft (through their Windows OS) with Intel (through their processors) in creating a de facto standard for a personal computer platform.
WLAN	Wireless Local Area Network
XaaS	Everything as a Service
XML	Extensible Markup Language
XMPP	Extensible Messaging and Presence Protocol
XNC	Extensible Network Controller
YANG	Yet Another Next Generation data modeling language

APPENDIX

Blacklist Application

B

The source code included in this appendix may be downloaded from: www.tallac.com/SDN/get-started/.

B.1 MESSAGELISTENER

```
//==================================================================
//   MessageListener class for receiving openflow messages
//   from Floodlight controller.
//==================================================================

public class MessageListener implements IOFMessageListener
{
    private static final MessageListener INSTANCE =
                                            new MessageListener();

    private static IFloodlightProviderService mProvider;

    private MessageListener() {}  // private constructor

    public static MessageListener getInstance() {
        return INSTANCE;
    }

    //- - - - - - - - - - - - - - - - - - - - - - - - - - - - - - -
    public void init(final FloodlightModuleContext context)
    {
        if( mProvider != null ) throw new RuntimeException(
                "BlackList Message listener already initialized" );

        mProvider = context.getServiceImpl(
                                IFloodlightProviderService.class );
    }
```

```
//- - - - - - - - - - - - - - - - - - - - - - - - - - - - - -
public void startUp()
{
    // Register class as MessageListener for PACKET_IN messages.
    mProvider.addOFMessageListener( OFType.PACKET_IN, this );
}

//- - - - - - - - - - - - - - - - - - - - - - - - - - - - - -
@Override
public String getName() { return BlackListModule.NAME; }

//- - - - - - - - - - - - - - - - - - - - - - - - - - - - - -
@Override
public boolean isCallbackOrderingPrereq( final OFType type,
                                         final String name )
{
    return( type.equals( OFType.PACKET_IN ) &&
            ( name.equals("topology") ||
              name.equals("devicemanager") ) );
}

//- - - - - - - - - - - - - - - - - - - - - - - - - - - - - -
@Override
public boolean isCallbackOrderingPostreq( final OFType type,
                                          final String name )
{
    return( type.equals( OFType.PACKET_IN ) &&
            name.equals( "forwarding" ) );
}

//- - - - - - - - - - - - - - - - - - - - - - - - - - - - - -
@Override
public Command receive( final IOFSwitch         ofSwitch,
                        final OFMessage         msg,
                        final FloodlightContext context )
{
    switch( msg.getType() )
    {
    case PACKET_IN:  // Handle incoming packets here

       // Create packethandler object for receiving packet in
       PacketHandler ph = new PacketHandler( ofSwitch,
                                             msg, context);

       // Invoke processPacket() method of our packet handler
       // and return the value returned to us by processPacket
       return ph.processPacket();
```

```
        default: break;  // If not a PACKET_IN, just return

        }

        return Command.CONTINUE;
    }

}
```

B.2 PACKETHANDLER

```
//=================================================================
//   PacketHandler class for processing packets receives
//   from Floodlight controller.
//=================================================================

public class PacketHandler
{
    public static final short   TYPE_IPv4 = 0x0800;
    public static final short   TYPE_8021Q = (short) 0x8100;

    private final IOFSwitch         mOfSwitch;
    private final OFPacketIn        mPacketIn;
    private final FloodlightContext mContext;
    private       boolean           isDnsPacket;

    //- - - - - - - - - - - - - - - - - - - - - - - - - - - - - - -
    public PacketHandler( final IOFSwitch         ofSwitch,
                          final OFMessage         msg,
                          final FloodlightContext context )
    {
        mOfSwitch   = ofSwitch;
        mPacketIn   = (OFPacketIn) msg;
        mContext    = context;
        isDnsPacket = false;
    }

    //- - - - - - - - - - - - - - - - - - - - - - - - - - - - - - -
    public Command processPacket()
    {
        // First, get the OFMatch object from the incoming packet
        final OFMatch ofMatch = new OFMatch();
        ofMatch.loadFromPacket( mPacketIn.getPacketData(),
                                mPacketIn.getInPort()     );

        // If the packet isn't IPv4, ignore.
        if( ofMatch.getDataLayerType() != Ethernet.TYPE_IPv4 )
        {
            return Command.CONTINUE;
```

```
}

//-- First handle all IP packets _ _ _ _ _ _ _ _
// We have an IPv4 packet, so check the
// destination IPv4 address against IPv4 blacklist.
try
{
    // Get the IP address
    InetAddress ipAddr = InetAddress.getByAddress(
              IPv4.toIPv4AddressBytes(
                    ofMatch.getNetworkDestination() ) );

    // Check the IP address against our blacklist
    if( BlacklistMgr.getInstance()
                    .checkIpv4Blacklist( ipAddr ) )
    {
        // It's on the blacklist, so update stats...
        StatisticsMgr.getInstance()
                        .updateIpv4Stats( mOfSwitch,
                                          ofMatch,
                                          ipAddr    );

        // ... and drop the packet so it doesn't
        // go through to the destination.
        FlowMgr.getInstance().dropPacket( mOfSwitch,
                                          mContext,
                                          mPacketIn );

        return Command.STOP; // Done with this packet,
                             // don't let somebody else
                             // change our DROP
    }

}

catch( UnknownHostException e1 )
{
    // If we had an error with something, bad IP or some
    return Command.CONTINUE;
}

//-- Now handle DNS packets _ _ _ _ _ _ _ _

// Is it DNS?
if( ofMatch.getNetworkProtocol() == IPv4.PROTOCOL_UDP &&
    ofMatch.getTransportDestination()
                        == FlowMgr.DNS_QUERY_DEST_PORT )
{
```

```
// Prepare data structure to hold DNS hostnames
// we extract from the request
final byte[] pkt = mPacketIn.getPacketData();
Collection<String> domainNames;

isDnsPacket = true;

// Get the domain names from the DNS request
try
{
    domainNames = parseDnsPacket( pkt );
}
catch( IOException e )  // Got here if there was an
                        // exception in parsing the
                        // domain names.
{
    return Command.CONTINUE; // Just return and
                             // allow other apps
                             // to handle the
                             // request.
}

// If there were not any domain names,
// no checking required, just forward the
// packet and return.
if( domainNames == null ) {
    forwardPacket();
    return Command.STOP;
}

// If there are domain names, proces them.
for( String domainName : domainNames )
{
    //  If the current domainName is in
    //  the blacklist, drop the packet and return.
    if( BlacklistMgr.getInstance()
                .checkDnsBlacklist( domainName ) )
    {
        // Update statistics about the dropped packet.
        StatisticsMgr.getInstance()
                        .updateDnsStats( mOfSwitch,
                                         ofMatch,
                                         domainName );

        // Drop the packet.
        FlowMgr.getInstance().dropPacket( mOfSwitch,
                                          mContext,
                                          mPacketIn );
```

```
                return Command.STOP; // Note that we are
                                     // dropping the whole
                                     // DNS request, even if
                                     // only one hostname is bad.
            }
        }
    }

    // If we made it here, everything is okay, so call the
    // method to forward the packet and set up flows for
    // the IP destination, if appropriate.
    forwardPacket();
    return Command.STOP;,
}

//- - - - - - - - - - - - - - - - - - - - - - - - - - - - - - - -
private void forwardPacket()
{
    // Get the output port for this destination IP address.
    short outputPort = FlowMgr.getInstance()
            .getOutputPort( mOfSwitch, mContext, mPacketIn );

    // If we can't get a valid output port for this
    // destination IP address, we have to drop it.
    if( outputPort == OFPort.OFPP_NONE.getValue() )
            FlowMgr.getInstance().dropPacket( mOfSwitch,
                                              mContext,
                                              mPacketIn );

    // Else if we should flood the packet, do so.
    else if( outputPort == OFPort.OFPP_FLOOD.getValue() ) {
            FlowMgr.getInstance().floodPacket( mOfSwitch,
                                               mContext,
                                               mPacketIn );
    }

    // Else we have a port to send this packet out on, so do it.
    else
    {
        final List<OFAction> actions = new ArrayList<OFAction>();

        // Add the action for forward the packet out outputPort
        actions.add( new OFActionOutput( outputPort ) );

        // Note that for DNS requests,
        // we don't need to set flows up on the switch.
        // Otherwise we must set flows so that subsequent
```

```
            // packets to this IP dest will get forwarded
            // locally w/o the controller.
            if( !isDnsPacket )
               FlowMgr.getInstance().createDataStreamFlow( mOfSwitch,
                                                           mContext,
                                                           mPacketIn,
                                                           actions );

            // In all cases, we have the switch forward the packet.
            FlowMgr.getInstance().sendPacketOut( mOfSwitch,
                                                 mContext,
                                                 mPacketIn,
                                                 actions    );

        }

    }

    //- - - - - - - - - - - - - - - - - - - - - - - - - - - - - - - - - - - -
    private Collection<String> parseDnsPacket(byte[] pkt) throws IOException
    {
        // Code to parse DNS are return
        // a collection of hostnames
        // that were in the DNS request
    }

}
```

B.3 FLOWMANAGER

```
//=================================================================
//   FlowManager class for handling interactions with Floodlight
//   regarding setting and unsetting flows
//=================================================================

public class FlowMgr
{
    private static final FlowMgr INSTANCE = new FlowMgr();

    private static IFloodlightProviderService mProvider;
    private static ITopologyService mTopology;

    public static final short PRIORITY_NORMAL      = 10;
    public static final short PRIORITY_IP_PACKETS  = 1000;
    public static final short PRIORITY_DNS_PACKETS = 2000;
    public static final short PRIORITY_IP_FLOWS    = 1500;
```

```
public static final short PRIORITY_ARP_PACKETS = 1500;

public static final short IP_FLOW_IDLE_TIMEOUT = 15;
public static final short NO_IDLE_TIMEOUT = 0;
public static final int BUFFER_ID_NONE = 0xffffffff;

public static final short DNS_QUERY_DEST_PORT = 53;

//- - - - - - - - - - - - - - - - - - - - - - - - - - - - - - - - - - -
private FlowMgr()
{
    // private constructor - prevent external instantiation
}

//- - - - - - - - - - - - - - - - - - - - - - - - - - - - - - - - - - -
public static FlowMgr getInstance()
{
    return INSTANCE;
}

//- - - - - - - - - - - - - - - - - - - - - - - - - - - - - - - - - - -
public void init( final FloodlightModuleContext context )
{
    mProvider = context.getServiceImpl(IFloodlightProviderService.class);
    mTopology = context.getServiceImpl(ITopologyService.class);
}

//- - - - - - - - - - - - - - - - - - - - - - - - - - - - - - - - - - -
public void setDefaultFlows(final IOFSwitch ofSwitch)
{
    // Note: this method is called whenever a switch is
    //     discovered by Floodlight, in our SwitchListener
    //     class (not included in this appendix).

    // Set the intitial 'static' or 'proactive' flows
    setDnsQueryFlow(ofSwitch);
    setIpFlow(ofSwitch);
    setArpFlow(ofSwitch);
}

//- - - - - - - - - - - - - - - - - - - - - - - - - - - - - - - - - - -
public void sendPacketOut( final IOFSwitch          ofSwitch,
                           final FloodlightContext cntx,
                           final OFPacketIn        packetIn,
                           final List<OFAction>    actions)
{
    // Create a packet out from factory.
    final OFPacketOut packetOut = (OFPacketOut) mProvider
```

```
                                    .getOFMessageFactory()
                                    .getMessage(OFType.PACKET_OUT);

        // Set the actions based on what has been passed to us.
        packetOut.setActions(actions);

        // Calculate and set the action length.
        int actionsLength = 0;
        for (final OFAction action : actions)
            { actionsLength += action.getLengthU(); }
        packetOut.setActionsLength((short) actionsLength);

        // Set the length based on what we've calculated.
        short poLength = (short) (packetOut.getActionsLength()
                                  + OFPacketOut.MINIMUM_LENGTH);

        // Set the buffer and in port based on the packet in.
        packetOut.setBufferId( packetIn.getBufferId() );
        packetOut.setInPort(   packetIn.getInPort() );

        // If the buffer ID is not present, copy and send back
        // the complete packet including payload to the switch.
        if (packetIn.getBufferId() == OFPacketOut.BUFFER_ID_NONE)
        {
            final byte[] packetData = packetIn.getPacketData();
            poLength += packetData.length;
            packetOut.setPacketData(packetData);
        }

        // Set the complete length of the packet we are sending.
        packetOut.setLength( poLength );

        // Now we actually send out the packet.
        try
        {
            ofSwitch.write( packetOut, cntx );
            ofSwitch.flush();
        }
        catch (final IOException e)
        {
            // Handle errors in sending packet out.
        }
    }

    //- - - - - - - - - - - - - - - - - - - - - - - - - - - - - - - -
    public void dropPacket( final IOFSwitch         ofSwitch,
                            final FloodlightContext cntx,
                                  OFPacketIn        packetIn)
```

```
{
    LOG.debug("Drop packet");

    final List<OFAction> flActions = new ArrayList<OFAction>();
    sendPacketOut( ofSwitch, cntx, packetIn, flActions );
}

//- - - - - - - - - - - - - - - - - - - - - - - - - - - - - - - - - - - - -
public void createDataStreamFlow(
                          final IOFSwitch         ofSwitch,
                          final FloodlightContext context,
                          final OFPacketIn        packetIn,
                                List<OFAction>    actions)
{
    final OFMatch match = new OFMatch();
    match.loadFromPacket( packetIn.getPacketData(),
                          packetIn.getInPort() );

    // Ignore packet if it is an ARP, or has not source/dest, or is not IPv4.
    if( ( match.getDataLayerType() == Ethernet.TYPE_ARP) ||
        ( match.getNetworkDestination() == 0 )           ||
        ( match.getNetworkSource() == 0 )                ||
        ( match.getDataLayerType() != Ethernet.TYPE_IPv4 ) )
      return;

    // Set up the wildcard object for IP address.
    match.setWildcards( allExclude( OFMatch.OFPFW_NW_DST_MASK,
                                    OFMatch.OFPFW_DL_TYPE ) );

    // Send out the data stream flow mod message
    sendFlowModMessage( ofSwitch,
                        OFFlowMod.OFPFC_ADD,
                        match,
                        actions,
                        PRIORITY_IP_FLOWS,
                        IP_FLOW_IDLE_TIMEOUT,
                        packetIn.getBufferId() );

}

//- - - - - - - - - - - - - - - - - - - - - - - - - - - - - - - - - - - - -
private void deleteFlow( final IOFSwitch ofSwitch,
                         final OFMatch   match )
{
    // Remember that an empty action list means 'drop'
    final List<OFAction> actions = new ArrayList<OFAction>();

    // Send out our empty action list.
```

```
    sendFlowModMessage( ofSwitch,
                        OFFlowMod.OFPFC_DELETE,
                        match,
                        actions,
                        PRIORITY_IP_FLOWS,
                        IP_FLOW_IDLE_TIMEOUT,
                        BUFFER_ID_NONE );
}



//- - - - - - - - - - - - - - - - - - - - - - - - - - - - - - - - - - - -
public void floodPacket(final IOFSwitch ofSwitch,
                        final FloodlightContext context,
                        final OFPacketIn packetIn)
{
    // Create action flood/all
    final List<OFAction> actions = new ArrayList<OFAction>();

    // If the switch supports the 'FLOOD' action...
    if (ofSwitch.hasAttribute( IOFSwitch.PROP_SUPPORTS_OFPP_FLOOD) )
    {
        actions.add(new OFActionOutput(
                              OFPort.OFPP_FLOOD.getValue()));
    }
    // ...otherwise tell it to send it to 'ALL'.
    else
    {
        actions.add( new OFActionOutput(
                              OFPort.OFPP_ALL.getValue() ) );
    }

    // Call our method to send the packet out.
    sendPacketOut( ofSwitch,
                   context,
                   packetIn,
                   actions );
}

//- - - - - - - - - - - - - - - - - - - - - - - - - - - - - - - - - - - -
public short getOutputPort( final IOFSwitch ofSwitch,
                            final FloodlightContext context,
                            final OFPacketIn packetIn )
{
    // return the output port for sending out the packet
    // that has been approved
}
```

```
//- - - - - - - - - - - - - - - - - - - - - - - - - - - - - - - - - - - -
private static int allExclude(final int... flags)
{
    // Utility routine for assistance in setting wildcard

    int wildcard = OFPFW_ALL;
    for( final int flag : flags ) { wildcard &= ~flag; }

    return wildcard;
}

//- - - - - - - - - - - - - - - - - - - - - - - - - - - - - - - - - - - -
private void sendFlowModMessage( final IOFSwitch      ofSwitch,
                                 final short          command,
                                 final OFMatch        ofMatch,
                                 final List<OFAction> actions,
                                 final short          priority,
                                 final short          idleTimeout,
                                 final int            bufferId )
{
    // Get a flow modification message from factory.
    final OFFlowMod ofm = (OFFlowMod) mProvider
                                 .getOFMessageFactory()
                                 .getMessage(OFType.FLOW_MOD);

    // Set our new flow mod object with the values that have
    // been passed to us.
    ofm.setCommand( command ).setIdleTimeout( idleTimeout )
                             .setPriority( priority )
                             .setMatch( ofMatch.clone() )
                             .setBufferId( bufferId )
                             .setOutPort( OFPort.OFPP_NONE )
                             .setActions( actions )
                             .setXid( ofSwitch
                                .getNextTransactionId() );

    // Calculate the length of the request, and set it.
    int actionsLength = 0;
    for( final OFAction action : actions ) { actionsLength += action.getLengthU(); }
    ofm.setLengthU(OFFlowMod.MINIMUM_LENGTH + actionsLength);

    // Now send out the flow mod message we have created.
    try
    {
        ofSwitch.write( ofm, null );
        ofSwitch.flush();
    }
    catch (final IOException e)
```

```
        {
            // Handle errors with the request
        }
    }

    //- - - - - - - - - - - - - - - - - - - - - - - - - - - - - - - - - - -
    private void setDnsQueryFlow( final IOFSwitch ofSwitch )
    {
        // Create match object to only match DNS requests
        OFMatch ofMatch = new OFMatch();
        ofMatch.setWildcards( allExclude( OFPFW_TP_DST,
                                          OFPFW_NW_PROTO,
                                          OFPFW_DL_TYPE ) )
                    .setDataLayerType( Ethernet.TYPE_IPv4 )
                    .setNetworkProtocol( IPv4.PROTOCOL_UDP )
                    .setTransportDestination( DNS_QUERY_DEST_PORT );

        // Create output action to forward to controller.
        OFActionOutput ofAction  = new OFActionOutput(
                                 OFPort.OFPP_CONTROLLER.getValue(),
                                 (short) 65535                     );

        // Create our action list and add this action to it
        List<OFAction> ofActions = new ArrayList<OFAction>();
        ofActions.add(ofAction);

        sendFlowModMessage( ofSwitch,
                            OFFlowMod.OFPFC_ADD,
                            ofMatch,
                            ofActions,
                            PRIORITY_DNS_PACKETS,
                            NO_IDLE_TIMEOUT,
                            BUFFER_ID_NONE );
    }

    //- - - - - - - - - - - - - - - - - - - - - - - - - - - - - - - - - - -
    private void setIpFlow( final IOFSwitch ofSwitch )
    {
        // Create match object to only match all IPv4 packets
        OFMatch ofMatch = new OFMatch();
        ofMatch.setWildcards( allExclude( OFPFW_DL_TYPE ) )
              .setDataLayerType( Ethernet.TYPE_IPv4 );

        // Create output action to forward to controller.
        OFActionOutput ofAction  = new OFActionOutput(
                                 OFPort.OFPP_CONTROLLER.getValue(),
                                 (short) 65535                     );
```

```
        // Create our action list and add this action to it.
        List<OFAction> ofActions = new ArrayList<OFAction>();
        ofActions.add(ofAction);

        // Send this flow modification message to the switch.
        sendFlowModMessage( ofSwitch,
                            OFFlowMod.OFPFC_ADD,
                            ofMatch,
                            ofActions,
                            PRIORITY_IP_PACKETS,
                            NO_IDLE_TIMEOUT,
                            BUFFER_ID_NONE );
    }


    //- - - - - - - - - - - - - - - - - - - - - - - - - - - - - - - - - - -
    private void setArpFlow( final IOFSwitch ofSwitch )
    {
        // Create match object match arp packets
        OFMatch ofMatch = new OFMatch();
        ofMatch.setWildcards( allExclude(OFPFW_DL_TYPE) )
                            .setDataLayerType( Ethernet.TYPE_ARP );

        // Create output action to forward normally
        OFActionOutput ofAction  = new OFActionOutput(
                                      OFPort.OFPP_NORMAL.getValue(),
                                      (short) 65535                );

        // Create our action list and add this action to it.
        List<OFAction> ofActions = new ArrayList<OFAction>();
        ofActions.add(ofAction);

        // Send this flow modification message to the switch.
        sendFlowModMessage( ofSwitch,
                            OFFlowMod.OFPFC_ADD,
                            ofMatch,
                            ofActions,
                            PRIORITY_ARP_PACKETS,
                            NO_IDLE_TIMEOUT,
                            BUFFER_ID_NONE );
    }

}
```

Index

B

C

D

E

F

G

H

N

O

P

Q

R

T

W

X

Y